NFPA® 70B

Recommended Practice for

Electrical Equipment Maintenance

2013 Edition

This edition of NFPA 70B, *Recommended Practice for Electrical Equipment Maintenance,* was prepared by the Technical Committee on Electrical Equipment Maintenance and released by the Technical Correlating Committee on National Electrical Code®. It was issued by the Standards Council on November 27, 2012, with an effective date of December 17, 2012, and supersedes all previous editions.

This edition of NFPA 70B was approved as an American National Standard on December 17, 2012.

Origin and Development of NFPA 70B

In the fall of 1967, the Board of Directors of the National Fire Protection Association authorized the formation of an Ad Hoc Committee on Electrical Equipment Maintenance to determine the need for the development of a suitable document on this subject. The purpose of the document would be to give recommendations on the maintenance of various types of electrical installations, apparatus, and equipment usually found in industrial and large commercial-type installations. A diversified group representing various interests and organizations was invited to participate.

At a meeting of the Ad Hoc Committee held January 10, 1968, in New York, with 31 representatives attending, it was pointed out that several requests had been made to the National Electrical Code Committee to include maintenance recommendations in the *National Electrical Code*®. The subject had been discussed by the Correlating Committee of the National Electrical Code Committee, and the decision was made that the Code was not the proper document in which to cover the maintenance of electrical equipment. However, the committee recognized that "lack-of-maintenance" frequently resulted in serious injuries and fatalities as well as high monetary damage. Electrical equipment maintenance was a subject requiring attention.

It was noted at that time that electrical safety information broke down logically into four main subdivisions: (1) design or product standards, (2) installation standards (as covered by the *National Electrical Code* and the *National Electrical Safety Code*), (3) maintenance recommendations, and (4) use instructions. The NFPA had not yet started work on *NFPA 70E®, Standard for Electrical Safety in the Workplace.* The problem was to explore whether something more should be done in the interest of electrical safety on the maintenance of electrical equipment and what form activity in this field should take.

It was recognized that much had been done to enunciate maintenance needs for specific types of equipment by the equipment manufacturers, and that guidance was available on the general subject from a number of sources. Therefore, it was determined that bringing together that guidance in the form of general guidelines in a single document under the NFPA procedure was advantageous. The stature of the document would be enhanced by becoming associated with the *National Electrical Code.* To this end, a tentative scope was drafted for presentation to the Board of Directors of the National Fire Protection Association with a recommendation that an NFPA Committee on Electrical Equipment Maintenance be authorized.

On June 27, 1968, the NFPA Board of Directors authorized the establishment of an NFPA Committee on Electrical Equipment Maintenance with the following scope: "To develop suitable texts relating to preventive maintenance of electrical systems and equipment used in industrial-type applications with the view of reducing loss of life and property. The purpose is to correlate generally applicable procedures for preventive maintenance that have broad application to the more common classes of industrial electrical systems and equipment without duplicating or superseding instructions that manufacturers normally provide. Reports to the Association through the Correlating Committee of the National Electrical Code Committee."

The committee was formed, and an organizational meeting was held December 12, 1968, in Boston. Twenty-nine members or representatives attended. The *Recommended Practice for Electrical Equipment Maintenance* represented the cumulative effort of the entire committee.

In 1973, the committee developed Part II, which became Chapters 5 through 15, and a new addition in the Appendix, "How to Instruct."

In 1976, the committee developed the chapters on Electronic Equipment, Ground-Fault Protection, Wiring Devices, and Maintenance of Electrical Equipment Subject to Long Intervals Between Shutdowns, and new additions in the Appendix, "NEMA Configurations," and "Long-Term Maintenance Guidelines."

In the 1983 edition, the committee developed the chapters on Deenergizing and Grounding of Equipment to Provide Protection for Electrical Maintenance Personnel and on Cable Tray Systems and added Appendix I, "Equipment Storage and Maintenance During Construction."

In the 1987 edition, the committee reorganized and reformatted the former Chapter 7, now Chapter 9, to include distribution transformers as well as power transformers.

In the 1990 edition, the committee developed the chapter on Uninterruptible Power Supply (UPS) Systems. This new chapter was developed with a significant contribution from several nonmembers who joined the Ad Hoc Technical Subcommittee to produce this new material. The committee recognized and extended its appreciation to Mr. Robert Adams, Liebert Corporation; Mr. Russ Grose, Liebert Corporation; and Mr. Ronald Mundt, U.S. Army Corps of Engineers. The chapter on Testing and Test Methods was amended by the addition of diagrams of different wave shapes for detecting problems in motors and generators using surge testing.

In the 1994 edition, the committee added three new chapters to cover power system studies that include short circuit studies, coordination studies, load-flow studies, and reliability studies; power quality and information pertaining to harmonics, the problems created by harmonic distortion, causes of harmonic distortion, harmonic surveying and testing, and recommended solutions to harmonic problems; and vibration analysis pertaining to rotating machinery. The third new chapter included a table on suggested vibration limits and a vibration severity chart for various-sized machines. To comply with the NFPA *Manual of Style*, a new Referenced Publications chapter was created. Also, the Bibliography was moved to Appendix J.

For the 1998 edition, the chapter on Power Quality was rewritten and expanded. In addition, the committee updated and revised maintenance techniques for stationary batteries and infrared inspections. Special handling and disposal considerations were introduced, and employee training was focused to emphasize workplace safety.

In the 2002 edition, the committee made several enhancements. The document was restructured to comply with the *Manual of Style for NFPA Technical Committee Documents*. The document scope was revised to include preventive maintenance for electronic and communications equipment. A new chapter added for Grounding provided definitions, symptoms, inspection, testing techniques, and solutions to grounding issues. A new section for Gas Insulated Substations was added to Chapter 8 to address the maintenance issues on industrial sites resulting from regulatory changes in the electrical utility industry. Charts were added in Chapter 11 for motor control troubleshooting for motor controllers, switchboards, and panelboards. Chapter 27 was enhanced with information on the latest technology on voltage fluctuation. A new annex was added that specified maintenance intervals for electrical equipment.

The 2006 edition of NFPA 70B included a significant change concerning safety. Whereas safety precautions and information in previous editions were dispersed throughout the individual equipment chapters, an entire new chapter on safety was written and placed up front in the document to provide more complete and updated coverage, as well as to emphasize the importance of safety. A series of updated test forms appeared in Annex F, revised testing schedules in Chapter 21, and maintenance of supervisory control and data acquisition (SCADA) systems in Chapter 30. An important part of maintenance is having a properly installed system with baseline performance data, and so a new chapter on commissioning the electrical system at a new facility was added. Because the industry trend is going from routine maintenance to reliability-centered maintenance (RCM), a chapter on how to apply RCM and an extensive annex with detailed reliability data on many types of electrical equipment also was added. Finally, revisions were made to update information on equipment cleaning, disconnects, busways, vibration testing, lamps, power quality, and rework and recertification of equipment.

Several significant changes have been made to the 2010 edition of NFPA 70B. Most noticeable is the reorganization of the document chapters and annexes. This effort groups like topics and equipment into a more logical arrangement. Major topic and equipment groupings used in this reorganization include: introduction, overview of EPM, electrical systems issues, testing and monitoring, switchgear, cables and wiring, static apparatus, rotating apparatus, and specific-purpose equipment. The extensive annex material was also reorganized using three major groupings that are general information, forms and diagrams, and maintenance. This reorganization effort is obvious by reviewing the table of contents in the 2010 edition.

In addition to the reorganization of the document, Chapter 11 (formerly Chapter 21) on testing and test methods is where the majority of provisions covering test procedures now reside. The testing procedures have been organized based on equipment type, and testing procedures formerly located in the individual equipment chapters have been centralized into Chapter 11. The revised provisions reflect current industry testing practice.

Responding to frequent questions coming from electrical equipment owners and maintainers are new provisions covering emergency preparedness and electrical system and equipment restoration (EPnSR) in Chapter 6. Procedures for emergency shutdown and post-emergency procedures are contained in Chapter 6 and in related annex material. Chapter 6 also contains new material covering outsourcing of electrical equipment maintenance. The requirements on personnel safety in Chapter 7 have been revised to correlate with and directly reference *NFPA 70E, Standard for Electrical Safety in the Workplace.*

Several other significant changes include reorganized provisions on maintaining SCADA systems, new material on data collection methods, new forms for conducting power quality surveys, and new information on failure mode effects and criticality analysis (FMECA) to support reliability centered maintenance (RCM). Significant material supporting reliability centered maintenance has been added in Annex N (formerly Annex K).

In the 2013 edition, the references, bibliography, and informational references were updated. In Chapter 3, new definitions were added for *arc flash hazard* and *arc flash hazard analysis,* both extracted from the 2012 edition of *NFPA 70E, Standard for Electrical Safety in the Workplace.* In Chapter 6, the section regarding procedures for post-emergency actions was deleted except for the title and a pointer to the new chapter on disaster recovery. A new section on counterfeit components, devices, tools, and equipment was also added to Chapter 6. A new section regarding arc-flash hazard analysis studies was added to Chapter 9. Chapter 11 included several upgrades, including the section on acceptance tests, the section on field testing of circuit breakers including solid state trip units, information regarding instantaneous trip tolerances for field testing, the section on battery tests, the section on cables, and a new section regarding a test or calibration decal system including a new figure. In Chapter 14 information was added regarding luminaire grounding. The section in Chapter 15 on stationary batteries and battery chargers was revised. In Chapter 19 the visual inspection and electrical testing sections were revised, and an inspection and testing records section was added. In Chapter 23 a new section regarding efficiency was added to the lamp and ballast section, and a new section regarding light emitting diode (LED) lamps was added with several new figures. A general upgrade was made to Chapter 29. Four new chapters were added: one on disaster recovery, one on photo voltaic systems, one on electrical vehicle charging systems, and one on wind power electrical systems and associated equipment. In Annex A, a new section was added and existing sections were renumbered where required.

Technical Correlating Committee on National Electrical Code®

Michael J. Johnston, *Chair*
National Electrical Contractors Association, MD [IM]

Jean O'Connor, *Recording Secretary*
National Fire Protection Association, MA

Mark Earley, *Administrative Secretary*
National Fire Protection Association, MA

James E. Brunssen, Telcordia, NJ [UT]
Rep. Alliance for Telecommunications Industry Solutions
Merton W. Bunker, Jr., U.S. Department of State, VA [U]
William R. Drake, Actuant Electrical, CA [M]
William T. Fiske, Intertek Testing Services, NY [RT]
Palmer L. Hickman, National Joint Apprentice and Training Committee, MD [L]
Rep. International Brotherhood of Electrical Workers
David L. Hittinger, Independent Electrical Contractors of Greater Cincinnati, OH [IM]
Rep. Independent Electrical Contractors, Inc.
Daniel J. Kissane, Legrand/Pass & Seymour, NY [M]
Rep. National Electrical Manufacturers Association
John R. Kovacik, Underwriters Laboratories Inc., IL [RT]
Neil F. LaBrake, Jr., National Grid, NY [UT]
Rep. Electric Light and Power Group/EEI
Danny Liggett, The DuPont Company, TX [U]
Rep. American Chemistry Council
Richard P. Owen, Oakdale, MN [E]
Rep. International Association of Electrical Inspectors

Alternates

Thomas L. Adams, Engineering Consultant, IL [UT]
(Alt. to N. F. LaBrake, Jr.)
Lawrence S. Ayer, Biz Com Electric, Inc., OH [IM]
(Alt. to D. L. Hittinger)
James T. Dollard, Jr., IBEW Local Union 98, PA [L]
(Alt. to P. L. Hickman)
Stanley J. Folz, Morse Electric Company, NV [IM]
(Alt. to M. J. Johnston)
Ernest J. Gallo, Telcordia Technologies, Inc., NJ [UT]
(Alt. to J. E. Brunssen)
Robert A. McCullough, Tuckerton, NJ [E]
(Alt. to R. P. Owen)
Michael E. McNeil, FMC Bio Polymer, ME [U]
(Alt. to D. Liggett)
Mark C. Ode, Underwriters Laboratories Inc., AZ [RT]
(Alt. to J. R. Kovacik)
Vincent J. Saporita, Cooper Bussmann, MO [M]
(Alt. to D. J. Kissane)

Nonvoting

Richard G. Biermann, Biermann Electric Company, Inc., IA [IM]
(Member Emeritus)
Timothy J. Pope, Canadian Standards Association, Canada [SE]
D. Harold Ware, Libra Electric Company, OK [IM]
(Member Emeritus)

Mark W. Earley, NFPA Staff Liaison

This list represents the membership at the time the Committee was balloted on the final text of this edition. Since that time, changes in the membership may have occurred. A key to classifications is found at the back of the document.

NOTE: Membership on a committee shall not in and of itself constitute an endorsement of the Association or any document developed by the committee on which the member serves.

Committee Scope: This Committee shall have primary responsibility for documents on minimizing the risk of electricity as a source of electric shock and as a potential ignition source of fires and explosions. It shall also be responsible for text to minimize the propagation of fire and explosions due to electrical installations.

Technical Committee on Electrical Equipment Maintenance

Alan Manche, *Chair*
Square D Company/Schneider Electric, KY [M]
Rep. National Electrical Manufacturers Association

Richard Bingham, Dranetz-BMI, NJ [M]
Thomas H. Bishop, Electrical Apparatus Service Association, MO [IM]
Timothy Crnko, Cooper Bussmann, MO [M]
Jeffrey Hall, Underwriters Laboratories Inc., NC [RT]
Palmer L. Hickman, National Joint Apprentice and Training Committee, MD [L]
Rep. International Brotherhood of Electrical Workers
Mark C. Horne, Georgia Power Company, GA [U]
Rep. Electric Light and Power Group/EEI
Ahmad A. Moshiri, Liebert Corporation, OH [M]
Ronald K. Mundt, U.S. Army Corps of Engineers, VA [U]
Erik G. Olsen, Chubb Group of Insurance Companies, NJ [I]
Kenneth J. Rempe, Siemens Industry Inc., GA [M]
Rep. National Electrical Manufacturers Association
Robert Urdinola, U.S. Department of State, DC [U]
Michael Velvikis, High Voltage Maintenance Corporation, WI [IM]
Rep. InterNational Electrical Testing Association
James R. White, Shermco Industries, TX [IM]
Bruce G. Wyman, Mount Snow Ltd., VT [U]

Alternates

Leonard Fiume, National Grid, NY [U]
(Alt. to M. C. Horne)
David Goodrich, Liebert Corporation, OH [M]
(Alt. to A. A. Moshiri)
David Huffman, Power Systems Testing Company, CA [IM]
(Alt. to M. Velvikis)
Christopher E. Kelly, JATC for Nassau & Suffolk Counties, NY [L]
(Alt. to P. L. Hickman)
Greg T. Nienaber, Connector Manufacturing Company, OH [M]
(Alt. to K. J. Rempe)
Ron Widup, Shermco Industries, TX [IM]
(Alt. to J. R. White)

Nonvoting

Albert J. Reed, Macungie, PA
(Member Emeritus)

Michael Fontaine, NFPA Staff Liaison

This list represents the membership at the time the Committee was balloted on the final text of this edition. Since that time, changes in the membership may have occurred. A key to classifications is found at the back of the document.

NOTE: Membership on a committee shall not in and of itself constitute an endorsement of the Association or any document developed by the committee on which the member serves.

Committee Scope: This Committee shall have the primary responsibility for documents relating to preventive maintenance of electrical, electronic, and communications systems and equipment used in industrial and commercial type applications with the view of: (1) reducing loss of life and property, and (2) improving reliability, performance, and efficiency in a cost-effective manner. The purpose is to provide generally applicable procedures for preventive maintenance that have broad application to the more common classes of industrial and commercial systems and equipment without duplicating or superseding instructions that manufacturers normally provide. This Committee shall report to Technical Correlating Committee of the National Electrical Code.

Contents

NFPA 70B

Recommended Practice for

Electrical Equipment Maintenance

2013 Edition

IMPORTANT NOTE: This NFPA document is made available for use subject to important notices and legal disclaimers. These notices and disclaimers appear in all publications containing this document and may be found under the heading "Important Notices and Disclaimers Concerning NFPA Documents." They can also be obtained on request from NFPA or viewed at www.nfpa.org/disclaimers.

NOTICE: An asterisk (*) following the number or letter designating a paragraph indicates that explanatory material on the paragraph can be found in Annex A.

Changes other than editorial are indicated by a vertical rule beside the paragraph, table, or figure in which the change occurred. These rules are included as an aid to the user in identifying changes from the previous edition. Where one or more complete paragraphs have been deleted, the deletion is indicated by a bullet (•) between the paragraphs that remain.

A reference in brackets [ ] following a section or paragraph indicates material that has been extracted from another NFPA document. As an aid to the user, the complete title and edition of the source documents for extracts in the recommendations sections of this document are given in Chapter 2 and those for extracts in the informational sections are given in Annex D. Extracted text may be edited for consistency and style and may include the revision of internal paragraph references and other references as appropriate. Requests for interpretations or revisions of extracted text should be sent to the technical committee responsible for the source document.

Information on referenced publications can be found in Chapter 2 and Annex D.

Chapter 1 Administration

1.1 Scope.

1.1.1 This recommended practice applies to preventive maintenance for electrical, electronic, and communication systems and equipment and is not intended to duplicate or supersede instructions that manufacturers normally provide. Systems and equipment covered are typical of those installed in industrial plants, institutional and commercial buildings, and large multifamily residential complexes.

1.1.2 Consumer appliances and equipment intended primarily for use in the home are not included.

1.2 Purpose. The purpose of this recommended practice is to reduce hazards to life and property that can result from failure or malfunction of industrial-type electrical systems and equipment.

1.2.1 Chapters 4, 5, and 6 of these recommendations for an effective electrical preventive maintenance (EPM) program have been prepared with the intent of providing a better understanding of benefits, both direct and intangible, that can be derived from a well-administered EPM program.

1.2.2 This recommended practice explains the function, requirements, and economic considerations that can be used to establish such an EPM program.

Chapter 2 Referenced Publications

2.1 General. The documents or portions thereof listed in this chapter are referenced within this recommended practice and should be considered part of the recommendations of this document.

2.2 NFPA Publications. The documents or portions thereof listed in this chapter are referenced within this recommended practice and should be considered part of the recommendations of this document.

NFPA 70®, *National Electrical Code®*, 2011 edition.

NFPA 70E®, *Standard for Electrical Safety in the Workplace®*, 2012 edition.

NFPA 110, *Standard for Emergency and Standby Power Systems,* 2013 edition.

NFPA 496, *Standard for Purged and Pressurized Enclosures for Electrical Equipment,* 2013 edition.

NFPA 780, *Standard for the Installation of Lightning Protection Systems,* 2011 edition.

NFPA 1600®, *Standard on Disaster/Emergency Management and Business Continuity Programs,* 2013 edition.

2.3 Other Publications.

2.3.1 ASTM Publications. American Society for Testing and Materials, 100 Barr Harbor Drive, West Conshohocken, PA 19428-2959.

ASTM D 92, *Standard Test Method for Flash and Fire Points by Cleveland Open Cup Tester,* 2012a.

ASTM D 445, *Standard Test Method for Kinematic Viscosity of Transparent and Opaque Liquids and Calculation of Dynamic Viscosity,* 2012.

ASTM D 664, *Standard Test Method for Acid Number of Petroleum Products by Potentiometric Titration,* 2011a.

ASTM D 877, *Standard Test Method for Dielectric Breakdown Voltage of Insulating Liquids Using Disk Electrodes,* 2002 (Rev. 2007).

ASTM D 923, *Standard Practices for Sampling Electrical Insulating Liquids,* 2007.

ASTM D 924, *Standard Test Method for Dissipation Factor (or Power Factor) and Relative Permittivity (Dielectric Constant) of Electrical Insulating Liquids,* 2008.

ASTM D 971, *Standard Test Method for Interfacial Tension of Oil Against Water by the Ring Method,* 2012.

ASTM D 974, *Standard Test Methods for Acid and Base Number by Color-Indicator Titration,* 2012.

ASTM D 1298, *Standard Test Method for Density, Relative Density (Specific Gravity), or API Gravity of Crude Petroleum and Liquid Petroleum Products by Hydrometer Method,* 2012b.

ASTM D 1500, *Standard Test Method for ASTM Color of Petroleum Products (ASTM Color Scale),* 2007.

ASTM D 1524, *Standard Test Method for Visual Examination of Used Electrical Insulating Oils of Petroleum Origin in the Field,* 1994 (Rev. 2010).

ASTM D 1533, *Standard Test Method for Water in Insulating Liquids by Coulometric Karl Fischer Titration,* 2012.

ASTM D 1816, *Standard Test Method for Dielectric Breakdown Voltage of Insulating Oils of Petroleum Origin Using VDE Electrodes,* 2012.

ASTM D 2129, *Standard Test Method for Color of Clear Electrical Insulating Liquids (Platinum-Cobalt Scale)*, 2005 (Rev. 2010).

ASTM D 2285, *Standard Test Method for Interfacial Tension of Electrical Insulating Oils of Petroleum Origin Against Water by the Drop-Weight Method*, 1999 (withdrawn 2008).

ASTM D 2472, *Standard Specification for Sulfur Hexafluoride*, 2000 (Rev. 2006).

ASTM D 3284, *Standard Practice for Combustible Gases in the Gas Space of Electrical Apparatus Using Portable Meters*, 2005 (Rev. 2011).

ASTM D 3612, *Standard Test Method for Analysis of Gases Dissolved in Electrical Insulating Oil by Gas Chromatography*, 2002 (Rev. 2009).

2.3.2 EASA Publications. Electrical Apparatus Service Association, 1331 Baur Blvd, St. Louis, MO 63132.

ANSI/EASAAR100, *Recommended Practice for the Repair of Rotating Electrical Apparatus*, 2010.

2.3.3 IEEE Publications. Institute of Electrical and Electronics Engineers, Three Park Avenue, 17th Floor, New York, NY 10016-5997.

ANSI/IEEE 43, *Recommended Practice for Testing Insulation Resistance of Rotating Machinery*, 2000.

ANSI/IEEE 80, *Guide for Safety in AC Substation Grounding*, 2000.

IEEE 81, *Guide for Measuring Earth Resistivity, Ground Impedance and Earth Surface Potentials of a Ground System*, 2012.

ANSI/IEEE 95, *Recommended Practice for Insulation Testing of AC Electric Machinery (2300 V and Above) with High Direct Voltage*, 2002.

ANSI/IEEE 141, *Recommended Practice for Electric Power Distribution for Industrial Plants (IEEE Red Book)*, 1993 (Rev. 1999).

ANSI/IEEE 142, *Recommended Practice for Grounding of Industrial and Commercial Power Systems (IEEE Green Book)*, 2007.

ANSI/IEEE 241, *Recommended Practice for Electric Power Systems in Commercial Buildings (IEEE Gray Book)*, 1990.

ANSI/IEEE 242, *Recommended Practice for Protection and Coordination of Industrial and Commercial Power Systems (IEEE Buff Book)*, 2001.

ANSI/IEEE 399, *Recommended Practice for Industrial and Commercial Power Systems Analysis (IEEE Brown Book)*, 1997.

ANSI/IEEE 400, *Guide for Field Testing and Evaluation of the Insulation of Shielded Power Cable Systems*, 2012.

IEEE 400.1, *Guide for Field Testing of Laminated Dielectric, Shielded Power Cable Systems Rated 5kV and Above with High Direct Current Voltage*, 2007.

IEEE 400.2, *Guide for Field Testing of Shielded Power Cable Systems Using Very Low Frequency (VLF)*, 2004.

IEEE 400.3, *Guide for Partial Discharge Testing of Shielded Power Cable Systems in a Field Environment*, 2006.

ANSI/IEEE 446, *Recommended Practice for Emergency and Standby Power Systems for Industrial and Commercial Applications (IEEE Orange Book)*, 1995 (Rev. 2000).

ANSI/IEEE 450, *Recommended Practice for Maintenance, Testing and Replacement of Vented Lead-Acid Batteries for Stationary Applications*, 2010.

ANSI/IEEE 493, *Recommended Practice for the Design of Reliable Industrial and Commercial Power Systems (IEEE Gold Book)*, 2007.

ANSI/IEEE 519, *Recommended Practices and Requirements for Harmonic Control in Electrical Power Systems*, 1992.

IEEE 637, *Guide for Reclamation of Insulating Oil and Criteria for Its Use*, 1985.

ANSI/IEEE 1100, *Recommended Practice for Powering and Grounding Electronic Equipment (IEEE Emerald Book)*, 2005.

IEEE 1106, *Recommended Practice for Installation, Maintenance, Testing and Replacement of Vented Nickel-Cadmium Batteries for Stationary Applications*, 2005.

ANSI/IEEE 1125, *Guide for Moisture Measurement and Control in SF_6 Gas-Insulated Equipment*, 1993 (Rev. 2000).

IEEE 1145, *IEEE Recommended Practice for the Installation and Maintenance of Nickel-Cadmium Batteries for Phtovoltaic (PV) Systems*, 199 (withdrawn).

IEEE 1159, *Recommended Practice on Monitoring Electric Power Quality*, 2009.

IEEE 1188, *Recommended Practice for Maintenance, Testing and Replacement of Valve-Regulated Lead Acid (VRLA) Batteries for Stationary Applications*, 2005.

IEEE 1584™, *Guide for Performing Arc Flesh Hazards Calculations*, 2002 (with Amendment 1 and 2).

IEEE 1657, *IEEE Recommended Practice for Personnel Qualifications for Installation and Maintenance of Stationary Batteries*, 2009.

IEEE 3007.2, *IEEE Recommended Practice for the Maintenance of Industrial and Commercial Power Systems*, 20010.

ANSI/Accredited Standards Committee C2, *National Electrical Safety Code®(NESC®)*, 2012.

ANSI/IEEE C37.13, *Standard for Low-Voltage AC Power Circuit Breakers Used in Enclosures*, 2008.

IEEE C37.20.1, *Standard for Metal-Enclosed Low-Voltage Power Circuit Breaker Switchgear*, 2002.

IEEE C37.23, *Standard for Metal-Enclosed Bus.*, 2003.

IEEE C37.122.1, *IEEE Guide for Gas-Insulated Substations*, 1993 (Rev. 2002).

ANSI/IEEE C57.104, *Guide for the Interpretation of Gases Generated in Oil-Immersed Transformers*, 2008.

ANSI/IEEE C57.106, *Guide for Acceptance and Maintenance of Insulating Oil in Equipment*, 2006.

ANSI/IEEE C57.110, *Recommended Practice for Establishing Liquid-Filled and Dry-Type Power and Distribution Transformer Capability When Supplying Nonsinusoidal Load Currents*, 2008.

ANSI/IEEE C57.111, *Guide for Acceptance of Silicone Insulating Fluid and Its Maintenance in Transformers*, 1989 (Rev. 1995).

ANSI/IEEE C57.121, *Guide for Acceptance and Maintenance of Less-Flammable Hydrocarbon Fluid in Transformers*, 1998.

2.3.4 ITI Publications. Information Technology Industry Council, 1250 Eye Street NW, Suite 200, Washington, DC 20005. 202-737-8880. http://www.itic.org.

ITI (CBEMA) Curve Application Note, 2000.

2.3.5 NEMA Publications. National Electrical Manufacturers Association, 1300 North 17th Street, Suite 1847, Rosslyn, VA 22209.

NEMA AB 1, *Molded Case Circuit Breakers, and Molded Case Switches, and Circuit Breaker Enclosures.* (Equivalent to the current edition of UL 489.)

ANSI/NEMA AB 4, *Guidelines for Inspection and Preventive Maintenance of Molded-Case Circuit Breakers Used in Commercial and Industrial Applications,* 2009.

ANSI/NEMA C84.1, *Electric Power Systems and Equipment, Voltage Ratings (60 Hertz),* 2011.

ANSI/NEMA KS 3, *Guidelines for Inspection and Preventive Maintenance of Switches Used in Commercial and Industrial Applications,* 2010.

NEMA MG 1, *Motors and Generators,* 2011.

ANSI/NEMA PB 2.1, *General Instructions for Proper Handling, Installation, Operation, and Maintenance of Dead Front Distribution Switchboards Rated 600 Volts or Less,* 2007.

Guidelines for Handling Water-Damaged Electrical Equipment, NEMA, www.nema.org/stds/water-damaged.cfm.

NEMA SG 6, *Power Switching Equipment,* 2000 (Rescinded).

NEMA WD 6, *Dimensions for Wiring Devices,* 2002 (Rev. 2008).

Evaluating Water-Damaged Electrical Equipment, NEMA 2011.

2.3.6 NETA Publications. InterNational Electrical Testing Association, P.O. Box 687, Morrison, CO 80465.

ANSI/NETA ATS, *Standard for Acceptance Testing Specifications for Electrical Power Distribution Equipment and Systems,* 2009.

ANSI/NETA MTS, *Standard for Maintenance Testing Specifications for Electrical Power Distribution Equipment and Systems,* 2011.

2.3.7 UL Publications. Underwriters Laboratories Inc., 333 Pfingsten Road, Northbrook, IL 60062-2096.

ANSI/UL 489, *Molded-Case Circuit Breakers, Molded-Case Switches and Circuit Breaker Enclosures,* Eleventh edition, 2009, (Rev. 2011).

ANSI/UL 943, *Standard for Ground-Fault Circuit Interrupters,* 2006 (Rev. 2010).

UL 1436, *Outlet Circuit Testers and Similar Indicating Devices,* 2004.

2.3.8 U.S. Government Publications. U.S. Government Printing Office, Washington, DC 20402-9328.

Federal Emergency Management Agency (FEMA), FEMA P-348, *Protecting Building Utilities from Flood Damage,* 1999 updated 2012.

Title 29, *Code of Federal Regulations,* Part 1910.

Title 29, *Code of Federal Regulations,* Part 1926.

Title 29, *Code of Federal Regulations,* Part 1910.94(a), "Occupational Health and Environmental Control — Ventilation."

Title 29, *Code of Federal Regulations,* Part 1910.146, "Permit-Required Confined Spaces."

Title 29, *Code of Federal Regulations,* Part 1910.242(b), "Hand and Portable Powered Tools and Other Hand Held Equipment."

Title 29, *Code of Federal Regulations,* Part 1910.269, "Electric Power Generation, Transmission, and Distribution," Paragraph (e), Enclosed Spaces.

Title 29, *Code of Federal Regulations,* Part 1910.331 through Part 1910.335, "Safety Related Work Practices."

Title 40, *Code of Federal Regulations,* Part 761, "Protection of Environment — Polychlorinated Biphenyls (PCBs) Manufacturing, Processing, Distribution in Commerce, and Use Prohibitions."

TM 5-694, *Commissioning of Electrical Systems for Command, Control, Communications, Computer, Intelligence, Surveillance, and Reconnaissance (C4ISR) Facilities,* 2006.

TM 5-698-1, *Reliability/Availability of Electrical and Mechanical Systems for Command, Control, Communications, Computer, Intelligence, Surveillance, and Reconnaissance (C4ISR) Facilities,* 2003.

TM 5-698-2, *Reliability-Centered Maintenance (RCM) for Command, Control, Communications, Computer, Intelligence, Surveillance, and Reconnaissance (C4ISR) Facilities,* 2003.

TM 5-698-3, *Reliability Primer for Command, Control, Communications, Computer, Intelligence, Surveillance, and Reconnaissance (C4ISR) Facilities,* 2003.

Toxic Substances Control Act, Environmental Protection Agency, www.epa.gov/agriculture/lsca.html.

U.S. General Services Administration and U.S. Department of Energy, *Building Commissioning Guide,* 1998.

2.3.9 Other Publications.

Merriam-Webster's Collegiate Dictionary, 11th edition, Merriam-Webster, Inc., Springfield, MA, 2003.

ABB Power T & D Company, Inc., *Instruction Book PC-2000 for Wecosol ™ Fluid-Filled Primary and Secondary Unit Substation Transformers.*

Penn-Union Catalog, http://www.penn-union.com/Services/Literature.

PowerTest Annual Technical Conference, Flood Repair of Electrical Equipment, Pat Beisert, Shermco Industries, March 12, 2009.

Square D Catalog, Schneider Electric, www.schneider-electric.com/us.

Square D Services, *Procedures for Startup and Commissioning of Electrical Equipment,* PDF available at http://static.schneider-electric.us/docs/Electrical%20Distribution/Services/New%20Installation%20Services/O18OIB0001.pdf

2.4 References for Extracts in Recommendations Sections.

NFPA 70®, National Electrical Code®, 2011 edition.

NFPA 70E®, Standard for Electrical Safety in the Workplace®, 2012 edition.

Chapter 3 Definitions

3.1 General. The definitions contained in this chapter apply to the terms used in this recommended practice. Where terms are not defined in this chapter or within another chapter, they should be defined using their ordinarily accepted meanings within the context in which they are used. *Merriam-Webster's Collegiate Dictionary,* 11th edition, is the source for the ordinarily accepted meaning.

3.2 NFPA Official Definitions.

3.2.1* Approved. Acceptable to the authority having jurisdiction.

3.2.2* Authority Having Jurisdiction (AHJ). An organization, office, or individual responsible for enforcing the requirements of a code or standard, or for approving equipment, materials, an installation, or a procedure.

3.2.3* Listed. Equipment, materials, or services included in a list published by an organization that is acceptable to the authority having jurisdiction and concerned with evaluation of products or services, that maintains periodic inspection of production of listed equipment or materials or periodic evaluation of services, and whose listing states that either the equipment, material, or service meets appropriate designated standards or has been tested and found suitable for a specified purpose.

3.2.4 Recommended Practice. A document that is similar in content and structure to a code or standard but that contains only nonmandatory provisions using the word "should" to indicate recommendations in the body of the text.

3.2.5 Should. Indicates a recommendation or that which is advised but not required.

3.3 General Definitions.

3.3.1 Arc Flash Hazard. A dangerous condition associated with the possible release of energy caused by an electric arc. [**70E,** 2012]

3.3.2 Arc Flash Hazard Analysis. A study investigating a worker's potential exposure to arc flash energy, conducted for the purpose of injury prevention and the determination of safe work practices, arc flash boundary, and the appropriate levels of personal protective equipment (PPE). [**70E,** 2012]

3.3.3 Bonding (Bonded). The permanent joining of metallic parts to form an electrically conductive path that will ensure electrical continuity and the capacity to conduct safely any current likely to be imposed. The "permanent joining" can be accomplished by the normal devices used to fasten clean, noncorroded parts together. Machine screws, bolts, brackets, or retainers necessary to allow equipment to function properly are items typically employed for this purpose. While welding and brazing can also be utilized, these preclude easy disassembly, and welding can increase rather than decrease resistance across joints. Metallic parts that are permanently joined to form an electrically conductive path that will ensure electrical continuity and the capacity to conduct safely any current likely to be imposed are bonded.

3.3.4 Bonding Jumper. A reliable conductor to ensure the required electrical conductivity between metal parts required to be electrically connected. This conductor can be solid or stranded or braided, and connected by compatible fittings to separate parts to provide this electrically conductive path. The bonding jumper can also be a screw or a bolt. This bonding jumper can be used alone or in conjunction with other electrically conductive paths. It generally is associated with the equipment-grounding path, but might or might not be electrically linked for a lowest impedance path.

3.3.5 Case (Enclosure) Ground. See 3.3.44, Grounding Terminal.

3.3.6 Central Grounding Point. The location where the interconnected parts of the grounding system are connected in a common enclosure. The central grounding point provides a common connection point for termination of the feeder or branch-circuit equipment-grounding conductors.

3.3.7 Commissioning. A qualitative and quantitative process used to: (1) develop procedures to verify and document functional system-level and component-level requirements; (2) develop a testing and operational tune-up (system and component final adjustment) plan; (3) determine and record baseline information for operation and maintenance procedures; (4) evaluate initial system performance results and measurements.

3.3.8 Common Mode Noise. See 3.3.53.1.

3.3.9 Concurrent Maintenance. The testing, troubleshooting, repair, and/or replacement of a component or subsystem while redundant component(s) or subsystem(s) are serving the load, where the ability to perform concurrent maintenance is critical to attaining the specified reliability/availability criteria for the system or facility.

3.3.10 Continuous Duty. See 3.3.16.1.

3.3.11 Coordination (Selective). Localization of an overcurrent condition to restrict outages to the circuit or equipment affected, accomplished by the choice of overcurrent protective devices and their ratings or settings. [**70,** 2008]

3.3.12 Coordination Study. A system planning process used to assist in selecting and setting protective devices to improve power system reliability.

3.3.13* Corona. An electrical discharge phenomenon occurring in gaseous substances, such as air.

3.3.14 Counterpoise. A conductor or system of conductors arranged beneath the transmission/distribution supply line; located on, above, or most frequently below the surface of the earth; and connected to the grounding system of the towers or poles supporting the line. (This conductor(s) might or might not be the continuous length of the supply path. It is often used to provide a lower surge impedance path to earth for lightning protection when there is a transition from overhead supply conductors to underground insulated cable.) Counterpoise is also used in communication systems, where it is a system of conductors, physically elevated above and insulated from the ground, forming a lower system of conductors of an antenna. Note that the purpose of a counterpoise is to provide a relatively high capacitance and thus a relatively low impedance path to earth. The counterpoise is sometimes used in medium- and low-frequency applications where it would be more difficult to provide an effective ground connection. Sometimes counterpoise is confused with equipotential plane. See also 3.3.29, Equipotential Plane.

3.3.15 Down Conductor. A conductor from a lightning protection system to earth ground designed to provide a low impedance path for the current from a lightning strike and/or dissipate the charge buildup that precedes a lightning strike. This conductor typically goes from the air terminals to earth. Due to the very high currents at very high frequencies, the impedance of the entire system is very critical. Normal wiring conductors are not suitable for the down conductor. Typically, they are braided conductors. There might be certain instances where additional investigation about the interconnection between the lightning and the grounding electrode system is warranted.

3.3.16 Duty.

3.3.16.1 ***Continuous Duty.*** Operation at a substantially constant load for an indefinitely long time.

3.3.16.2 ***Intermittent Duty.*** Operation for alternate intervals of (1) load and no load; (2) load and rest; and (3) load, no load, and rest.

3.3.16.3 ***Periodic Duty.*** Intermittent operation in which the load conditions are regularly recurrent.

3.3.16.4 ***Short-Time Duty.*** Operation at a substantially constant load for a short and definitely specified time.

3.3.16.5 ***Varying Duty.*** Operation at loads, and for intervals of time, both of which might be subject to wide variation.

3.3.17 Earth Grounding. The intentional connection to earth through a grounding electrode of sufficiently low impedance to minimize damage to electrical components and prevent an electric shock that can occur from a superimposed voltage from lightning and voltage transients. In addition, earth grounding helps prevent the buildup of static charges on equipment and material. It also establishes a common voltage reference point to enable the proper performance of sensitive electronic and communications equipment.

3.3.18 Earthing. An IEC term for *ground.* See 3.3.30, Ground.

3.3.19 Effective Grounding Path. The path to ground from circuits, equipment, and metal enclosures for conductors shall (1) be permanent and electrically continuous, (2) have capacity to conduct safely any fault current likely to be imposed on it, and (3) have sufficiently low impedance to limit the voltage to ground and to facilitate the operation of the circuit protection devices. The earth should not be used as the sole equipment-grounding conductor.

3.3.20 Effectively Grounded (as applied to equipment or structures). Intentionally connected to earth (or some conducting body in place of earth) through a ground connection or connections of sufficiently low impedance and having sufficient current-carrying capacity to prevent the buildup of voltages that might result in undue hazards to connected equipment or to persons.

3.3.21 Effectively Grounded (as applied to systems). This is defined by ratios of impedance values that must be within prescribed limits.

3.3.22 Electrical Equipment. A general term applied to the material, fittings, devices, fixtures, and apparatus that are part of, or are used in connection with, an electrical installation and includes the electrical power-generating system; substations; distribution systems; utilization equipment; and associated control, protective, and monitoring devices.

3.3.23* Electrical Preventive Maintenance (EPM). A managed program of inspecting, testing, analyzing, and servicing electrical systems and equipment with the purpose of maintaining safe operations and production by reducing or eliminating system interruptions and equipment breakdowns.

3.3.24 Electrostatic Discharge (ESD) Grounding. The conductive path created to reduce or dissipate the electrostatic charge where it builds up as a result of equipment operation or induced from an electrostatically charged person or material coming in contact with the equipment. Also referred to as *static grounding.*

3.3.25 Equipment Bonding Jumper. The connection between two or more portions of the equipment-grounding conductor.

3.3.26 Equipment Ground. An ambiguous term that can mean either case ground, equipment-grounding conductor or equipment bonding jumper; hence, use of this term should be avoided.

3.3.27 Equipment-Grounding Conductor. The conductor used to connect the noncurrent-carrying metal parts of equipment, raceways, and other enclosures to the system grounded conductor, the grounding electrode conductor, or both, at the service equipment or at the source of a separately derived system.

3.3.28 Equipotential Bonding. Electrical connection putting various exposed conductive parts and extraneous conductive parts at a substantially equal potential.

3.3.29 Equipotential Plane. (1) (as applied to livestock) An area accessible to livestock where a wire mesh or other conductive elements are embedded in concrete, are bonded to all metal structures and fixed nonelectrical metal equipment that might become energized, and are connected to the electrical grounding system to prevent a difference in voltage from developing within the plane. (2) (as applied to equipment) A mass or masses of conducting material that, when bonded together, provide a uniformly low impedance to current flow over a large range of frequencies. Sometimes the equipotential plane is confused with counterpoise.

3.3.30 Ground. The earth. [**70,** 2008]

3.3.30.1 ***Lightning Ground.*** See 3.3.43, Grounding Electrode System.

3.3.30.2 ***Noise(less) Ground.*** The supplemental equipment-grounding electrode installed at machines, or the isolated equipment-grounding conductor, intended to reduce electrical noise.

3.3.30.3 ***Personnel Protective Ground.*** Bonding jumper that is intentionally installed to ground deenergized, normally ungrounded circuit conductors when personnel are working on them, to minimize voltage differences between different parts of the equipment and personnel, so as to protect against shock hazard and/or equipment damage.

3.3.30.4 ***Safety Ground.*** See 3.3.30.3, Personnel Protective Ground.

3.3.31 Ground Fault. Unintentional contact between an ungrounded conductor and earth or conductive body that serves in place of earth. Within a facility, this is typically a fault between a current-carrying conductor and the equipment-grounding path that results in the operation of the overcurrent protection.

3.3.32 Ground Leakage Current. Current that is introduced into the grounding conductor by normal equipment operation, such as capacitive coupling. Many RFI/EMI filters in electronic equipment have capacitors from current-carrying conductors to the equipment-grounding conductor to shunt noise emitted from or injected into their power supplies. While there are relatively low current level limits imposed by regulatory agencies (e.g., UL specifies maximum 3.5 mA, hospital equipment 0.5 mA), not all equipment is listed. Even with listed equipment, the sum of the current from a large quantity of such equipment in a facility can result in significant ground currents.

3.3.33 Ground Loop. Multiple intentional or unintentional connections from a conductive path to ground or the conductive body that serves in place of earth. Current will flow in the ground loop if there is voltage difference between the connection nodes. Regrounding of the grounded circuit conductor (neutral) beyond the service point will result in ground loops. This might or might not be harmful depending on the application.

3.3.34 Ground Resistance/Impedance Measurement. The use of special test equipment to measure the grounding electrode resistance or impedance to earth at a single frequency at or near power line frequency.

3.3.35 Ground Well. See 3.3.43, Grounding Electrode System.

3.3.36 Grounded (Grounding). Connected (connecting) to ground or to a conductive body that extends the ground connection. [**70,** 2008]

3.3.37 Grounded Conductor. A system or circuit conductor that is intentionally grounded. This intentional grounding to earth or some conducting body that serves in place of earth takes place at the premises service location or at a separately derived source. Control circuit transformers are permitted to have a secondary conductor bonded to a metallic surface that is in turn bonded to the supply equipment-grounding conductor. Examples of grounded system conductors would be a grounded system neutral conductor (three phase or split phase) or a grounded phase conductor of a 3-phase, three-wire, delta system.

3.3.38* Ground-Fault Circuit Interrupter (GFCI). A device intended for the protection of personnel that functions to deenergize a circuit or portion thereof within an established period of time when a current to ground exceeds the values established for a Class A device. Note: Class A Ground-Fault Circuit Interrupters trip when the current to ground has a value in the range of 4 mA to 6 mA. For further information, see UL 943, *Standard for Ground-Fault Circuit Interrupters.* [**70,** 2008]

3.3.39* Ground-Fault Protection of Equipment (GFP). A system intended to provide protection of equipment from damaging line-to-ground fault currents by operating to cause a disconnecting means to open all ungrounded conductors of the faulted circuit. This protection is provided at current levels less than those required to protect conductors from damage through the operation of a supply circuit overcurrent device. [**70,** 2008]

3.3.40 Grounding.

3.3.40.1 ***Multipoint Grounding.*** Multipoint grounding consists of interconnecting primary and secondary neutrals of the transformer. The secondary and primary neutral are common, and they both utilize the same grounding electrode that connects the system to earth.

3.3.40.2 ***Single Point Grounding.*** The single-point grounding of a transformer means connecting the secondary side of the transformer to earth ground through one or more grounding electrodes. This connection should be made at any point on the separately derived system from the source to the first system-disconnecting means or overcurrent device.

3.3.40.3 ***System Grounding.*** The intentional connection of an electrical supply system to its associated grounding electrode(s).

3.3.41 Grounding Electrode. A conductive body deliberately inserted into earth to make electrical connection to earth. Typical grounding electrodes include the following: (1) The nearest effectively grounded metal member of the building structure (2) The nearest effectively grounded metal water pipe, but only if the connection to the grounding electrode conductor is within 5 ft of the point of entrance of the water pipe to the building (3) Any metal underground structure that is effectively grounded (4) Concrete encased electrode in the foundation or footing (e.g., Ufer ground) (5) Ground ring completely encircling the building or structure (6) Made electrodes (e.g., ground rods or ground wells) (7) Conductive grid or mat used in substations

3.3.42 Grounding Electrode Conductor. The conductor used to connect the grounding electrode to the equipment-grounding conductor, to the grounded conductor, or to both, of the circuit at the service equipment or at the source of a separately derived system. This conductor must be connected to provide the lowest impedance to earth for surge current due to lightning, switching activities from either or both of the supply and load side, and to reduce touch potentials when equipment insulation failures occur.

3.3.43 Grounding Electrode System. The interconnection of grounding electrodes.

3.3.44 Grounding Terminal. A terminal, lug, or other provision provided on some equipment cases (enclosures) to connect the conductive portion of the enclosure to the equipment-grounding conductor.

3.3.45 Grounding-Type Receptacle. A receptacle with a dedicated terminal that is to be connected to the equipment grounding conductor.

3.3.46 Harmonics. Those voltages or currents whose frequencies are integer multiples of the fundamental frequency.

3.3.47 Interharmonics. Not all frequencies that occur on an electrical power system are integer multiples of the fundamental frequency (usually 60 Hz), as are harmonics. Some loads draw currents that result in voltages that are between harmonic frequencies or less than the fundamental frequency. These frequencies are referred to as interharmonics and can be made of discrete frequencies or as a wide-band spectrum. A special category of these interharmonics is called subharmonics, in which the frequencies involved are less than the fundamental power line frequency.

3.3.48 Intermittent Duty. See 3.3.16.2.

3.3.49 Isolated Equipment-Grounding Conductor. An insulated equipment-grounding conductor that has one intentional connection to the equipment-grounding system. The isolated equipment-grounding conductor is typically connected to an equipment-grounding terminal either in the facility's service enclosure or in the first applicable enclosure of a separately derived system. The isolated equipment-grounding conductor should be connected to the equipment-grounding system within the circuits' derived system.

3.3.50 Lightning Ground. See 3.3.30.1.

3.3.51 Long Duration Undervoltage. A decrease of the supply voltage to less than 90 percent of the nominal voltage for a time duration greater than 1 minute. *[See IEEE 1159, Recommended Practice on Monitoring Electric Power Quality, Table 4-2.]*

3.3.52 Multipoint Grounding. Multipoint grounding consists of interconnecting primary and secondary neutrals of the transformer. The secondary and primary neutral are common, and they both utilize the same grounding electrode that connects the system to earth.

3.3.53 Noise. Undesirable electrical signals in an electrical or electronic circuit.

3.3.53.1 ***Common Mode Noise.*** Undesirable electrical signals that exist between a circuit conductor and the grounding conductor.

3.3.53.2 ***Transverse Mode Noise.*** Undesirable electrical signals that exist between a pair of circuit conductors. These signals are sometimes referred to as normal or differential mode noise.

3.3.54 Noise(less) Ground. See 3.3.30.2.

3.3.55 Periodic Duty. See 3.3.16.3.

3.3.56 Personnel Protective Ground. See 3.3.30.3.

3.3.57 Power Transformers. Determines the type of transformer and is defined as those larger than 500 kVA, while distribution transformers are those 500 kVA or smaller.

3.3.58 Protective Bonding Circuit. See 3.3.28, Equipotential Bonding.

3.3.59 Protective Conductor. A conductor required by some measures for protection against electric shock for electrically connecting any of the following parts: exposed conductive parts, extraneous conductive parts, or main (grounding) earthing terminal. Also identified in some instances as the protective external (PE) conductor. *(See also 14.1.6.15, Equipment-Grounding Conductor.)*

3.3.60 Protective Ground. See 3.3.28, Equipotential Bonding.

3.3.61 Qualified Person. One who has the skills and knowledge related to the construction and operation of the electrical equipment and installations and has received safety training on the hazards involved.

3.3.62 RFI/EMI Grounding. See 3.3.32, Ground Leakage Current.

3.3.63 Safety Ground. See 3.3.30.3, Personnel Protective Ground.

3.3.64 Sag. A decrease to between 10 percent and 90 percent of the normal voltage at the power frequency for durations of 0.5 cycle to 1 minute. (If the voltage drops below 10 percent of the normal voltage, then this is classified as an interruption.) It is further classified into three categories: (1) instantaneous — 0.5 cycle to 30 cycles; (2) momentary — 30 cycles to 3 seconds; and (3) temporary — 3 seconds to 1 minute.

3.3.65 Separately Derived System. A premises wiring system whose power is derived from a battery, a solar photovoltaic system, or from a generator, transformer, or converter windings, and that has no direct electrical connection, including a solidly connected grounded circuit conductor, to supply conductors originating in another system. Equipment-grounding conductors are not supply conductors and are to be interconnected.

3.3.66 Shield Ground. Intentional grounding of one or both ends of the shield of a cable.

3.3.66.1 ***Data Communications Cables.*** The shield of data communication cables can be connected to the equipment-grounding conductor at either one end of the cable (single end) or at both ends (double ended). When both ends of a shield are grounded, another shield should be provided inside the outer shield and that one single end grounded.

3.3.66.2 ***Shield Ground, Power Cables.*** The shield of power cables can be connected to the equipment-grounding conductor at either one end of the cable (single end) or at both ends (double ended). Shielding will ensure uniform dielectric stress along the length of the cable. When grounded at both ends, cable derating might be necessary because of heat due to ground loop current.

3.3.67 Short-Time Duty. See 3.3.16.4.

3.3.68 Single-Point Grounding. See 3.3.40.2.

3.3.69 Substation Ground. Grounding electrode system (grid) in a substation. See 14.1.6.28, Grounding Electrode System.

3.3.70 Survey. The collection of accurate data on the electrical system and the evaluation of this data to obtain the necessary information for developing the EPM program. The systems and equipment covered in specific parts of the survey should be based on logical divisions of the electrical system.

3.3.71 Sustained Voltage Interruption. The loss of the supply voltage to less than 10 percent on one or more phases for a period greater than 1 minute.

3.3.72 Swell. An increase to between 110 percent and 180 percent in normal voltage at the power frequency durations from 0.5 cycle to 1 minute. It is further classified into three categories: (1) instantaneous — 0.5 cycle to 30 cycles; (2) momentary — 30 cycles to 3 seconds; and (3) temporary — 3 seconds to 1 minute.

3.3.73 System Grounding. See 3.3.40.3.

3.3.74 Transformer. A device for changing energy in an alternating current system from one voltage to another; usually includes two or more insulated coils on an iron core.

3.3.75 Transients. Transients (formerly referred to as surges, spikes, or impulses) are very short duration, high amplitude excursions outside of the limits of the normal voltage and current waveform. Waveshapes of the excursions are usually unidirectional pulses or decaying amplitude, high frequency oscillations. Durations range from fractions of a microsecond to milliseconds, and the maximum duration is in the order of one half-cycle of the power frequency. Instantaneous amplitudes of voltage transients can reach thousands of volts.

3.3.76 Transverse Mode Noise. See 3.3.53.2.

3.3.77 Unbalanced Voltages. Unequal voltage values on 3-phase circuits that can exist anywhere on the power distribution system.

3.3.78 Varying Duty. See 3.3.16.5.

Chapter 4 Why an Effective Electrical Preventive Maintenance (EPM) Program Pays Dividends

4.1 Why EPM?

4.1.1 Electrical equipment deterioration is normal, and equipment failure is inevitable. However, equipment failure can be delayed through appropriate EPM. As soon as new equipment is installed, a process of normal deterioration begins. Unchecked, the deterioration process can cause malfunction or an electrical failure. Deterioration can be accelerated by factors such as a hostile environment, overload, or severe duty cycle. An effective EPM program identifies and recognizes these factors and provides measures for coping with them.

4.1.2 In addition to normal deterioration, other potential causes of equipment degradation can be detected and corrected through EPM. Among these are load changes or additions, circuit alterations, improperly set or improperly selected protective devices, and changing voltage conditions.

4.1.3 Without an EPM program, management assumes a greatly increased risk of a serious electrical failure and its consequences.

4.2 Value and Benefits of a Properly Administered EPM Program.

4.2.1 A well-administered EPM program reduces accidents, saves lives, and minimizes costly breakdowns and unplanned shutdowns of production equipment. Impending troubles can be identified — and solutions applied — before they become major problems requiring more expensive, time-consuming solutions.

4.2.2 Benefits of an effective EPM program fall into two general categories. Direct, measurable economic benefits are derived from reduced cost of repairs and reduced equipment downtime. Less measurable but very real benefits result from improved safety. To understand fully how personnel and equipment safety are served by an EPM program, the mechanics of the program — inspection, testing, and repair procedures — should be understood. Such an understanding explains other intangible benefits such as improved employee morale, better workmanship and increased productivity, reduced absenteeism, reduced interruption of production, and improved insurance considerations. Improved morale comes with employee awareness of a conscious management effort to promote safety by reducing the likelihood of electrical injuries or fatalities, electrical explosions, and fires. Reduced personnel injuries and property loss claims can help keep insurance premiums at favorable rates.

4.2.3 Some of the benefits that result from improved safety are difficult to measure. However, direct and measurable economic benefits can be documented by equipment repair cost and equipment downtime records after an EPM program has been implemented.

4.2.4 Dependability can be engineered and built into equipment, but effective maintenance is required to keep it dependable. Experience shows that equipment is reduced when it is covered by an EPM program. In many cases, the investment in EPM is small compared with the cost of accidents, equipment repair, and the production losses associated with unexpected outages.

4.2.5 Careful planning is the key to the economic success of an EPM program. With proper planning, maintenance costs can be held to a practical minimum, while production is maintained at a practical maximum.

4.2.6 An EPM program requires the support of top management, because top management provides the funds that are required to initiate and maintain the program. The maintenance of industrial electrical equipment is essentially a matter of business economics. Maintenance costs can be placed in either of two basic categories: preventive maintenance or breakdown repairs. The money spent for preventive maintenance will be reflected as less money required for breakdown repairs. An effective EPM program holds the sum of these two expenditures to a minimum. Figure 4.2.6 is a typical curve illustrating this principle. According to this curve, as the interval of time between EPM inspections increases, the cost of the EPM diminishes and the cost of breakdown repairs and replacement of failed equipment increases. The lowest total annual expense is realized by maintaining an inspection frequency that keeps the sum of repair/replacement and EPM costs at a minimum.

4.2.7 An EPM program is a form of protection against accidents, lost production, and loss of profit. An EPM program enables management to place a monetary value on the cost of such protection. An effective EPM program satisfies an important part of management's responsibility for keeping costs down and production up.

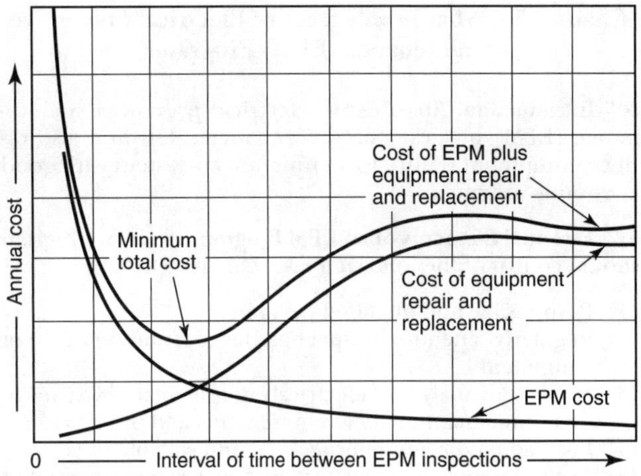

FIGURE 4.2.6 Effect of EPM Inspection Frequency on Overall Costs.

4.2.8* Insurance statistics document the high cost of inadequate electrical maintenance.

4.3 EPM and Energy Conservation. Energy conservation is one of the worthwhile benefits associated with an EPM program, saving money and vital resources. Equipment that is well maintained operates more efficiently and utilizes less energy.

4.4 Case Histories.

4.4.1 In one industrial plant, the failure of a transformer caused a total plant shutdown. Contamination of the transformer's insulating oil caused the failure. The contamination went undetected because the oil had not been tested for several years. Fire damage and equipment replacement costs amounted to $50,000 (U.S.), exclusive of the cost of plant downtime. This amount would have paid for the cost of operating an EPM program covering the entire plant's electrical distribution system for several years.

4.4.2 In another industrial plant, damage amounting to $100,000 (U.S.) was attributed to the failure of the main switchgear. Fouling by dirt, gummy deposits, and iron filings caused the failure. The cost of this failure would have supported a comprehensive EPM program covering all of the plant's electrical distribution system for several years.

4.4.3 McCormick Place, a large exhibition hall in Chicago, was destroyed by a fire believed to have been started because of a defective extension cord serving a display booth. Direct property loss was $60 million (U.S.), and loss of the facility cost an additional $100 million (U.S.) to the economy in the Chicago area. This fire might have been prevented if a program had been in effect to ensure that worn cords were replaced, that only heavy-duty cords were used, and that cords and their supply circuits were not overloaded.

4.4.4 The failure of a large motor shut down an entire industrial plant for 12 days. The cause of the failure was overheating resulting from dust-plugged cooling ducts. An EPM inspection would have detected the clogged ducts and averted the failure and accompanying plant outage.

Chapter 5 What Is an Effective Electrical Preventive Maintenance (EPM) Program?

5.1 Introduction. An effective electrical preventive maintenance (EPM) program should enhance safety and also reduce equipment failure to a minimum consistent with good economic judgment.

5.2 Essential Elements of an EPM Program. An EPM program should consist of the following essential elements:

(1) Responsible and qualified personnel
(2) Regularly scheduled inspection, testing, and servicing of equipment
(3) Survey and analysis of electrical equipment and systems to determine maintenance requirements and priorities
(4) Programmed routine inspections and suitable tests
(5) Accurate analysis of inspection and test reports so that proper corrective measures can be prescribed
(6) Performance of necessary work
(7) Concise but complete records

5.3 Planning an EPM Program. The following factors should be considered in the planning of an EPM program.

(1) *Personnel Safety*: Will an equipment failure endanger or threaten the safety of any personnel? What can be done to ensure personnel safety?
(2) *Equipment Loss*: Is installed equipment — both electrical and mechanical — complex or so unique that required repairs would be unusually expensive?
(3) *Production Economics*: Will breakdown repairs or replacement of failed equipment require extensive downtime? How many production dollars will be lost in the event of an equipment failure? Which equipment is most vital to production?

5.4 Personnel.

5.4.1 A well-qualified individual should be in charge of the program. (*See 6.1.3.2.*)

5.4.2 Personnel assigned to electrical preventive maintenance duties should be selected from the most technically qualified personnel in the plant.

5.4.3 Where in-plant personnel are not qualified, a technically competent maintenance contractor should be employed.

5.5 Survey and Analysis.

5.5.1 Survey and analysis should cover equipment and systems that have been determined to be essential in accordance with a priority plan.

5.5.2 Regardless of the size of the program being contemplated, the EPM supervisor should determine the scope of the work to be done and where to begin.

5.5.3 All electrical equipment — such as motors, transformers, circuit breakers, and controls — should receive a thorough inspection and evaluation to permit the EPM supervisor to make a qualified judgment as to how, where, and when each piece of equipment should fit into the program.

5.5.4 In addition to determining the equipment's physical condition, the survey should determine if the equipment is operating within its rating, and how the load level could affect the frequency of maintenance.

5.5.5 It should be stressed that environmental or operating conditions of a specific installation should be considered and might dictate a different frequency of maintenance.

5.5.6 In the course of the survey, the condition of electrical protective devices such as fuses, circuit breakers, protective relays, and motor overload relays should be checked. These devices are the safety valves of an electrical system, and their proper operation ensures the safety of personnel, protection of equipment, and reduction of economic loss.

5.5.7 After the survey has been completed, data should be evaluated to determine equipment condition. Equipment condition will reveal repair work to be done, as well as determine the nature and frequency of required inspections and tests.

5.6 Programmed Inspections. Inspection and testing procedures should be carefully tailored to requirements. In some plants, regularly scheduled tests will call for scheduled outages of production or process equipment. In such cases, close coordination between maintenance and production personnel is necessary.

5.6.1 Analysis of Inspection and Test Reports. Analysis of inspection and test reports should be followed by implementation of appropriate corrective measures. Follow-through with necessary repairs, replacement, and adjustment is the end purpose of an effective EPM program.

5.6.2 Records.

5.6.2.1 Records should be accurate and contain all vital information.

5.6.2.2 Care should be taken to ensure that all relevant information becomes part of the record.

5.6.3 EPM Support Procedures.

5.6.3.1 Design for Ease of Maintenance. Effective electrical preventive maintenance begins with good design. In the design of new facilities, a conscious effort to ensure optimum maintainability is recommended. Dual circuits, tie circuits, auxiliary power sources, and drawout protective devices make it easier to schedule maintenance and to perform maintenance work with minimum interruption of production. Other effective design techniques include equipment rooms to provide environmental protection, grouping of equipment for more convenience and accessibility, and standardization of equipment and components.

5.6.4 Training for Safety and Technical Skills.

5.6.4.1 Training Requirements.

5.6.4.1.1 All employees who face a risk of electrical hazard should be trained to understand the specific hazards and related injuries associated with electrical energy.

5.6.4.1.2 All employees should be trained in safety-related work practices and required procedures as necessary to provide protection from electrical hazards associated with their jobs or task assignments.

5.6.4.1.3 Refresher training should be provided as required.

5.6.4.2 Type of Training. The training can be in the classroom, on the job, or both. The type of training should be determined by the needs of the employee.

5.6.4.3 Emergency Procedures. Employees working on or near exposed energized electrical conductors or circuit parts should be instructed regularly and be familiar with methods of first aid and emergency procedures, such as approved methods of resuscitation, release of victims from contact with exposed energized conductors or circuit parts, and any other emergency procedures that are related to their work and necessary for their safety.

5.6.4.4 Training Scope. Employees should be trained and knowledgeable in the following:

(1) Construction and operation of equipment
(2) Specific work method
(3) Electrical hazards that can be present with respect to specific equipment or work method
(4) Proper use of special precautionary techniques, personal protective equipment, insulating and shielding materials, and insulated tools and test equipment
(5) Skills and techniques necessary to distinguish exposed, energized parts from other parts of electrical equipment
(6) Skills and techniques necessary to determine the nominal voltage of exposed energized parts
(7) Decision-making process necessary to determine the degree and extent of hazard
(8) Job planning necessary to perform the task safely
(9) Self-discipline necessary to maintain a safe work environment

5.6.4.5 Recordkeeping. Records of training should be maintained for each employee.

5.6.5 Outside Service Firms. Some maintenance and testing operations, such as relay and circuit-breaker inspection and testing, require specialized skills and special equipment. In small organizations, it might be impractical to develop the skills and acquire the equipment needed for this type of work. In such cases, it might be advisable to contract the work to firms that specialize in providing such services.

5.6.6 Tools and Instruments. Proper tools and instruments are an important part of an EPM program, and safety protective gear is an essential part of the necessary equipment. Proper tools, instruments, and other equipment should be used to ensure maximum safety and productivity from the maintenance crew. There should be adequate storage facilities for tools and test equipment that are common to each maintenance work group. Where specialized instruments and test equipment are needed only occasionally, they can be rented from a variety of sources.

Chapter 6 Planning and Developing an Electrical Preventive Maintenance (EPM) Program

6.1 Introduction.

6.1.1 The purpose of an EPM program is to reduce hazard to life and property resulting from the failure or malfunction of electrical systems and equipment. This chapter explains the planning and development considerations that can be used to establish such a program.

6.1.2 The following four basic steps should be taken in the planning and development of an EPM program:

(1) Compile a listing of all equipment and systems.
(2) Determine which equipment and systems are most critical.
(3) Develop a system for monitoring.
(4) Determine the internal and/or external personnel needed to implement and maintain the EPM program.

6.1.3 A single individual should have the overall responsibility for EPM program implementation.

6.1.3.1 The individual responsible for the EPM program should be given the authority to perform the job and should have the cooperation of management, production, and other departments whose operations might affect the EPM program.

6.1.3.2 Ideally, the person designated to head the EPM program should have the following qualifications:

(1) *Technical competence.* The person should, by education, training, and experience, be well-rounded in all aspects of electrical maintenance.
(2) *Administrative and supervisory skills.* The person should be skilled in the planning and development of long-range objectives to achieve specific results and should be able to command respect and solicit the cooperation of all persons involved in the program.

6.1.4 The maintenance supervisor should have open lines of communication with design supervision. Frequently, an unsafe installation or one that requires excessive maintenance can be traced to improper design or construction methods or misapplication of hardware.

6.1.5 The work center of each maintenance work group should be conveniently located. This work center should contain the following:

(1) Copies of all the inspection and testing procedures for that zone
(2) Copies of previous reports
(3) Single-line diagrams
(4) Schematic diagrams
(5) Records of complete nameplate data
(6) Vendors' catalogs
(7) Facility stores' catalogs
(8) Supplies of report forms

6.1.6 In a continuously operating facility, running inspections (inspections made with equipment operating) play a vital role in the continuity of service. The development of running inspection procedures varies with the type of operation. Running inspection procedures should be as thorough as practicable within the limits of safety and the skill of the maintenance personnel. These procedures should be reviewed to keep them current. Each failure of electrical equipment, be it an electrical or a mechanical failure, should be reviewed against the running inspection procedure to determine if some other inspection technique would have indicated the impending failure. If so, the procedure should be modified to reflect the findings.

6.1.7 When the electrical maintenance supervisor initiates corrective action, the maintenance personnel should be so informed. The maintenance personnel who found the condition will then realize their importance in the EPM program. However, if nothing is done, individual motivation and the EPM program might be affected adversely.

6.2 Survey of Electrical Installation.

6.2.1 Data Collection.

6.2.1.1 The first step in organizing a survey should be to examine available resources. Will the available personnel permit the survey of an entire system, process, or building, or should it be divided into segments?

6.2.1.2 Where the project will be divided into segments, a priority should be assigned to each segment. Segments found to be related should be identified before the actual work commences.

6.2.1.3 The third step should be the assembling of all documentation. This might necessitate a search of desks, cabinets, computers, and such, and might also require that manufacturers be contacted, to replace lost documents. All of the documents should be centralized, controlled, and maintained. The documentation should include recommended practices and procedures for some or all of the following:

(1) Installation
(2) Disassembly/assembly (interconnections)
(3) Wiring diagrams, schematics, bills of materials
(4) Operation (set-up and adjustment)
(5) Maintenance (including parts list and recommended spares)
(6) Software program (if applicable)
(7) Troubleshooting

6.2.2 Diagrams and Data. The availability of up-to-date, accurate, and complete diagrams is the foundation of a successful EPM program. The diagrams discussed in 6.2.2.1 through 6.2.2.8.2 are some of those in common use.

6.2.2.1 Single-line diagrams should show all electrical equipment in the power system and give all pertinent ratings. In making this type of diagram, it is basic that voltage, frequency, phase, and normal operating position be included. No less important, but perhaps less obvious, are items such as transformer impedance, available short-circuit current, all overcurrent protective device types, ampere ratings, settings, and interrupting ratings. Other items include current and potential transformers and their ratios, surge capacitors, and protective relays. If one diagram cannot cover all the equipment involved, additional diagrams, appropriately noted on the main diagram, can be drawn.

6.2.2.2 Some managers have the misconception that these engineering studies are part of the initial facility design, after which the subject can be forgotten. Engineering studies such as short-circuit coordination and arc flash studies are important and should be updated periodically based on a number of factors including changes in the supply capacity of the source of power, changes in size or percent impedance of transformers, changes in conductor size, addition of motors, and changes in system operating conditions.

6.2.2.2.1 In the course of periodic maintenance testing of protective equipment, such as relays and series or shunt-trip devices, equipment settings should be evaluated. Along with the proper sizing of fuses, this evaluation is part of the coordination study.

6.2.2.2.2 It is desirable to develop a computerized short-circuit study to improve accuracy and reduce engineering time. Should resources not be available within the facility organization, the short-circuit study can be performed on a contract basis.

6.2.2.2.3 Fuses are rated on the basis of their current-carrying and interrupting capacities. These ratings should be determined and recorded. Other protective devices are usually adjustable as to pickup point and time-current characteristics. The settings of such protective devices should be determined by engineering studies, verified by electrical tests, and recorded for future reference.

6.2.2.2.4 Personnel performing the tests should be trained and qualified. Various organizations and manufacturers of power and test equipment periodically schedule seminars in which participants are taught the principles of maintenance and testing of electrical protective devices.

6.2.2.2.5 The available short-circuit current data is used as follows:

(1) To determine the momentary and interrupting ratings of circuit breakers and fuses
(2) To determine the short-circuit current ratings for equipment
(3) As required information to conduct the following:
    (a) Overcurrent protective device coordination studies
    (b) Arc flash hazard analysis

6.2.2.2.6 Additional guidance on electrical systems can be found in Chapter 28.

6.2.2.3 Circuit-routing diagrams, cable maps, or raceway layouts should show the physical location of conductors. In addition to voltage, such diagrams should also indicate the type of raceway, number and size of conductors, and type of insulation.

6.2.2.3.1 Where control conductors or conductors of different systems are contained within the same raceway, the identification appropriate to each conductor should be noted.

6.2.2.3.2 The location of taps, headers, and pull boxes should be shown on the circuit routing diagrams.

6.2.2.3.3 Access points for raceways should be noted.

6.2.2.4 Layout diagrams, plot plans, equipment location plans, or facility maps should show the physical layout (and in some cases, the elevations) of all equipment in place.

6.2.2.4.1 Switching equipment, transformers, control panels, mains, and feeders should be identified.

6.2.2.4.2 Voltage and current ratings should be shown for each piece of equipment.

6.2.2.5 Schematic diagrams should be arranged for simplicity and ease of understanding circuits without regard for the actual physical location of any components. The schematic should always be drawn with switches and contacts shown in a deenergized position.

6.2.2.6 Wiring diagrams, like schematics, should show all components in the circuit and arranged in their actual physical location. Wiring diagrams should identify all equipment parts and devices by standard methods, symbols, and markings. Of particular value is the designation of terminals and terminal strips with their appropriate numbers, letters, or colors.

6.2.2.7 An effective EPM program should have manufacturers' service manuals and instructions. These manuals should include recommended practices and procedures.

6.2.2.8 Electrical Equipment Installation Change. The documentation of the changes that result from engineering decisions, planned revisions, and so on, should be the responsibility of the engineering group that initiates the revisions.

6.2.2.8.1 Periodically, changes occur as a result of an EPM program. The EPM program might also uncover undocumented practices or installations.

6.2.2.8.2 A responsibility of those responsible for the EPM program is to highlight these changes, note them in an appropriate manner, and formally submit the revisions to the organization responsible for the maintenance of the documentation.

6.2.3 System Diagrams. System diagrams should be provided to complete the data being assembled. The importance of the system determines the extent of information shown. The information can be shown on the most appropriate type of diagram but should include the same basic information, source and type of power, conductor and raceway information, and switching and protective devices with their physical locations. It is vital to show where the system might interface with another system, such as with emergency power; hydraulic, pneumatic, or mechanical systems; security and fire-alarm systems; and monitoring and control systems. Some of the more common of these are described in 6.2.3.1 through 6.2.3.3.

6.2.3.1 Lighting System Diagrams. Lighting system diagrams (normal and emergency) can terminate at the branch circuit panelboard, listing the number of fixtures, type and lamp size for each area, and design lighting level. The diagram should show watchman night lighting lights and probably an automatic transfer switch to the emergency power system.

6.2.3.2 Heating, Ventilation, and Air-Conditioning. Ventilation systems normally comprise the heating, cooling, and air-filtering system. Basic information, including motor and fan sizes, motor or pneumatically operated dampers, and so on, should be shown. Additionally, many safety features can be involved to ensure that fans start before the process — airflow switches to shut down an operation on loss of ventilation and other interlocks of similar nature. Each of these should be identified with respect to type, function, physical location, and operating limits. Heating and air-conditioning systems are usually manufactured and installed as a unit, furnished with diagrams and operating and maintenance manuals. This information should be updated as the system is changed or modified. Because these systems are often critical to the facility operation, additional equipment might have been incorporated — for example, humidity, lint, and dust control for textile, electronic, and similar processes, and corrosive and flammable vapor control for chemical and related industries. Invariably, these systems interface with other electrical or nonelectrical systems; pneumatic or electromechanical operation of dampers, valves, and so on, electric operation for normal and abnormal temperature control, and manual control stations for emergency smoke removal are just a few. There might be others, and all should be shown and complete information given for each.

6.2.3.3 Control and Monitoring. Control and monitoring system diagrams should be provided to describe how these complicated systems function. They usually are in the form of a schematic diagram and can refer to specific wiring diagrams. Maximum benefit can be obtained only when every switching device is shown, its function is indicated, and it is identified for ease in finding a replacement. These devices often involve interfaces with other systems, whether electromechanical (heating or cooling medium) pumps and valves, electro-pneumatic temperature and damper controls, or safety and emergency operations. A sequence-of-operation chart and a list of safety precautions should be included to promote the safety of personnel and equipment. Understanding these complex circuits is best accomplished by breaking down the circuits into their natural functions, such as heating, cooling, process, or humidity controls. The knowledge of how each function relates to another enables the maintenance personnel to have a better understanding of the entire system and thus perform assignments more efficiently.

6.2.4 Emergency Procedures. Emergency procedures should list, step by step, the action to be taken in case of emergency or for the safe shutdown or start-up of equipment or systems. *(See Section 6.9 for details.)* Optimum use of these procedures is made when they are readily available near the area of the equipment or systems. If an emergency could make some areas of the facility inaccessible, the associated emergency procedures should be located outside the potentially affected area. Some possible items to consider for inclusion in the emergency procedures are interlock types and locations, interconnections with other systems, and tagging procedures of the equipment or systems. Accurate single-line diagrams posted in strategic places are particularly helpful in emergency situations. The production of such diagrams in anticipation of an emergency is essential to a complete EPM program. Diagrams are a particularly important training tool in developing a state of preparedness. Complete and up-to-date diagrams provide a quick review of the emergency plan. During an actual emergency, when time is of the essence, they provide a simple, quick reference guide.

6.2.5 Test and Maintenance Equipment.

6.2.5.1 All maintenance work requires the use of proper tools and equipment to properly perform the task to be done. In addition to their ordinary tools, maintenance personnel (such as carpenters, pipe fitters, and machinists) use special tools or equipment based on the nature of the work to be performed. The electrician is no exception, but for EPM, special-use tools should be readily available. The size of the facility, the nature of its operations, and the extent of its maintenance, repair, and test facilities are all factors that determine the use frequency of the equipment. Economics seldom justify purchasing an infrequently used, expensive tool when it can be rented. However, a corporation having a number of facilities in the area might well justify common ownership of the same device for joint use, making it quickly available at any time to any facility. Typical examples might be high-current test equipment, infrared thermography equipment, or a ground-fault locator.

6.2.5.2 Because a certain amount of mechanical maintenance is often a part of the EPM program being conducted on associated equipment, the electrical maintenance personnel should have ready access to such items as the following:

(1) Assorted lubrication tools and equipment
(2) Various types and sizes of wrenches
(3) Nonmetallic hammers and blocks to protect against injury to machined surfaces
(4) Feeler gauges to function as inside- and outside-diameter measuring gauges
(5) Instruments for measuring torque, tension, compression, vibration, and speed
(6) Standard and special mirrors with light sources for visual inspection
(7) Industrial-type portable blowers and vacuums having insulated nozzles for removal of dust and foreign matter
(8) Nontoxic, nonflammable cleaning solvents
(9) Clean, lint-free wiping cloths

6.2.5.3 The use of well-maintained safety equipment is essential and should be mandatory for work on energized electrical conductors or circuit parts. Prior to performing maintenance on energized electrical conductors or circuit parts, *NFPA 70E, Standard for Electrical Safety in the Workplace*, should be used to identify the degree of personal protective equipment (PPE) required. Some of the more important equipment that should be provided includes the following:

(1) Heavy leather gloves
(2) Insulating gloves, mats, blankets, baskets, boots, jackets, and coats

(3) Insulated hand tools such as screwdrivers and pliers
(4) Nonmetallic hard hats with suitable arc-rated face protection
(5) Poles with hooks and hot sticks to safely open isolating switches

6.2.5.3.1 A statiscope is recommended to indicate the presence of high voltage on certain types of equipment.

6.2.5.4 Portable electric lighting should be provided, particularly in emergencies involving the power supply. Suitable extension cords should be provided.

6.2.5.4.1 Portable electric lighting used for maintenance areas that are normally wet or where personnel will be working within grounded metal structures such as drums, tanks, and vessels should be operated at an appropriate low voltage from an isolating transformer or through a GFCI device. The aim is to limit the exposure of personnel to hazardous current levels, or limit the voltage.

6.2.5.5 Portable meters and instruments are necessary for testing and troubleshooting, especially on circuits of 600 volts or less. These include general-purpose volt meters, voltohmmeters, and clamp-on-type ammeters with multiscale ranges. In addition to conventional instruments, recording meters are useful for measuring magnitudes and fluctuations of current, voltage, power factor, watts, volt-amperes, and transients versus time values. These instruments are a definite aid in defining specific electrical problems and determining if equipment malfunction is due to abnormal electrical conditions. Other valuable test equipment includes devices to measure the insulation resistance of motors and similar equipment in the megohm range and similar instruments in the low range for determining ground resistance, lightning protection systems, and grounding systems. Continuity testers are particularly valuable for checking control circuits and for circuit identification.

6.2.5.6 Special instruments can be used to test the impedance of the grounding circuit conductor or the grounding path of energized low-voltage distribution systems and equipment. These instruments can be used to test the equipment-grounding circuit path of electrical equipment.

6.2.5.7 Insulation resistance-measuring equipment should be used to indicate insulation baseline values at the time equipment is put into service. Later measurements might indicate any deterioration trend of the insulation values of the equipment. High-potential ac and dc testers are used effectively to indicate dielectric strength and insulation resistance of the insulation, respectively. It should be recognized that the possibility of breakdown under test due to concealed weakness is always present. High-potential testing should be performed with caution and only by qualified operators.

6.2.5.8 Portable ground-fault locators can be used to test ungrounded power systems. Such devices will indicate ground location while the power system is energized. They thus provide a valuable aid for safe operation by indicating where to take corrective steps before an insulation breakdown occurs on another phase.

6.2.5.9 Receptacle circuit testers are devices that, by a pattern of lights, indicate some types of incorrect wiring of 15- and 20-ampere, 125-volt grounding-type receptacles.

CAUTION: Although these test devices can provide useful and easily acquired information, some have limitations, and the test results should be used with caution. For example, a high-resistance ground can give a correct wiring display, as can some multiple wiring errors. An incorrect display can be considered a valid indication that there is an incorrect situation, but a correct wiring display should not be accepted without further investigation.

6.3 Identification of Critical Equipment.

6.3.1 Equipment (electric or otherwise) should be considered critical if its failure to operate normally and under complete control will cause a serious threat to people, property, or the product. Electric power, like process steam, water, and so forth, might be essential to the operation of a machine, but unless loss of one or more of these supplies causes the machine to become hazardous to people, property, or production, that machine might not be critical. The combined knowledge and experience of several people might be needed to make this determination. In a small facility, the facility engineer or master mechanic working with the operating superintendent should be able to make this determination.

6.3.1.1 A large operation should use a team comprising the following personnel:

(1) The electrical foreman or superintendent
(2) Production personnel thoroughly familiar with the operation capabilities of the equipment and the effect its loss will have on final production
(3) The senior maintenance person who is generally familiar with the maintenance and repair history of the equipment or process
(4) A technical person knowledgeable in the theoretical fundamentals of the process and its hazards (in a chemical plant, a chemist; in a mine, a geologist; etc.)
(5) A safety engineer or the person responsible for the overall security of the plant and its personnel against fire and accidents of all kinds

6.3.1.2 The team should review the entire plant or each of its operating segments in detail, considering each unit of equipment as related to the entire operation and the effect of its loss on safety and production.

6.3.2 There are entire systems that might be critical by their very nature. Depending on the size and complexity of the operation, a plant can contain any or all of the following examples: emergency power, emergency lighting, fire-alarm systems, fire pumps, and certain communications systems. There should be a clear determination in establishing whether a system is critical and in having the proper amount of emphasis placed on its maintenance.

6.3.3 More difficult to identify are the parts of a system that are critical because of the function of the utilization equipment and its associated hardware. Some examples are as follows:

(1) The cooling water source of an exothermic reactor might have associated with it some electrical equipment such as a drive motor, solenoid valves, controls, or the like. Failure of the cooling water might allow the exothermic reaction to go beyond the stable point and overpressurize and destroy the vessel.
(2) A process furnace recirculating fan drive motor or fan might fail, nullifying the effects of temperature-sensing points and thus allowing hot spots to develop, with serious side reactions.
(3) The failure of gas analysis equipment and interlocks in a drying oven or annealing furnace might allow the

atmosphere in the drying oven or furnace to become flammable, with the possibility of an explosion.
(4) The failure of any of the safety combustion controls on a large firebox, such as a boiler or an incinerator, can cause a serious explosion.
(5) Two paralleled pump motors might be needed to provide the total requirements of a continuous process. Failure of either motor can cause a complete shutdown, rather than simply reduce production.

6.3.4 There are parts of a system that are critical because they reduce the widespread effect of a fault in electrical equipment. The determination of these parts should be primarily the responsibility of the electrical person on the team. Among the things that fall into this category are the following:

(1) Source overcurrent protective devices, such as circuit breakers or fuses, including the relays, control circuits, and coordination of trip characteristics of the devices
(2) Automatic bus transfer switches or other transfer switches that would supply critical loads with power from the emergency power source if the primary source failed; includes instrument power supplies as well as load power supplies

6.3.5 Parts of the control system are critical because they monitor the process and automatically shut down equipment or take other action to prevent catastrophe. These items are the interlocks, cutout devices, or shutdown devices installed throughout the plant or operation. Each interlock or shutdown device should be considered carefully by the entire team to establish whether it is a critical shutdown or a "convenience" shutdown. The maintenance group should thoroughly understand which shutdowns are critical and which are convenience. Critical shutdown devices are normally characterized by their use in sensing an abnormal condition and might have a sensing device separate from the normal control device. They probably have separate, final, or end devices that cause action to take place. Once the critical shutdown systems have been determined, they should be distinctly identified on drawings, on records, and on the hardware itself. Some examples of critical shutdown devices are overspeed trips; high or low temperature, pressure, flow, or level trips; low-lube-oil pressure trips; pressure-relief valves; overcurrent trips; and low-voltage trips.

6.3.6 Some parts of a system are critical because they alert operating personnel to dangerous, out-of-control, or abnormal conditions. These are normally referred to as alarms. Like shutdown devices, alarms fall into at least three categories: (1) those that signify a true pending catastrophe, (2) those that indicate out-of-control conditions, and (3) those that indicate the end of an operation or similar condition. The entire team should consider each alarm in the system with the same thoroughness with which they have considered the shutdown circuits. A truly critical alarm should be characterized by its separate sensing device, a separate readout device, and, preferably, separate circuitry and power source. The maintenance department should thoroughly understand the critical level of each alarm. The critical alarms and their significance should be distinctly marked on drawings, in records, and on the operating unit. For an alarm to be critical does not necessarily mean that it is complex or related to complex action. A simple valve position indicator can be one of the most critical alarms in an operating unit.

6.4 Establishment of a Systematic Program. The purpose of any inspection and testing program is to establish the condition of equipment to determine what work should be done and to verify that it will continue to function until the next scheduled servicing occurs. Inspection and testing are best done in conjunction with routine maintenance. In this way, many minor items that require no special tools, training, or equipment can be corrected as they are found. The inspection and testing program is probably the most important function of a maintenance department in that it establishes what should be done to keep the system in service to perform the function for which it is required.

6.4.1 Atmosphere or Environment.

6.4.1.1 The atmosphere or environment in which electrical equipment is located has a definite effect on its operating capabilities and the degree of maintenance required. An ideal environment is one in which the air is (1) clean or filtered to remove dust, harmful vapor, excess moisture, and so on; (2) maintained in the temperature range of 15°C to 29°C (60°F to 85°F); and (3) in the range of 40 percent to 70 percent humidity. Under such conditions, the need for maintenance will be minimized. Where these conditions are not maintained, the performance of electrical equipment could be adversely affected. Good housekeeping contributes to a good environment and reduced maintenance.

6.4.1.2 Dust can foul cooling passages and thus reduce the capabilities of motors, transformers, switchgear, and so on, by raising their operating temperatures above rated limits, decreasing operating efficiencies, and increasing fire hazard. Similarly, chemicals and vapors can coat and reduce the heat transfer capabilities of heating and cooling equipment. Chemicals, dusts, and vapors can be highly flammable, explosive, or conductive, increasing the hazard of fire, explosion, ground faults, and short circuits. Chemicals and corrosive vapors can cause high contact resistance that will decrease contact life and increase contact power losses with possible fire hazard or false overload conditions due to excess heat. Large temperature changes combined with high humidity can cause condensation problems, malfunction of operating and safety devices, and lubrication problems. High ambient temperatures in areas where thermally sensitive protective equipment is located can cause such protective equipment to operate below its intended operating point. Preferably, both the electrical apparatus and its protective equipment should be located within the same ambient temperature. Where the ambient-temperature difference between equipment and its protective device is extreme, compensation in the protective equipment should be made.

6.4.1.3 Electrical equipment installed in hazardous (classified) locations as described in *NFPA 70, National Electrical Code,* requires special maintenance considerations. *(See Section 27.2.)*

6.4.2 Load Conditions.

6.4.2.1 Equipment is designed and rated to perform satisfactorily when subjected to specific operating and load conditions. A motor designed for safe continuous operation at rated load might not be satisfactory for frequent intermittent operation, which can produce excessive winding temperatures or mechanical trouble. The resistance grid or transformer of a reduced-voltage starter will overheat if left in the starting position. So-called "jogging" or "inching" service imposes severe demands on equipment such as motors, starters, and controls. Each type of duty influences the type of equipment used and the extent of maintenance required. The five most common types of duty are defined in *NFPA 70, National Electrical Code,* and they are repeated in 6.4.2.2.

6.4.2.2 The following definitions can be found in Chapter 3 and are unique to this chapter:

(1) Continuous duty *(See 3.3.16.1.)*
(2) Intermittent duty *(See 3.3.16.2.)*
(3) Periodic duty *(See 3.3.16.3.)*
(4) Short-time duty *(See 3.3.16.4.)*
(5) Varying duty *(See 3.3.16.5.)*

6.4.2.3 Some devices used in establishing a proper maintenance period are running-time meters (to measure total "on" or "use" time); counters to measure number of starts, stops, or load-on, load-off, and rest periods; and recording ammeters to graphically record load and no-load conditions. These devices can be applied to any system or equipment and will help classify the duty. They will help establish a proper frequency of preventive maintenance.

6.4.2.4 Safety and limit controls are devices whose sole function is to ensure that values remain within the safe design level of the system. Because these devices function only during an abnormal situation in which an undesirable or unsafe condition is reached, each device should be periodically and carefully inspected, checked, and tested to be certain that it is in reliable operating condition.

6.4.3 Wherever practical, a history of each electrical system should be developed for all equipment or parts of a system vital to a facility's operation, production, or process. The record should include all pertinent information for proper operation and maintenance. This information is useful in developing repair cost trends, items replaced, design changes or modifications, significant trouble or failure patterns, and replacement parts or devices that should be stocked. System and equipment information should include the following:

(1) Types of electrical equipment, such as motors, starters, contactors, heaters, relays
(2) Types of mechanical equipment, such as valves, controls, and so on, and driven equipment, such as pumps, compressors, fans, and whether they are direct, geared, or belt driven
(3) Nameplate data
(4) Equipment use
(5) Installation date
(6) Available replacement parts
(7) Maintenance test and inspection dates: type and frequency of lubrication; electrical inspections, test, and repair; mechanical inspections, test, and repair; replacement parts list with manufacturer's identification; electrical and mechanical drawings for assembly, repair, and operation

6.4.4 Inspection Frequency. Those pieces of equipment found to be critical should require the most frequent inspections and tests. Depending on the degree of reliability required, other items can be inspected and tested much less frequently.

6.4.4.1* Manufacturers' service manuals should have a recommended frequency of inspection. The frequency given is based on standard or usual operating conditions and environments. It would be impossible for a manufacturer to list all combinations of environmental and operating conditions. However, a manufacturer's service manual is a good basis from which to begin considering the frequency for inspection and testing.

6.4.4.2 There are several points to consider in establishing the initial frequency of inspections and tests. Electrical equipment located in a separate air-conditioned control room or switch room certainly would not be considered normal, so the inspection interval might be extended 30 percent. However, if the equipment is located near another unit or operating plant that discharges dust or corrosive vapors, this time might be reduced by as much as 50 percent.

6.4.4.3 Continuously operating units with steady loads or with less than the rated full load tend to operate much longer and more reliably than intermittently operated or standby units. For this reason, the interval between inspections might be extended 10 to 20 percent for continuously operating equipment and possibly reduced by 20 to 40 percent for standby or infrequently operated equipment.

6.4.4.4 Once the initial frequency for inspection and tests has been established, this frequency should be adhered to for at least four maintenance cycles unless undue failures occur. For equipment that has unexpected failures, the interval between inspections should be reduced by 50 percent as soon as the trouble occurs. On the other hand, after four cycles of inspections have been completed, a pattern should have developed. If equipment consistently goes through more than two inspections without requiring service, the inspection period can be extended by 50 percent. Loss of production due to an emergency shutdown is almost always more expensive than loss of production due to a planned shutdown. Accordingly, the interval between inspections should be planned to avoid the diminishing returns of either too long or too short an interval.

6.4.4.5 Adjustment in the interval between inspections should continue until the optimum interval is reached. This adjustment time can be minimized and the optimum interval approximated more closely initially by providing the person responsible for establishing the first interval with as much pertinent history and technology as possible.

6.4.4.6 The frequency of inspection for similar equipment operating under differing conditions can differ widely. Typical examples are as follows:

(1) In a continuously operating plant using less than full design load factor and located in a favorable environment, the high-voltage oil circuit breakers might need an inspection only every 2 years. On the other hand, an electrolytic process plant using similar oil circuit breakers for controlling furnaces might find it necessary to inspect and service them as frequently as every 7 to 10 days.
(2) An emergency generator to provide power for noncritical loads can be tested on a monthly basis. Yet the same generator in another plant having processes sensitive to explosion on loss of power might need to be tested during each shift.

6.5 Methods and Procedures.

6.5.1 General.

6.5.1.1 If a system is to operate without failure, not only should the discrete components of the system be maintained, but the connections between these components also should be covered by a thorough set of methods and procedures. Overlooking this important link in the system causes many facilities to suffer high losses every year.

6.5.1.2 Other areas where the maintenance department should develop its own procedures are shutdown safeguards, interlocks, and alarms. Although the individual pieces of equipment can have testing and calibrating procedures furnished by the manufacturer, the equipment application is probably unique, so the system should have an inspection and testing procedure developed.

6.5.2 Forms and Reports.

6.5.2.1 A variety of forms can be included with the inspection, testing, and repair (IT&R) procedure; these forms should be detailed and direct, yet simple and durable enough to be used in the field.

6.5.2.2 The reports should go in the master file maintained by first-line electrical maintenance supervision. A document control system should be in place to assure current documents are utilized by personnel. If reports to production or engineering are needed, they should be separate, and inspection reports should not be used. See Annex H for a set of forms that might be applicable.

6.5.2.3 The IT&R procedure file for a piece of equipment should list the following items:

(1) All the special tools, materials, and equipment necessary to do the job
(2) The estimated or actual average time to do the job
(3) Appropriate references to technical manuals
(4) Previous work done on the equipment
(5) Points for special attention indicated by previous IT&R
(6) References to unusual incidents reported by production that might be associated with the equipment
(7) If major work was predicted at the last IT&R, copies of the repair work orders and parts references

6.5.2.4 Special precautions relative to operation, such as the following, should be part of the IT&R document:

(1) What other equipment is affected and in what way?
(2) Who has to be informed that the IT&R is going to be done?
(3) How long will the equipment be out of service if all goes well? How long if major problems are uncovered?

6.5.3 Planning.

6.5.3.1 After the IT&R procedures have been developed and the frequency has been established (even though preliminary), the task of scheduling should be handled. Scheduling in a continuous-process plant (as opposed to a batch-process plant) is most critically affected by availability of equipment in blocks consistent with maintenance personnel capabilities. In general, facilities should be shut down in whole or in part on some regular basis for overall maintenance and repair. Some of the electrical maintenance items should be done at this time. IT&R that could be done while equipment is in service should be done prior to shutdown. Only work that needs to be done during shutdown should be scheduled at that time, to limit personnel requirements and limit downtime.

6.5.3.2 The very exercise of scheduling IT&R will point out design weaknesses that require excessive personnel during critical shutdown periods or that require excessive downtime to do the job with the personnel available. Once these weaknesses have been uncovered, consideration can be given to rectifying them.

6.5.3.3 Availability of spare equipment affects scheduling in many ways. Older facilities might have installed spares for a major part of the equipment, or the facility might be made up of many parallel lines so that they can be shut down, one at a time, without seriously curtailing operations. This concept is particularly adaptable to electrical distribution. The use of a circuit breaker and a transfer bus can extend the interval between total shutdown on a main transformer station from once a year to once in 5 years or more.

6.5.3.4 Many continuous-process plants use a large single-process line with no installed spares. This method of operation requires ongoing inspections and tests. Downtime in such plants is particularly costly, so it is desirable to build as much monitoring into the electrical systems as possible.

6.5.3.5 Planning running inspections can vary from a simple desk calendar to a computer program. Any program for scheduling should have at least the following four facets:

(1) A reminder to order parts and equipment with sufficient lead time to have them on the job when needed
(2) The date and man-hours estimate to do the job
(3) A check to see that the job has been completed
(4) Identifying if additional parts will be needed for the next IT&R and when they should be ordered

6.5.3.6 Planning shutdown IT&R is governed by the time between shutdowns, established by the limitations of the process or production units involved. Reliability of electrical equipment can and should be built in to correspond to almost any length of time.

6.5.3.7 Small plants should use, in an abbreviated form, the following shutdown recommendations of a large facility IT&R:

(1) Know how many personnel-shifts the work will take.
(2) Know how many persons will be available.
(3) Inform production of how many shifts the electrical maintenance will require.
(4) Have all the necessary tools, materials, and spare parts assembled on the job site. Overage is better than shortage.
(5) Plan the work so that each person is used to best suit his or her skills.
(6) Plan what each person will be doing during each hour of the shutdown. Allow sufficient off time so that if a job is not finished as scheduled, the person working on that job can be held over without becoming overtired for the next shift. This procedure will allow the schedule to be kept.
(7) Additional clerical people during shutdown IT&R will make the job go more smoothly, help prevent omission of some important function, and allow an easier transition back to normal.
(8) Supply copies of the electrical group plan to the overall shutdown coordinator so it can be incorporated into the overall plan. The overall plan should be presented in a form that is easy to use by all levels of supervision. In a large, complex operation, a critical path program or some similar program should be used.

6.5.3.8 Automatic shutdown systems and alarm systems that have been determined as critical should be designed and maintained so that nuisance tripping does not reduce operator confidence. Loss of operator confidence could cause these systems to be bypassed and the intended safety lost. Maintenance of these systems should prove that each shutdown and alarm operation was valid and was caused by an unsafe condition.

6.5.3.9 A good electrical preventive maintenance program should identify the hierarchy of job criticality so it is clear to first-line supervision which EPM can be delayed to make personnel available for emergency breakdown repair.

6.5.4 Analysis of Safety Procedures.

6.5.4.1 It is beyond the scope of this recommended practice to cover the details of safety procedures for each IT&R activity. Manufacturers' instructions typically contain safety procedures required in using their test equipment. See Chapter 7 for more detailed information on personnel safety.

6.5.4.2 The electrical test equipment should be inspected in accordance with vendor recommendations before the job is started. Any unsafe condition should be corrected before proceeding.

6.5.4.3 The people performing the IT&R should be briefed to be sure that all facets of safety before, during, and after the IT&R are understood. It is important that all personal protective equipment (PPE) is in good condition and available and used when required.

6.5.4.4 Screens, ropes, guards, and signs needed to protect people other than the IT&R team should be provided and used.

6.5.4.5 A procedure should be developed, understood, and used for leaving the test site in a safe condition when unattended at times such as breaks or overnight work.

6.5.4.6 A procedure should be developed, understood, and used to ensure safety to and from the process before, during, and after the IT&R. The process or other operation should be put in a safe condition for the IT&R by the operating people before the work is started. The procedure should include checks to ensure that the unit is ready for operation after the IT&R is completed and before the operation is restarted.

6.5.5 Records.

6.5.5.1 General. Sufficient records should be kept by maintenance management to evaluate overall EPM results. Analysis of the records should guide budget planning for EPM and breakdown repair.

6.5.5.2 Records of Cost. Figures should be kept to show the total cost of each unplanned outage. This should be the actual cost plus an estimated cost of the business interruption. This figure is a powerful indicator for the guidance of expenditures for EPM.

6.5.5.3 Records Kept by First-Line Supervisor of EPM. Of the many approaches to this phase of the program, the following approach is a typical one that fulfills the minimum requirements.

6.5.5.3.1 Inspection Schedule. The first-line supervisor should maintain, in some easy-to-use form, a schedule of inspections so that he or she can plan personnel requirements.

6.5.5.3.2 Work Order Log. An active log should be kept of unfinished work orders. A greater probability of breakdown may be indicated by a large number of outstanding work orders resulting from the inspection function.

6.5.5.3.3 Unusual Event Log. As the name implies, this log lists unusual events that affect the electrical system in any way. This record is derived from reports of operating and other personnel and is a good tool for finding likely problems. Near misses can be recorded and recognition given for averting trouble.

6.5.6 Emergency Procedures. It should be recognized that properly trained electrical maintenance personnel have the potential to make an important contribution in the emergency situations that are most likely to occur. However, most such situations will also involve other crafts and disciplines, such as operating personnel and mechanics. Emergency procedure for each anticipated emergency situation should be developed. Qualified personnel of each discipline should be involved, detailing steps to be followed, sequence of steps, and assignment of responsibility. The total procedure should then be performed periodically as an emergency drill to ensure that all involved personnel are kept thoroughly familiar with the tasks they are to perform.

6.6 Maintenance of Imported Electrical Equipment. Imported equipment can pose additional maintenance considerations, and the original equipment manufacturer's documentation and local codes and standards should be referenced for any special needs or requirements.

6.6.1 Timely delivery of replacement parts cannot be taken for granted. Suppliers should be identified, and any replacement parts issues should be reflected in the facility spare parts inventory. In addition to considering possible delays in delivery of replacement parts, knowledgeable outside sources of engineering services for the imported equipment should be established.

6.6.2 Parts catalogs, maintenance manuals, and drawings should be available in the language of the user. Documents created in a different language and then translated should not be presumed to be understandable. Unclear portions of the translation should be identified as soon as literature is received to ensure that material will be fully understood later, when actual maintenance must be performed.

6.7 Maintenance of Electrical Equipment for Use in Hazardous (Classified) Locations. (*See Section 27.2.*)

6.8 Outsourcing of Electrical Equipment Maintenance.

6.8.1 General. This section describes the process for a facility to request the services of qualified contractors to perform maintenance on electrical equipment.

6.8.2 Contract Elements. Elements in a contract for outsourcing electrical maintenance service are to include, but are not limited to, the following:

(1) Define project scope of work, what is included and not included, along with equipment specifications on any new or replacement parts, and the time period(s) in which the activities are to be performed.
(2) Determine if it is a performance-based or detailed (step-by-step) specification.
(3) Determine which safety and maintenance codes and standards are to be followed, including appropriate permits.
(4) Determine methodology for pricing: lump sum or unit price.
(5) Determine the qualifications of potential contractors and develop and maintain a list of such.
(6) Obtain appropriate liability, insurance coverage, and warranty information.
(7) Assemble the appropriate up-to-date and accurate facility and equipment specific documents, such as, but not limited to, the following:
    (a) Facility one line diagrams
    (b) Facility layout drawings showing location of substations and major facility electrical equipment
    (c) Facility equipment list (if facility drawings show equipment, facility drawings may be used in lieu of specific equipment lists)
    (d) Equipment manufacturers' requirements (these include equipment service manuals, equipment drawings, etc.)

(e) Arc flash hazard analysis, short circuit analysis, and time-current coordination studies

(8) Conduct a pre-bid/negotiation walk through with the potential contractor.
(9) Conduct a post-work walk-through to verify proper completion of scope of work, and review written report from the contractor on findings and recommendations, as applicable.

6.8.3 Sample Forms. Sample forms for electrical maintenance service are found in Annex H.

6.9 Emergency Preparedness and Electrical System and Equipment Restoration (EPnSR).

6.9.1 Introduction.

6.9.1.1 A plan should be developed for a safe and orderly shut down of the facility in event of an emergency, and for post-emergency actions required to restore normal operations in the facility. This includes, but is not limited to, the following:

(1) The information provided in 6.2.4.
(2) Preparation of a primary contact matrix, as illustrated in Annex J, which should be updated yearly, prominently posted, and visible to maintenance personnel during an emergency.
(3) Spare parts stocking, including the determination of which ones should be stocked and determination of accessible storage location.

6.9.1.2 For further information on disaster and emergency management, see *NFPA 1600, Standard on Disaster/Emergency Management and Business Continuity Programs* and *NEMA Guidelines for Handling Water-Damaged Electrical Equipment.*

6.9.2 Procedure for Emergency Shutdown.

6.9.2.1 A procedure should be developed for the shutdown of the electrical system and incorporated in the overall plant shutdown and evacuation plan.

6.9.2.2 In the absence of notification by the authority having jurisdiction (AHJ), the determination to cease operations is based on when the qualified person decides that the facility personnel can no longer safely and properly operate and maintain the equipment and the facility.

6.9.2.3 Notify the appropriate authorities that the facility has been shut down and evacuated.

6.9.3 Procedure for Post-Emergency Actions. See Chapter 32, Electrical Disaster Recovery.

• **6.9.4 Training.** Refer to 6.5.6 for additional information about emergency procedures.

6.10 Counterfeit Components, Devices, Tools, and Equipment.

6.10.1 When the maintenance of electrical equipment requires the replacement of existing components, devices, tools, or equipment, care should be exercised to minimize the inadvertent use or installation of counterfeit goods.

6.10.2 Products should be purchased from an authorized vendor.

6.10.3 Careful visual inspection of the goods and packaging can distinguish counterfeit goods from those of the legitimate manufacturer. The product and/or packaging could contain grammatical or spelling errors, missing or improper certification marks, or lack of applicable safety warnings and instructions.

6.10.4 If it is suspected that the goods are counterfeit, contact the manufacturer or the nationally recognized testing laboratory (NRTL).

Chapter 7 Personnel Safety

7.1 Introduction.

7.1.1 Personnel safety is a primary consideration in system design and in establishing safety-related work practices where performing preventative maintenance for electrical, electronic, and communication systems and equipment. Maintenance should be performed only by qualified persons trained in safe maintenance practices and the special considerations necessary to maintain electrical equipment. Safe work practices should be instituted and followed to prevent injury to those who are performing tasks as well as others who might be exposed to the hazards. Among the hazards associated with working on energized electrical conductors or circuit parts are shock, arc flash, and arc blast, any of which, may result in severe injury or death. Preventive maintenance should be performed only when equipment is in an electrically safe work condition.

7.1.2 *NFPA 70E, Standard for Electrical Safety in the Workplace,* ANSI/Accredited Standards Committee C2, *National Electrical Safety Code,* and OSHA 29 CFR 1926 and 1910 are among the references that should be utilized for the development of programs and procedures associated with maintenance activities, and are necessary to be used in conjunction with this document.

7.1.3 Chapter 1 of *NFPA 70E* covers electrical safety-related work practices and procedures for employees who work on or near exposed energized electrical conductors or circuit parts in workplaces that are included in the scope of that standard. These practices and procedures are intended to provide for employee safety relative to electrical hazards in the workplace. All maintenance personnel should confirm that the requirements of *NFPA 70E* are adhered to when performing electrical maintenance procedures.

7.1.3.1 Below is a listing of some of the considerations as required by *NFPA 70E* in Article 110:

(1) Employee and employer responsibilities *(see Section 110.3)*
(2) Host and contract relationship *(see Section 110.5)*
(3) Training: emergency procedures and qualified and unqualified procedures *(see Section 110.6)*
(4) Electrical safety program *(see Section 110.7)*
(5) Working while exposed to electrical hazards *(see Section 110.8)*
(6) Use of equipment *(see Section 110.9)*

7.1.3.2 Below is a listing of some of the considerations as required by *NFPA 70E* in Article 120:

(1) Process of achieving an electrically safe work condition *(see Section 120.1)*
(2) Deenergized conductors or circuit parts that have lockout/tagout devices applied *(see Section 120.2)*
(3) Temporary protective grounding equipment *(see Section 120.3)*

7.1.3.3 Below is a listing of some of the considerations as required by *NFPA 70E* in Article 130:

(1) Justification for work *(see Section 130.1)*

(2) Approach boundaries to energized electrical conductors or circuit parts *(see Section 130.2)*
(3) Arc flash hazard analysis *(see Section 130.3)*
(4) Test instruments and equipment use *(see Section 130.4)*
(5) Work within the limited approach boundary of uninsulated overhead lines *(see Section 130.5)*
(6) Other precautions for personnel activities *(see Section 130.6)*
(7) Personal and other protective equipment *(see Section 130.7)*

7.2 Grounding of Equipment to Provide Protection for Electrical Maintenance Personnel.

7.2.1 Personnel working on, or in close proximity to, deenergized lines or conductors in electrical equipment should be protected against shock hazard and flash burns that could occur if the circuit were to be inadvertently reenergized. Sound judgment should be exercised in deciding the extent of protection to be provided and determining the type of protective equipment and procedures that should be applied. The extent of protection that should be provided will be dictated by specific circumstances.

7.2.2 The following possible conditions and occurrences should be considered in determining the type and extent of protection to be provided:

(1) Induced voltages from adjacent energized conductors, which can be appreciably increased when high fault currents flow in adjacent circuits
(2) Switching errors causing inadvertent reenergizing of the circuit
(3) Any unusual condition that might bring an energized conductor into electrical contact with the deenergized circuit
(4) Extremely high voltages caused by direct or nearby lightning strikes
(5) Stored charges from capacitors or other equipment

7.2.3 Providing proper protection begins with establishing an electrically safe work condition.

7.2.4 In spite of all precautions, deenergized circuits can be inadvertently reenergized. When this occurs, adequate grounding is the only protection for personnel working on those circuits. For this reason, it is especially important that adequate grounding procedures be established and rigidly enforced.

7.2.4.1 A variety of terms are used to identify the grounding of deenergized electrical equipment to permit personnel to safely perform work on it without using special insulated tools. Some of these terms are *safety grounding, temporary grounding,* and *personnel protective grounding.* Throughout this chapter, the word *grounding* is used to refer to this activity; it does not refer to permanent grounding of system neutrals or non-current-carrying metal parts of electrical equipment.

7.2.4.2 Grounding equipment consists mainly of special heavy-duty clamps that are connected to cables of adequate capacity for the system fault current. This current, which might well be in excess of 100,000 amperes, will flow until the circuit overcurrent protective devices operate to deenergize the conductors. The grounding equipment should not be larger than necessary, because bulkiness and weight hinder personnel connecting them to the conductors, especially when they are working with hot-line sticks. Selection of grounding equipment should take the provisions of 7.2.4.2.1 through 7.2.4.2.5 into consideration.

7.2.4.2.1 Grounding clamps should be of proper size to fit the conductors and have adequate capacity for the fault current. An inadequate clamp can melt or be blown off under fault conditions. Hot-line clamps should not be used for grounding deenergized conductors because they are not designed to carry the high current that would flow if the circuit were to be inadvertently reenergized. They are intended to be used only for connecting tap conductors to energized overhead lines by means of hot-line sticks and are designed to carry only normal load current. If hot-line clamps are used for grounding, high fault current could melt or blow them off without operating the overcurrent protective devices to deenergize the conductors, thereby exposing personnel to lethal voltages and arc burns.

7.2.4.2.2 Grounding cables should be of adequate capacity, which, in some instances, might require two or more to be paralleled. Three factors contribute to adequate capacity: (1) terminal strength, which largely depends on the ferrules installed on the cable ends; (2) size to carry maximum current without melting; and (3) low resistance to keep the voltage drop across the areas in which the personnel are working at a safe level during any period of inadvertent reenergization.

7.2.4.2.3 Solid metal-to-metal connections are essential between grounding clamps and the deenergized conductors. Conductors are often corroded and are sometimes covered with paint. Ground clamps should have serrated jaws because it is often impractical to clean the conductors. The clamps should be slightly tightened in place, given a slight rotation on the conductors to provide cleaning action by the serrated jaws, and then securely tightened. Ground clamps that attach to the steel tower, switchgear, or station ground bus are equipped with pointed or cupped set screws that should be tightened to ensure penetration through corrosion and paint to provide adequate connections.

7.2.4.2.4 Grounding cables should be no longer than is necessary to keep resistance as low as possible and to minimize slack in cables to prevent their violent movement under fault conditions. If the circuit should be inadvertently reenergized, the fault current and resultant magnetic forces could cause severe and dangerous movement of slack grounding cables in the area where personnel are working. Proper routing of grounding cables to avoid excessive slack is essential for personnel safety.

7.2.4.2.5 Grounding cables should be connected between phases to the grounded structure and to the system neutral (when available) to minimize the voltage drop across the work area if inadvertent reenergization should occur. The arrangement is shown in Figure 7.2.4.2.5 with the equivalent electrical diagram.

7.2.4.3 In Figure 7.2.4.2.5, electrical equivalent diagram, it can be presumed that the resistance of the worker's body (Rm) is 500 ohms. The worker is in parallel with only the resistance of a single cable (Rj), which can be on the order of 0.001 ohm. Rg is the ground resistance of the switchgear or structure area. If a 1000-ampere current should flow in the circuit grounded in this manner, the worker in Figure 7.2.4.2.5 would be subjected to only about 1 volt imposed across the work area; therefore, the current flow through the worker's body would be negligible.

7.2.4.4 Connecting the phase conductors together with short cables and clamps of adequate capacity, as shown in Figure 7.2.4.2.5, minimizes resistance between phases for fast action of the circuit overcurrent protective devices to deenergize the circuit, if it is inadvertently reenergized. The short down-lead cable between the jumpered phase

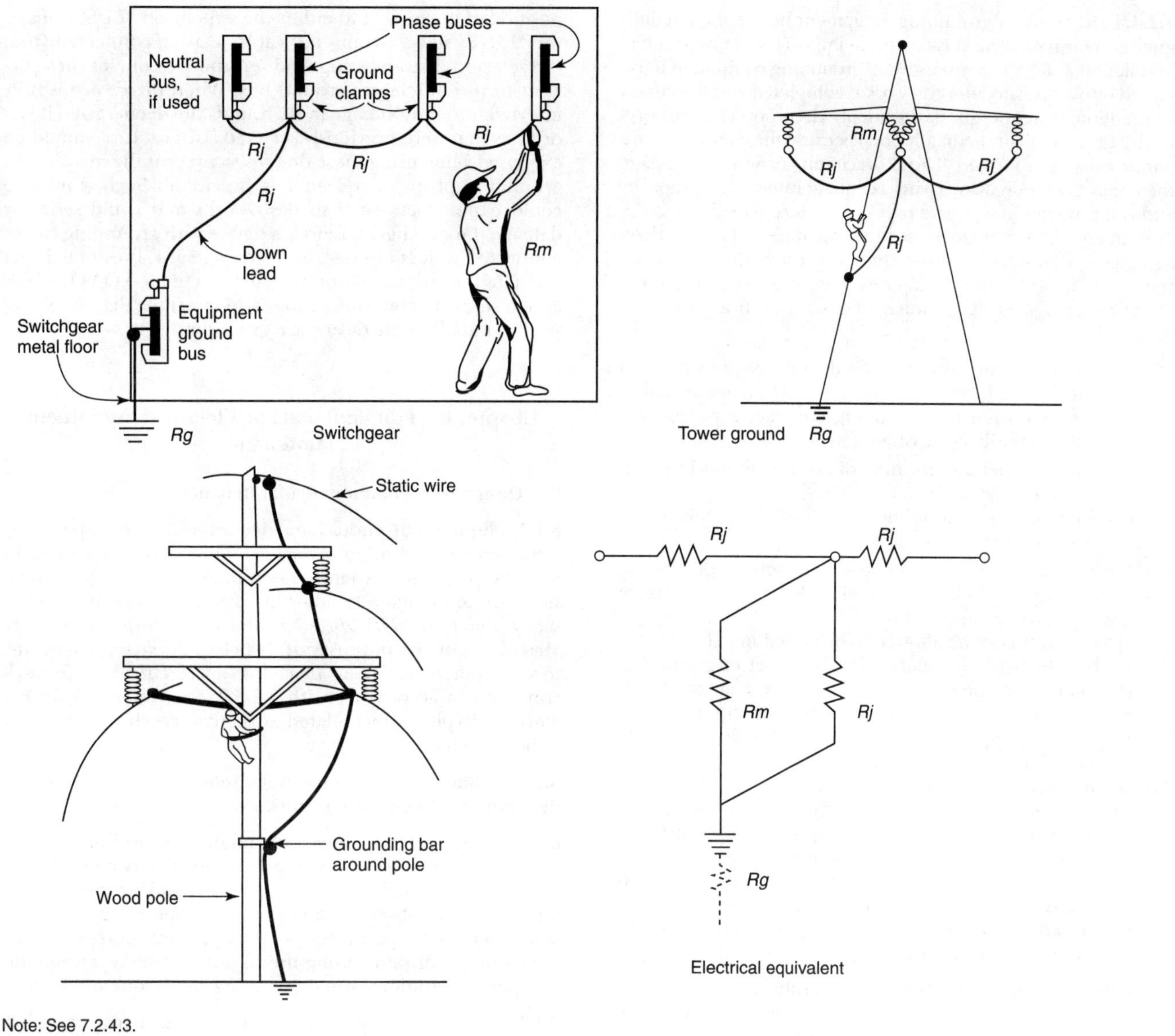

Note: See 7.2.4.3.

FIGURE 7.2.4.2.5 Preferred Grounding Arrangement.

conductors and the grounded tower or switchgear ground bus reduces resistance to ground and the amount of cable that can move violently in the work area during high current flow. If there is a system neutral conductor at the work location, a cable should also be connected to that conductor for more complete protection and to ensure lowest resistance in the ground return path to the source. Figure 7.2.4.2.5 shows buses and a person working inside switchgear; the same conditions would apply to personnel on overhead line towers and outdoor substation steel structures. Someone working on such properly grounded areas is in parallel with a minimum of resistance so he or she would be exposed to minimum voltage drop in the event of current flow in the system, and the low resistance would cause rapid operation of the fuses or circuit breakers, thus minimizing the time the person is exposed to the voltage drop.

7.2.4.5 Prior to installation, grounding equipment should be inspected for broken strands in the conductors, loose connections to the clamp terminals, and defective clamp mechanisms. Defective equipment should not be used.

7.2.4.6 Grounding equipment should be installed at each point where work is being performed on deenergized equipment. Often it is advisable to install grounding equipment on each side of a work point or at each end of a deenergized circuit.

7.2.4.7 One end of the grounding down lead should be connected to the metal structure or ground bus of the switchgear before the other end is connected to a phase conductor of the deenergized equipment. Then, and only then, should the grounding cables be connected between phase conductors.

7.2.4.8 When grounding equipment is removed, the above installation procedure should be reversed by first disconnecting the cables between phases, then disconnecting the down lead from the phase conductor, and, finally, disconnecting the down lead from the metal structure or ground bus.

7.2.4.9 Removal of grounding equipment before the circuit is intentionally reenergized is equally as important as was its initial installation, but for other reasons. If grounding equipment is forgotten or overlooked after the work is completed and the circuit is intentionally reenergized, the supply circuit overcurrent protective devices will immediately open because the conductors are jumpered and grounded. The short-circuit current can damage the contacts of a breaker having adequate interrupting capacity and can cause an inadequate breaker or fuses to explode. If the grounding cables are inadequate, they can melt and initiate damaging power arcs. A procedure should be established to ensure removal of all grounding equipment before the circuit is intentionally reenergized. Recommendations for such a procedure are as follows:

(1) An identification number should be assigned to each grounding equipment set, and all sets that are available for use by all parties, including contractor personnel, should be rigidly controlled.
   (a) The number and location of each set that is installed should be recorded.
   (b) That number should be crossed off the record when each set is removed.
(2) Before the circuit is reenergized, all sets of grounding equipment should be accounted for by number to ensure that all have been removed.
(3) Doors should not be allowed to be closed nor should covers be allowed to be replaced where a set of grounding equipment has been installed inside switchgear. If it is necessary to do so to conceal grounding equipment, a highly visible sign should be placed on the door or cover to remind personnel that a ground is inside.
(4) Before reenergizing the circuit, personnel should inspect interiors of equipment to verify that all grounding sets, including small ones used in testing potential transformers, relays, and so on, have been removed.
(5) Before the circuit is reenergized, all conductors should be tested with a megohmmeter to ascertain if any are grounded. If so, the cause should be determined and corrective action taken.

7.2.4.10 Use of insulated hot-line sticks, rubber gloves, or similar protective equipment by personnel is advisable while installing grounding equipment on ungrounded, deenergized overhead line conductors and also while removing the grounding equipment.

7.2.4.11 Data available from grounding-equipment manufacturers should be referred to for ampacities of cables and clamps and for detailed application information.

7.2.4.12 In some instances, specialized grounding equipment might be required, such as traveling grounds on new overhead line conductors being strung adjacent to energized circuits.

7.2.4.13 Drawout-type grounding and testing devices are available for insertion into some models of switchgear to temporarily replace circuit breakers. These devices provide a positive and convenient grounding means for switchgear buses or associated circuits by connecting to the switchgear buses or line stabs in the same manner as drawout breakers. One such device has two sets of primary disconnecting stabs: the set designated "BUS" connects to the switchgear bus stabs, and the other set, designated "LINE," connects to the switchgear supply line or load circuit stabs. Another type of grounding device has only one set of primary disconnecting stabs that can be positioned to connect to either the switchgear "BUS" stabs or the "LINE" stabs. Grounding cables can be connected from the selected disconnecting stud terminals in one of these devices to the switchgear ground bus. When the device is fully inserted into the switchgear, it grounds the deenergized buses or lines that were previously selected. Utmost care should be exercised when using these devices to prevent the inadvertent grounding of an energized bus or circuit. Such a mistake could expose personnel to flash burns and could seriously damage the switchgear. Before a device with grounding cables connected to it is inserted into switchgear, it is essential that the stabs that are to be grounded are tested for NO VOLTAGE and to verify that only the proper and matching disconnecting stud terminals in the device are grounded.

Chapter 8 Fundamentals of Electrical Equipment Maintenance

8.1 Design to Accommodate Maintenance.

8.1.1 Equipment should be deenergized for inspections, tests, repairs, and other servicing. Where maintenance tasks must be performed when the equipment is energized, provisions are to be made to allow maintenance to be performed safely. Refer to *NFPA 70E, Standard for Electrical Safety in the Workplace*. For the purposes of this chapter, *deenergized* means the equipment has been placed in an electrically safe work condition in accordance with 7.1.3.2. See Chapter 7 for examples of typical safety-related work practices that might need to be implemented.

8.1.1.1 Many maintenance tasks require equipment to be deenergized for effective results.

8.1.1.2 Other maintenance tasks might specifically require or permit equipment to be energized and in service while the tasks are performed. Examples include removing transformer oil for analysis, observing and recording operating characteristics such as temperatures, load conditions, corona, noise, lamp output, or performing thermographic surveys while the equipment is under normal operating conditions and load.

8.1.1.3 Coordinating maintenance with planned production outages and providing system flexibility with redundant equipment and processes are two recommended means to avoid major disruptions of operations. An example of flexibility is a selective radial distribution system incorporating double-ended low-voltage substations. This system permits maintenance and testing of equipment such as the primary feeders, transformers, and main, and tie circuit breakers during periods of light loads.

8.1.2 Larger production equipment, such as air compressors, air-conditioning units, and pumps, that can be difficult to repair or replace quickly is often installed in multiples to provide reserve capacity. Redundant equipment and systems enable maintenance to be performed economically without costly premium time and ensures continuous production in the event of a breakdown.

8.1.3 Selection of equipment that is adequate for present and projected load growth is a prime factor in reducing maintenance costs. Overloaded equipment or equipment not suited for the application will have a short service life and will be costly to maintain. Abnormal conditions, such as a corrosive atmosphere, excessive temperature, high humidity, abrasive or conducting particles, and frequent starting and stopping, require special consideration in the selection and location of the equipment in order to minimize maintenance costs.

8.1.4 Installation costs without sufficient regard and planning for efficient and economic maintenance influence system design. Within a few years, the added cost of performing maintenance, plus production loss from forced outages due to lack of maintenance, will more than offset the savings in initial cost. As equipment ages and is possibly worked harder or becomes more critical to facility operations, scheduling outages to perform accelerated maintenance could become a major challenge.

8.2 Scheduling Maintenance.

8.2.1 In larger facilities, routine maintenance scheduling is often done by a computerized maintenance planning program that generates work orders for projects to be accomplished on a daily, weekly, or monthly basis. In smaller facilities, the maintenance schedule is oftentimes not formally structured and relies on facility maintenance personnel to perform the required maintenance tasks. An effective maintenance program requires a positive mechanism for scheduling and recording of work that needs to be accomplished.

8.2.2 Maintenance outages, particularly in facilities that operate 24 hours a day, 7 days a week, are difficult to schedule; however, there are areas that can be relieved with a nominal investment. For example, low-voltage power circuit breakers should be inspected on an annual basis and tested under simulated overload and fault conditions every 3 to 5 years. An investment in a few spare circuit breakers, one or two of each make and size in use, would allow them to be inspected, and tested at a more convenient time. The in-service breakers could then be exchanged with spares at the opportune time, with negligible production downtime.

8.2.3 The scope of the maintenance work should be confined to the limited time and available personnel. Contracting maintenance to qualified electrical personnel can relieve these and other support tasks associated with preventive maintenance. Electrical contractors who specialize in this type of work have trained technicians along with the proper tools and equipment. Many of them carry inventories of spare electrical parts and equipment.

8.2.4 It is necessary to establish intervals for performing specific tasks when scheduling maintenance. The following considerations should be reviewed during development of a routine maintenance schedule:

(1) Potential of equipment failure to endanger or threaten personnel safety *(see Section 5.3)*
(2) Manufacturer's recommended service and maintenance practices and procedures *(See 6.4.4.1.)*
(3) Operating environment *(see Section 5.6, 6.4.1, and 6.4.3)*
(4) Operating load conditions and equipment rating *(see 5.6.3, 5.6.4, 6.4.2, and 6.4.4.4)*
(5) Unusually expensive equipment repairs *(see Section 5.3)*
(6) Failure and repair of equipment causing extensive downtime and lost production dollars *(see section 5.3)*
(7) Equipment condition *(see 6.3.3 and 6.3.5)*
(8) Production and operating schedules *(see 6.1.6 and 6.5.3)*
(9) Ability to take equipment out of service *(see 6.1.6)*
(10) Failure history *(see 6.1.6, and 6.4.4.5)*
(11) Inspection history *(see 6.4.4.5)*

8.2.4.1 A guide for maintenance intervals is included in Annex L.

8.3 Equipment Safety.

8.3.1 Destructive energy, capable of disintegrating an entire switchgear or switchboard assembly in a very short period of time, can be released during a low-voltage phase-to-phase or phase-to-ground sustained arcing fault. The fault current, in the order of thousands of amperes, multiplied by the arc voltage drop multiplied by the duration of the arc (in seconds) is a measure of the energy released (watt-seconds).

8.3.2 Equipment safety demands sensitive and effective protection. The equipment protective device should be capable of immediately sensing an abnormality and causing it to be isolated with the least possible damage and disturbance to the system. The degree of sensitivity and speed of response is vital to the effectiveness of the protection.

8.3.3 The protective device, such as a fuse, relay, sensor, transducer, etc. generally responds to abnormal conditions. Ideally, the device should not be applied or set to respond to normal load excursions, yet it should have the ability to function on a low-level fault. This can be a difficult situation to monitor and control without properly applied protective devices. For example, unless ground-fault protection is utilized, a phase-to-ground fault could be less than normal load current and may not be sensed as a problem until catastrophic failure of the equipment causes a fault trip from the upstream protective device.

8.4 Protective Scheme.

8.4.1 Proper application of circuit protection, as developed in a short-circuit and coordination study, is typically an engineering function and therefore recognized as a facet of system design. Maintenance functions include, but are not limited to, the following:

(1) Assuring that this designed protection system remains in operation.
(2) Applying the engineered settings and periodic testing, lubricating, and cleaning of the protective devices, relays, and trip elements.
(3) Verifying the proper type and ampere rating of the fuses used within the system.

8.4.2 In larger facilities, interpretation of the short-circuit and coordination study is generally made by facility engineering, and the settings and test points for the adjustable protective devices are furnished to the maintenance department, as are the type and ampere rating of the fuses. While the maintenance personnel need not be able to formulate the engineering study, they should be able to interpret the time-current curves and understand the anticipated performance of the protective device being tested.

8.4.3 An up-to-date short-circuit and coordination study is essential for the safety of personnel and equipment. As a function of the study, the momentary and interrupting rating requirements of the protective devices should be analyzed and verification made that the circuit breaker or fuse will safely interrupt a fault during fault conditions.

8.4.3.1 Additionally, the study should provide the application of the protective device to realize minimum equipment damage and the least disturbance to the system in the event of a fault by properly clearing downstream devices nearest to the point of a fault.

8.5 Acceptance Testing. The initial acceptance testing of the electrical system is part of design and facility construction and typically not part of maintenance. However, the acceptance test data does provide important benchmarks for subsequent maintenance testing. The acceptance testing should be witnessed by the owner's representative, and a copy of the test reports should be forwarded to the facility engineer for inclusion into the maintenance records. See Chapter 31 for more detailed information on commissioning and acceptance testing.

8.6 Guidelines and Impact of Additions/Rework to Retrofitting Equipment.

8.6.1 Rework, remanufacturing, or retrofitting of equipment typically involves replacement or refurbishing of major components of equipment or systems.

8.6.2 Repairs or modifications not authorized by the original equipment manufacturer might void the equipment warranties and third-party certifications.

8.6.3 Equipment can be reconditioned under rebuild programs, provided the reconditioning follows established guidelines.

8.6.4 The rework, remanufacturing, or retrofitting process can be conducted by the original manufacturer or by another party with sufficient facilities, technical knowledge, and manufacturing skills (as evaluated by an accepted certification organization). Safety certifications should be maintained for repaired or rebuilt equipment.

8.6.5 Refurbished or remanufactured equipment should be marked to identify it as such.

8.7 Equipment Cleaning.

8.7.1 General. When cleaning equipment, the method used should be determined by the type of contamination to be removed and whether the apparatus is to be returned to service immediately. Drying is necessary after using a solvent or water. Insulation should be tested to determine if it has been properly cleaned. Enclosure and substation room filters should be cleaned at regular intervals and replaced if they are damaged or clogged. Loose hardware, dust, and debris should be removed from equipment enclosures. When properly cleaned, new or unusual wear or loss of parts can be detected during subsequent maintenance operations.

8.7.2 Methods of Cleaning.

8.7.2.1 Wiping off dirt with a clean, dry, lint-free cloth or soft brush is usually satisfactory if the apparatus is small, the surfaces to be cleaned are accessible, and only dry dirt is to be removed. Lint-free rags should be used so lint will not adhere to the insulation and act as a further dirt-collecting agent. Care should be used to avoid damage to delicate parts.

8.7.2.2 To remove loose dust, dirt, and particles, suction cleaning methods should be used.

8.7.2.3 Where dirt cannot be removed by wiping or vacuuming, compressed-air blowing might be necessary.

8.7.2.3.1 If compressed air is used, protection should be provided against injury to workers' faces and eyes from flying debris and to their lungs from dust inhalation. The use of compressed air should comply with OSHA regulations in 29 CFR 1910.242(b),"Hand and Portable Powered Tools and Other Hand Held Equipment," including limiting air pressure for such cleaning to less than a gauge pressure of 208.85 kPa (30 psi) and the provision of effective chip guarding and appropriate personal protective equipment.

8.7.2.3.2 Care should be exercised as compressed air can cause contaminants to become airborne, which can compromise the integrity of insulation surfaces or affect the mechanical operation of nearby equipment. Provisions should be made to remove the equipment to a suitable location for cleaning or to cover other equipment and guard it from cross contamination. Air should be dry and directed in a manner to avoid further blockage of ventilation ducts and recesses in insulation surfaces.

8.7.2.3.3 Protection might also be needed against contamination of other equipment if the insulation is cleaned in place with compressed air. If feasible, equipment should be removed to a suitable location for cleaning, or other exposed equipment should be covered before cleaning to keep the debris from entering exposed equipment.

8.7.2.4 Accumulated dirt, oil, or grease might require a solvent to remove it. A lint-free cloth barely moistened (not wet) with a nonflammable solvent can be used for wiping. Solvents used for cleaning of electrical equipment should be selected carefully to ensure compatibility with materials being cleaned. Liquid cleaners, including spray cleaners, are not recommended unless solvent compatibility is verified with the equipment manufacturer, as residues could cause damage, interfere with electrical or mechanical functions, or compromise the integrity of insulation surfaces.

8.7.2.5 Some equipment could require cleaning by nonconductive abrasive blasting.

8.7.2.5.1 Shot blasting should not be used.

> **CAUTION:** Cleaning with abrasives or abrasive blasting methods can create a hazard to personnel and equipment.

8.7.2.5.2 Abrasive blasting operations should comply with OSHA regulations in 29 CFR 1910.94(a), "Occupational Health and Environmental Control — Ventilation." Protection should be provided against injury to workers' faces and eyes from abrasives and flying debris and to their lungs from dust inhalation.

8.7.2.6 Airborne asbestos fibers can endanger health and are subject to government regulations. Knowledge of government regulations related to the handling of asbestos is required before handling asbestos and other such materials. (Copies of the *Toxic Substances Control Act* as defined in the U.S. Code of Federal Regulations can be obtained from the U.S. Environmental Protection Agency.)

8.7.2.7 If sweeping of an electrical equipment room is required, a sweeping compound should be used to limit the amount of dirt and dust becoming airborne. During mopping, the mop bucket should be kept as far as practical from the electrical equipment.

8.8 Special Handling and Disposal Considerations.

8.8.1 The handling and disposal of certain electrical equipment, components, and materials can present special maintenance obligations. Examples of such materials are given in 8.8.1.1 through 8.8.1.9.

8.8.1.1 Asbestos. Asbestos-containing materials can be present in equipment such as wire, switches, circuit protectors, panelboards, and circuit breakers, particularly in various arc chute constructions. *(See 8.7.2.6.)*

8.8.1.2 Polychlorinated Biphenyls (PCBs). PCBs were used as noncombustible dielectric fluids in transformers, capacitors, cables, and fluorescent ballasts. Although PCBs are no longer manufactured in the United States and are no longer put in new equipment, PCBs might still exist in older transformers, power capacitors, oil-insulated cables, and fluorescent lighting ballasts. Unless verified and labeled as PCB free, the device should be carefully handled until verified PCB free. *(See 21.2.1.3.)*

8.8.1.3 Lead. The disposal of paper-insulated, lead-covered cables can be an environmental concern. Abandoning a lead product in the ground, such as lead-covered cable, is prohibited in some jurisdictions. If left abandoned, the lead can leach soluble lead salts into the environment.

8.8.1.4 Mineral Oil. Mineral oil is a petroleum product, therefore, the disposal of ordinary transformer oil can be an environmental concern. Spent oil should be sent to a manufacturer or processor for recycling. In the United States, certain quantities of oil spills require state and regional EPA notification.

8.8.1.5 Tetrachlorethylene. Some transformers contain tetrachlorethylene, a toxic substance. Where possible, recycling should be considered.

8.8.1.6 Trichloroethane. Vapors from trichloroethane, sometimes used as an electrical cleaning and degreasing solvent, are toxic and present an environmental threat. Handling and disposal of the liquid require special precautions, as trichloroethane is an ozone-depleting chemical. Many jurisdictions have banned the use of trichloroethane products.

8.8.1.7 Mercury Vapor and Phosphor Coatings. Fluorescent lamps and similar gas tubes could contain mercury vapor and phosphor coatings. If the tube breaks, these materials can escape into the environment. The disposal of large quantities of tubes warrants capturing these materials.

8.8.1.8 Radioactive Materials. Devices containing radioactive materials require special precautions.

8.8.1.9 Other Harmful Agents. Hazards presented by materials and processes should be reviewed whenever changes are planned. For example, a substitute cleaning agent might be more hazardous than the original cleaner, and special handling precautions might be needed for the new cleaner. Or, because of a planned change in operations, a fabric filter might soon be collecting a toxic substance, and new procedures for filter replacement and disposal might be needed.

8.8.2 Those responsible for establishing and sustaining maintenance programs should keep abreast of relevant material-handling and disposal issues, including knowledge of toxic substances, environmental threats, and the latest technologies for waste handling and salvage. Testing might be required to determine the presence of toxic substances.

8.8.3 Health and environmental issues, governmental regulations, and salvage values should all be addressed in disposal-planning programs.

8.9 Supervisory Control and Data Acquisition (SCADA) Systems.

8.9.1 General. A comprehensive maintenance program is critical to attaining long-term reliable performance of supervisory control and data acquisition (SCADA) systems. Periodic device calibration, preventive maintenance, and testing allow potential problems to be identified before they can cause a failure of intended mission or function. Prompt corrective maintenance ensures reliability by minimizing downtime of redundant components.

8.9.2 Special Term. The following special term is used in this chapter.

8.9.2.1 Concurrent Maintenance. The testing, troubleshooting, repair, and/or replacement of a component or subsystem while redundant component(s) or subsystem(s) are serving the load, where the ability to perform concurrent maintenance is critical to attaining the specified reliability/availability criteria for the system or facility.

8.9.3 Preventive Maintenance. The SCADA system should be part of the overall preventive maintenance program for the facility. The recommended maintenance activities and frequencies can be found in Annex L for the various components of SCADA. Preventive maintenance schedules for SCADA components and subsystems should be coordinated with those for the mechanical/electrical systems that they serve, to minimize overall schedule downtime.

8.9.4 Testing. Many components of SCADA systems, such as dead-bus relays, are not required to function under normal system operating modes. For this reason, the system should be tested periodically under actual or simulated contingency conditions. Periodic system testing procedures can duplicate or be derived from the recommended functional performance testing procedures of individual components, as provided by the manufacturers.

8.9.5 Concurrent Maintenance. aintenance should be scheduled to occur during maintenance of associated equipment.

8.10 Lubrication. Lubrication is the application of grease or oil to the bearings of motors, rotating shafts, circuit breaker mechanisms, gears, and so forth. It also includes light lubrication to door hinges or other sliding surfaces on the equipment. Some special parts are identified as being prelubricated for life and should require no further lubrication.

8.10.1 Proper application of lubricants, compatible with existing lubricants on the equipment, is paramount if the equipment is to operate as intended. Manufacturer's instruction bulletins and maintenance procedures should be referenced before applying any lubricants to electrical equipment. Caution should be exercised in the use of products identified as penetrating solvents because they may not be acceptable as lubricants in accordance with manufacturers specifications.

Chapter 9 System Studies

9.1 Introduction. Electrical studies are an integral part of system design, operations, and maintenance. These engineering studies generally cover the following areas:

(1) Short-circuit studies
(2) Coordination studies
(3) Load-flow studies
(4) Reliability studies
(5) Arc flash hazard analysis study

9.1.1 Copies of single-line diagrams and system study data should be given to the facility maintenance department. It is critical to efficient, safe system operation that the maintenance department keep the single-line diagrams current and discuss significant changes with the facility engineering department or consulting electrical engineer. It should be noted, however, that the information required for system studies is highly specialized, and outside help might be necessary.

9.2 Short-Circuit Studies.

9.2.1 Short circuits or fault currents represent a significant amount of destructive energy that can be released into electrical systems under abnormal conditions. During normal system operation, electrical energy is controlled and does useful work. However, under fault conditions, short-circuit currents can cause serious damage to electrical systems and equipment and create the potential for serious injury to personnel. Short-circuit currents can approach values as large as several hundred thousands of amperes.

9.2.1.1 During short-circuit conditions, thermal energy and magnetic forces are released into the electrical system. The thermal energy can cause insulation and conductor melting as well as explosions contributing to major equipment burndowns. Magnetic forces can bend bus bars and cause violent conductor whipping and distortion. These conditions have grim consequences on electrical systems, equipment, and personnel.

9.2.1.2 Protecting electrical systems against damage during short-circuit faults is required in *NFPA 70, National Electrical Code,* Sections 110.9 and 110.10. Additional information on short-circuit currents can be found in ANSI/IEEE 242, *Recommended Practice for Protection and Coordination of Industrial and Commercial Power Systems (IEEE Buff Book),* ANSI/IEEE 141, *Recommended Practice for Electric Power Distribution for Industrial Plants (IEEE Red Book),* ANSI/IEEE 241, *Recommended Practice for Electric Power Systems in Commercial Buildings (IEEE Gray Book),* and ANSI/IEEE 399, *Recommended Practice for Industrial and Commercial Power Systems Analysis (IEEE Brown Book).*

9.2.2 Baseline short-circuit studies should be performed when the facility electrical system is designed. They should be updated when a major modification or renovation takes place, but no more frequently than every 5 years. A copy of the most recent study should be kept with other important maintenance documents.

9.2.2.1 The following are some of the conditions that might require an update of the baseline short-circuit study:

(1) A change by the utility
(2) A change in the primary or secondary system configuration within the facility
(3) A change in the transformer size (kVA) or impedance (percent Z)
(4) A change in conductor lengths or sizes
(5) A change in the motors connected to the system

9.2.2.2 A periodic review of the electrical system configuration and equipment ratings should be checked against the permanent records. Specific attention should be paid to the physical changes in equipment, including changes in type and quantity. Significant changes should be communicated to the maintenance supervisor, the facility engineering department, or the electrical engineer.

9.2.2.3 A comprehensive treatment of short-circuit currents is beyond the scope of this document. However, there is a simple method to determine the maximum available short-circuit current at the transformer secondary terminals. This value can be calculated by multiplying the transformer full load amperes by 100, and dividing the product by the percent impedance of the transformer.

9.2.2.3.1 Figure 9.2.2.3.1 shows an example: 500 kVA transformer, 3-phase, 480 V primary, 208 Y/120 V secondary, 2 percent Z.

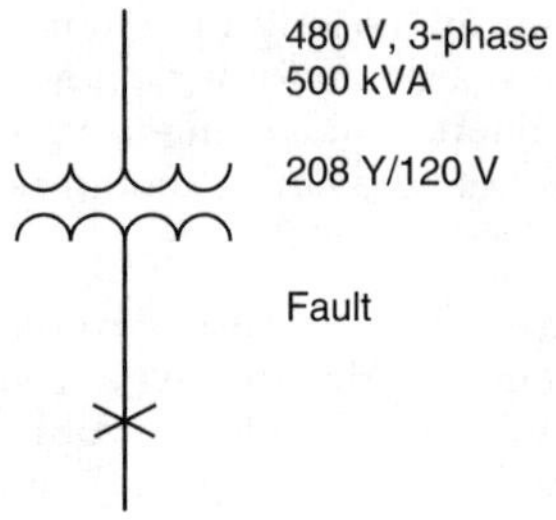

Transformer secondary full-load amperes = 1388
2 percent Z
Maximum short-circuit amperes

$$Isc = FLA \times \frac{100}{\%Z}$$

$$Isc = 1388\ A \times \frac{100}{2}$$

= 69,400 maximum short-circuit amperes

FIGURE 9.2.2.3.1 Example Calculation of Maximum Available Short-Circuit Current at the Transformer Secondary Terminals.

9.2.2.3.2 There are several computer programs commercially available to conduct thorough short-circuit calculation studies.

9.2.2.4 When modifications to the electrical system increase the value of available short-circuit amperes, a review of overcurrent protection device interrupting ratings and equipment withstand ratings should take place. This might require replacing overcurrent protective devices with devices having higher interrupting ratings or installing current-limiting devices such as current-limiting fuses, current-limiting circuit breakers, or current-limiting reactors. For silicon control rectifier (SCR) or diode input devices, change of the source impedance can affect equipment performance. Proper operation of this equipment depends on maintaining the source impedance within the rated range of the device. The solutions to these engineering problems are the responsibility of the maintenance supervisor, the facility engineering department, or the electrical engineer.

9.3 Coordination Studies.

9.3.1 A coordination study, sometimes called a selectivity study, is done to improve power system reliability. *[See 3.3.11, Coordination (Selective).]*

9.3.1.1 Improper coordination can cause unnecessary power outages. For example, branch-circuit faults can open multiple upstream overcurrent devices. This process can escalate and cause major blackouts, resulting in the loss of production. Blackouts also affect personnel safety.

9.3.1.2 *NFPA 70, National Electrical Code,* and various IEEE standards contain the requirements and suggested practices to coordinate electrical systems. The IEEE standards include ANSI/IEEE 242, *Recommended Practice for Protection and Coordination of Industrial and Commercial Power Systems (IEEE Buff Book),* ANSI/IEEE 141, *Recommended Practice for Electric Power Distribution for Industrial Plants (IEEE Red Book),* ANSI/IEEE 241, *Recommended Practice for Electric Power Systems in Commercial Buildings (IEEE Gray Book),* and ANSI/IEEE 399, *Recommended Practice for Industrial and Commercial Power Systems Analysis (IEEE Brown Book).*

9.3.2 A baseline coordination study is generally made when the electrical system is designed. A copy of the study should be kept with other important facility maintenance documents.

9.3.3 Changes affecting the coordination of overcurrent devices in the electrical system include the following:

(1) A change in the available short-circuit current
(2) Replacing overcurrent devices with devices having different ratings or operating characteristics
(3) Adjusting the settings on circuit breakers or relays
(4) Changes in the electrical system configuration
(5) Inadequate maintenance, testing, and calibration

9.3.4 The facility electrical system should be periodically reviewed for configuration changes, available short-circuit current changes, changes in fuse class or rating, changes in circuit-breaker type or ratings, and changes in adjustable trip settings on circuit breakers and relays.

9.3.4.1 Any changes noted in the coordinated performance of overcurrent protective devices should be reported to the maintenance supervisor, the facility engineering department, or the consulting electrical engineer.

9.3.4.2 Time–current curves should be kept up to date. Usually this is the responsibility of the facility engineering department or the consulting electrical engineer. However, it is vitally important for facility maintenance to observe and communicate coordination information to the maintenance supervisor, facility engineering department, or consulting electrical engineer.

9.4 Load-Flow Studies.

9.4.1 Load-flow studies show the direction and amount of power flowing from available sources to every load. By means of such a study, the voltage, current, power, reactive power, and power factor at each point in the system can be determined.

9.4.1.1 This information is necessary before changes to the system can be planned and will assist in determining the operating configuration. This study also helps determine losses in the system. ANSI/IEEE 399, *Recommended Practice for Industrial and Commercial Power Systems Analysis (IEEE Brown Book)*, provides more detailed information.

9.4.1.2 Load-flow studies should be done during the design phase of an electrical distribution system. This is called the baseline load-flow study. The study should be kept current and revised whenever significant increases or changes to the electrical system are completed.

9.4.1.3 Some of the events that result in load-flow changes include changing motors, motor horsepower, transformer size, or impedance; operating configurations not planned for in the existing study; adding or removing power-factor correction capacitors; and adding or removing loads.

9.4.2 It is important that the system single-line diagrams and operating configurations (both normal and emergency) be kept current along with the load-flow study.

9.4.3 Some signs that indicate a need to review a load-flow study include unbalanced voltages, voltage levels outside the equipment rating, inability of motors to accelerate to full load, motor starters dropping off line when other loads are energized, or other signs of voltage drop. Additional signs also include poor system power factor, transformer or circuit overloading during normal system operation, and unacceptable overloading when the system is operated in the emergency configuration.

9.4.4 When changes to the electrical system are made, the maintenance department should note the changes on their copy of the single-line diagram. Significant changes, as mentioned in 9.4.1.3, should be reviewed with the maintenance supervisor, facility engineering department, or the consulting electrical engineer to determine if changes are necessary to the single-line diagram.

9.4.5 Data Collection Methods. In order to conduct short circuit, coordination, and arc flash studies, specific data should be collected. This data should be included on a single line diagram: utility company points of contact, and data records for equipment such as, but not limited to transformers, cables, overhead lines, fuses, medium voltage breakers, reclosers, capacitor banks, low voltage breakers, disconnects, generators, and motors. This information should be developed for each type of operating conditions. Typical data collection forms are included in Figure H.47 through Figure H.49.

9.4.5.1 Utility information should at least include the minimum and maximum short circuit MVA and the X/R ratio at the service point; point of contact name, address, and telephone number in addition to a facility point of contact, address, and telephone number.

9.4.5.2 Transformer data records should include location, rated kVA, maximum kVA, primary voltage, secondary voltage, impedance in percent, type of primary and secondary connection, ground impedance, and if appropriate, the voltage tap.

9.4.5.3 Cable data should include "to" and "from," rated volts, nominal volts, single conductor or three conductor cable, the number of conductors per phase, the neutral size, copper or aluminum, and length in feet.

9.4.5.4 Raceway material (magnetic or nonmagnetic) should be noted.

9.4.5.5 Overhead line information should include "to" and "from," connection configuration, nominal volts, number of lines, lines per phase, ground size, type of cable (material), and length in feet.

9.4.5.6 Medium voltage breaker information should include location of the breaker, manufacturer, type, rated volts, interrupting current, interrupting time (cycles), close/latch amps and for the associated relays the manufacturer/type, time delay range and existing tap, time dial, instantaneous range and existing tap, and CT ratio.

9.4.5.7 Recloser information should include location, CT ratio, nominal volts, manufacturer, type, BIL, continuous current rating, interrupting rating, minimum trip, operational sequence, reclosing times (if available), and tripping curves (if available).

9.4.5.8 Low voltage information for the breaker should include: location, manufacturer, type, rated volts, frame rating, and interrupting rating; for the trip device: manufacturer, type, long time delay range and bands available, short time delay range and bands available, instantaneous range, and ground range and bands available.

9.4.5.9 Generator information should include location, type, kVA rating, generated volts, rated current, rpm, wiring connection (e.g., delta or wye), system ground, subtransient impedance, ground impedance, and power factor.

9.4.5.10 Motor information should include location, type, horsepower, rated volts, full load amps, rpm, code letter, locked rotor amps, power factor, and starter type.

9.4.5.11 Capacitor bank information should include the location, kVAR rating, rated volts, and wiring connection (e.g., delta or wye).

9.4.5.12 Fuse information should include the location, voltage rating, interruption rating, fuse type or class, manufacturer, and manufacturer's part number.

9.5 Reliability Studies.

9.5.1 A reliability study is conducted on facility electrical systems to identify equipment and circuit configurations that can lead to unplanned outages.

9.5.1.1 The study methods are based on probability theory. The computed reliability of alternative system designs as well as the selection and maintenance of components can be made to determine the most economical system improvements. A complete study considering all the alternatives to improve system performance add technical credibility to budgetary requests for capital improvements.

9.5.1.2 An immediate benefit from this investigation is the listing of all system components with their failure modes, frequencies, and consequences. This allows weakness in component selection to be identified prior to calculation of risk indices.

9.5.1.3 ANSI/IEEE 399, *Recommended Practice for Industrial and Commercial Power Systems Analysis (IEEE Brown Book)*, Chapter 12, provides more detailed information. In addition, there are publications that deal with reliability calculations, including TM 5-698-1, *Reliability/Availability of Electrical and Mechanical Systems for Command, Control, Communications, Computer, Intelligence, Surveillance, and Reconnaissance (C4ISR) Facilities*; TM 5-698-2, *Reliability-Centered Maintenance (RCM) for Command, Control, Communications, Computer, Intelligence, Surveillance, and Reconnaissance (C4ISR) Facilities*; and TM 5-698-3, *Reliability Primer for Command, Control, Communications, Computer, Intelligence, Surveillance, and Reconnaissance (C4ISR) Facilities.*

9.5.2 A reliability study can be conducted when alternative systems, components, or technologies are being considered to improve reliability. Changes affecting the reliability of an electrical system or component can include one or more of the following:

(1) System design
(2) Reliability of the power source
(3) Equipment selection
(4) Quality of maintenance
(5) Age of equipment
(6) Equipment operating environment
(7) Availability of spare parts

9.5.2.1 Generally, the existing system design cannot be significantly altered; however, it is possible to meet with the utility and discuss methods for increasing the reliability of service. The selection of reliable equipment and the need for additional maintenance can be evaluated from an economic standpoint. The age of equipment and the environment in which it is operated affects the probability of equipment failure. Spare parts should be monitored and inspected periodically to ensure that they will be available when needed. The study should be kept current and revised whenever a significant change to the electrical system has been made.

9.5.3 A reliability study begins with the system configuration documented by a single-line diagram. Reliability numerics are applied to a system model identifying system outages based on component downtime and system interactions. A failure modes and effect analysis (FMEA) is used to generate a list of events that can lead to system interruption and includes the probability of each event and its consequences. An example of an FMEA table for a facility's electrical equipment is shown in Table 9.5.3. The frequency of failures per year can be obtained from ANSI/IEEE 493, *Recommended Practice for the Design of Reliable Industrial and Commercial Power Systems (IEEE Gold Book).*

Table 9.5.3 Sample FMEA Table

System/ Component	Failure Mode	Frequency per Year	Consequence ($1000)
Breaker B1	Internal fault	0.0036	150
Transformer T1	Winding failure	0.0062	260
Motor M1	Stator damage	0.0762	225

9.5.4 The information in Table 9.5.3 can be analyzed using event-tree analysis or by computing a system reliability index. The event tree is used to further break down each system or component failure into a series of possible scenarios, each with an assigned probability. The outcome is a range of consequences for each event tree.

9.5.4.1 A system reliability index assigns a number (usually expressed in hours down per year) for each system configuration. The calculations for alternative system configurations can be redone until an acceptable downtime per year is obtained. For details on how to conduct a reliability study and how to obtain the number of hours down per year, see Chapter 30.

9.6 Arc Flash Hazard Analysis Studies.

9.6.1 An arc flash hazard analysis study is conducted on facility electrical systems to determine the following for each designated piece of electrical equipment:

(1) Incident energy exposure at working distance
(2) Arc flash boundary
(3) Appropriate arc-rated personal protective equipment required within the arc flash boundary

9.6.1.1 An arc flash hazard analysis study is an important consideration for electrical safe work practices. Refer to Chapter 3 for definitions of *arc flash hazard* and *arc flash hazard analysis.* Refer to *NFPA 70E* for use of the incident energy analysis method or the hazard/risk category (HRC) method.

9.6.2 The benefit of an arc flash hazard analysis is being able to provide the necessary information to a qualified electrical worker so that proper safe work practices can be followed if the worker has to work on or near electrical equipment not in an electrically safe work condition.

9.6.3 The available short-circuit current and the total clearing time at each designated piece of electrical equipment is needed to perform an arc flash hazard analysis. *NFPA 70E* and OSHA provides the requirements. *IEEE 1584, Guide for Performing Arc Flash Hazards Calculations,* provides suggested calculation methods.

9.6.4 Where the result of the arc flash hazard analysis at a designated piece of equipment is greater than what is appropriate for the available PPE, a means to reduce the hazard level should be implemented.

9.6.5 The arc flash hazard analysis study results are field marked by a label on the equipment. The documentation for the arc flash hazard analysis should be retained for reference and use as needed.

9.6.6 The arc flash hazard analysis should be repeated if there are changes that occur that affect the arc flash hazard, such as changes in the available short-circuit current or in the overcurrent protective devices.

Chapter 10 Power Quality

10.1 Introduction.

10.1.1 Special Terms. The following special terms are used in this chapter.

10.1.1.1 Bonding (Bonded). The permanent joining of metallic parts to form an electrically conductive path that will ensure electrical continuity and the capacity to conduct safely any current likely to be imposed. The "permanent joining" can be accomplished by the normal devices used to fasten clean, noncorroded parts together. Machine screws, bolts, brackets, or retainers necessary to allow equipment to function properly are items typically employed for this purpose. While welding and brazing can also be utilized, these preclude easy disassembly, and welding can increase rather than decrease resistance across joints. Metallic parts that are permanently joined to form an electrically conductive path that will ensure electrical continuity and the capacity to conduct safely any current likely to be imposed are bonded.

10.1.1.2 Bonding Jumper. A reliable conductor to ensure the required electrical conductivity between metal parts required to be electrically connected. This conductor can be solid or stranded or braided, and connected by compatible fittings to separate parts to provide this electrically conductive path. The bonding jumper can also be a screw or a bolt. This bonding jumper can be used alone or in conjunction with other electrically conductive paths. It generally is associated with the equipment-grounding path, but might or might not be electrically linked for a lowest impedance path.

10.1.1.3 Central Grounding Point. The location where the interconnected parts of the grounding system are connected in a common enclosure. The central grounding point provides a common connection point for termination of the feeder or branch-circuit equipment-grounding conductors.

10.1.1.4 Common Mode Noise. Undesirable electrical signals that exist between a circuit conductor and the grounding conductor.

10.1.1.5 Equipment-Grounding Conductor. The conductor used to connect the noncurrent-carrying metal parts of equipment, raceways, and other enclosures to the system grounded conductor, the grounding electrode conductor, or both, at the service equipment or at the source of a separately derived system.

10.1.1.6 Grounded Conductor. A system or circuit conductor that is intentionally grounded. This intentional grounding to earth or some conducting body that serves in place of earth takes place at the premises service location or at a separately derived source. Control circuit transformers are permitted to have a secondary conductor bonded to a metallic surface that is in turn bonded to the supply equipment-grounding conductor. Examples of grounded system conductors would be a grounded system neutral conductor (3 phase or split phase) or a grounded phase conductor of a 3-phase, 3-wire, delta system.

10.1.1.7 Grounding Electrode Conductor. The conductor used to connect the grounding electrode to the equipment-grounding conductor, to the grounded conductor, or to both, of the circuit at the service equipment or at the source of a separately derived system. This conductor must be connected to provide the lowest impedance to earth for surge current due to lightning, switching activities from either or both of the supply and load side, and to reduce touch potentials when equipment insulation failures occur.

10.1.1.8 Harmonics. Those voltages or currents whose frequencies are integer multiples of the fundamental frequency.

10.1.1.9 Interharmonics. Not all frequencies that occur on an electrical power system are integer multiples of the fundamental frequency (usually 60 Hz), as are harmonics. Some loads draw currents that result in voltages that are between harmonic frequencies or less than the fundamental frequency. These frequencies are referred to as interharmonics and can be made of discrete frequencies or as a wide-band spectrum. A special category of these interharmonics is called subharmonics, in which the frequencies involved are less than the fundamental power line frequency.

10.1.1.10 Long-Duration Undervoltage. A decrease of the supply voltage to less than 90 percent of the nominal voltage for a time duration greater than 1 minute. *(See IEEE 1159, Recommended Practice on Monitoring Electric Power Quality, Table 4-2.)*

10.1.1.11 Multipoint Grounding. Multipoint grounding consists of interconnecting primary and secondary neutrals of the transformer. The secondary and primary neutral are common, and they both utilize the same grounding electrode that connects the system to earth.

10.1.1.11.1 These provide corresponding neutral circuit conductors in both the primary and secondary single-phase and wye-connected windings. This provides a low impedance path between each system and allows ground current disturbances to flow freely between them with little or no attenuation. Although there are advantages to these "wye-wye" systems, they can contribute to a common mode noise problem.

10.1.1.11.2 Multipoint grounding can also be found with systems where one or both windings are delta connected.

10.1.1.11.3 The primary and secondary windings are only casually interconnected, and this provides significant impedance to any current flow between them, since there are no corresponding circuit conductors that can be directly connected together. Grounding a circuit conductor at any point up to the service entrance disconnect location of the premises is permitted. Multipoint grounding of separately derived systems is not permitted, and single-phase 2-wire, single-phase 3-wire (split-phase), or delta-wye multiphase systems are recommended.

10.1.1.12 Sag. A decrease to between 10 percent and 90 percent of the normal voltage at the power frequency for durations of 0.5 cycle to 1 minute. (If the voltage drops below 10 percent of the normal voltage, then this is classified as an interruption.) It is further classified into three categories: (1) instantaneous — 0.5 cycle to 30 cycles; (2) momentary — 30 cycles to 3 seconds; and (3) temporary — 3 seconds to 1 minute.

10.1.1.13 Separately Derived System. A premises wiring system whose power is derived from a battery, a solar photovoltaic system, or from a generator, transformer, or converter windings, and that has no direct electrical connection, including a solidly connected grounded circuit conductor, to supply conductors originating in another system. Equipment-grounding conductors are not supply conductors and are to be interconnected.

10.1.1.14 Sustained Voltage Interruption. The loss of the supply voltage to less than 10 percent on one or more phases for a period greater than 1 minute. *(See IEEE 1159, Recommended Practice on Monitoring Electric Power Quality, Table 4-2.)*

10.1.1.15 Swell. An increase to between 110 percent and 180 percent in normal voltage at the power frequency durations from 0.5 cycle to 1 minute. It is further classified into three categories: (1) instantaneous — 0.5 cycle to 30 cycles; (2) momentary — 30 cycles to 3 seconds; and (3) temporary — 3 seconds to 1 minute. *(See IEEE 1159, Recommended Practice on Monitoring Electric Power Quality, Table 4-2.)*

10.1.1.16 Transients. Transients (formerly referred to as surges, spikes, or impulses) are very short duration, high amplitude excursions outside of the limits of the normal voltage and current waveform. Waveshapes of the excursions are usually unidirectional pulses or decaying amplitude, high frequency oscillations. Durations range from fractions of a microsecond to milliseconds, and the maximum duration is on the order of one half-cycle of the power frequency. Instantaneous amplitudes of voltage transients can reach thousands of volts.

10.1.1.17 Transverse Mode Noise. Undesirable electrical signals that exist between a pair of circuit conductors. These signals are sometimes referred to as normal or differential mode noise.

10.1.1.18 Unbalanced Voltages. Unequal voltage values on three-phase circuits that can exist anywhere on the power distribution system.

10.1.2 General.

10.1.2.1 Power quality addresses deviations and interruptions from the pure, ideal power supply. Alternating-current (ac) power used to run equipment often consists of distorted, nonsinusoidal waveforms (nonlinear); waveforms in the three phases of a 3-phase circuit commonly differs slightly in size and shape; and circuit voltage can change as the load on the circuit changes.

10.1.2.2 Historically, most equipment has been moderately tolerant of typical power quality problems. Some equipment with electronic components is more susceptible to power quality problems. Some equipment conducts current during only part of the power frequency cycle. These are typically called nonlinear loads and are sources of harmonic currents. Rectified input switch–mode power supplies; arcing devices, including fluorescent lamps; and other nonlinear devices affect waveforms and cause a decrease in power quality.

10.1.2.3 Power quality problems are frequently caused by equipment or conditions on the customer's premises. Power quality problems are less frequently caused by utility generating, transmission, or distribution equipment. However, off-site equipment belonging to neighbors and line exposures such as from capacitor switching, lightning, vehicles, contaminants, and wildlife can create problems that are carried by the utility to its customers.

10.1.2.4 Poor power quality can cause electrical faults, jeopardize personal safety, damage or reduce the life of electrical and electronic equipment, cause an increased fire hazard, and reduce equipment performance and productivity. Poor power quality can also affect data and communications. Maintaining the quality of the entire process might require monitoring of the process transducers and associated communication systems, as well as the electrical supply.

10.1.2.5 While power quality problems are often identified by maintenance personnel, diagnosing problems and finding solutions can be difficult. Some solutions require knowledge of electrical engineering, testing, and specialized equipment. A solution might require a custom-engineered approach, not merely equipment repair, upgrade, or replacement.

10.1.2.6 Power quality disturbances include the following:

(1) Harmonics imposed on the fundamental sine wave
(2) Voltage transients
(3) Voltage sags and swells
(4) Long-duration undervoltage and sustained voltage interruptions
(5) Unbalanced voltages and single phasing (partial interruption)
(6) Inadvertent and inadequate grounding
(7) Electrical noise
(8) Interharmonics
(9) Voltage fluctuations (light flicker)

10.1.2.6.1 For common power system disturbances, a waveform of a sag cleared by the supply line circuit breaker, and a waveform with transient from power factor capacitor switching, see Figure 10.1.2.6.1(a), Figure 10.1.2.6.1(b), and Figure 10.1.2.6.1(c).

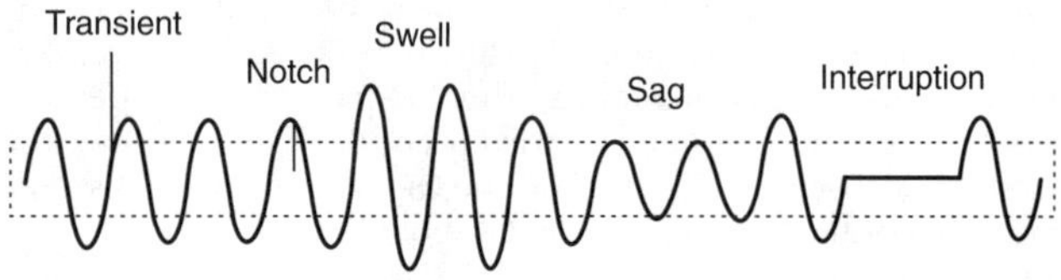

FIGURE 10.1.2.6.1(a) Common Power System Disturbances.

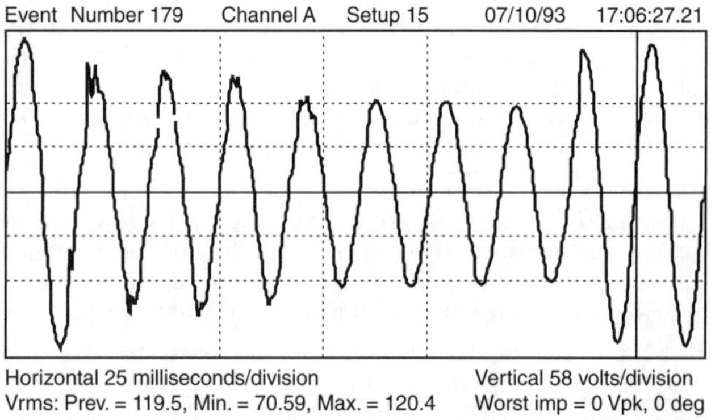

FIGURE 10.1.2.6.1(b) Waveform of Sag Cleared by Supply Line Breaker Operation.

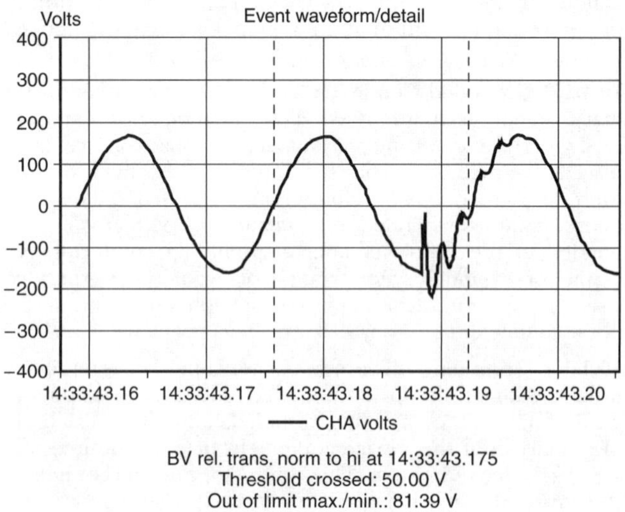

FIGURE 10.1.2.6.1(c) Waveform with Transient from Power Factor Capacitor Switching.

10.2 Harmonics.

10.2.1 Introduction.

10.2.1.1 The fundamental frequency (usually 60 Hz) is the predominant, intended frequency of a power system. Harmonics are identified by their harmonic number. For example, with a 60 Hz fundamental frequency, 120 Hz is the second harmonic, 180 Hz the third harmonic, and 300 Hz the fifth harmonic.

10.2.1.2 Harmonics distort and change the rms magnitude of the waveform. *(See Figure 10.2.1.2.)*

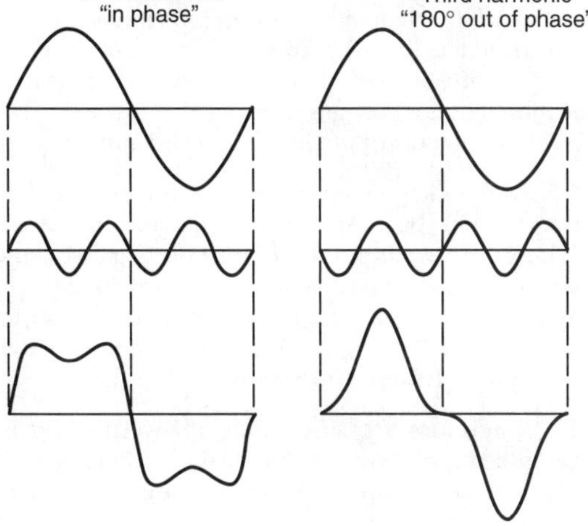

FIGURE 10.2.1.2 Harmonics and the Fundamental Waveform.

10.2.1.3 Harmonics imposed on the power system are typically calculated from measurements over either 10 or 12 cycles, depending on whether the fundamental frequency is 50 Hz or 60 Hz, respectively, and are usually expressed as a percentage of the fundamental voltage or current. For example, the total harmonic distortion (THD) of a voltage waveform is stated as a percentage and can be defined as 100 times the square root of the ratio of the sum of the squares of the rms amplitudes of the individual harmonics, or root mean square (rms), divided by the square of the voltage at the fundamental frequency and is represented by the following formula:

$$THD = \left(\frac{100 \times \sqrt{\sum V_h^2}}{V_f} \right) \%$$

where:
THD = total harmonic distortion (percent)
V_h = rms voltage of the individual harmonic
V_f = rms voltage of the fundamental frequency

10.2.1.3.1 This value should be considered related to the maximum load capacity when used for current THD. For example, a circuit may have 1A of current flowing on a 30A circuit, where 50 percent of the current is harmonic current. A load of 0.5A of harmonic current is normally not a problem on a 30A circuit. However, if the situation was 50 percent current THD on a circuit and the total current was 25A, this may cause problems. Instead, Total Demand Distortion (TDD) should be used. See ANSI/IEEE 519, *Recommended Practices and Requirements for Harmonic Control in Electrical Power Systems.*

10.2.1.4 Line voltage notching is a form of harmonic distortion, as shown in Figure 10.2.1.4. In many cases, line voltage notching caused by phase-controlled rectifiers can be more of a problem than current harmonics. Commutation notches can affect the performance of electronic equipment.

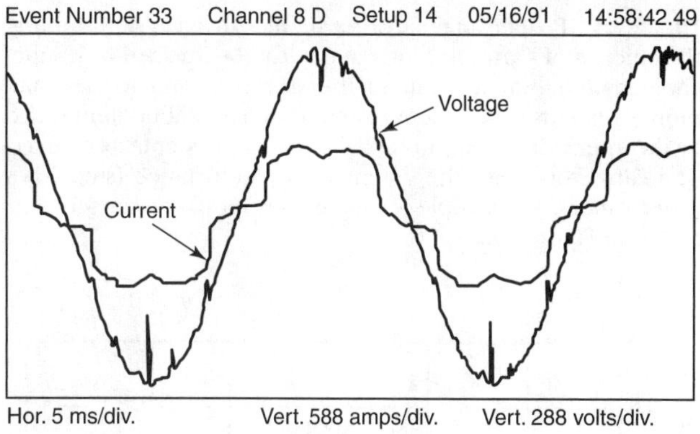

FIGURE 10.2.1.4 Line Voltage Notching Harmonic Distortion.

10.2.2 Harmonic Symptoms and Effects.

10.2.2.1 Harmonics are caused by nonlinear loads in which the current waveform does not conform to the waveform of the impressed voltage. Some of the symptoms and effects of nonlinear loads are comparable to other, more readily recognized symptoms such as overloading and can be difficult to diagnose.

10.2.2.1.1 Problems created by harmonics might include the following:

(1) Excessive neutral current
(2) Overheating of transformers, motors, generators, solenoid coils, and lighting ballasts
(3) Nuisance operation of protective devices
(4) Unexplained blowing of fuses on power-factor correction capacitors
(5) Unusual audible noise in electrical switchgear
(6) Voltage and current waveform distortion that results in misoperation or failure of solid-state electronic equipment
(7) Audible noise interference on telephone circuits
(8) Loss of data on computer systems
(9) False operation of facility distribution power line carrier control systems such as lights, clocks, and load shedding
(10) Failure of uninterruptible power supply (UPS) systems to properly transfer
(11) Shaft voltages and currents on electric motors causing bearing failure if the bearings are not insulated

10.2.2.1.2 The neutrals of 3-phase, 4-wire systems are especially susceptible to harmonic problems. On such circuits, each phase is displaced by 120 electrical degrees from adjacent phases. If the load is balanced and no harmonics are present, the phase currents cancel vectorially, and the neutral current is zero. However, odd triplen harmonics, such as the third, ninth, and fifteenth, are additive rather than subtractive in the neutral and do not cancel.

10.2.2.1.3 For example, in a 3-phase, 4-wire system, if there are 30 amperes of triplen harmonic current present in a 100-ampere phase current, 90 amperes of triplen harmonic current will flow in the neutral. Neutral current will be higher if the phase currents are unbalanced.

10.2.2.2 Capacitors do not create harmonics, but capacitor failures and blown fuses on power-factor improvement capacitors are often attributable to harmonics. This situation occurs because capacitance can combine with circuit inductance to establish a resonant condition when harmonics are present. Resonance can cause high voltages to appear across elements of the power system and can cause high currents to flow.

10.2.2.2.1 Proper analysis of harmonics involves determining the amount of harmonic current that can be injected by nonlinear loads and then determining the system response to these harmonic currents. The system response usually will be dominated by the interaction of shunt capacitor banks (power-factor correction capacitors) with the system source inductance (step-down transformer). An example of frequency response characteristic is shown in Figure 10.2.2.2.1.

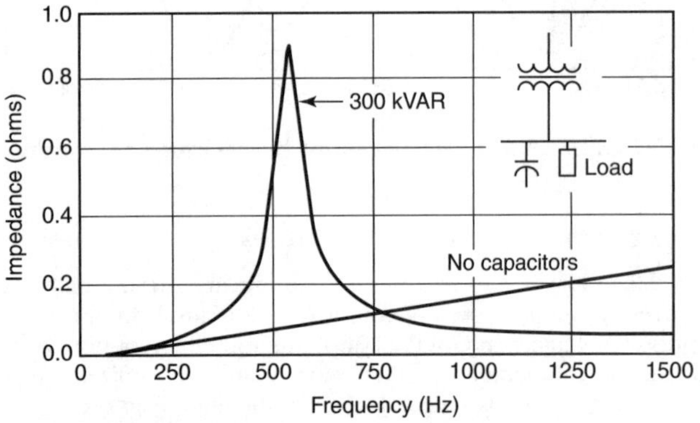

FIGURE 10.2.2.2.1 Example of Frequency Response Characteristic for a 1500 kVA, 13.8/0.48 kV, 6.0 Percent Transformer and a 300 kVAR, 480 V Capacitor Bank.

10.2.2.2.2 The parallel resonance occurs at the frequency where the shunt capacitive reactance is equal to the inductive source reactance and can be expressed in terms of the 60 Hz values as follows:

$$h = \sqrt{\frac{X_c}{X_{sc}}} = \sqrt{\frac{kVA_{sc}}{kVARcap}}$$

where:

h = resonant frequency as a multiple of the fundamental frequency
X_c = shunt capacitive reactance of capacitor
X_{sc} = short-circuit reactance of source
kVA_{sc} = short-circuit kVA of source
$kVARcap$ = total capacitor kVAR

10.2.2.2.3* This simple relationship provides an excellent first check of whether harmonics are likely to be a problem. An overvoltage condition problem occurs when this parallel resonance moves close to the fifth or seventh harmonic, since these are the largest harmonic current components in most nonlinear loads that contain 3-phase, full wave rectifiers. However, the eleventh and thirteenth harmonics can also be a problem when nonlinear loads are a large percentage of the total load. If a parallel resonance exists at one of the characteristic harmonics, the harmonic currents injected by the nonlinear loads are magnified, and high magnitudes of voltage distortion occur.

10.2.2.3 Harmonics can cause overheating, overvoltage, and excessive noise in transformers. Overheating is a compound effect of increased winding I^2R losses due to both excessive current and skin effect and increased eddy current and hysteresis losses in the transformer core.

10.2.2.3.1 On a 3-phase delta-wye-connected transformer, third harmonics generated by the transformer secondary loads are reflected into the primary in the form of circulating currents in the delta-connected primary. It is therefore especially important to use a true rms-reading ammeter when checking a transformer's secondary line, neutral, and, where practical and safe, primary winding current, for possible overload.

10.2.2.3.2 Transformers with relatively high impedance are susceptible to overvoltage and core saturation in the presence of harmonics, causing increased current flow and resultant additional heating. Higher voltages can cause excessive 60 Hz hum, and harmonics can contribute higher-pitched audible noise.

10.2.2.3.3 Motors are also subject to overheating in the presence of harmonics, because of skin effect and increased iron losses. Where fifth harmonics are present, negative-sequence currents will also flow in opposition to the current necessary to develop the torque required for rotation. This counter torque contributes to overheating, and in some extreme cases can result in pulsating torque and excessive vibration.

10.2.2.4 Generators are susceptible to overheating in the presence of harmonics for essentially the same reasons as are motors. Generators equipped with solid-state controls can also operate erratically in the presence of harmonics, especially if the controls incorporate zero-crossing sensing circuits. Generators can cause harmonics due to internal construction. The type of generating winding pitch can determine the magnitude and types of harmonics generated.

10.2.2.4.1 Generators operating in parallel should have the same winding pitch to minimize problems. Where generators operate in parallel with a common neutral, third harmonic currents can circulate between the machines and cause overheating. High resistance grounding of these generators can adequately limit the harmonic current.

10.2.2.5 As is the case with all equipment operating on the principle of electromagnetic induction, ferromagnetic ballasts will also develop excessive heating where harmonics are present. The presence of harmonics can contribute to inaccurate readings (high or low) of induction disc meters.

10.2.2.6 Electrical panels and cables can also exhibit overheating because of excessive neutral current; excessive heating might be detected as discoloration. Harmonics can also cause conductor-insulation failure because of voltage stress and corona discharge.

10.2.2.7 Instrumentation transformers such as current and potential transformers can transfer harmonics from a primary to a secondary, resulting in misoperation of instrumentation, protective relaying, and control circuits.

10.2.2.8 Computers and other computer-type equipment such as programmable logic controllers (PLCs) are susceptible to harmonic-distorted waveforms, and the possibility of harmonics should be investigated on circuits serving such equipment where neutral-to-ground voltages in excess of 2 volts are measured at the equipment. Harmonic effects on such equipment can range from data errors, process controls operating out of sequence, and erratic operation of production robots or machine tools, through total failure of the electronic equipment.

10.2.2.8.1 In addition to polluting data and control signal transmissions, harmonics can be a source of audible noise on telephone communication circuits. Audible noise can be induced in communication conductors that run in proximity to conductors containing harmonic currents, such as with integral hp adjustable speed drives or other equipment that has power rectifiers without proper filters. For this reason, voice communication lines should be shielded or rerouted for wider separation. ANSI/IEEE 519, *Recommended Practices and Requirements for Harmonic Control in Electrical Power Systems,* and ANSI/IEEE 1100, *Recommended Practice for Powering and Grounding Sensitive Electronic Equipment (IEEE Emerald Book),* contain additional information on the telephone interference factor (TIF).

10.2.3 Causes of Harmonic Distortion. Harmonics from neighboring utility customers can be introduced to the premises by the incoming utility supply. Harmonics originate on the premises in most cases.

10.2.3.1 All equipment operating on the principle of ferromagnetics produces harmonics when operating in the saturation region of the magnetic core. This equipment includes transformers, motors, generators, induction heaters, solenoid coils, lifting magnets, and iron-core arc-discharge lighting ballasts. The extent to which harmonics are generated varies with the type of equipment.

10.2.3.2 Arc-producing equipment, such as welding machines and arc furnaces, also develops harmonics. Arc-discharge lamps produce harmonics over and above those introduced by the lamp ballast.

10.2.3.3 The most significant contributor to harmonics is often electronic equipment, especially equipment that utilizes a rectified-input switching-mode power supply. The wave-chopping characteristic operation of thyristors, silicon-controlled rectifiers, transistors, and diodes develops current waveforms that do not conform to the applied voltage waveform and therefore develops harmonics. Included among electronic equipment that is rich in harmonic generation are welders, battery chargers, rectifiers, ac and dc adjustable-speed motor drives, electronic lighting ballasts, computers, printers, reproducing machines, and programmable logic controllers.

10.2.4 Harmonic Surveying and Testing.

10.2.4.1 Where harmonics are suspected as the cause of problems, it is necessary to determine the magnitude of the harmonic frequencies and their contribution to THD. This information will define the extent of the harmonic problem, provide clues as to causes of the harmonics, and provide the data needed to engineer solutions. It will also permit calculation of transformer derating factors in accordance with ANSI/IEEE C57.110, *Recommended Practice for Establishing Transformer Capability When Supplying Nonsinusoidal Load Currents.*

10.2.4.2 The extent of harmonic surveying and testing will vary widely depending on the severity of the problem, available resources, and the facility's particular needs. A simple test can be performed to confirm or refute the existence of harmonics.

10.2.4.2.1 Where the test is conducted by taking current readings with an average responding ammeter and a true rms responding ammeter, the average responding instrument generally yields a lower reading than the rms responding meter if harmonics are present.

10.2.4.3 The presence of odd triplen harmonics (third, ninth, fifteenth, and so on) can be readily determined on 4-wire "wye" circuits by measuring neutral current with a true rms responding ammeter and comparing it with current to be expected on the basis of rms phase currents. Neutral-to-ground voltages in excess of 2 volts measured at the equipment can also indicate the presence of triplen harmonics. Where such readings determine that triplen harmonics are present, analysis should be undertaken to determine their specific frequencies and magnitudes. Instruments available for harmonic analysis include oscilloscopes, harmonic analyzers, and spectrum analyzers.

10.2.4.4 Oscilloscopes readily permit visual observation of the waveform to determine if it deviates from a sine wave or if line voltage notching exists.

CAUTION: Because one side of the oscilloscope probe might be common to the case, a line isolation device should be used between the probe and the line voltage being measured.

10.2.4.4.1 Harmonic analyzers measure the contribution of harmonic voltage and current at each frequency and calculate THD. Harmonic analyzers are available in a broad range of sophistication, with some also measuring circuit parameters such as kW, kVA, and power factor and some determining transformer derating factors and telephone interference factors. Spectrum analyzers provide detailed waveform analysis indicating the harmonic frequencies imposed on the fundamental.

10.2.4.5 Total harmonic distortion can vary significantly with load. Therefore, readings should be taken under different load conditions. Measurements should be taken to determine the location and extent of harmonics. Readings should be taken on all phases and neutrals of 3-phase, 4-wire systems and especially on 3-phase circuits that serve single-phase loads.

10.2.4.6 Voltage measurements on low-voltage circuits can be taken easily by connecting instrumentation directly to the measured point. High voltages, however, require the use of instrument potential transformers (PTs). The PTs should be dedicated to the test; existing bus PTs serving relaying and bus instrumentation should not be used. The measurement instrument should have an input impedance of at least 100 kilohms, and the instrument manufacturer's connection and operating instructions should always be followed.

10.2.4.7 Each facility differs in its tolerance to harmonic distortion. There are guidelines that can be followed to determine if harmonics are within acceptable limits. ANSI/IEEE 519, *Recommended Practices and Requirements for Harmonic Control in Electrical Power Systems,* is one such standard. Other information sources include the following:

(1) ANSI/IEEE C57.110, *Recommended Practice for Establishing Transformer Capability When Supplying Nonsinusoidal Load Currents*
(2) ANSI/IEEE 1100, *Recommended Practice for Powering and Grounding Sensitive Electronic Equipment (IEEE Emerald Book)*

10.2.5 Recommended Solutions to Harmonic Problems. Solutions to harmonic problems are unique in nature and depend on the results of the harmonic survey and testing and subsequent analysis. Recommended solutions include any or all of the following:

(1) Derating of existing equipment
(2) Replacement of existing equipment with higher rated equipment

(3) Use of delta-wye- or delta-delta-connected transformers as appropriate
(4) Use of equipment specifically rated for harmonic circuits
(5) Better selection and application of protective and metering devices
(6) Use of rms-sensing protective devices
(7) Balancing of single-phase loads on 3-phase systems
(8) Use of 3-phase rectifiers instead of single-phase rectifiers
(9) Relocating power-factor improvement capacitors
(10) Shielding of conductors and electronic equipment
(11) Isolation of harmonic-sensitive loads
(12) Use of filters to block or shunt off harmonics
(13) Specification of new equipment for low harmonic content
(14) Periodic surveys and power-system adjustments/modifications as might be indicated by survey results
(15) Increased neutral conductor size
(16) Replacement or repair of harmonic producing equipment
(17) Utilization of a motor or generator with an insulated bearing

10.2.5.1 In some cases, solutions can be engineered and implemented with in-house personnel; in other cases, it might be necessary to engage personnel with specialized expertise and equipment.

10.3 Transients (Surges).

10.3.1 Introduction. Transient current is proportional to the transient voltage and the system impedance. System impedance includes source impedance and transient impedance. For rating transient protective devices, transient energy is usually expressed in joules (watt-seconds).

10.3.2 Transient Symptoms and Effects.

10.3.2.1 The effect and the severity of the transient depend on magnitude, duration, and frequency. Low-energy transients can cause equipment to malfunction. High-energy transients can damage equipment. When transient-sensitive equipment or transient-producing equipment is installed, problems previously not encountered with existing equipment can occur. In addition, if transient-sensitive equipment and transient-producing equipment are moved electrically closer to each other, problems can result.

10.3.2.2 Within electrical systems without transient voltage protection, transient voltages are limited by flashover of clearances. When the transient reaches breakdown voltage, an arc is established through the air or across the surface of insulation, limiting the maximum transient voltage on the system. Typically, in low-voltage (1000 volts or less) distribution, the maximum transient is limited to about 6 kV for indoor systems and to about 10 kV to 20 kV for outdoor systems. The transient voltage can be limited to lesser values by surge protective devices.

10.3.2.3 Problems associated with transients include the following:

(1) Unusual equipment damage due to insulation failures or arc-over, even with proven maintenance practices
(2) Damage to electronic equipment components due to their inability to withstand transient voltages
(3) Total failure, lock-up, or misoperation of computer or other microprocessor-based equipment

10.3.3 Causes of Transients.

10.3.3.1 Transient voltages in low-voltage ac power circuits usually originate from lightning effects (direct or indirect) on the power system or from switching operation.

10.3.3.2 Lightning strikes can cause severe transients because of the very high voltages and currents. Lightning can enter the electrical circuit directly or can be induced by nearby strikes. This might also produce a transient on the grounded and grounding systems.

10.3.3.3 Transients can be caused by the switching of inductive or capacitive loads, such as motors, ballasts, transformers, or capacitor banks. Arcing contacts can also cause transients.

10.3.3.4 Transients can result from abnormal conditions on the power system, such as phase-to-phase or phase-to-ground short circuits.

10.3.4 Transient Monitoring.

10.3.4.1 Monitoring can be used to determine the presence of transients. Storage-type, high-bandwidth oscilloscopes with high-voltage capability can be used, but more information can be obtained from the use of power disturbance analyzers specifically designed for transient and other types of power-quality problems. Monitoring might be required over an extended period of time, due to the characteristics of transients, which vary as loads and system configurations change.

10.3.4.2 Monitoring is often performed at specific locations where a sensitive load is connected or is to be connected. Other devices on the monitored circuit, such as the power quality monitor itself, can contain surge-protection devices that limit transients and distort the results. If possible, use an alternative power source for powering monitoring equipment.

10.3.4.3 Monitoring can be required phase-to-phase, phase-to-ground, phase-to-neutral, and neutral-to-ground to develop a complete profile of the system.

10.3.5 Recommended Solutions to Transient Problems.

10.3.5.1 The following are devices intended for the suppression of transients: surge arresters, surge capacitors, surge protectors, inductive reactors, and surge suppressors. Proper grounding of all circuits intended to be grounded is required for correct operation of these devices. The manufacturer's instructions should be followed when any of these devices is installed. Engineering evaluation might be required to select the proper type and rating of these devices.

10.3.5.2 Surge arresters are intended to be installed ahead of the service entrance equipment for limiting transient voltage by discharging or bypassing transient currents to ground. They typically provide protection for the effects of lightning.

10.3.5.3 Surge capacitors are placed in a circuit to slow the transient voltage rise time. By spreading out the voltage increase over a longer time span, less electrical stress occurs to equipment subjected to the transient.

10.3.5.4 Surge protectors are gas-tube devices or assemblies composed of one or more gas-tube devices. They are used for low-voltage applications (up to 1000 volts rms or 1200 volts dc.)

10.3.5.5 A transient voltage surge suppressor (TVSS) is a device intended for installation on the load side of the main overcurrent protection in circuits not exceeding 600 volts rms. These devices comprise any combination of linear or nonlinear circuit elements (i.e., varistors, avalanche diodes, and gas tubes) and are intended for limiting transient voltages by diverting or limiting surge current. Filters for electromagnetic interference (EMI) reduction can also be incorporated within TVSS devices.

10.3.5.6 The following are typical locations where transient protection is installed: power circuits at the service entrance, communication circuits entering the building, computer room power, and at susceptible loads.

10.3.5.7 Inductive reactors placed in series with the supply circuit can attenuate off-site transients. Inductive reactors placed in series with noisy equipment can localize transients to the equipment.

10.4 Voltage Sags and Swells.

10.4.1 Sags and swells are the most common types of power quality disturbances. Millions of dollars are lost in productivity each year in the United States due to these disturbances. A simple understanding of their causes will help obtain effective solutions to minimize these disturbances.

10.4.1.1 Different equipment might require a different susceptibility curve. *(See Figure 10.4.1.1 for an example of a curve.)*

10.4.1.2 Sags are the most common type of voltage disturbances. Typically, sags occur twice as often as swells.

10.4.2 Symptoms of Sags and Swells.

10.4.2.1 Introduction. The effects of a sag are more noticeable than those of a swell. A sag duration longer than three cycles is visible because lighting output is reduced. Sags often are not distinguishable from momentary interruptions, and the effects to the equipment can be the same. Sensitive equipment, such as computers, can experience intermittent lockups or garbled data. Even relays and contactors in motor starters can be sensitive to voltage sags, resulting in shutdown of a process when they drop out.

10.4.2.2 Sophisticated Equipment. Equipment used in industrial plants (e.g., process controllers, programmable logic controllers, adjustable-speed drives, and robotics) is increasingly sensitive to voltage sags as the equipment becomes more complex. For example, computers are faster, operate at lower logic voltages, and have less power supply ride-through, making them much more vulnerable to voltage sags.

10.4.2.3 Loss of Memory. Voltage sags can cause the loss of stored data in programmable electronic systems.

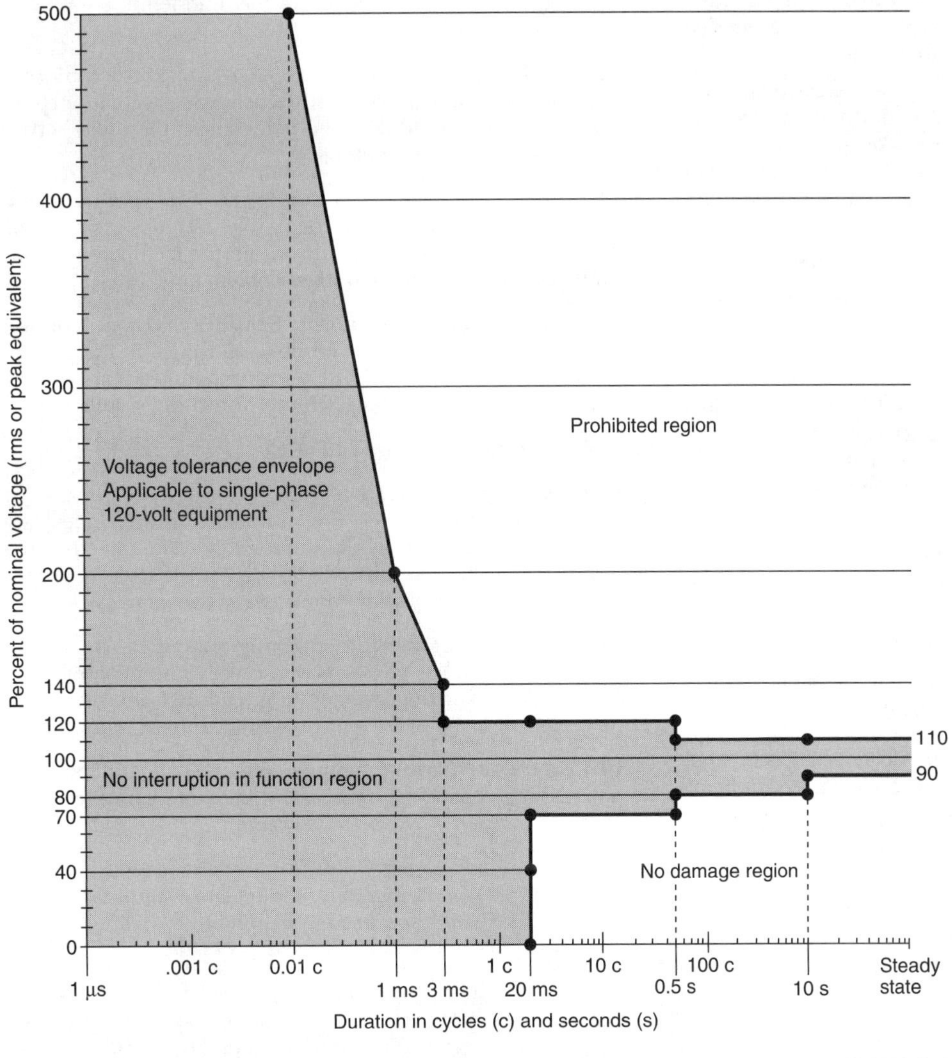

FIGURE 10.4.1.1 ITI Curve. [Courtesy of Information Technology Industry Council (ITI).] For proper use of this figure, refer to complete text of ITI (CBEMA) Curve Application Note.

10.4.2.4 Equipment Shutdowns. Motor contactors and electromechanical relays can drop out with a sag to 70 percent of rated voltage lasting 1 cycle or longer. High-intensity discharge (HID) lamps generally require restriking for sags below 80 percent. The ride-through of some adjustable-speed drives varies from 0.05 second to 0.5 second. Remote I/O units of some programmable logic controllers trip on a reduction to 90 percent of rated voltage for just a few cycles.

10.4.2.5 Component Breakdown. The effects of a swell can be more physically destructive to the equipment than those of a sag. The overvoltage can cause breakdown of components in the power supplies of the equipment. This can be a gradual, cumulative effect.

10.4.3 Causes of Voltage Sags and Swells. Sags and swells occur in utility transmission and distribution systems and facility power distribution systems. A common, underlying cause in all three areas is a sudden change of current flow.

10.4.3.1 Sag Causes — Transmission Systems.

10.4.3.1.1 Outside Sources. Severe weather, construction accidents, transportation accidents, or animals can cause faults that result in sags. Lightning is a common cause of faults on overhead transmission and distribution lines. A fault can occur by lightning directly striking a phase conductor or striking a nearby grounded object, such as a transmission shield wire or tower.

10.4.3.1.2 Sag Duration. Transmission system voltage sags normally are shorter in duration than utility distribution sags. The fault-clearing mechanisms (the relay/breaker schemes) are designed to react faster for transmission faults. Sags to 75 percent in a facility have been caused from transmission faults 300 miles away.

10.4.3.2 Sag Causes — Utility Distribution Systems.

10.4.3.2.1 Outside Causes. In addition to transmission system causes, contact with tree limbs and motor vehicle accidents can result in voltage sags. Such faults can be 3-phase, line-to-line or line-to-ground. The 3-phase faults are the most severe but are relatively uncommon. Single line-to-ground faults on the utility system are a common cause of voltage sags in an industrial plant. A fault on a single feeder can result in an interruption to loads on that feeder, as well as a sag on the other feeders.

10.4.3.2.2 Sag Duration. Typically, distribution system sags are 6 cycles to 20 cycles. Repeated sags can occur with reclosing on the same fault. Depending on the number of reclosures, feeders can experience several voltage sags in succession.

10.4.3.3 Sag Causes — Facility Power Systems.

10.4.3.3.1 Sudden increases in the current demand within a facility can cause sags until the large current demand decreases. The sudden increases can be the result of fault conditions in the building or the startup of large inductive loads, such as motors. In one large-scale study *(see Section 10.10)*, 50 percent or more of the sags and swells recorded were caused by load equipment in the same building.

10.4.3.3.2 A voltage sag from energization of large high-current-demand motors, can last for 30 cycles.

10.4.3.3.3 A utility fault usually creates a more severe sag than a motor start sag. The sag will last until the fault is cleared or removed.

10.4.3.4 Swell Causes.

10.4.3.4.1 Swells are less common than voltage sags and are usually associated with system fault conditions. A swell can occur due to a single line-to-ground fault on the system, which can result in a temporary voltage rise on the unfaulted phases.

10.4.3.4.2 Swells can also be generated by sudden load decreases. The abrupt interruption of current can generate a large voltage, proportional to the inductance and the rate of change of the current. Switching on a large capacitor bank can also cause a swell, although it more often causes a high-frequency transient.

10.4.4 Monitoring and Testing for Sags and Swells.

10.4.4.1 Equipment. Different types of monitoring equipment are available to monitor sags and swells. These range from event indicators that visually indicate that a sag or swell has occurred to monitors that provide a cycle-by-cycle graph of the disturbance and record the minimum/maximum values, duration, and time of occurrence.

10.4.4.2 Finding the Source. Data on the timing and magnitude of the sag or swell can often identify the source of the initiating condition. If the phase current levels of the load did not change prior to the voltage sag, the source is more likely to be upstream. When the magnitude of the sag is severe, it is likely that the source was close by. A power-factor correction capacitor being switched on can result in an oscillatory transient followed by a swell.

10.4.4.3 Initial Placement. Unless there is significant information pointing to the source of the disturbance, it is common practice to begin monitoring at the point where the utility service connects to the facility equipment.

10.4.4.4 Other Locations. If the source of the disturbance is determined to be internal to the facility, then placing multiple monitors on the various circuits within the facility most likely would identify the source of the problem quickly.

10.4.4.5 Monitoring Instrument Sensitivity. Power monitoring instruments are quite sensitive, and outside factors can influence their accuracy. Long measurement leads are susceptible to RFI/EMI pickup, which can distort the results.

10.4.5 Solutions for Sags and Swells.

10.4.5.1 A transformer tap change can be used to raise or lower the nominal voltage level and make the system less susceptible to sags or swells. Automatic solid-state tap-changing transformers that are controlled by electronic sensing circuits can react relatively quickly (one cycle to three cycles).

10.4.5.2 Different transformer configurations can be used to minimize the effects of events that cause sags and swells. For example, a delta-delta configuration tends to hold voltage levels higher than a delta-wye or wye-delta configuration.

10.4.5.3 Fault current limiters, zero voltage independent pole closing capacitor switches, and high-energy surge arresters can be added to the electric system.

10.4.5.4 Ferroresonant transformers, also called constant-voltage transformers, can handle most short-duration voltage sags. They provide excellent regulation but have limited overload capacity and poor efficiency at low loads.

10.4.5.5 Magnetic controlled voltage regulators use transformers, inductors, and capacitors to synthesize 3-phase voltage outputs. Enough energy is stored in the capacitors to ride through 1 cycle. The overall response time is relatively slow (3 cycles to 10 cycles).

10.4.5.6 A UPS can provide isolation from power line disturbances, in addition to providing ride-through during a sag. *(See Chapter 28.)*

10.4.5.7 A static transfer switch is capable of transferring the supply voltage from one voltage source to another within a quarter-cycle.

10.5 Long-Duration Undervoltages and Sustained Voltage Interruptions.

10.5.1 Normal Supply Voltage Variations. Variations in the normal supply voltage are to be expected because loads on the supply system and plant distribution system are not constant. Electric utilities, equipment manufacturers, and end users have established standards for steady-state operating voltage limits that accommodate these variations. Facility utilization equipment can be designed and rated to operate within the range of supply system voltage while allowing for voltage drop in the plant system. *[See ANSI /NEMA C84.1, Electric Power Systems and Equipment, Voltage Ratings (60 Hertz).]*

10.5.1.1 Electric Utilities. Electric utilities can be required by regulatory commissions to maintain service voltages within prescribed limits for the various types of service. Plant electrical people should be aware of any required service voltage limits for their type of service. The utility generally works with the customer to ensure that the service voltage remains within the required limitations or within standard design limits where there are no required limitations.

10.5.1.2 As the system load varies, the utility automatic voltage-regulating equipment maintains the service voltage within the required range. When the serving utility's electrical system is severely stressed, the utility can implement a load-reduction strategy by reducing the voltage on its distribution lines, typically up to 5 percent. During these periods, the service voltage can be near the lower limit of the required range. As a result, a long-term undervoltage condition can exist at plant utilization equipment. It is strongly recommended that plant distribution system voltage drops be kept to a reasonable level.

10.5.2 Symptoms of Long-Duration Undervoltage. Undervoltage might not be readily apparent. Depending on the length and magnitude of the undervoltage, there can be a detrimental effect on electric and electronic equipment. Equipment such as induction motors might run hotter. Electronic equipment such as computers or microprocessor-based devices might function erratically.

10.5.3 Causes of Long-Duration Undervoltage. A long-duration undervoltage can originate on the electric utility system or on the plant electrical system. The utility system can be stressed due to line or equipment failure or system load conditions exceeding the supply capability. The plant electrical system or connected loads can result in unacceptable voltage drops even though the voltage is normal at the service point.

10.5.4 Monitoring and Testing of Long-Duration Undervoltages. Because the occurrence of a long-duration undervoltage might not be obvious, and damage to equipment and systems can result, an appropriate monitoring system is recommended where reliability is vital.

10.5.4.1 The monitoring system can consist of a sophisticated warning scheme with visual and audible alarms at appropriate locations. Alternatively, it can simply be a voltage-sensing relay located at the facility service entrance or at sensitive equipment with alarms placed in appropriate locations.

10.5.5 Solutions for Long-Duration Undervoltages. When a long-duration undervoltage occurs, costly or sensitive equipment should be disconnected to prevent possible damage. If it is necessary to keep the equipment or system in operation, an alternative power supply should be provided.

10.5.6 Symptoms of a Sustained Voltage Interruption. A sustained voltage interruption is obvious because electric power is unavailable for an extended period of time except for equipment served by an alternative power source.

10.5.7 Causes of Sustained Voltage Interruption. Sustained voltage interruptions are caused by power system disruptions such as power lines going down in a storm, the utility's distribution transformer failing, a fault condition causing a circuit protective device to open, or plant wiring problems.

10.5.8 Solutions for Sustained Voltage Interruptions. Solutions include generator sets, multiple power sources, and battery banks.

10.6 Unbalanced Voltages and Single Phasing. *(See 3.3.77, Unbalanced Voltages.)*

10.6.1 Percentage Limitations. On 3-phase circuits, unbalanced voltages can cause serious problems, particularly to motors, transformers, and other inductive devices.

10.6.1.1 Single phasing, which is the complete loss of a phase, is the worst-case voltage unbalance condition for a 3-phase circuit.

10.6.1.2 The National Electrical Manufacturers Association (NEMA) in MG1, *Motors and Generators*, part 14.36, defines voltage unbalance as follows: percent unbalance = 100 × (maximum voltage deviation from the average voltage/average voltage).

10.6.1.3 NEMA states that polyphase motors shall operate successfully under running conditions at rated load when the voltage unbalance at the motor terminals does not exceed 1 percent. Also, operation of a motor with more than 5 percent unbalance condition is not recommended and will probably result in damage to the motor.

10.6.1.4 For example, with line-to-line voltages of 460, 467, and 450, the average is 459, the maximum deviation from average is 9, and the percent unbalance equals 100 × (9/459) = 1.96 percent, which exceeds the 1 percent limit.

10.6.2 Causes of Unbalanced Voltages.

10.6.2.1 Unbalanced voltages usually occur because of variations in the load. When phases are unequally loaded, unbalanced voltages will result because of different impedances.

10.6.2.2 Symptoms and causes of unbalanced voltages include the following:

(1) Unequal impedance in conductors of power supply wiring
(2) Unbalanced distribution of single-phase loads such as lighting
(3) Heavy reactive single-phase loads such as welders
(4) Unbalanced incoming utility supply
(5) Unequal transformer tap settings
(6) Large single-phase load on the system
(7) Open phase on the primary of a 3-phase transformer
(8) Open delta-connected transformer banks
(9) A blown fuse on a 3-phase bank of power factor correction capacitors

10.6.3 Symptoms of Unbalanced Voltages.

10.6.3.1 General. The most common symptoms of unbalanced voltages are improper operation of or damage to electric motors, power supply wiring, transformers, and generators.

10.6.3.2 Phase Current Unbalance at Motor Terminals. Unbalanced voltages at motor terminals can cause phase current unbalance to range from 6 to 10 times the voltage unbalance for a fully loaded motor. As an example, if a voltage unbalance is 2 percent, then current unbalance could be anywhere from 12 percent to 20 percent. This causes motor overcurrent, resulting in excessive heat, which shortens motor life.

10.6.3.2.1 Speed and Torque at Motor Terminals. The unbalanced voltages at the motor terminals will also cause speed and torque to be reduced. If the voltage unbalance is great enough, the reduced torque capability might not be adequate for the application. Noise and vibration levels can also increase as a result of voltage unbalance.

10.6.3.3 Motor Heating and Losses. Insulation life is approximately halved for every 18°F (10°C) increase in winding temperature. Table 10.6.3.3 illustrates the typical percentage increases in motor losses and heating for various levels of voltage unbalance.

Table 10.6.3.3 Voltage Unbalance Versus Temperature Rise at Average Voltage of 230

Percent Unbalanced Voltage	Percent Unbalanced Current	Increased Temperature Rise (°C)
0.3	0.4	0
2.3	17.7	30
5.4	40	40

10.6.3.3.1 The motor often continues to operate with unbalanced voltages; however, its efficiency is reduced. This reduction of efficiency is caused by both increased current (I) and increased resistance (R) due to heating. Essentially, this means that as the resulting losses increase, the heating intensifies rapidly. This can lead to a condition of uncontrollable heat rise, called thermal runaway, which results in a rapid deterioration of the winding insulation, ending in winding failure.

10.6.3.4 Motor Operation Under Single-Phase Condition. Single-phase operation of a 3-phase motor will cause overheating due to excessive current and decreased output capability. If the motor is at or near full load when single phasing occurs, it will not develop enough torque and therefore will stall. This results in high currents, causing an extremely rapid temperature rise. If motor protection is not adequate, the stator winding will fail, and the rotor might be damaged or destroyed.

10.6.3.4.1 Standard (thermal, bimetallic, eutectic alloy) overload relays normally are relied on to provide protection against single phasing where properly selected and applied. Protective relays or other devices can provide supplemental single-phasing protection.

10.6.4 Monitoring and Testing.

10.6.4.1 Measuring. The first step in testing for unbalanced voltages should be to measure line-to-line voltages at the machine terminals. If the motor starter is close by, the tests can be made at load or "T" terminals in the starter. The current in each supply phase should be measured to check for current unbalance.

10.6.4.2 Detecting Single Phasing.

10.6.4.2.1 Single phasing should be suspected when a motor fails to start. The voltage should be checked for balanced line-to-line voltages.

10.6.4.2.2 If the motor is running, the voltage and the current in each phase of the circuit should be measured. One phase will carry zero current when a single-phasing condition exists.

10.6.5 Solutions for Unbalanced Voltages.

10.6.5.1 Unbalanced voltages should be corrected; unbalance caused by excessively unequal load distribution among phases can be corrected by balancing the loads. Also, checking for a blown fuse on a 3-phase bank of power-factor correction capacitors is recommended.

10.6.5.2 When voltage unbalance exceeds 1 percent, the motor should be derated as indicated by the curve in Figure 14-1 of NEMA MG 1, *Motors and Generators*.

10.6.5.3 Automatic Voltage Regulator (AVR). AVRs can be used on a per phase basis to correct under- and overvoltage, as well as voltage unbalance. The AVR can compensate for voltage unbalance, provided that the input voltage to the AVR is within its range of magnitude.

10.6.5.4 Relays. Negative sequence voltage relays can detect single phasing, phase-voltage unbalance, and reversal of supply phase rotation. Reverse phase or phase sequence relays provide limited single-phasing protection by preventing the starting of a motor with one phase of the system open.

10.6.5.5 Transformer tap settings should be checked; unequal power transformer tap settings can be a cause of voltage unbalance. This condition should be checked prior to taking other steps.

10.6.5.6 An unsymmetrical transformer bank should be replaced. For example, an open delta bank can be replaced with a three-transformer bank.

10.7 Symptoms — Grounding.

10.7.1 General. If the equipment ground conductor and the service neutral are not electrically connected to the central grounding point, noise voltages can develop between them and appear as common mode noise.

10.7.1.1 Wiring without an equipment ground conductor and without electrically continuous conduit can produce common mode noise.

10.7.1.2 Ground loops are undesirable because they create a path for noise currents to flow.

10.7.2 Monitoring and Testing — Grounding. The electrical connection to earth can be measured using the three-point system referred to in ANSI/IEEE 142, *Recommended Practice for Grounding of Industrial and Commercial Power Systems (IEEE Green Book)*. Minimizing the impedance between the equipment grounded conductor and the grounding conductor is recommended, as follows:

(1) A visual inspection should be made to verify the integrity of the grounding and bonding conductors and associated connections.
(2) An impedance test should be performed on the equipment-grounding conductor.
(3) Voltage should be measured between the equipment-grounding conductor and the grounded conductor.
(4) A check should be made for abnormal currents on the equipment-grounding conductor.

10.7.3 Solutions — Grounding.

10.7.3.1 The grounded conductor should be connected to the equipment-grounding conductor only as permitted by Article 250 of *NFPA 70, National Electrical Code.*

10.7.3.2 Isolated Equipment Ground. One solution is to install an "isolated ground" receptacle (identified by orange color or an orange triangle) in which the equipment-grounding terminal is insulated from the mounting strap. An insulated equipment-grounding conductor is then connected from the grounding terminal of the receptacle in accordance with Article 250 of *NFPA 70, National Electrical Code.* The insulated equipment-grounding conductor is connected to the applicable derived system or service grounding terminal only at the power source.

10.7.3.3 Isolation Transformer. An isolation transformer has separate primary and secondary windings with an interwinding shield that has its own grounding connection. The bonding jumper between the equipment-grounding conductor and the secondary grounded conductor provides protection from common mode electrical noise.

10.7.3.4 Signal Circuit Isolation. Breaking the ground loop current path minimizes ground currents on signal circuits. This can be accomplished by one or more of the following:

(1) Grounding at a single point per system *[See ANSI/IEEE 142, Recommended Practice for Grounding of Industrial and Commercial Power Systems (IEEE Green Book), and ANSI/IEEE 1100, Recommended Practice for Powering and Grounding Sensitive Electronic Equipment (IEEE Emerald Book).]*
(2) Fiber-optic transmission over completely nonconducting path
(3) Optical isolators
(4) Signal circuit isolation transformers in signal circuit or power circuits

10.8 Noise in Electrical and Electronic Systems.

10.8.1 Introduction. Noise is undesirable electrical signals in an electrical or electronic circuit. It can be random or continuous in nature. Noise can occur at any frequency and amplitude. Noise can be introduced into a circuit from a multitude of sources and can manifest itself in equipment malfunction or data corruption.

10.8.1.1 Common Mode Noise. In a 3-phase system, common mode noise is seen equally and in phase between any phase conductor or neutral and the grounding conductor. Neutral-to-ground voltage or ground current can be a result of common mode noise. *(See 3.3.53.1, Common Mode Noise.)*

10.8.1.2 Transverse Mode Noise. In a 3-phase system, transverse mode noise occurs in phase on all 3-phase conductors and neutral. *(See 3.3.53.2, Transverse Mode Noise.)*

10.8.1.3 Interference. Interference that is electromagnetically coupled into a wiring system is called electromagnetic interference (EMI). Interference that is capacitively coupled into a wiring system is called radio frequency interference (RFI). Interference can appear as transverse or common mode noise.

10.8.2 Symptoms.

10.8.2.1 Electrical noise is present in all circuits to a certain degree. It might or might not present a problem. Unlike sags or swells, electrical noise does not normally destroy equipment. It does not cause circuit breakers to trip, unless the noise affects shunt trip or undervoltage release controls.

10.8.2.2 Electrical noise usually manifests itself in the form of data corruption or unexplained equipment malfunction. For example, electrical noise can create "hum" in a telephone system or "snow" on a video image or can cause a computer to lock up.

10.8.3 Causes. Electrical noise is a by-product of the normal operation of electric equipment. The type and sources of noise are as diverse and numerous as the number of facilities that contain power systems and include the following:

(1) Transformers generate magnetic fields that can influence adjacent pieces of equipment.
(2) Long cable runs between interconnected pieces of computer equipment can act as an antenna to a local radio station.
(3) Any piece of electronic equipment that contains a switch mode power supply will introduce electrical noise to some degree into both the building wiring system and the air.
(4) Poor electrical wiring connections can create electrical noise. A loose connection can vibrate, creating an arc at the connection, resulting in noise.

10.8.4 Monitoring and Testing. Locating the sources, frequency, and amplitude of noise can be a difficult and time-consuming task. Troubleshooting becomes increasingly difficult because multiple sources of noise might be present. Determining the amplitude and frequency of the noise signal is essential in identifying the source. Typically, several different types of test equipment can be required to isolate the nature of the noise, including the following:

(1) Spectrum analyzer — capable of measuring a wide range of frequencies
(2) Conducted RFI/EMI recorder — capable of measuring noise levels superimposed on the voltage waveform
(3) Radiated RFI/EMI recorder — capable of measuring electrical noise levels present in the air
(4) Digital storage oscilloscope with line decoupler
(5) Power-quality monitor

10.8.5 Solutions.

10.8.5.1 Elimination. Once the nature of noise disturbance is determined, the best solution is to isolate and eliminate the source. Unfortunately, the source of the noise cannot always be located or the offending piece of equipment removed. In these cases, the noise should be attenuated or filtered out of the system. Some methods of attenuating or filtering out noise are listed below.

10.8.5.2 Signal Reference Grid. This is a special type of grounding system designed for data processing installations. When properly installed, it provides the lowest possible ground impedance across the widest spectrum of frequencies. The grid places the entire data processing ground system at a common potential.

10.8.5.3 Isolation Transformers. Transformers equipped with multiple electrostatic shields can significantly attenuate transverse and common mode noise.

10.8.5.4 Filters. Filtering can be low-pass, high-pass, band-pass, or notch type. Once the frequency and amplitude of the noise signal is determined, a filter can be tuned to "trap" the unwanted noise signal.

10.8.5.5 Signal Cable. The use of twisted pair and shields in low-power signal cables can effectively reduce noise.

10.8.5.6 Shielding. Plane shielding mounted on walls, floors, or ceilings can reduce radiated noise if properly grounded.

10.9 Interharmonics.

10.9.1 Symptoms.

10.9.1.1 The flickering of lights is often a result of subharmonics that occur below 24 Hz, which is observable. Around 9 Hz, as little as a 0.25 percent variation in the rms voltage can be detected in some types of lighting.

10.9.1.2 The misoperation of equipment that occurs because of harmonics, such as the overheating of transformers and the misoperation of control devices, can also occur with interharmonics. In addition, CRT flicker, overload of conventional series tuned filters, overload of outlet strip filters, communications interference, and CT saturation can result from interharmonics. *(See 10.2.2 for additional information.)*

10.9.2 Causes. Operation of loads that draw current or have mechanical processes that are not synchronized to the power line frequency can result in interharmonic voltages and currents. Examples are cycloconverters, static frequency converters, subsynchronous converter cascades, induction motors, arc furnaces, and arc welders. Arc furnaces, which draw very large arcing currents during the melting stage, can generate interharmonics over a wide range of frequencies.

10.9.3 Monitoring and Testing. Harmonic analyzers that use conventional Fast Fourier Transform (FFT)–based harmonic analysis might not be fully effective in determining the presence of interharmonics. The energy of the interharmonic often will be split between two adjacent harmonic values. Spectrum analyzers or harmonic analyzers with interharmonic capabilities are recommended.

10.9.3.1 A flickermeter is a special type of meter for measuring the presence of voltage fluctuations that can result in light flicker.

10.9.4 Solutions. The solutions for minimizing the effects of interharmonics often are similar to those used with harmonics *(see 10.2.5)*. These solutions include filtering, impedance reduction, derating of transformers and motors, and isolation of sensitive equipment.

10.10 Voltage Fluctuations and Flicker.

10.10.1 Explanation of Voltage Fluctuations and Flicker.

10.10.1.1 Voltage fluctuations are variations in the rms voltage that are less than those that would be considered a sag. They generally do not cause equipment to malfunction but often result in light flicker.

10.10.1.2 Flicker is the change in light output from a lamp, caused by the fluctuation of the supply voltage in the frequency range of 0.5 Hz to 30 Hz, where as little as a quarter of a percent voltage fluctuation at nine times per second can be perceived. Figure 10.10.1.2 is an example of voltage fluctuation on a sine wave.

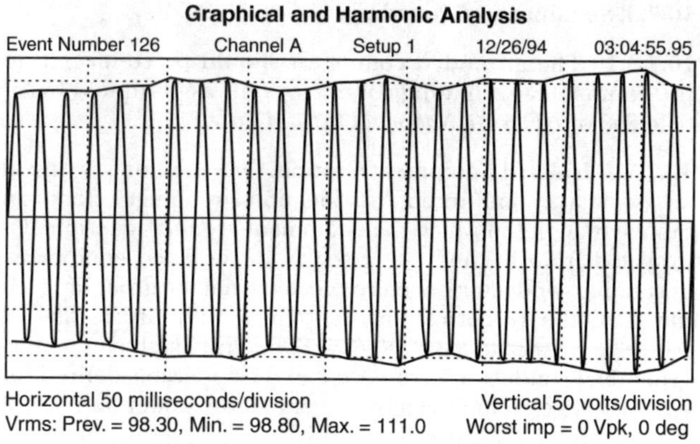

FIGURE 10.10.1.2 Example of a Flicker Sine Wave. (Courtesy of Dranetz Technologies, Inc.)

10.10.2 Symptoms.

10.10.2.1 The effects of the voltage fluctuations normally are perceived as an annoyance and distraction when the lights flicker. They can induce discomfort in the form of nausea or headaches. They usually are not severe enough to disrupt most manufacturing processes, though they can cause variations in the processes.

10.10.2.2 Factors affecting severity include the following:

(1) Frequency of voltage fluctuations (how often it occurs)
(2) Magnitude of voltage fluctuations (how much of a change)
(3) Type of lighting (incandescent, fluorescent, HID)
(4) Ambient light level
(5) Amount of the surface illuminated
(6) Type of activity and the eye-brain characteristics of the individual person

10.10.3 Causes.

10.10.3.1 The voltage fluctuations or modulations of the rms envelope follow the same basic rules as the rms variations that result in sags (dips) or swells. They usually are the result of a change in load current, which causes a change in the voltage drop across the source impedance, which then results in a change in the voltage supply.

10.10.3.2 The frequency of the voltage fluctuation can be the direct result of the frequency of the current draw by the load as the result of folding back of higher frequencies modulating with 50/60 Hz fundamental or their harmonics, which produces sidebands around the fundamental or harmonic frequency.

10.10.3.3 Common sources of voltage fluctuations include the following:

(1) Lamp dimmers
(2) Resistance welding machines
(3) Rolling mills
(4) Large electric motors with variable loads
(5) Arc furnaces and arc welders
(6) Switching on and off of PF correction capacitors
(7) Medical imaging machines (X-ray, MRI, CAT scan)
(8) Large-capacity copy machines
(9) Electric motor starts
(10) Household appliances

10.10.4 Monitoring and Testing.

10.10.4.1 The perception of the light flickering is measured using a flickermeter, which measures two parameters: *Pst*, the short-term perception index, and *Plt*, the long-term perception index.

10.10.4.1.1 A value of 1 should be assigned to the lower bound of the observable flicker perception curve for 60 W incandescent light bulbs.

10.10.4.1.2 The larger the number, the more perceptible the flicker is.

10.10.4.2 Monitoring for voltage fluctuations generally begins at the point of common coupling. The change in current versus the change in voltage helps to determine whether the monitoring point is upstream or downstream from the source. Some flickermeters and some power quality analyzers have the capability to capture the waveforms and other calculations that help determine the source.

10.10.5 Solutions. Three solutions to minimize the effects of voltage fluctuation on lighting are reducing the magnitude and/or changing the frequency of the load current; reducing the source impedance; and changing the type of lighting, such as changing from an incandescent lamp to a fluorescent lamp.

10.11 Power Quality Audit.

10.11.1 Power quality audits are overall studies of a number of the previously described PQ phenomena in a single survey, rather than individual measurements for a specific phenomena. These audits are often done as follows:

(1) Before installing new equipment into a facility, as part of commissioning of a new facility
(2) As part of a routine maintenance program to look for changes or trends that can indicate problems that can be proactively corrected before failure

10.11.2 The audit can be conducted as follows:

(1) At the point of common coupling
(2) At a distribution panel
(3) And/or at a point of utilization (load)

10.11.2.1 Conducting the audit for troubleshooting purposes typically begins at the point of utilization of the equipment that experienced a problem. Overall site audits typically begin at the point of common coupling, or the local power source, such as the UPS output. These should be conducted at several times during the year, such as in the summer, when the HVAC load is high; in the winter, when heating systems are on and humidity might be lower; or when there has been a change to the electrical infrastructure or loads.

10.11.3 Use of power quality monitors that can simultaneously collect data in all of the categories of interest can save time and provide a more complete picture, especially the interaction between various parameters; for example, if the voltage sags were preceded by transients or rapid voltage changes. They can also detect other incipient problems, such as unintended paths of current. Typical parameters to monitor include those found in Table 10.11.3.

10.11.4 The report generated from all of this data often includes items described in Annex H, Forms in Figure H.36 to Figure H.46. It is important to include information, such as the setup of the PQ monitor, as well as data showing "normal" and "abnormal" conditions. For example, knowing that the voltage harmonic distortion was always below 3 percent during all times of the day and all months of the year is valuable, as well as the rms variation that caused the process to trip off-line.

10.12 Power Quality References. See Annex C for these references.

Table 10.11.3 Typical Power Quality Parameters to be Monitored

	TIMEPLOTS with minimum/maximum/average	
1	Voltage rms	L-N, L-L, N-G where applicable
2	Current rms	Phase, Neutral, Residual, Net
3	Voltage harmonic distortion	Vthd, selective harmonics as % of fundamental
4	Current harmonic distortion	Selective harmonics in amps
5	Voltage unbalance	Either rms deviation from average, or negative sequence/positive sequence method
6	Current unbalance	Either rms deviation from average, or negative sequence/positive sequence method
7	Frequency	
	PQ DISTURBANCES	
8	Transients	
9	RMS variations	Sags, swells, interruptions: with magnitude/duration, rms envelope, waveforms
10	Activity chart	Time of day of disturbances
	DEMAND & ENERGY TIMEPLOTS with min/max/avg	
11	Watts	Per phase and total
12	Volt-amperes	Per phase and total
13	Volt-ampere-reactive	Per phase Fundamental VARs per phase and total
14	Demand	
15	Energy	

Chapter 11 Testing and Test Methods

11.1 Introduction. This chapter covers the tests ordinarily used in the field to determine the condition of various elements of an electrical power distribution system. The data obtained in these tests provide information that is used as follows:

(1) To determine whether any corrective maintenance or replacement is necessary or desirable
(2) To ascertain the ability of the element to continue to perform its design function adequately
(3) To chart the gradual deterioration of the equipment over its service life

11.2 Acceptance Tests and Maintenance Tests.

11.2.1 Acceptance Tests. Acceptance tests are tests that are performed on new equipment, at the factory, on-site and after installation, prior to energization. These tests determine whether a piece of equipment is in compliance with the purchase specification and design intent and also establish test benchmarks that can be used as references during future tests.

11.2.1.1 Acceptance tests at the factory are valuable in ensuring the equipment was appropriately designed and manufactured and can be appropriately configured in the field to comply with the operational check to be performed in the field.

11.2.1.2 Acceptance tests on-site are also valuable in ensuring that the equipment has not been damaged during shipment or installation. In addition to the tests that are performed, an acceptance program should include a comprehensive visual inspection and an operational check of all circuitry, accessory devices, and the overall system.

11.2.2 Routine Maintenance Tests. Routine maintenance tests are tests that are performed at regular intervals over the service life of equipment. These tests normally are performed concurrently with preventive maintenance on the equipment.

11.2.3 Special Maintenance Tests. Special maintenance tests are tests performed on equipment that is thought or known to be defective or equipment that has been subjected to conditions that possibly could adversely affect its condition or operating characteristics. Examples of special maintenance tests are cable fault–locating tests or tests performed on a circuit breaker that has interrupted a high level of fault current.

11.2.4 Pretest Circuit Analysis. An analysis of the circuit to be tested should be made prior to the testing to assess the potential meaning of the test results.

11.3 As-Found and As-Left Tests.

11.3.1 As-Found Tests. As-found tests are tests performed on equipment on receipt or after it has been taken out of service for maintenance but before any maintenance work is performed.

11.3.2 As-Left Tests. As-left tests are tests performed on equipment after preventive or corrective maintenance and immediately prior to placing the equipment back in service.

11.3.3 Correlation of As-Found and As-Left Tests. When equipment is taken out of service for maintenance, performance of both an as-found and an as-left test is recommended. The as-found tests will show any deterioration or defects in the equipment since the last maintenance period and, in addition, will indicate whether corrective maintenance or special procedures should be taken during the maintenance process. The as-left tests will indicate the degree of improvement in the equipment during the maintenance process and will also serve as a benchmark for comparison with the as-found tests during the next maintenance cycle.

11.4 Frequency of Tests. Most routine testing can best be performed concurrently with routine preventive maintenance, because a single outage will serve to allow both procedures. For that reason, the frequency of testing generally coincides with the frequency of maintenance. The optimum cycle depends on the use to which the equipment is put and the operating and environmental conditions of the equipment. In general, this cycle can range from 6 months to 3 years, depending on conditions and equipment use. The difficulty of obtaining an outage should never be a factor in determining the frequency of testing and maintenance. Equipment for which an outage is difficult to obtain is usually the equipment that is most vital in the operation of the electrical system. Consequently, a failure of this equipment would most likely create the most problems relative to the continued successful operation of the system. In addition to routine testing, tests should be performed any time equipment has been subjected to conditions that possibly could have caused it to be unable to continue to perform its design function properly.

11.5 Special Precautions and Safety.

11.5.1 Many tests on electrical equipment involve the use of high voltages and currents that are dangerous, both from the standpoint of being life hazards to personnel and because they are capable of damaging or destroying the equipment under test. Adequate safety rules should be instituted and practiced to prevent injury to personnel, both personnel who are performing the tests and personnel who might be exposed to the hazard. Also, the test procedures used should be designed to ensure that no intentional damage to equipment results from the testing process.

11.5.2 It should be recognized that, as the name implies, overpotential or high-potential testing is intended to stress the insulation structure above that of normal system voltage. The purpose of the test is to establish the integrity of the insulation to withstand voltage transients associated with switching and lightning surges and hence reduce the probability of in-service equipment failures. Direct voltage over-potential testing is generally considered a controlled, nondestructive test in that an experienced operator, utilizing a suitable test set, can often detect marginal insulation from the behavior of measured current. It is therefore possible, in many cases, to detect questionable insulation and plan for replacement without actually breaking it down under test. Unfortunately, some insulations might break down with no warning. Plans for coping with this possibility should be included in the test schedule.

11.5.3 Low-voltage insulation testing generally can be done at the beginning of the planned maintenance shutdown. In the event of an insulation failure under test, maximum time would be available for repair prior to the scheduled plant start-up. Equipment found in wet or dirty condition should be cleaned and dried before high-potential testing is done, since a breakdown could damage the equipment.

11.5.4 Low-voltage circuit breakers, which require very high interrupting ratings, are available with integral current-limiting fuses. Although the fuse size is selected to override without damage to the time–current operating characteristic of the series trip device, it is desirable to bypass or remove the fuse prior to applying simulated overload and fault current.

11.6 Qualifications of Test Operators. If a testing program is to provide meaningful information relative to the condition of the equipment under test, the person evaluating the test data should be assured that the test was conducted in a proper manner and that all the conditions that could affect the evaluation of the tests were considered and any pertinent factors reported. The test operator, therefore, should be thoroughly familiar with the test equipment used in the type of test to be performed and also should be sufficiently experienced to be able to detect any equipment abnormalities or questionable data during the performance of the tests.

11.7 Test Equipment. It is important in any test program to use the proper equipment to perform the required tests. In general, any test equipment used for the calibration of other equipment should have an accuracy at least twice the accuracy of the equipment under test. The test equipment should be maintained in good condition and should be used only by qualified test operators. All test equipment should be calibrated at regular intervals to ensure the validity of the data obtained. In order to get valid test results, it might be necessary to regulate the power input to the test equipment for proper waveform and frequency and to eliminate voltage surges.

11.8 Forms. If a testing and maintenance program is to provide optimum benefits, all testing data and maintenance actions should be recorded on test circuit diagrams and forms that are complete and comprehensive. It is often useful to record both test data and maintenance information on the same form. A storage and filing system should be set up for these forms that will provide efficient and rapid retrieval of information regarding previous testing and maintenance on a piece of equipment. A well-designed form also serves as a guide or a checklist of inspection requirements. Samples of typical forms are included in Annex H and are summarized as follows:

(1) Figure H.1, Typical Work Order Request Form
(2) Figure H.2, Typical Air Circuit Breaker Inspection Record
(3) Figure H.3, Typical Air Circuit Breaker Test and Inspection Report
(4) Figure H.4, Typical Medium-Voltage Vacuum Breaker Form
(5) Figure H.5, Typical Oil Circuit Breaker Test Report
(6) Figure H.6, Typical Disconnect Switch Test Report
(7) Figure H.7, Typical Low-Voltage Circuit Breaker 5-Year Tests Form
(8) Figure H.8, Typical Electrical Switchgear–Associated Equipment Inspection Record
(9) Figure H.9, Typical Current or Potential Transformer Ratio Test Report
(10) Figure H.10, Typical Overload Relay Test Report
(11) Figure H.11, Typical Ground-Fault System Test Report
(12) Figure H.12, Typical Instrument/Meter Calibration and Test Report
(13) Figure H.13, Typical Watt-Hour Meter Test Sheet
(14) Figure H.14, Typical Panelboard/Circuit Breaker Test Report
(15) Figure H.15, Typical Transformer Test and Inspection Report
(16) Figure H.16, Typical Transformer (Dry Type) Inspection Record
(17) Figure H.17, Typical Transformer (Liquid Filled) Inspection Record
(18) Figure H.18, Typical Transformer Oil Sample Report
(19) Figure H.19, Typical Transformer Oil Trending Report
(20) Figure H.20, Typical Transformer Insulation Resistance Record
(21) Figure H.21, Typical Battery Record
(22) Figure H.22, Typical Engine Generator Set Inspection Checklist
(23) Figure H.23, Typical Automatic Transfer Switch Report
(24) Figure H.24, Typical Uninterruptible Power Supply Inspection Checklist
(25) Figure H.25, Typical Back-Up Power System Inspection Checklist
(26) Figure H.26, Typical Insulation Resistance–Dielectric Absorption Test Sheet for Power Cable
(27) Figure H.27, Typical Cable Test Sheet
(28) Figure H.28, Typical Insulation Resistance Test Record
(29) Figure H.29, Typical Insulation Resistance Test Record for Rotating Machinery
(30) Figure H.30, Typical Motor Test Information Form
(31) Figure H.31, Typical Ground System Resistance Test Report
(32) Figure H.32, Typical Ground Test Inspection — Health Care Facilities
(33) Figure H.33, Typical Line Isolation Monitor Test Data — Health Care Facilities
(34) Figure H.34, Typical Torque Value Record
(35) Figure H.35, Typical Main Power Energization Checklist
(36) Figure H.36, Instructions to Contractor
(37) Figure H.37, Project Scope of Work Template
(38) Figure H.38, Project Scope of Work Form
(39) Figure H.39, Project Scope of Work Modification Form
(40) Figure H.40, Cover and Contents
(41) Figure H.41, Point of Contact
(42) Figure H.42, Power Distribution Unit (PDU) Survey
(43) Figure H.43, Generator Set Survey
(44) Figure H.44, Electrical Panel Survey
(45) Figure H.45, Inverter Survey
(46) Figure H.46, Building Lightning Protection Survey
(47) Figure H.47, Rectifier Survey
(48) Figure H.48, Electrical Panel Survey
(49) Figure H.49, Transfer Switches Survey
(50) Figure H.50, Power Transformers Survey
(51) Figure H.51, Uninterruptible Power System Survey
(52) Figure H.52, Low Voltage Breaker Data Record
(53) Figure H.53, Recloser Data Record
(54) Figure H.54, Generator Data Record

11.9 Insulation Testing.

11.9.1 Introduction.

11.9.1.1 General. Insulation is the material between points of different potential in an electrical system that prevents the flow of electricity between those points. Insulation materials can be in the gaseous, liquid, or solid form. A vacuum is also a commonly used insulation medium. The failure of the insulation system is the most common cause of problems in electrical equipment. This is true on both high-voltage and low-voltage systems. Insulation tests are tests used to determine the quality or condition of the insulation systems of electrical equipment. Both alternating current and direct current are used in insulation testing.

11.9.1.2 Reasons for Insulation Failure. Liquid and solid insulating materials with organic content are subject to natural deterioration due to aging. This natural deterioration is accelerated by excessive heat and moisture. Heat, moisture, and dirt are the principal causes of all insulation failures. Insulation can also fail due to chemical attack, mechanical damage, sunlight, and excessive voltage stresses.

11.9.2 Direct-Current (dc) Testing — Components of Test Current.

11.9.2.1 When a dc potential is applied across an insulation, the resultant current flow is composed of several components as follows:

(1) Capacitance-charging current
(2) Dielectric-absorption current
(3) Surface leakage current
(4) Partial discharge (corona current)
(5) Volumetric leakage current

11.9.2.2 The capacitance-charging current and the dielectric-absorption current decrease as the time of application of the voltage increases. The test readings of resistance or current should not be taken until these two currents have decreased to a low value and will not significantly affect the reading. The time lapse between the application of voltage and the taking of the reading should be reported as part of the test data. The surface leakage current is caused by conduction on the surface of the insulation between the points where the conductor emerges from the insulation and points of ground potential. This current is not desired in the test results (except for as-found tests) and can be eliminated by carefully cleaning the leakage paths described. Corona current occurs only at high values of test voltage. This current is caused by the overstressing of air at sharp corners or points on the conductor. This current is not desired in the test results and can be eliminated by installing stress-control shielding at such points during the test. Volumetric leakage current is the current that flows through the volume insulation itself. It is the current that is of primary interest in the evaluation of the condition of the insulation.

11.9.2.3 Insulation Resistance Testing. In an insulation resistance test, an applied voltage, from 100 volts to 5000 volts, supplied from a source of constant potential, is applied across the insulation. The usual potential source is a megohmmeter, either hand or power operated, which indicates the insulation resistance directly on a scale calibrated in megohms. The quality of the insulation is evaluated based on the level of the insulation resistance.

11.9.2.3.1 The insulation resistance of many types of insulation varies with temperature, so the data obtained should be corrected to the standard temperature for the class of equipment under test. Published charts are available for this purpose.

11.9.2.3.2 The megohm value of insulation resistance obtained is inversely proportional to the volume of insulation being tested. For example, a cable 304.8 m (1000 ft) long would be expected to have one-tenth the insulation resistance of a cable 30.48 m (100 ft) long if all other conditions were identical.

11.9.2.3.3 The insulation resistance test is relatively easy to perform and is a useful test used on all types and classes of electrical equipment. Its main value lies in the charting of data from periodic tests, corrected for temperature, over the life of the equipment so that deteriorative trends might be detected.

11.9.2.4 Dielectric Absorption.

11.9.2.4.1 In a dielectric-absorption test, a voltage supplied from a source of constant potential is applied across the insulation. The range of voltages used is much higher than the insulation resistance test and can exceed 100,000 volts. The potential source can be either a megohmmeter, as described in 11.9.2.3, or a high-voltage power supply with an ammeter indicating the current being drawn by the specimen under test. The voltage is applied for an extended period of time, from 5 minutes to 15 minutes, and periodic readings are taken of the insulation resistance or leakage current.

11.9.2.4.2 The test data are evaluated on the basis that if an insulation is in good condition, its apparent insulation resistance will increase as the test progresses. Unlike the insulation resistance test, the dielectric-absorption test results are independent of the volume and the temperature of the insulation under test.

11.9.2.5 Polarization Index. The polarization index is a specialized application of the dielectric-absorption test. The index is the ratio of insulation resistance at two different times after voltage application, usually the ratio of the insulation resistance at 10 minutes to the insulation resistance at 1 minute. The use of polarization-index testing is usually confined to rotating machines, cables, and transformers. A polarization index less than 1.0 indicates that the equipment needs maintenance before being placed in service. References are available for polarization indexes for various types of equipment.

11.9.2.6 Direct-Current (dc) Overpotential Testing.

11.9.2.6.1 General. A dc overpotential test consists of applying voltage across an insulation at or above the dc equivalent of the 60 Hz operating crest voltage. This test can be applied either as a dielectric-absorption test or a step-voltage test. A dc overpotential test is an appropriate method for an acceptance test for most equipment.

> **CAUTION:** It is strongly recommended that dc overpotential testing should not be performed as a maintenance test on extruded insulated power cables because of the possibility of damage to the cable.

11.9.2.6.1.1 Dielectric-Absorption Test. When applied as a dielectric-absorption test, the maximum voltage is applied gradually over a period from 60 seconds to 90 seconds. The maximum voltage is then held for 5 minutes with leakage-current readings being taken each minute.

11.9.2.6.1.2 Step-Voltage Test. When applied as a step-voltage test, the maximum voltage is applied in a number of equal increments, usually not fewer than eight, with each voltage step being held for an equal interval of time. The time interval between steps should be long enough to allow the leakage current to reach stability, approximately 1 or 2 minutes. A leakage-current reading is taken at the end of each interval before the voltage is raised to the next level. A linear increase in leakage current is expected, and it should stabilize or decrease from the initial value at each step. A plot of test voltage versus leakage current or insulation resistance is drawn as the test progresses. A nonlinear increase in leakage current can indicate imminent failure, and the test should be discontinued. After the maximum test voltage is reached, a dielectric-absorption test can be performed at that voltage, usually for a 5-minute period.

11.9.2.6.1.3 Proper Discharge. At the end of each test, the test equipment control should be turned to zero voltage and the voltage should be monitored. When the voltage is reduced to 20 percent or lower of the maximum test voltage, the metallic components should be grounded in accordance with test procedures or for at least 30 minutes.

11.9.2.6.2 Arrangement Before Testing. Before equipment insulation is tested, it should be cleaned, inspected, and repaired as necessary to minimize leakage currents. Surge arresters should be disconnected.

11.9.3 Alternating-Current (ac) Testing.

11.9.3.1 High-Potential Testing. An ac high-potential test is made at voltages above the normal system voltage for a short time, such as 1 minute. The test voltages to be used vary depending on whether the device or circuit is low or high voltage, whether it is a primary or control circuit, and whether it was tested at the factory or in the field. Manufacturers' instructions and the applicable standards should be consulted for the proper values.

11.9.3.2 Insulation Power-Factor or Dissipation-Factor Testing. When power-factor or dissipation-factor testing is performed, the criteria in 11.9.3.2.1 and 11.9.3.2.2 should be utilized.

11.9.3.2.1 General. The power factor or dissipation factor of an insulation is the cosine of the angle between the charging current vector and the impressed voltage vector when the insulation system is energized with an ac voltage. In other words, it is a measure of the energy component of the charging current. The term *power-factor testing* means any testing performed to determine the power factor or dissipation factor of an insulation system. For low values of power factor, the dissipation factor can be assumed to be the same as the power factor. Power-factor or dissipation-factor testing is a useful tool in evaluating the quality of insulation in power, distribution, and instrument transformers; circuit breakers; rotating machines; cables; regulators; and insulating liquids. The equipment to be tested should be isolated from the rest of the system, if practical, and all bushings or terminations should be cleaned and dried. The test should be conducted when the relative humidity is below 70 percent and when the insulation system is at a temperature above 0°C (32°F). Data obtained at relative humidity above 70 percent can be interpreted to recognize the higher humidity.

11.9.3.2.2 Test Equipment. The test equipment used should be such that the power factor or dissipation factor can be read directly or such that the charging volt-amperes and the dielectric losses can be read separately so that a ratio can be computed.

11.9.3.2.2.1 The test equipment should also have sufficient electromagnetic interference cancellation devices or shielding to give meaningful test results even when used in an area of strong interference, such as an energized substation.

11.9.3.2.2.2 The test equipment should be able to produce and maintain a sinusoidal wave shape while performing the test at 60 Hz and should be of sufficient capacity and voltage range to perform the test at a minimum voltage of 2500 volts or the operating voltage of the equipment under test, whichever is lower, but in no case less than 500 volts.

11.9.3.2.3 Data Evaluation. Evaluation of the data obtained should be based on the following:

(1) Industry standards for the particular type of equipment tested
(2) Correlation of data obtained with test data from other similar units tested
(3) Comparison of data with previous test data on the same equipment (if available)

11.10 Low-Voltage Circuit Breakers.

11.10.1 Low-Voltage Circuit Breakers — General. Low-voltage circuit breakers generally can be divided into the following two categories depending on the applicable industry design standards:

(1) Insulated-case/molded-case circuit breakers are designed, tested, and evaluated in accordance with ANSI/UL 489, *Molded-Case Circuit Breakers, Molded-Case Switches and Circuit Breaker Enclosures.*
(2) Low-voltage power circuit breakers are designed, tested, and evaluated in accordance with NEMA SG 6, *Power Switching Equipment,* and ANSI/IEEE C37.13, *Standard for Low-Voltage AC Power Circuit Breakers Used in Enclosures.*

11.10.2 Field Testing in General.

11.10.2.1 The procedures outlined in Sections 2 and 3 of the NEMA publication listed in 11.10.1(2) are intended for checking the condition and basic electrical operation of circuit breakers, but they should not be considered as calibration tests or comparisons to laboratory tests. Section 3 outlines factors to be considered if laboratory accuracy is to be approached. If evaluation indicates operation outside acceptable limits, the circuit breaker should be removed and either replaced or repaired. Refer to the manufacturer's instruction manual to understand appropriate maintenance procedures. If checking indicates maloperation, the circuit breaker should be removed and sent to the manufacturer for investigation and test. Checking the condition and basic electrical operation of circuit breakers can be accomplished by performing field testing, but these tests should not be considered as calibration tests or comparisons to laboratory tests. The applicable industry field evaluation standards include the following:

(1) For insulated-case/molded-case circuit breakers, inspection and preventative maintenance performed in accordance with NEMA AB 4, *Guidelines for Inspection and Preventive Maintenance of Molded-Case Circuit Breakers Used in Commercial and Industrial Applications*; ANSI/NETA ATS, *Standard for Acceptance Testing Specifications for Electrical Power Distribution Equipment and Systems*; and ANSI/NETA MTS, *Standard for Maintenance Testing Specifications for Electrical Power Distribution Equipment and Systems*
(2) For low-voltage power circuit breakers, evaluation in accordance with ANSI/IEEE C37.13, *Standard for Low-Voltage AC Power Circuit Breakers Used in Enclosures*; ANSI/NETA ATS, *Standard for Acceptance Testing Specifications for Electrical Power Distribution Equipment and Systems*; and ANSI/NETA MTS, *Standard for Maintenance Testing Specifications for Electrical Power Distribution Equipment and Systems*

11.10.2.2 Where field testing is required, it is recommended that a qualified field service team be employed and that instructions be followed as recommended by the appropriate standard and manufacturer's instructions. If the evaluation of the circuit breaker indicates results that differ significantly from the recommended values, the circuit breaker should be removed and sent to the manufacturer for investigation and test. It is not advisable that repairs be attempted in the field.

11.10.2.3 The testing of electromechanical trip devices or solid-state devices by the primary injection method requires the use of a high-current test set capable of producing sufficient current at low voltage to operate each of the elements of the trip device. This test should have means of adjusting the amount of current applied to the trip device and a cycle and second timer to measure the amount of time to trip the breaker at each current setting. At least one test should be made in the range of each element of the trip device. The long time-delay element ordinarily should be tested at approximately 300 percent of its setting. The short time-delay element should be tested at 150 percent to

200 percent of its setting. The instantaneous element should be tested for pickup. For the test of the instantaneous element, the applied current should be symmetrical without an asymmetrical offset, or random errors will be introduced. As-found and as-left tests should be performed if any need of adjustments is found.

11.10.2.4 The applicable industry field evaluation standards include the following:

(1) ANSI/NETA ATS, *Standard for Acceptance Testing Specifications for Electrical Power Distribution Equipment and Systems*
(2) ANSI/NETA MTS, *Standard for Maintenance Testing Specifications for Electrical Power Distribution Equipment and Systems*

11.10.3 Assistance. Where needed, manufacturers, electrical contractors, and other competent service organizations generally provide field-test services; some are equipped to perform field tests on any make of unit. Such service is more practicable where accurate tests are required and for all tests on circuit breakers of 600-ampere capacity and above. This is, in part, due to the need for special heavy loading equipment and the difficulty of making suitable testing connections.

11.10.4 Field Testing of Circuit Breakers Employing Solid-State Trip Devices.

11.10.4.1 Insulation Resistance Test. Insulation resistance tests are performed to ensure acceptable insulation resistance between each phase conductor and ground. The equipment used to conduct these tests should be used to create an overvoltage to verify the insulation integrity. Review of the markings and instructional material for the circuit breaker should be done before conducting an insulation resistance test. It might be necessary to remove the rating plug or other connections before performing the test to prevent the trip system from being damaged.

11.10.4.2 Testing the Tripping System. Solid-state trip units are designed to operate on low-level currents obtained from the current transformers mounted on the phase conductors. Primary injection test sets will test the entire tripping system, which validates the measurement functions and interconnectivity of sensing and trip devices. Secondary injection tests validate the functionality of the trip unit and circuit breaker opening, however will not test the entire tripping system. Because these breakers have unique design characteristics, the manufacturers should be consulted for available test kits and testing instructions. Attempted field repair of the solid-state trip units should be avoided. Any suspected malfunction should be referred to a competent service group.

11.10.5 Insulated-Case/Molded-Case Circuit-Breaker Testing. When performing insulated-case/molded-case circuit-breaker testing, the criteria in 11.10.5.1 through 11.10.5.3 should be utilized.

11.10.5.1 Insulated-Case/Molded-Case Circuit Breakers — General.

11.10.5.1.1 Insulated-case/molded-case circuit breakers are available in a wide variety of sizes, shapes, and ratings. Voltage ratings, by standard definitions, are limited to 600 volts, although special applications have been made to 1000 volts. Current ratings are available from 10 amperes through 4000 amperes. Insulated-case/molded-case circuit breakers can be categorized generally by the types of trip units employed as described in Section 17.5.

11.10.5.1.2 Electrical testing should be performed in a manner and with the type of equipment required by the type of trip unit employed.

11.10.5.1.3 For further information on insulated-case/molded-case circuit breakers, see Chapter 17, Insulated-Case/Molded-Case Circuit Breakers.

11.10.5.1.4 Insulation Resistance Tests. Measure insulation resistance tests for 1 minute on each pole, phase-to-phase and phase-to-ground, with the circuit breaker closed and across each open pole. Apply voltage in accordance with manufacturer's published data. Insulation resistance values should be in accordance with manufacturer's published data. For further information on insulation resistance testing see 11.9.2.3, Insulation Resistance Testing.

11.10.5.1.5 Contact/Pole Resistance or Millivolt Drop Tests. Measure contact/pole resistance or millivolt drop. These tests are used to test the quality of the contacts. The contact resistance or millivolt drop should be kept as low as possible to reduce power losses at the contacts with the resultant localized heating, which will shorten the life of both the contacts and nearby insulation. Microhm or millivolt drop values should not exceed the high levels of the normal range as indicated in the manufacturer's published data.

11.10.5.2 Testing Thermal-Magnetic Trip Units.

11.10.5.2.1 The electrical testing of thermal-magnetic trip units in circuit breakers can be divided into three steps:

(1) Overload of individual poles at 300 percent of trip rating
(2) Verification of test procedures
(3) Verification of manufacturer's published data

11.10.5.2.2 Complete and detailed instructions for testing insulated-case/molded-case circuit breakers in accordance with the steps in 11.10.5.2.1 are outlined in detail in NEMA AB 4, *Guidelines for Inspection and Preventive Maintenance of Molded-Case Circuit Breakers Used in Commercial and Industrial Applications.* Individual manufacturers also publish recommended testing procedures as well as time–current characteristic tripping curves.

11.10.5.2.3 When circuit-breaker tripping characteristics are tested, it is recommended that the inverse time trip (thermal or long time-delay element) tests be performed on individual poles at 300 percent of rated current.

11.10.5.2.4 The reaction of the circuit breaker to this overload is indicative of its reaction throughout its entire overcurrent tripping range. This load is chosen as the test point because it is relatively easy to generate the required current in the field, and the wattage per pole from line to load is large enough that the dissipation of heat in the nonactive pole spaces is minor and does not affect the test results appreciably.

11.10.5.2.5 Values for Inverse Time Trip (Thermal or Long Time-Delay Element) Data. Table 11.10.5.2.5 outlines the current and trip-time values as recommended by NEMA. The minimum/maximum range of values in Table 11.10.5.2.5 was developed to encompass most brands. For more specific values, refer to the manufacturer's data for the circuit breaker being tested.

11.10.5.2.6 Testing the instantaneous (magnetic) element of a trip unit requires the use of elaborate constant rate-of-rise test equipment coupled with accurate current-monitoring instrumentation, preferably digital readout, for accurate confirmation of manufacturers' test results. Unless this type of equipment is available, it is recommended that these breakers be referred to the manufacturer, electrical contractor, or other competent service organization when calibration is required. Instantaneous

Table 11.10.5.2.5 Values for Inverse Time Trip Test (at 300 Percent of Rated Continuous Current of Circuit Breaker)

Rated Current in Amperes	Maximum Trip Time in Seconds*	
	≤250 V	251–600 V
0–30	50	70
31–50	80	100
51–100	140	160
101–150	200	250
151–225	230	275
226–400	300	350
401–600	—	450
601–800	—	500
801–1000	—	600
1001–1200	—	700
1201–1600	—	775
1601–2000	—	800
2001–2500	—	850
2501–5000	—	900
6000	—	1000

*For integrally fused circuit breakers, trip times may be substantially longer if tested with the fuses replaced by solid links (shorting bars).

pickup values of insulated-case/molded-case circuit breakers should fall within the manufacturer's published tolerances. In the absence of manufacturer's published tolerances, refer to Table 11.10.5.2.6 with values as recommended in Table 4 of ANSI/NEMA AB4, *Guidelines for Inspection and Preventive Maintenance of Molded Case Circuit Breakers Used in Commercial and Industrial Applications.*]

11.10.5.3 Testing Instantaneous-Only Circuit Breakers. The testing of instantaneous-only circuit breakers requires the use of elaborate constant rate-of-rise test equipment coupled with accurate current-monitoring instrumentation, preferably digital readout, for accurate confirmation of manufacturers' test results. Unless this type of equipment is available, it is recommended that these breakers be referred to the manufacturer, electrical contractor, or other competent service organization when calibration is required. Instantaneous pickup values of insulated-case/molded-case circuit breakers should fall within the manufacturer's published tolerances.

Table 11.10.5.2.6 Instantaneous Trip Tolerances for Field Testing of Circuit Breakers

Breaker Type	Tolerances of Settings	Tolerances of Manufacturers' Published Trip Range	
		High Side	Low Side
Electronic trip units*	+30%	—	—
	−30%	—	—
Adjustable*	+40%	—	—
	−30%	—	—
Nonadjustable†	—	+25%	−25%

*Tolerances are based on variations from the nominal settings.
†Tolerances are based on variations from the manufacturer's published trip band (i.e., −25 percent below the low side of the band, +25 percent above the high side of the band).

11.10.5.4 Testing Solid-State Trip Units. Breakers employing solid-state trip units offer testing opportunities not readily available in thermal-magnetic or instantaneous only trip units. These devices have two or more of the following elements.

11.10.5.4.1 Long Time-Delay Element. This element is designed to operate on overloads between its pickup setting and the pickup of a short time delay or an instantaneous element. The long time-delay pickup adjustment is generally within the range of 80 percent to 160 percent of the trip-device rating. Settings higher than solid-state trip-device ampere rating do not increase the continuous-current rating of the trip device, and in no event is the rating increased beyond the breaker frame size. The operating time of this element ranges from seconds to minutes. Determine long-time pickup and delay. Long-time pickup values should be as specified, and the trip characteristic should not exceed manufacturer's published time–current characteristic tolerance band, including adjustment factors.

11.10.5.4.2 Short Time-Delay Element. This element has a time delay measured in cycles and is used to protect against moderate fault currents and short circuits. This element usually can be adjusted to pick up within the range of 250 percent to 1000 percent of the trip-device rating. Determine short-time pickup and delay. Short-time pickup values should be as specified, and the trip characteristic should not exceed manufacturer's published time–current tolerance band.

11.10.5.4.3 Instantaneous Element. This element has no intentional time delay and is used to protect against heavy fault currents and short circuits. The pickup settings for this type of element usually range from 500 percent to 1500 percent of the trip-device rating. Determine instantaneous pickup. Instantaneous pickup values should be within the tolerances of manufacturer's published data.

11.10.5.4.4 Ground-Fault Element. This element is available only on solid-state devices and is used to protect against ground-fault currents at levels below those that would be sensed otherwise. Determine ground-fault pickup and delay. Ground-fault pickup values should be as specified, and the trip characteristic should not exceed manufacturer's published time–current tolerance band.

11.10.6 Low-Voltage Power Circuit-Breaker Testing. When low-voltage circuit-breaker testing is performed, the criteria in 11.10.6.3 through 11.10.6.3.4 should be utilized.

11.10.6.1 Insulation Resistance Tests. Measure insulation resistance tests for 1 minute on each pole, phase-to-phase and phase-to-ground with the circuit breaker closed and across each open pole. Apply voltage in accordance with manufacturer's published data. Insulation resistance values should be in accordance with the manufacturer's published data. For further information on insulation resistance testing, see 11.9.2.3, Insulation Resistance Testing.

11.10.6.2 Contact/Pole Resistance or Millivolt Drop Tests. Measure contact/pole resistance or millivolt drop. These tests are used to test the quality of the contacts. The contact resistance or millivolt drop should be kept as low as possible to reduce power losses at the contacts with the resultant localized heating, which will shorten the life of both the contacts and nearby insulation. Microhm or millivolt drop values should not exceed the high levels of the normal range as indicated in the manufacturer's published data.

11.10.6.3 Overcurrent Trip Device. Most low-voltage power circuit breakers are equipped with overcurrent trip devices that sense overload of fault currents and trip the breaker. These devices can be either electromechanical or solid state and usually have two or more of the following types of elements.

11.10.6.3.1 Long Time-Delay Element. This element is designed to operate on overloads between its pickup setting and the pickup of a short time delay or an instantaneous element. The electromechanical long time-delay pickup adjustment is generally within the range of 80 percent to 160 percent of the trip-device rating. Settings higher than an electromechanical trip-device ampere rating do not increase the continuous-current rating of the trip device, and in no event is the rating increased beyond the breaker frame size. The operating time of this element ranges from seconds to minutes. Determine long-time pickup and delay. Long-time pickup values should be as specified, and the trip characteristic shall not exceed manufacturer's published time–current characteristic tolerance band, including adjustment factors.

11.10.6.3.2 Short Time-Delay Element. This element has a time delay measured in cycles and is used to protect against moderate fault currents and short circuits. This element usually can be adjusted to pick up within the range of 250 percent to 1000 percent of the trip-device rating. Determine short-time pickup and delay. Short-time pickup values should be as specified, and the trip characteristic should not exceed manufacturer's published time–current tolerance band.

11.10.6.3.3 Instantaneous Element. This element has no intentional time delay and is used to protect against heavy fault currents and short circuits. The pickup settings for this type of element usually range from 500 percent to 1500 percent of the trip-device rating. Determine instantaneous pickup. Instantaneous pickup values should be within the tolerances of manufacturer's published data.

11.10.6.3.4 Ground-Fault Element. This element is available only on solid-state devices and is used to protect against ground-fault currents at levels below those that would be sensed otherwise. Determine ground-fault pickup and delay. Ground-fault pickup values should be as specified, and the trip characteristic should not exceed manufacturer's published time–current tolerance band.

11.10.6.4 Low-Voltage Power Circuit Breakers in General. For further information on low-voltage air circuit breakers see Section 15.4, Air Circuit Breakers.

11.11 Transformer Tests.

11.11.1 Field Testing in General.

11.11.1.1 The applicable industry field evaluation standards include the following:

(1) ANSI/NETA ATS, *Standard for Acceptance Testing Specifications for Electrical Power Distribution Equipment and Systems*
(2) ANSI/NETA MTS, *Standard for Maintenance Testing Specifications for Electrical Power Distribution Equipment and Systems*

11.11.1.2 For further information on transformers, see Chapter 21, Power and Distribution Transformers.

11.11.1.3 In general, insulation tests, such as power-factor testing, and insulation resistance testing, and diagnostic tests, such as turns-ratio testing, winding resistance testing, and exciting-current testing, are the major maintenance tests for all transformers. In addition, liquid-filled transformers should be tested to determine the quality of the insulating liquid.

11.11.2 Turns-RatioTests.

11.11.2.1 The turns-ratio test is used to determine the number of turns in one winding of a transformer in relation to the number of turns in the other windings of the same phase of the transformer. The polarity test determines the vectoral relationships of the various transformer windings. The turns-ratio test is used as both an acceptance and a maintenance test, while the polarity test is primarily an acceptance test.

11.11.2.2 The tests are applicable to all power, distribution, and instrument transformers. The test equipment used ordinarily is a turns-ratio test set designed for the purpose, although, if not available, two voltmeters or two ammeters (for current transformers only) can be used. If the two-meter method is used, the instruments should be at least of the 0.25 percent full-scale accuracy type.

11.11.2.3 When a turns-ratio test is performed, the ratio should be determined for all no-load taps (maintenance tests should be performed at the designated tap position). If the transformer is equipped with a load-tap changer (LTC), the ratio should be determined for each LTC position. If the transformer has both an LTC and a no-load-tap changer, the ratio should be determined for each position of the LTC to one position of the no-load-tap changer and vice versa. This test is useful in determining whether a transformer has any shorted turns or improper connections and, on acceptance testing, verifying nameplate information.

11.11.3 Insulation Resistance Tests. Perform insulation resistance tests winding-to-winding and winding-to-ground. Apply voltage in accordance with manufacturer's published data. Calculate polarization index. Minimum insulation resistance values of transformer insulation should be in accordance with manufacturer's published data. The polarization index shall be compared to previously obtained results and should not be less than 1.0. For further information on insulation resistance testing, see 11.9.2.3, Insulation Resistance Testing, and 11.9.2.5, Polarization Index.

11.11.4 Power-Factor or Dissipation-Factor Tests.

11.11.4.1 Power-factor or dissipation-factor of the following should be obtained:

(1) Each winding with respect to ground
(2) Each winding with respect to every other winding

For further information on insulation power-factor testing see 11.9.3.2, Insulation Power-Factor or Dissipation-Factor Testing.

11.11.4.2 The high voltage side (CH) and low-voltage side (CL) power-factor or dissipation-factor values will vary due to support insulators and bus work utilized on dry type transformers. Consult transformer manufacturer's or test equipment manufacturer's data for additional information.

11.11.4.3 Maximum power-factor/dissipation-factor values of liquid-filled transformers corrected to 20°C should be in accordance with the transformer manufacturer's published data.

11.11.4.4 In addition to the provisions of 11.11.4, tests should be made of each bushing with a rated voltage above 600 volts, using either the power factor or capacitance tap if the bushing is so equipped or a "hot-collar" test using a test electrode around the outside shell of the bushing.

11.11.5 Excitation-Current Tests. Excitation-current tests are used to detect short-circuited turn-to-turn insulation, short-circuited core laminations, loosening of the core clamping, or improper winding connections. Perform excitation-current tests in accordance with the test equipment manufacturer's published data. A typical excitation-current test data pattern for a three-legged core transformer is two similar current readings and one lower current reading.

11.11.6 Core Insulation Tests. On dry type transformers, measure the core insulation resistance at 500 volts dc if the core is insulated and if the core ground strap is removable. Core insulation resistance values should be comparable to previously obtained results but not less than one megohm at 500 volts dc.

11.11.7 Winding Resistance Tests. When a winding resistance test is performed, the resistance should be determined for all taps (maintenance tests should be performed at the designated tap position). Temperature-corrected winding-resistance values should compare within 1 percent of the factory or previously obtained results.

11.11.8 Liquid Maintenance and Analysis.

11.11.8.1 Liquid Analysis.

11.11.8.1.1 For insulating oils, the tests routinely performed are dielectric breakdown, acidity, color, power factor, interfacial tension, and visual examination. These tests are covered in Section 11.19. For other insulating liquids, the manufacturers' recommendations should be followed.

11.11.8.1.2 Tests can also be performed to determine levels of PCBs. Test results might require service or replacement of the transformer tested as specified by government regulations. (*See 21.2.1.3.*)

11.11.8.1.3 Samples should never be taken from energized transformers except by means of an external sampling valve. If the transformer has no external sampling valve, the unit should first be deenergized and a sample taken internally. (*See ASTM D 923, Standard Practices for Sampling Electrical Insulating Liquids.*)

11.11.8.2 Maintenance. If any test indicates that an insulating liquid is not in satisfactory condition, the liquid can be restored by reconditioning or reclaiming, or it can be completely replaced. Reconditioning is the removal of moisture and solid materials by mechanical means such as filter presses, centrifuges, or vacuum dehydrators. Reclaiming is the removal of acidic and colloidal contaminants and products of oxidation by chemical and absorbent means such as processes involving fuller's earth, either alone or in combination with other substances. Replacing the liquid involves draining, flushing, testing, and proper disposal of materials removed.

11.11.9 Fault-Gas Analysis. (*See ASTM D 3284, Standard Practice for Combustible Gases in the Gas Space of Electrical Apparatus Using Portable Meters.*) The determination of the percentage of combustible gases present in the nitrogen cap of sealed, pressurized oil-filled transformers can provide information as to the likelihood of incipient faults in the transformer. When arcing or excessive heating occurs below the top surface of the oil, insulation decomposition can occur. Some of the products of the decomposition are combustible gases that rise to the top of the oil and mix with the nitrogen above. A small sample of nitrogen is removed from the transformer and analyzed. The test set has a direct reading scale calibrated in percentages of combustible gas. Ordinarily, the nitrogen cap in a transformer has less than 0.5 percent combustible content. As a problem develops over a period of time, the combustible content can rise to 10 or 15 percent. A suggested evaluation of the test results is shown in Table 11.11.9.

Table 11.11.9 Sample Test Results Evaluation

Percentage of Combustible Gas	Gas Evaluation
0.0 to 1.0	No reason for concern. Make tests at regularly scheduled intervals.
1.0 to 2.0	Indication of contamination or slight incipient fault. Make more frequent readings and watch trends.
2.0 to 5.0	Begin more frequent readings immediately. Prepare to investigate cause by internal inspection.
over 5.0	Remove transformers from service and make internal inspection.

11.11.10 Dissolved-Gas-in-Oil Analysis. A refinement of the fault-gas analysis is the dissolved-gas-in-oil test. (See ASTM D 3612, *Standard Test Method for Analysis of Gases Dissolved in Electrical Insulating Oil by Gas Chromatography.*) In this test, an oil sample is withdrawn from the transformer, and the dissolved gases are extracted from the oil. A portion of the gases is then subjected to chromatographic analysis, which determines the exact gases present and the amount of each. Different types of incipient faults have different patterns of gas evolution. With this test, the nature of the problems can often be diagnosed. (*See ANSI/IEEE C57.104, Guide for the Interpretation of Gases Generated in Oil-Immersed Transformers.*)

11.11.11* Metals-in-Oil Analysis on Large Power Transformers. An oil sample should be withdrawn from the transformer and analyzed by a qualified laboratory that may use one of several methods to determine the exact metals present, including atomic absorption spectroscopy.

11.12 Protective Relays.

11.12.1 Introduction.

11.12.1.1 When performing protective relays testing, the criteria in 11.12.1.1.11 through 11.12.1.1.3 should be utilized.

11.12.1.1.1 (*See Caution in 15.9.7.1.*) Protective relays are used in conjunction with medium-voltage circuit breakers (above 600 volts) to sense abnormalities and cause the trouble to be isolated with minimum disturbance to the electrical system and with the least damage to the equipment at fault. They have the accuracy and sophistication demanded by the protective requirements of the primary feeder circuits and larger electrical equipment. Protective relays designed to be responsive to an abnormal excursion in current, voltage, frequency, phase-angle, direction of current or power flow, and so on, and with varying operation characteristics, are commercially available. Each relay application requires custom engineering to satisfy the parameters of its particular intended function in the system.

11.12.1.1.2 The more common protective relay is of the electromechanical type. That is, some mechanical element such as an induction disk, an induction cylinder, or a magnetic plunger is caused to move in response to an abnormal change in a parameter of the electrical system. The movement can cause a contact in the control circuit to operate, tripping the related circuit breaker. Protective relays should be acceptance tested prior to being placed in service and should be tested periodically thereafter to ensure reliable performance. In a normal industrial application, periodic testing should be done at least every 2 years.

11.12.1.1.3 The various facets involved in testing protective relays include the following:

(1) The technician should understand the construction, operation, and testing of the particular relay.
(2) The manufacturer's instruction bulletin, as identified on the nameplate of the relay, should be available.
(3) The technician should be given the settings to be applied to each particular relay and the test points. This information is often furnished on a time–current curve of the coordination study displaying the characteristics of the relay.
(4) A test instrument, suitable to accurately accommodate the various acceptance and periodic maintenance tests described in the manufacturer's instruction manual, should be available.
(5) Most protective relays can be isolated for testing while the electrical system is in normal operation. However, an operation of the breaker is recommended to ascertain that the operation of the relay contacts will trigger the intended reaction, such as to trip the associated circuit breaker.

11.12.1.2 For further information on protective relays, see 15.9.7, Protective Relays and Metering Devices.

11.12.2 Testing Procedure. When protective relays testing is performed, the procedures listed in 11.12.2.1 through 11.12.2.10 should be followed.

11.12.2.1 Inspection. If recommended or desirable, each relay should be removed from its case for a thorough inspection and cleaning. If the circuit is in service, one relay at a time should be removed so as not to totally disable the protection. The areas of inspection are detailed in the manufacturer's instruction manual. These areas generally consist of inspection for loose screws, friction in moving parts, iron filings between the induction disk and the permanent magnet, and any evidence of distress with the relay. The fine silver contacts should be cleaned only with a burnishing tool.

11.12.2.2 Settings. Prescribed settings should be applied, or it should be ascertained that they have been applied to the relay.

11.12.2.3 Pickup Test. In the case of a time-overcurrent relay, its contacts should eventually creep to a closed position with a magnitude of current introduced in its induction coil equal to the tap setting. The pickup is adjusted by means of the restraining spiral-spring adjusting ring. A pickup test on a voltage relay is made in much the same manner.

11.12.2.4 Timing Test. In the case of a time-overcurrent relay, one or more timing tests are made at anywhere from 2 to 10 times the tap setting to verify the time–current characteristic of the relay. One timing point should be specified in the prescribed settings. Tests should be made with the relay in its panel and case, and the time test run at the calibration setting.

11.12.2.5 Time Delay Settings. For example, in the case of one particular overcurrent relay having a 5-ampere tap setting, the timing test could be specified as "25 amperes at 0.4 second." It could be seen from the family of curves in the manufacturer's instruction manual for that relay that the test should result in a time-dial setting of approximately 1.6.

11.12.2.6 Relays to Be Tested. A timing test should be made on most types of relays.

11.12.2.7 Instantaneous Test. Some protective relays are instantaneous in operation or might have a separate instantaneous element. In this context, the term *instantaneous* means "having no intentional time delay." If used, the specified pickup on the instantaneous element should be set by test. Again referring to the relay used in the example given in 11.12.2.5, at two times pickup, its instantaneous element should have an operating time of between 0.016 second and 0.030 second.

11.12.2.8 Test of Target and Seal-In Unit. Most types of protective relays have a combination target and seal-in unit. The target indicates that the relay has operated. The seal-in unit is adjustable to pick up at either 0.2 ampere or 2.0 amperes. The setting for the seal-in unit should be specified with the relay settings.

11.12.2.9 Contact Verification. It should be verified by test that the contacts will seal in (hold in closed position) with the minimum specified direct current applied in the seal-in unit.

11.12.2.10 Test of Tripping Circuit. A test should be made, preferably at the time of testing the relays, to verify that operation of the relay contacts causes the breaker to trip.

11.12.3 Field Testing in General.

11.12.3.1 The applicable industry field evaluation standards include the following:

(1) ANSI/NETA ATS, *Standard for Acceptance Testing Specifications for Electrical Power Distribution Equipment and Systems*
(2) ANSI/NETA MTS, *Standard for Maintenance Testing Specifications for Electrical Power Distribution Equipment and Systems*

11.13 Grounding Systems.

11.13.1 Impedance Testing of Equipment Grounding Conductor.

11.13.1.1 This test is used to determine the integrity of the grounding path from the point of test back to the source panel or supply transformer. A low-impedance grounding path is necessary to facilitate operation of the overcurrent device under ground-fault conditions as well as to provide a zero voltage reference for reliable operation of computers and other microprocessor-based electronic equipment.

11.13.1.2 Instruments are available to measure the impedance of the grounding path. When using these instruments, the user should remember that, although a high-impedance value is an indication of a problem, for example, a loose connection or excessive conductor length, a low-impedance readout does not necessarily indicate the adequacy of the grounding path.

11.13.1.3 A grounding path that is found to have a low impedance by the use of relatively low test currents might not have sufficient capacity to handle large ground faults. Visual examinations and actual checking for tightness of connections are still necessary to determine the adequacy of the grounding path.

11.13.1.4 Impedance tests can be performed reliably on circuits where an equipment-grounding conductor is not connected to other parallel paths. These equipment-grounding conductors can be in nonmetallic sheathed cable, circuits installed in nonmetallic conduits and fittings, flexible cords, and systems using an isolated ground.

11.13.1.5 Ground loop or grounding conductor impedance cannot be measured reliably in situations where metallic conduits are used or where metallic boxes or equipment are attached to metal building frames or interconnected structures. Such situations create parallel paths for test currents that make it impossible to measure the impedance of the grounding conductor or even to detect an open or missing grounding conductor. Also, the impedance of a steel raceway varies somewhat unpredictably with the amount of current flowing through it. The relatively low test currents used during testing usually produce a higher impedance than that actually encountered by fault currents. However, this higher impedance tends to render the tests conservative, and the impedance values might still be acceptable.

11.13.2 Grounded Conductor Impedance Testing.

11.13.2.1 On solidly grounded low-voltage systems (600 volts or less) supplying microprocessor-based electronic equipment with switching power supplies, this test is used to determine the quality of the grounded conductor (neutral) from the point of test back to the source panel or supply transformer. These electronic loads can create harmonic currents in the neutral that can exceed the current in the phase conductors. A low-impedance neutral is necessary to minimize neutral-to-ground potentials and common-mode noise produced by these harmonic currents.

11.13.2.2 Some instruments used to perform the equipment ground-impedance tests in 11.13.1 can be used to perform grounded conductor (neutral) impedance tests.

11.13.3 Grounding-Electrode Resistance Testing.

11.13.3.1 Grounding-electrode resistance testing is used to determine the effectiveness and integrity of the grounding system. An adequate grounding system is necessary to (1) provide a discharge path for lightning, (2) prevent induced voltages caused by surges on power lines from damaging equipment connected to the power line, and (3) maintain a reference point of potential for instrumentation safety. Periodic testing is necessary because variations in soil resistivity are caused by changes in soil temperature, soil moisture, conductive salts in the soil, and corrosion of the ground connectors.

11.13.3.2 The test set used ordinarily is a ground-resistance test set, designed for the purpose, using the principle of the fall of potential of ac-circulated current from a test spot to the ground connection under test. This instrument is direct reading and calibrated in ohms of ground resistance.

11.13.4 Field Testing in General.

11.13.4.1 Where field testing is required, it is recommended that a qualified field service team be employed and that instructions be followed as recommended by appropriate standard and manufacturer's instructions.

11.13.4.2 The applicable industry field evaluation standards include the following:

(1) ANSI/NETA ATS, *Standard for Acceptance Testing Specifications for Electrical Power Distribution Equipment and Systems*
(2) ANSI/NETA MTS, *Standard for Maintenance Testing Specifications for Electrical Power Distribution Equipment and Systems*
(3) IEEE Standard 81, *Guide for Measuring Earth Resistivity, Ground Impedance and Earth Surface Potentials of a Ground System*

11.13.4.3 For further information on grounding see Chapter 14, Grounding.

11.14 Battery Testing.

11.14.1 Field Testing in General. The applicable industry field evaluation standards include the following:

(1) ANSI/NETA ATS, *Standard for Acceptance Testing Specifications for Electrical Power Distribution Equipment and Systems*
(2) ANSI/NETA MTS, *Standard for Maintenance Testing Specifications for Electrical Power Distribution Equipment and Systems*
(3) ANSI/IEEE 450, *Recommended Practice for Maintenance, Testing and Replacement of Vented Lead-Acid Batteries for Stationary Applications*
(4) IEEE 1188, *Recommended Practice for Maintenance, Testing and Replacement of Valve-Regulated Lead Acid (VRLA) Batteries for Stationary Applications*
(5) IEEE 1106, *Recommended Practice for Installation, Maintenance, Testing, and Replacement of Vented Nickel-Cadmium Batteries for Stationary Application*

11.14.2 Battery Tests.

11.14.2.1 Tests should be performed and results recorded to establish trends that can be used in predicting battery life. For lead-acid batteries, a pilot cell may be selected to obtain a representative temperature while the recommended voltage and specific gravity measurements are made.

11.14.2.2 If used, pilot cell voltage, specific gravity, and electrolyte temperature should be measured and recorded monthly in accordance with the established maintenance program *(see 15.9.4.2)*. Refer to the manufacturer's recommended practices for the use of pilot cells and for the range for float voltage applicable to a specific battery.

11.14.2.3 A performance test should be performed at intervals not greater than 25 percent of the expected service life as determined by the initial design, or as recommended by the manufacturer, depending on the load reliability requirements and environmental conditions of the installation. The frequency of battery tests should be increased to yearly when the battery reaches 85 percent of its service life or when it shows signs of deterioration.

11.14.2.4 Measure connection resistances of cell-to-cell and terminal connections in accordance with manufacturer's instructions and with an established maintenance program.

11.14.2.5 Batteries should be examined under full load with an infrared scanning device whenever a performance test is conducted. Infrared scanning can reveal problems such as abnormal temperature of a cell, a poor connection at a battery post, and a deteriorated link, strap, or conductor.

11.14.2.6 Measurements should be recorded for future reference along with log notations of the visual inspection and corrective action. An example of a battery record is included as Figure H.21. The record should be modified to correspond to the user's maintenance program per 15.9.4.2.

11.15 Switches.

11.15.1 Low-Voltage Air Switches.

11.15.1.1 Field Testing in General.

11.15.1.1.1 Where field testing is required, it is recommended that a qualified field service team be employed and that instructions be followed as recommended by appropriate standard and manufacturer's instructions. If evaluation of the switch indicates results that differ significantly from the recommended values, the switch should be removed and repaired.

11.15.1.1.2 The applicable industry field evaluation standards include the following:

(1) ANSI/NETA ATS, *Standard for Acceptance Testing Specifications for Electrical Power Distribution Equipment and Systems*
(2) ANSI/NETA MTS, *Standard for Maintenance Testing Specifications for Electrical Power Distribution Equipment and Systems*

11.15.1.1.3 For further information on air switches, see 15.1.4, Air-Disconnecting Switches.

11.15.1.2 Field Tests.

11.15.1.2.1 Insulation Resistance.

11.15.1.2.1.1 Measure insulation resistance on each pole, phase-to-phase and phase-to-ground with the switch closed and across each open pole for one minute. Apply voltage in accordance with manufacturer's published data. For further information on insulation resistance testing, see 11.9.2.3, Insulation Resistance Testing.

11.15.1.2.1.2 Insulation resistance values of the low voltage air switch should be in accordance with manufacturer's published data.

11.15.1.2.2 Contact/Pole Resistance Test.

11.15.1.2.2.1 Measure the contact resistance across each switchblade and fuseholder. This test is used to test the quality of the contacts. The contact resistance should be kept as low as possible to reduce power losses at the contacts with the resultant localized heating, which will shorten the life of both the contacts and nearby insulation.

11.15.1.2.2.2 Microhm values should not exceed the high levels of the normal range as indicated in the manufacturer's published data.

11.15.1.2.3 Fuse Resistance.

11.15.1.2.3.1 Measure Fuse Resistance.

11.15.1.2.3.2 Fuses in the same circuit should not have fuse-resistance values that deviate from each other by more than 15 percent. Refer to Section 11.18 for the proper fuse testing procedures.

11.15.1.2.3.3 For further information on fuses see Chapter 18, Fuses.

11.15.1.2.4 Heaters. Verify operation of heaters. Heaters should be operational.

11.15.2 Medium Voltage Air Switches, Metal Enclosed.

11.15.2.1 Field Testing in General.

11.15.2.1.1 Where field testing is required, it is recommended that a qualified field service team be employed and that instructions be followed as recommended by appropriate standard and manufacturer's instructions. If evaluation of the switch indicates results that differ significantly from the recommended values, the switch should be removed and repaired.

11.15.2.1.2 The applicable industry field evaluation standards include the following:

(1) ANSI/NETA ATS, *Standard for Acceptance Testing Specifications for Electrical Power Distribution Equipment and Systems*
(2) NETA *Maintenance Testing Specifications for Electrical Power Distribution Equipment and Systems*

11.15.2.1.3 For further information on air switches see 15.1.4, Air-Disconnecting Switches.

11.15.2.2 Field Tests.

11.15.2.2.1 Insulation Resistance.

11.15.2.2.1.1 Measure insulation resistance on each pole, phase-to-phase and phase-to-ground with the switch closed and across each open pole for one minute. Apply voltage in accordance with manufacturer's published data. For further information on insulation resistance testing see 11.9.2.3, Insulation Resistance Testing.

11.15.2.2.1.2 Insulation resistance values of the medium voltage air switch should be in accordance with manufacturer's published data.

11.15.2.2.2 Contact/Pole Resistance Test.

11.15.2.2.2.1 Measure the contact resistance across each switchblade and fuseholder. This test is used to test the quality of the contacts. The contact resistance should be kept as low as possible to reduce power losses at the contacts with the resultant localized heating, which will shorten the life of both the contacts and nearby insulation.

11.15.2.2.2.2 Microhm values should not exceed the high levels of the normal range as indicated in the manufacturer's published data.

11.15.2.2.3 Direct-Current (dc) Overpotential Test.

11.15.2.2.3.1 Measure leakage current on each phase with the switch closed and the poles not under test grounded. Test voltage should be in accordance with manufacturer's published data. For further information on dc overpotential testing, see 11.9.2.6, Direct Current (dc) Overpotential Testing.

11.15.2.2.3.2 If no evidence of distress or insulation failure is observed by the end of the total time of voltage application during the dc overpotential test, the switch is considered to have passed the test.

11.15.2.2.4 Fuse Resistance.

11.15.2.2.4.1 Measure Fuse Resistance.

11.15.2.2.4.2 Fuses in the same circuit should not have fuse-resistance values that deviate from each other by more than 15 percent. Refer to Section 11.18 for the proper fuse testing procedures.

11.15.2.2.4.3 For further information on fuses, see Chapter 18, Fuses.

11.15.2.2.5 Heaters. Verify operation of heaters. Heaters should be operational.

11.15.3 Medium and High Voltage Air Switches, Open.

11.15.3.1 Field Testing in General.

11.15.3.1.1 Where field testing is required, it is recommended that a qualified field service team be employed and that instructions be followed as recommended by appropriate standard and manufacturer's instructions. If evaluation of the switch indicates results that differ significantly from the recommended values, the switch should be removed and repaired.

11.15.3.1.2 The applicable industry field evaluation standards include the following:

(1) ANSI/NETA ATS, *Standard for Acceptance Testing Specifications for Electrical Power Distribution Equipment and Systems*
(2) ANSI/NETA MTS, *Standard for Maintenance Testing Specifications for Electrical Power Distribution Equipment and Systems*

11.15.3.1.3 For further information on air switches, see 15.1.4, Air-Disconnecting Switches.

11.15.3.2 Field Tests.

11.15.3.2.1 Insulation Resistance.

11.15.3.2.1.1 Measure insulation resistance on each pole, phase-to-phase and phase-to-ground with the switch closed and across each open pole for one minute. Apply voltage in accordance with manufacturer's published data. For further information on insulation resistance testing, see 11.9.2.3, Insulation Resistance Testing.

11.15.3.2.1.2 Insulation resistance values of the medium or high voltage air switch should be in accordance with manufacturer's published data.

11.15.3.2.2 Contact/Pole Resistance Test.

11.15.3.2.2.1 Measure the contact resistance across each switchblade and fuseholder. This test is used to test the quality of the contacts. The contact resistance should be kept as low as possible to reduce power losses at the contacts with the resultant localized heating, which will shorten the life of both the contacts and nearby insulation.

11.15.3.2.2.2 Microhm values should not exceed the high levels of the normal range as indicated in the manufacturer's published data.

11.15.3.2.3 Direct-Current (dc) Overpotential Test.

11.15.3.2.3.1 Measure leakage current on each phase with the switch closed and the poles not under test grounded. Test voltage should be in accordance with manufacturer's published data. For further information on dc overpotential testing, see 11.9.2.6, Direct Current (dc) Overpotential Testing.

11.15.3.2.3.2 If no evidence of distress or insulation failure is observed by the end of the total time of voltage application during the dc overpotential test, the switch is considered to have passed the test.

11.16 Medium and High Voltage Circuit Breakers.

11.16.1 Medium Voltage Air Circuit Breakers.

11.16.1.1 Field Testing in General.

11.16.1.1.1 Where field testing is required, it is recommended that a qualified field service team be employed and that instructions be followed as recommended by appropriate standard and manufacturer's instructions. If evaluation of the switch indicates results that differ significantly from the recommended values, the switch should be removed and repaired.

11.16.1.1.2 The applicable industry field evaluation standards include the following:

(1) ANSI/NETA ATS, *Standard for Acceptance Testing Specifications for Electrical Power Distribution Equipment and Systems*
(2) ANSI/NETA MTS, *Standard for Maintenance Testing Specifications for Electrical Power Distribution Equipment and Systems*

11.16.1.1.3 For further information on medium voltage air circuit breakers, see Section 15.4, Air Circuit Breakers.

11.16.1.2 Field Tests.

11.16.1.2.1 Insulation Resistance.

11.16.1.2.1.1 Measure insulation resistance on each pole, phase-to-phase and phase-to-ground with the switch closed and across each open pole for one minute. Apply voltage in accordance with manufacturer's published data. For further information on insulation resistance testing, see 11.9.2.3, Insulation Resistance Testing.

11.16.1.2.1.2 Insulation resistance values of the medium voltage air circuit breaker should be in accordance with manufacturer's published data.

11.16.1.2.2 Contact/Pole Resistance Test.

11.16.1.2.2.1 Measure the contact resistance across each pole. This test is used to test the quality of the contacts. The contact resistance should be kept as low as possible to reduce power losses at the contacts with the resultant localized heating, which will shorten the life of both the contacts and nearby insulation.

11.16.1.2.2.2 Microhm values should not exceed the high levels of the normal range as indicated in the manufacturer's published data.

11.16.1.2.3 Power Factor or Dissipation Factor Test.

11.16.1.2.3.1 On air circuit breakers, the power factor of the following should be obtained:

(1) Each line-side and load-side bushing assembly complete with stationary contacts and interrupters, with the circuit breaker open
(2) Each phase of the circuit breaker with the breaker closed
(3) Air magnetic circuit breakers should be tested both with and without arc chutes

11.16.1.2.3.2 Measure power factor or dissipation factor on each bushing. Tests should be made of each bushing using either the power factor or capacitance tap if the bushing is so equipped or a "hot collar" test using a test electrode around the outside shell of the bushing.

11.16.1.2.3.3 Power factor or dissipation factor values should be compared with previous test results of similar breakers or manufacturer's published data. Bushing values should be within 10 percent of nameplate rating for bushings or manufacturer's published data. Hot collar tests are evaluated on a milliampere/milliwatt loss basis and the results should be compared to values of similar bushings or manufacturer's published data.

11.16.1.2.3.4 For further information on power factor or dissipation factor testing, see 11.9.2.3, Insulation Power Factor or Dissipation Factor Testing.

11.16.1.2.4 Direct-Current (dc) Overpotential Test.

11.16.1.2.4.1 Measure leakage current on each phase with the circuit breaker closed and the poles not under test grounded. Test voltage should be in accordance with manufacturer's published data. For further information on dc overpotential testing, see 11.9.2.6, Direct-Current (dc) Overpotential Testing.

11.16.1.2.4.2 If no evidence of distress or insulation failure is observed by the end of the total time of voltage application during the dc overpotential test, the circuit breaker is considered to have passed the test.

11.16.1.2.5 Blowout Coils. Verify blowout coil circuit continuity. The blowout coil circuit should exhibit continuity.

11.16.1.2.6 Heaters. Verify operation of heaters. Heaters should be operational.

11.16.1.2.7 Time-Travel Analysis.

11.16.1.2.7.1 This test, used on medium- and high-voltage circuit breakers, provides information as to whether the operating mechanism of the circuit breaker is operating properly. All test instruments should be used in strict compliance with the manufacturer's instructions and recommendations. Failure to follow the manufacturer's instructions can result in injury to personnel and can produce meaningless data. The test presents in graphical form the position of the breaker contacts versus time. This test can be used to determine the opening and closing speeds of the breaker, the interval time for closing and tripping, and the contact bounce. The test provides information that can be used to detect problems such as weak accelerating springs, defective shock absorbers, dashpots, buffers, and closing mechanisms.

11.16.1.2.7.2 The test is performed by a mechanical device that is attached to the breaker. There are several types of devices available to perform this function. One device, a rotating drum with a chart attached, is temporarily connected to the chassis or tank of the breaker. A movable rod with a marking device attached is installed on the lift rod portion of the breaker. As the breaker is opened or closed, the marking device indicates the amount of contact travel on the chart as the drum rotates at a known speed. With another available device, a transducer is attached to the movable rod, and the breaker operation is recorded on an oscillograph.

11.16.1.2.7.3 Compare travel and velocity values to manufacturer's published data and previous test data.

11.16.2 Medium- and High-Voltage Oil Circuit Breakers.

11.16.2.1 Field Testing in General.

11.16.2.1.1 Where field testing is required, it is recommended that a qualified field service team be employed and that instructions be followed as recommended by appropriate standard and manufacturer's instructions. If evaluation of the switch indicates results that differ significantly from the recommended values, the switch should be removed and repaired.

11.16.2.1.2 The applicable industry field evaluation standards include the following:

(1) ANSI/NETA ATS, *Standard for Acceptance Testing Specifications for Electrical Power Distribution Equipment and Systems*
(2) ANSI/NETA MTS, *Standard for Maintenance Testing Specifications for Electrical Power Distribution Equipment and Systems*

11.16.2.1.3 For further information on medium-voltage oil circuit breakers, see Section 15.6, Oil Circuit Breakers.

11.16.2.2 Field Tests.

11.16.2.2.1 Insulation Resistance.

11.16.2.2.1.1 Measure insulation resistance on each pole, phase-to-phase and phase-to-ground with the switch closed and across each open pole for 1 minute. Apply voltage in accordance with manufacturer's published data. For further information on insulation resistance testing, see 11.9.2.3, Insulation Resistance Testing.

11.16.2.2.1.2 Insulation resistance values of the medium- or high-voltage oil circuit breaker should be in accordance with manufacturer's published data.

11.16.2.2.2 Contact/Pole Resistance Test.

11.16.2.2.2.1 Measure the contact resistance across each pole. This test is used to test the quality of the contacts. The contact resistance should be kept as low as possible to reduce power losses at the contacts with the resultant localized heating, which will shorten the life of both the contacts and nearby insulation.

11.16.2.2.2.2 Microhm values should not exceed the high levels of the normal range as indicated in the manufacturer's published data.

11.16.2.2.3 Power Factor or Dissipation Factor Test.

11.16.2.2.3.1 On oil circuit breakers, the power factor or dissipation factor should be obtained in both the open and closed positions. Determine the tank loss index.

11.16.2.2.3.2 Measure power factor or dissipation factor on each bushing. Tests should be made of each bushing using either the power factor or capacitance tap if the bushing is so equipped or a "hot collar" test using a test electrode around the outside shell of the bushing.

11.16.2.2.3.3 Power factor or dissipation factor values and tank loss index should be compared with previous test results of similar breakers or manufacturer's published data. Bushing values should be within 10 percent of nameplate rating for bushings or manufacturer's published data. Hot collar tests are evaluated on a milliampere/milliwatt loss basis and the results should be compared to values of similar bushings or manufacturer's published data.

11.16.2.2.3.4 The range of data and recommendations given in Table 11.16.2.2.3.4 should be taken as general information.

11.16.2.2.3.5 For further information on power factor or dissipation factor testing, see 11.9.3.2, Insulation Power Factor or Dissipation Factor Testing.

Table 11.16.2.2.3.4 Tank Loss Index Results in Watts (W)

Below −0.20 W	Between −10 W and −0.20 W	Between −10 W and +0.05 W	Between +0.05 W and +0.10 W	Above +0.10 W
Investigate immediately (Investigate)	Retest on a more frequent basis (Acceptable)	Normal for most breakers (Acceptable)	Retest on a more frequent basis (Acceptable)	Investigate immediately (Investigate)
Problem could be in the lift-rod guide assembly, contact assembly (interrupter), or upper portion of lift rod.			Problem could be in the lift rod, tank oil, tank liner, or auxiliary contact support insulation.	

11.16.2.2.4 Direct-Current (dc) Overpotential Test.

11.16.2.2.4.1 Measure leakage current on each phase with the circuit breaker closed and the poles not under test grounded. Test voltage should be in accordance with manufacturer's published data. For further information on dc overpotential testing, see 11.9.2.6, Direct Current (dc) Overpotential Testing.

11.16.2.2.4.2 If no evidence of distress or insulation failure is observed by the end of the total time of voltage application during the dc overpotential test, the circuit breaker is considered to have passed the test.

11.16.2.2.5 Heaters. Verify operation of heaters. Heaters should be operational.

11.16.2.2.6 Time-Travel Analysis.

11.16.2.2.6.1 This test, used on medium- and high-voltage circuit breakers, provides information as to whether the operating mechanism of the circuit breaker is operating properly. All test instruments should be used in strict compliance with the manufacturer's instructions and recommendations. Failure to follow the manufacturer's instructions can result in injury to personnel and can produce meaningless data. The test presents in graphical form the position of the breaker contacts versus time. This test can be used to determine the opening and closing speeds of the breaker, the interval time for closing and tripping, and the contact bounce. The test provides information that can be used to detect problems such as weak accelerating springs, defective shock absorbers, dashpots, buffers, and closing mechanisms.

11.16.2.2.6.2 The test is performed by a mechanical device that is attached to the breaker. There are several types of devices available to perform this function. One device, a rotating drum with a chart attached, is temporarily connected to the chassis or tank of the breaker. A movable rod with a marking device attached is installed on the lift rod portion of the breaker. As the breaker is opened or closed, the marking device indicates the amount of contact travel on the chart as the drum rotates at a known speed. With another available device, a transducer is attached to the movable rod, and the breaker operation is recorded on an oscillograph.

11.16.2.2.6.3 Compare travel and velocity values to manufacturer's published data and previous test data.

11.16.2.2.7 Insulating Fluid Tests. Remove a sample of insulating liquid and test for the following:

(1) Dielectric breakdown
(2) Color
(3) Power factor
(4) Interfacial tension
(5) Visual condition

For further information on insulating fluid tests, see Section 11.19, Insulating-Liquid Analysis.

11.16.3 Medium-Voltage Vacuum Breakers.

11.16.3.1 Field Testing in General.

11.16.3.1.1 Where field testing is required, it is recommended that a qualified field service team be employed and that instructions be followed as recommended by appropriate standard and manufacturer's instructions. If evaluation of the switch indicates results that differ significantly from the recommended values, the switch should be removed and repaired.

11.16.3.1.2 The applicable industry field evaluation standards include the following:

(1) ANSI/NETA ATS, *Standard for Acceptance Testing Specifications for Electrical Power Distribution Equipment and Systems*
(2) ANSI/NETA MTS, *Standard for Maintenance Testing Specifications for Electrical Power Distribution Equipment and Systems*

11.16.3.1.3 For further information on medium-voltage vacuum circuit breakers, see Section 15.5, Vacuum Circuit Breakers.

11.16.3.2 Field Tests.

11.16.3.2.1 Insulation Resistance.

11.16.3.2.1.1 Measure insulation resistance on each pole, phase-to-phase and phase-to-ground with the switch closed and across each open pole for one minute. Apply voltage in accordance with manufacturer's published data. For further information on insulation resistance testing, see 11.9.2.3, Insulation Resistance Testing.

11.16.3.2.1.2 Insulation resistance values of the low-voltage vacuum circuit breakers should be in accordance with manufacturer's published data.

11.16.3.2.2 Contact/Pole Resistance Test.

11.16.3.2.2.1 Measure the contact resistance across each pole. This test is used to test the quality of the contacts. The contact resistance should be kept as low as possible to reduce power losses at the contacts with the resultant localized heating, which will shorten the life of both the contacts and nearby insulation.

11.16.3.2.2.2 Microhm values should not exceed the high levels of the normal range as indicated in the manufacturer's published data.

11.16.3.2.3 Power Factor or Dissipation Factor Test.

11.16.3.2.3.1 On vacuum circuit breakers, the power factor of the following should be obtained:

(1) Each pole with breaker in the open position
(2) Each phase of the circuit breaker with the breaker closed

11.16.3.2.3.2 Measure power factor or dissipation factor on each bushing. Tests should be made of each bushing using either the power factor or capacitance tap if the bushing is so equipped or a "hot collar" test using a test electrode around the outside shell of the bushing.

11.16.3.2.3.3 For further information on power factor or dissipation factor testing, see 11.9.3.2, Insulation Power Factor or Dissipation Factor Testing.

11.16.3.2.4 Overpotential Test.

11.16.3.2.4.1 Measure leakage current on each phase with the circuit breaker closed and the poles not under test grounded. Test voltage (ac or dc) should be in accordance with manufacturer's published data.

11.16.3.2.4.2 If no evidence of distress or insulation failure is observed by the end of the total time of voltage application during the overpotential test, the circuit breaker is considered to have passed the test.

11.16.3.2.5 Vacuum Bottle Integrity Overpotential Test.

11.16.3.2.5.1 Measure leakage current on each phase across each vacuum bottle with the circuit breaker open. Test voltage should be in accordance with manufacturer's published data.

11.16.3.2.5.2 Do not exceed maximum voltage stipulated by the manufacturer for this test. Be aware that some dc test sets are half-wave rectified and can produce peak voltages in excess of the breaker manufacturer's recommended maximum.

11.16.3.2.5.3 Provide adequate barriers and protection against x-radiation during this test. The level of x-ray emission from a vacuum breaker with proper contact spacing and subjected to standard test voltages is extremely small and well below the maximum level permitted by standards. In view of the possibility that the contacts are out of adjustment or that the applied voltage is greater than prescribed, it is advisable that during the overvoltage test all personnel stand behind the front steel barrier and remain farther from the breaker than would otherwise be necessary for reasons of electrical safety. During the high-voltage test, the vapor shield inside the interrupter can acquire an electrostatic charge. This charge should be bled off immediately after the test.

11.16.3.2.5.4 Do not perform the vacuum bottle integrity test unless the contact displacement of each interrupter is within manufacturer's tolerance.

11.16.3.2.5.5 If no evidence of distress or insulation failure is observed by the end of the total time of voltage application during the vacuum bottle integrity test, the vacuum bottle is considered to have passed the test.

11.16.3.2.6 Time-Travel Analysis.

11.16.3.2.6.1 This test, used on medium-voltage circuit breakers, provides information as to whether the operating mechanism of the circuit breaker is operating properly. All test instruments should be used in strict compliance with the manufacturer's instructions and recommendations. Failure to follow the manufacturer's instructions can result in injury to personnel and can produce meaningless data. The test presents in graphical form the position of the breaker contacts versus time. This test can be used to determine the opening and closing speeds of the breaker, the interval time for closing and tripping, and the contact bounce. The test provides information that can be used to detect problems such as weak accelerating springs, defective shock absorbers, dashpots, buffers, and closing mechanisms.

11.16.3.2.6.2 The test is performed by a mechanical device that is attached to the breaker. There are several types of devices available to perform this function. One device, a rotating drum with a chart attached, is temporarily connected to the chassis or tank of the breaker. A movable rod with a marking device attached is installed on the lift rod portion of the breaker. As the breaker is opened or closed, the marking device indicates the amount of contact travel on the chart as the drum rotates at a known speed. With another available device, a transducer is attached to the movable rod, and the breaker operation is recorded on an oscillograph.

11.16.3.2.6.3 Compare travel and velocity values to manufacturer's published data and previous test data.

11.16.3.2.7 Heaters. Verify operation of heaters. Heaters should be operational.

11.16.4 SF_6 Circuit Breakers.

11.16.4.1 Field Testing in General.

11.16.4.1.1 Where field testing is required, it is recommended that a qualified field service team be employed and that instructions be followed as recommended by appropriate standard and manufacturer's instructions. If evaluation of the circuit breaker indicates results that differ significantly from the recommended values, the circuit breaker should be removed and repaired.

11.16.4.1.2 The applicable industry field evaluation standards include the following:

(1) ANSI/NETA ATS, *Standard for Acceptance Testing Specifications for Electrical Power Distribution Equipment and Systems*
(2) ANSI/NETA MTS, *Standard for Maintenance Testing Specifications for Electrical Power Distribution Equipment and Systems*

11.16.4.1.3 For further information on SF_6 breakers, see Section 15.8, Gas-Insulated Substations and Gas-Insulated Equipment.

11.16.4.2 Field Tests.

11.16.4.2.1 Insulation Resistance.

11.16.4.2.1.1 Measure insulation resistance on each pole, phase-to-phase and phase-to-ground with the circuit breaker closed and across each open pole for 1 minute. Apply voltage in accordance with manufacturer's published data. For further information on insulation resistance testing, see 11.9.2.3, Insulation Resistance Testing.

11.16.4.2.1.2 Insulation resistance values of SF_6 circuit breakers should be in accordance with manufacturer's published data.

11.16.4.2.2 Contact/Pole Resistance Test.

11.16.4.2.2.1 Measure the contact/pole resistance. This test is used to test the quality of the contacts. The contact resistance should be kept as low as possible to reduce power losses at the contacts with the resultant localized heating, which will shorten the life of both the contacts and nearby insulation.

11.16.4.2.2.2 Microhm values should not exceed the high levels of the normal range as indicated in the manufacturer's published data.

11.16.4.2.3 Power Factor or Dissipation Factor Test.

11.16.4.2.3.1 On SF_6 circuit breakers, the power factor of the following should be obtained:

(1) Each pole with the circuit breaker in the open position
(2) Each phase of the circuit breaker with the breaker closed

11.16.4.2.3.2 Measure power factor or dissipation factor on each bushing. Tests should be made of each bushing using either the power factor or capacitance tap if the bushing is so equipped or a "hot collar" test using a test electrode around the outside shell of the bushing.

11.16.4.2.3.3 Power factor or dissipation factor values should be compared with previous test results of similar breakers or manufacturer's published data. High losses or high power factor usually indicates problems with bushings, operating road, support insulation or the SF_6 gas. Bushing values should be within ten percent of nameplate rating for bushings or manufacturer's published data. Hot collar tests are evaluated on a milliampere/milliwatt loss basis and the results should be compared to values of similar bushings or manufacturer's published data.

11.16.4.2.3.4 For further information on power factor or dissipation factor testing, see 11.9.3.2, Insulation Power Factor or Dissipation Factor Testing.

11.16.4.2.4 Direct-Current (dc) Overpotential Test.

11.16.4.2.4.1 Measure leakage current on each phase with the circuit breaker closed and the poles not under test grounded. Test voltage should be in accordance with manufacturer's published data. For further information on dc overpotential testing, see 11.9.2.6, Direct-Current (dc) Overpotential Testing.

11.16.4.2.4.2 If no evidence of distress or insulation failure is observed by the end of the total time of voltage application during the dc overpotential test, the circuit breaker is considered to have passed the test.

11.16.4.2.5 Time-Travel Analysis.

11.16.4.2.5.1 This test, used on medium- and high-voltage circuit breakers, provides information as to whether the operating mechanism of the circuit breaker is operating properly. All test instruments should be used in strict compliance with the manufacturer's instructions and recommendations. Failure to follow the manufacturer's instructions can result in injury to personnel and can produce meaningless data. The test presents in graphical form the position of the breaker contacts versus time. This test can be used to determine the opening and closing speeds of the breaker, the interval time for closing and tripping, and the contact bounce. The test provides information that can be used to detect problems such as weak accelerating springs, defective shock absorbers, dashpots, buffers, and closing mechanisms.

11.16.4.2.5.2 The test is performed by a mechanical device that is attached to the breaker. There are several types of devices available to perform this function. One device, a rotating drum with a chart attached, is temporarily connected to the chassis or tank of the breaker. A movable rod with a marking device attached is installed on the lift rod portion of the breaker. As the breaker is opened or closed, the marking device indicates the amount of contact travel on the chart as the drum rotates at a known speed. With another available device, a transducer is attached to the movable rod, and the breaker operation is recorded on an oscillograph.

11.16.4.2.5.3 Compare travel and velocity values to manufacturer's published data and previous test data.

11.16.4.2.6 Heaters. Verify operation of heaters. Heaters should be operational.

11.17 Infrared Inspection.

11.17.1 Introduction. Infrared inspections of electrical systems are beneficial to reduce the number of costly and catastrophic equipment failures and unscheduled plant shutdowns.

11.17.1.1 Infrared inspections should be performed by qualified and trained personnel who have an understanding of infrared technology, electrical equipment maintenance, and the safety issues involved. Infrared inspections have uncovered a multitude of potentially dangerous situations. Proper diagnosis and remedial action of these situations have also helped to prevent numerous major losses.

11.17.1.2 The instruments most suitable for infrared inspections are of the type that use a scanning technique to produce an image of the equipment being inspected. These devices display a picture in which "hot spots" appear as bright or brighter spots.

11.17.1.3 Infrared surveys can be accomplished either by in-house teams or by a qualified outside contractor. The economics and effectiveness of the two alternatives should be carefully weighed. Many organizations find it preferable to obtain these surveys from qualified outside contractors. Because of outside contractors' more extensive experience, their findings and recommendations are likely to be more accurate, practical, and economical than those of a part-time in-house team.

11.17.1.4 Infrared surveys of electrical systems should not be viewed as replacement for visual inspections. Visual inspections or checks are still required on lightly loaded circuits or on circuits not energized or not carrying current at the time of the infrared survey (e.g., neutral connections).

11.17.2 Advantages of Infrared Inspections. Infrared inspections are advantageous to use in situations where electrical equipment cannot be deenergized and taken out of service or where plant production is affected. They can reduce typical visual examinations and tedious manual inspections and are especially effective in long-range detection situations.

11.17.2.1 Infrared detection can be accurate, reliable, and expedient to use in a variety of electrical installations. More important, it can be relatively inexpensive to use considering the savings often realized by preventing equipment damage and business interruptions.

11.17.2.2 Infrared inspections are considered a useful tool to evaluate previous repair work and proof test new electrical installations and new equipment still under warranty.

11.17.2.3 Regularly scheduled infrared inspections often require the readjustment of electrical maintenance priorities as well as detect trends in equipment performance that require periodic observation.

11.17.3 Disadvantages. There are some disadvantages to individual ownership of certain types of equipment. Scanning-type thermal imaging devices can be costly to purchase outright. Training is recommended for persons who operate scanning-type thermal imaging instruments.

11.17.3.1 Infrared inspections require special measures and analysis. Equipment enclosed for safety or reliability can be difficult to scan or to detect radiation from within. Special precautions, including the removal of access panels, might be necessary for satisfactory measurements. Weather can be a factor in the conduct of a survey of electrical systems located outdoors, for example, overhead electric open lines and substations. Rain and wind can produce abnormal cooling of defective conductors and components. Because the reflection of sun rays from bright surfaces can be misread as hot spots, infrared work on outdoor equipment might have to be performed at night. That, in turn, presents a problem, because electrical loads usually are lower at night, and the faulty connections and equipment might not overheat enough to enable detection. Shiny surfaces do not emit radiation energy efficiently and can be hot while appearing cool in the infrared image.

11.17.3.2 The handling of liquid nitrogen, argon, and other liquefied gases with their inherent hazards is a disadvantage of some infrared testing equipment.

11.17.4 Desirable Operational Features. The equipment display should be large and provide good resolution of hot spots. The equipment should provide color or black-and-white photographs to identify the exact location of the hot spot. The unit should be portable, easy to adjust, and approved for use in the atmosphere in which it is to be used. It should also have a cone of vision that gives enough detail to accurately identify the hot spot.

11.17.4.1 The unit should be designed so that the operator knows the degree of accuracy in the display. There should be easily operated checks to verify the accuracy of the display.

11.17.5 Inspection Frequency and Procedures. Routine infrared inspections of energized electrical systems should be performed annually prior to shutdown. More frequent infrared inspections, for example, quarterly or semiannually, should be performed where warranted by loss experience, installation of new electrical equipment, or changes in environmental, operational, or load conditions.

11.17.5.1 All critical electrical equipment as determined by Section 6.3 should be included in the infrared inspection.

11.17.5.2 Infrared surveys should be performed during periods of maximum possible loading but not less than 40 percent of rated load of the electrical equipment being inspected. The circuit-loading characteristics should be included as part of the documentation provided in 11.17.5.4.

11.17.5.3 Equipment enclosures should be opened for a direct view of components whenever possible. When opening the enclosure is impossible, such as in some busway systems, internal temperatures can be higher than the surface temperatures. Plastic and glass covers in electrical enclosures are not transparent to infrared radiation.

11.17.5.4 Infrared surveys should be documented as outlined in 6.5.2 and Section 11.8.

11.17.5.5 The electrical supervisor should be immediately notified of critical, impending faults so that corrective action can be taken before a failure occurs. Priorities should be established to correct other deficiencies.

11.17.5.6 Section 9 and Table 10.18 of the ANSI/NETA MTS, *Standard for Maintenance Testing Specifications for Electrical Power Distribution Equipment and Systems* suggest temperature benchmarks similar to those in the following list. The temperature differences in this list denote differences from the normal referenced temperature. The normal referenced temperature is determined by a qualified technician.

(1) Temperature differences of 1°C to 3°C indicate possible deficiency and warrant investigation.
(2) Temperature differences of 4°C to 15°C indicate deficiency; repairs should be made as time permits.
(3) Temperature differences of 16°C and above indicate major deficiency; repairs should be made immediately.

11.18 Fuses.

11.18.1 Fuses can be tested with a continuity tester to verify that the fuse is not open. Resistance readings can be taken using a sensitive 4-wire instrument such as a Kelvin bridge or micro-ohmmeter. Fuse resistance values should be compared against values recommended by the manufacturer.

11.18.2 Where manufacturers' data is not readily available, resistance deviations of more than 15 percent for identical fuses in the same circuit should be investigated.

11.19 Insulating-Liquid Analysis. Regular semiannual tests should be made on insulating oils and askarels. Samples should be taken from the equipment in accordance with ASTM D 923, *Standard Test Method for Sampling Electrical Insulating Liquids.* The maintenance tests most commonly performed on used insulating liquids, together with the appropriate ASTM test methods, are shown in Table 11.19(a) through Table 11.19(g).

Table 11.19(a) Suggested Limits for Class I Insulating Oil

		Mineral Oil[a]		
		Acceptable Values		
Test	**ASTM Method**	**69 kV and Below**	**Above 69 kV – Below 230 kV**	**230 kV and Above**
Dielectric breakdown, kV minimum[b]	D 877	26	26	26
Dielectric breakdown, kV minimum @ 1 mm (0.04 in.) gap	D 1816	23	28	30
Dielectric breakdown, kV minimum @ 2 mm (0.08 in.) gap	D 1816	40	47	50
Interfacial tension mN/m minimum	D 971 or D 2285	25	30	32
Neutralization number, mg KOH/g maximum	D 974	0.2	0.15	0.10
Water content, ppm maximum @ 60°C[c]	D 1533	35	25	20
Power factor at 25°C, %	D 924	0.5	0.5	0.5
Power factor at 100°C, %	D 924	5.0	5.0	5.0
Color[d]	D 1500	3.5	3.5	3.5
Visual condition	D 1524	Bright, clear and free of particles	Bright, clear and free of particles	Bright, clear and free of particles
Specific gravity (relative density) @ 15°C maximum[e]	D 1298	0.91	0.91	0.91

[a]ANSI/IEEE C57.106-2002 *Guide for Acceptance and Maintenance of Insulating Oil in Equipment,* Table 7.
[b]IEEE STD 637-1985 *Guide for Reclamation of Insulating Oil and Criteria for Its Use,* Table 1.
[c]ANSI/IEEE C57.106-2002 *Guide for Acceptance and Maintenance of Insulating Oil in Equipment,* Table 5.
[d]In the absence of consensus standards, NETA's Standard Review Council suggests these values.
[e]ANSI/IEEE C57.106 *Guide for Acceptance and Maintenance of Insulating Oil in Equipment,* Table 1.

Table 11.19(b) Suggested Limit for Less-Flammable Hydrocarbon Insulating Liquid

Test	ASTM Method	Acceptable Values
Dielectric breakdown voltage, kV minimum	D 877	24
Dielectric breakdown voltage for 0.04 in. gap, kV minimum	D 1816	34
Dielectric breakdown voltage for 0.08 in. gap, kV minimum	D 1816	24
Water content, ppm maximum	D 1533 B	35
Dissipation/power factor, 60 Hz, % max. @ 25°C	D 924	1.0
Fire point, °C, minimum	D 92	300
Interfacial tension, mN/m, 25°C	D 971	24
Neutralization number, mg KOH/g	D 664	0.2

ANSI/IEEE C57.121-1998 *Guide for Acceptance and Maintenance of Less-Flammable Hydrocarbon Fluid in Transformers,* Table 4.

The values in this table are considered typical for acceptable service-aged LFH fluids as a general class. If actual test analysis approaches the values shown, consult the fluid manufacturer for specific recommendations.

If the purpose of the HMWH installation is to comply with the NFPA 70, *National Electrical Code,* this value is the minimum for compliance with *NEC* Article 450.23.

Table 11.19(c) Suggested Limit for Service-Aged Silicone Insulating Liquid

Test	ASTM Method	Acceptable Values
Dielectric breakdown, kV minimum	D 877	25
Visual	D 2129	Colorless, clear, free of particles
Water content, ppm maximum	D 1533	100
Dissipation/power factor, 60 Hz, maximum @ 25°C	D 924	0.2
Viscosity, cSt @ 25°C	D 445	47.5–52.5
Fire point, °C, minimum	D 92	340
Neutralization number, mg KOH/g maximum	D 974	0.2

ANSI/IEEE C57.111-1989 (R1995) *Guide for Acceptance of Silicone Insulating Fluid and Its Maintenance in Transformers,* Table 3.

Table 11.19(d) Suggested Limits for Service-Aged Tetrachloroethylene Insulating Fluid

Test	ASTM Method	Acceptable Values
Dielectric breakdown, kV minimum	D 877	26
Visual	D 2129	Clear with purple iridescence
Water content, ppm maximum	D 1533	35
Dissipation/power factor, % maximum @ 25°C	D 924	12.0
Viscosity, cSt @ 25°C	D 445	0
Fire point, °C, minimum	D 92	—
Neutralization number, mg KOH/g maximum	D 974	0.25
Neutralization number, mg KOH/g maximum	D 664	—
Interfacial tension, mN/m minimum @ 25°C	D 971	—

Instruction Book PC-2000 for Wecosol™ Fluid-Filled Primary and Secondary Unit Substation Transformers, ABB Power T&D

Table 11.19(e) Insulating Fluid Limits

Table 100.4.1 Test Limits for New Insulating Oil Received in New Equipment

	Mineral Oil				
Test	**ASTM Method**	**≤69 kV and Below**	**>69 kV – < 230 kV**	**>230 kV – < 345 kV**	**>345 kV and Above**
Dielectric breakdown, kV minimum	D 877	30	30	30	
Dielectric breakdown, kV minimum @ 1 mm (0.04 in.) gap	D 1816	25	30	32	35
Dielectric breakdown, kV minimum @ 2 mm (0.08 in.) gap	D 1816	45	52	55	60
Interfacial tension mN/m minimum	D 971 or D 2285	38	38	38	38
Neutralization number, mg KOH/g maximum	D 974	0.015	0.015	0.015	0.015
Water content, ppm maximum	D 1533	20	10	10	10
Power factor at 25°C, %	D 924	0.05	0.05	0.05	0.05
Power factor at 100°C, %	D 924	0.40	0.40	0.30	0.30
Color	D 1500	1.0	1.0	1.0	1.0
Visual condition	D 1524	Bright and clear	Bright and clear	Bright and clear	Bright and clear

ANSI/IEEE C57.106-2002, *Guide for Acceptance and Maintenance of Insulating Oil in Equipment,* Tables 1, 2, and 3.

Table 11.19(f) Test Limits for Silicone Insulating Liquid in New Transformers

Table 100.4.2 Test Limits for Silicone Insulating Liquid in New Transformers

Test	ASTM Method	Acceptable Values
Dielectric breakdown, kV minimum	D 877	30
Visual	D 2129	Clear, free of particles
Water content, ppm maximum	D 1533	50
Dissipation/power factor, 60 Hz, % max. @ 25°C	D 924	0.1
Viscosity, cSt @ 25°C	D 445	47.5–52.5
Fire point, °C, minimum	D 92	340
Neutralization number, mg KOH/g max.	D 974	0.01

ANSI/IEEE C57.111-1989 (R1995), *Guide for Acceptance of Silicone Insulating Fluid and Its Maintenance in Transformers,* Table 2.

11.19.1 Field Testing in General.

11.19.1.1 The applicable industry field evaluation standards include the following:

(1) ANSI/NETA ATS, *Standard for Acceptance Testing Specifications for Electrical Power Distribution Equipment and Systems*
(2) ANSI/NETA MTS, *Standard for Maintenance Testing Specifications for Electrical Power Distribution Equipment and Systems*

11.19.1.2 For further information on transformer liquid maintenance, and analysis, see 11.11.8.

11.19.2 Types of Test.

11.19.2.1 Dielectric Breakdown Test. This test measures the electrical stress an insulating liquid can withstand without failure. It indicates if contaminants, particularly moisture and conducting particulate matter are in the fluid The dielectric breakdown voltage is expressed in kilovolts. Dielectric testing is performed by ASTM method D-877 (disk electrodes) or D-1816 (VDE cell).

11.19.2.2 Acid Neutralization Number. Measuring acidity provides a means of monitoring the progress of oxidation. Acidic compounds precede the formation of sludge in a transformer which is the end product of oxidation. The acid neutralization number is express in dynes/cm. Acid neutralization number testing is performed by ANSI/ASTM method D-974 or D-664.

11.19.2.3 Specific Gravity. Specific gravity of oil is the ratio of the weights of equal volumes of oil and water at the same temperature. The specific gravity is not significant in determining the quality of oil but is applicable in determining suitability for use in a specific situation. Specific gravity testing is performed by ANSI/ASTM method D-1298.

11.19.2.4 Interfacial Tension. Interfacial tension (IFT) determines the presence of polar contaminants in oil. Interfacial

Table 11.19(g) Typical Values for Less-Flammable Hydrocarbon Insulating Liquid in New Equipment

Table 100.4.3 Typical Values for Less-Flammable Hydrocarbon Insulating Liquid Received in New Equipment

ASTM Method	Test	Minimum	Results	Maximum
D 1816	Dielectric breakdown voltage for 0.08 in. gap, kV	40	34.5 kV class and below	
		50	Above 34.5 kV class	
		60	Desirable	
D 1816	Dielectric breakdown voltage for 0.04 in. gap, kV	20	34.5 kV class and below	
		25	Above 34.5 kV class	
		30	Desirable	
D 974	Neutralization number, mg KOH/g			0.03
D 877	Dielectric breakdown voltage, kV		30	
D 924	AC loss characteristic (dissipation factor), % 25°C 100°C			0.11
D 1533B	Water content, ppm			25
D 1524	Condition-visual		Clear	
D 92	Flash point (°C)		275	
D 92	Fire point (°C)		300a	—
D 971	Interfacial tension, mN/m, 25°C		38	
D 445	Kinematic viscosity, mm^2/s, (cSt), 40°C		1.0 × 102 (100)	1.3 × 102 (130)
D 1500	Color			L2.5

ANSI/IEEE C57.121-1998, *IEEE Guide for Acceptance and Maintenance of Less Flammable Hydrocarbon Fluid in Transformers*, Table 3.
The test limits shown in this table apply to less-flammable hydrocarbon fluids as a class. Specific typical values for each brand of fluid should be obtained from each fluid manufacturer.

tension together with the acid neutralization number is an indicator to monitor sludge development. Foreign substances such as dissolved varnishes and organic coating materials can also affect the IFT. The interfacial tension is expressed in dynes per centimeter. Interfacial tension testing is performed by ANSI/ASTM method D-971 (ring method) or D-2285 (drop weight).

11.19.2.5 Color. Monitoring the color of oil provides a rapid assessment of the oil quality. Insulating oils tend to darken due to oxidation and/or the presence of contamination. Color testing is performed by ANSI/ASTM method D-1500 (petroleum oils).

11.19.2.6 Visual Condition. The visual examination provides an assessment of undesirable materials suspended in oil. Visual condition is performed by ASTM method D-1524 (petroleum oils).

11.19.2.7 Water in Insulating Liquids. Water in oil can adversely affect the dielectric strength of an insulating fluid. Water content is expressed in parts per million. Water in insulating liquids test is performed by ASTM method D-1533.

11.19.2.8 Power Factor or Dissipation Factor. Power factor or dissipation factor is a measurement of the dielectric losses in an insulating fluid. It is a means of evaluating the quality of the insulating fluid. Power factor or dissipation factor tests are performed by ASTM method D-924.

11.19.3 Analysis.

11.19.3.1 Service-Aged Insulating Fluid Limits.

11.20 Rotating Machine Testing. On ac rotating machines with an external neutral connection, the stator neutral should be disconnected and a test of each winding with respect to the other two windings and to ground should be obtained.

11.20.1 Insulation Resistance Testing.

11.20.1.1 This testing procedure applies to armature and rotating or stationary field windings. A hand crank, rectifier, or battery-operated instrument is suitable for testing equipment rated to 600 volts. For equipment rated over 600 volts, a 1000-volt or 2500-volt motor-driven or rectifier-operated instrument is recommended for optimum test results. Operating machines should be tested immediately following shutdown when the windings are hot and dry. On large machines, the temperature should be recorded and converted to a base temperature in accordance with ANSI/IEEE 43 *Recommended Practice for Testing Insulation Resistance of Rotating Machinery*, paragraph 6.3, to provide continuity for comparative purposes. Voltage sources, lightning arresters, and capacitors or other potential low-insulation sources should always be disconnected before insulation measurements are made. Lead-in cables or buses and line side of circuit breakers or starters can be tested as a part of the circuit provided a satisfactory reading is obtained. If the insulation resistance is below the established minimum, the circuit components should be tested separately until the low insulation reading is located. Insulation resistance history based on tests conducted on new motors, after rewind and cleaning or from recorded data made under uniform conditions forms a useful basis for interpretation of a machine winding condition. When records of periodic tests are compared, *any persistent downward trend is an indication of insulation trouble*, even though the values might be higher than the recommended minimum safe values listed in Table 11.20.1.2.

11.20.1.2 Insulation resistance readings taken for purposes of correlation should be made at the end of a definite interval following the application of a definite test voltage. For purposes of standardization, 60-second applications are recommended where short-time single readings are to be made on windings and where comparisons with earlier and later data are to be made. Recommended minimum acceptable insulation values without further investigation are as shown in Table 11.20.1.2.

Table 11.20.1.2 Rotating Machine Insulation Testing

Rotating Machinery Voltage	Insulation Resistance (at 40°C)
1000 volts and below	2 megohms
Above 1000 volts	1 megohm per 1000 volts plus 1 megohm

11.20.2 Dielectric-Absorption Testing. A more complete and preferred test applies the voltage for 10 minutes or more to develop the dielectric-absorption characteristic. The curve obtained by plotting insulation resistance against time gives a good indication of moist or dirty windings. A steady rising curve is indicative of a clean, dry winding. A quickly flattening curve is the result of leakage current through or over the surface of the winding and is indicative of a moist or dirty winding. If facilities are not available for a 10-minute test, readings can be taken at 30 seconds and 60 seconds. The 60:30–second ratio or the 10:1–minute ratio serves as an indication of the winding condition. Table 11.20.2 should serve as a guide in interpreting these ratios.

Table 11.20.2 Dielectric-Absorption Testing

Condition	60:30–Second Ratio	10:1–Minute Ratio
Dangerous	—	Less than 1
Poor	Less than 1.1	Less than 1.5
Questionable	1.1 to 1.25	1.5 to 2
Fair	1.25 to 1.4	2 to 3
Good	1.4 to 1.6	3 to 4
Excellent	Above 1.6	Above 4

11.20.3 Overpotential Testing.

11.20.3.1 Overvoltage tests are performed during normal maintenance operations or after the servicing or repair of important machines. Such tests, made on all or parts of the circuit to ground, ensure that the insulation level is sufficiently high for continued safe operation. Both ac and dc test equipment are available. There is no conclusive evidence that one method is preferred over the other. However, where equipment using several insulating materials is tested, ac stresses the insulation more nearly to actual operating conditions than dc. Also, more comparable data have been accumulated, because ac testing has had a head start. However, the use of dc has several advantages and is rapidly gaining favor with increased usage. The test equipment is much smaller, lighter in weight, and lower in price. There is far less possibility of damage to equipment under test, and dc tests give more information than is obtainable with ac testing.

11.20.3.2 The test overvoltages that should be applied depend on the type of machine involved and the level of reliability required from the machines. However, the overvoltage should be of sufficient magnitude to search out weaknesses in the insulation that might cause failure. Standard over-potential test voltage when new is twice rated voltage plus 1000 volts ac. On older or repaired apparatus, tests are reduced to approximately 50 to 60 percent of the factory (new) test voltage. For dc tests, the ac test voltage is multiplied by a factor (1.7) to represent the ratio between the direct test voltage and alternating rms voltage. *[See ANSI/IEEE 95, Recommended Practice for Insulation Testing of AC Electric Machinery (2300 V and Above) with High Direct Voltage.]*

11.20.3.3 A high-potential test made to determine the condition of the insulation up to a predetermined voltage level is difficult to interpret. It is common practice to compare known good results against test specimens to determine what is acceptable and what fails the test. For a dc high-potential test, another criterion used is the shape of the leakage current plotted against voltage rise.

11.20.3.3.1 As long as the knee of the curve, which indicates impending breakdown (point c in Figure 11.20.3.3.1), does not occur below the maximum required test voltage, and as long as the shape of the curve is not too steep compared with that of similar equipment or prior test of the same equipment, the results can be considered satisfactory. It should be recognized that if the windings are clean and dry, overvoltage tests will not detect any defects in the end turns or in lead-in wire located away from the stator iron.

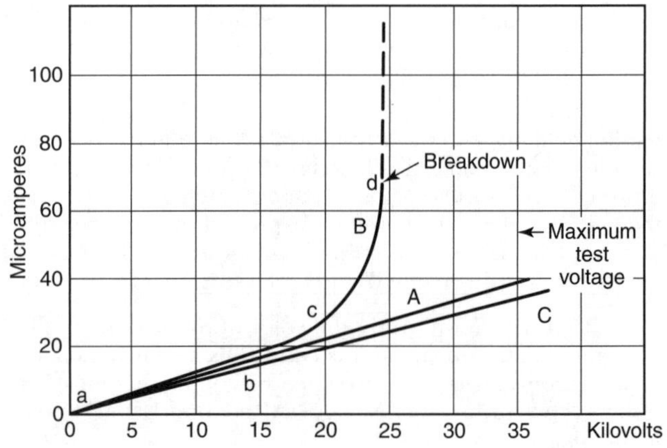

FIGURE 11.20.3.3.1 High-Potential Test.

11.20.4 Surge-Comparison Testing.

11.20.4.1 Surge-comparison testing can detect turn-to-turn, coil-to-coil, group-to-group, and phase-to-phase winding flaws that cannot be detected by insulation resistance, dielectric-absorption, or over-potential testing. Surge testing should not be undertaken until after the integrity of insulation to ground has been verified.

11.20.4.2 The surge testing principle is based on the premise that the impedances of all 3-phase windings of a 3-phase machine should be identical if there are no winding flaws. Each phase (A/B, B/C, C/A) is tested against the others to determine if there is a discrepancy in winding impedances.

11.20.4.3 The test instrument imposes identical, high-voltage, high-frequency pulses across two phases of the machine. The reflected decay voltages of the two windings are displayed and captured on an oscilloscope screen. If the winding impedances are identical, the reflected decay voltage signatures coincide and appear on the screen as a single trace. Two dissimilar traces indicate dissimilar impedances and a possible winding flaw. *(See Figure 11.20.4.3.)*

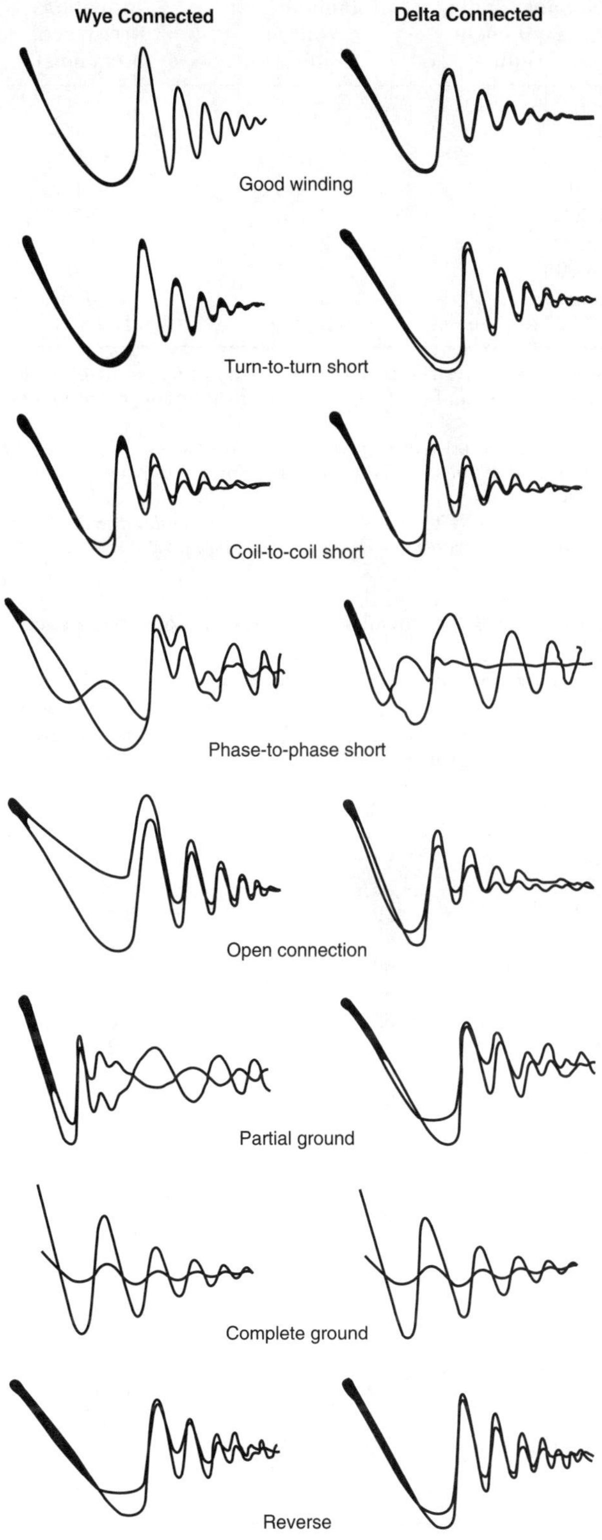

FIGURE 11.20.4.3 Wave Shapes for Winding Faults.

11.20.4.4 The testing and interpretation of results should be conducted by a trained individual.

11.20.5 Other Electrical Tests. More complex tests are not employed unless apparatus performance indicates the tests should be made and experienced testers are available with the test equipment. The other types of tests, depending on the need and desired results, include the following:

(1) Turn-to-turn insulation
(2) Slot discharge and corona
(3) Winding impedance test
(4) Power-factor value
(5) Core loss test

11.20.6 Vibration Testing. Chapter 26 contains information on common methods of measuring vibration.

11.21 Cables.

11.21.1 Cable Field Testing in General.

11.21.1.1 The applicable industry field evaluation standards for electrical power cables include the following:

(1) ANSI/NETA ATS, *Standard for Acceptance Testing Specifications for Electrical Power Distribution Equipment and Systems (ATS)*
(2) ANSI/NETA MTS, *Standard for Maintenance Testing Specifications for Electrical Power Distribution Equipment and Systems (MTS)*
(3) IEEE-400, *Guide for Field Testing and Evaluation of the Insulation of Shielded Power Cable Systems*
(4) IEEE-400.1, *Guide for Field Testing of Laminated Dielectric, Shielded Power Cable Systems Rated 5kV and Above with High Direct Current Voltage*
(5) IEEE-400.2, *Guide for Field Testing of Shielded Power Cable Systems Using Very Low Frequency (VLF)*
(6) IEEE-400.3, *Guide for Partial Discharge Testing of Shielded Power Cable Systems in a Field Environment*

11.21.1.2 For further information on cables, see Chapter 19, Power Cables.

11.21.1.3 Before equipment insulation is tested, it should be cleaned, inspected, and repaired as necessary to minimize leakage currents. The same action should be taken for cable terminations. Surge arresters, capacitors, or similar equipment should be disconnected.

11.21.1.4 When cables are being tested, all transformers, switches, fuse cutouts, switchgear, or other ancillary equipment should be disconnected wherever practicable. If significant leakage currents are encountered it will be known that those currents are probably in the cable insulation and not in equipment connected thereto. If such disconnection is impractical, it might be necessary to limit the maximum test voltage to the level that such equipment can withstand without damage.

11.21.1.5 High leakage currents in cables might be due to improper preparation before cable terminations or splices were installed, which could allow high surface leakage while under test.

11.21.2 Low-Voltage Cables.

11.21.2.1 Low-Voltage Cable Testing Methods.

11.21.2.1.1 Insulation Resistance.

11.21.2.1.1.1 Perform an insulation resistance test on each conductor with respect to ground and adjacent conductors.

Apply 500 volts dc for 300-volt rated cable and 1000 volts dc for 600-volt rated cable with a test duration of 1 minute.

11.21.2.1.1.2 Insulation resistance values should be in accordance with manufacturer's published data.

11.21.2.1.2 Cable Continuity. Perform continuity tests to ensure correct cable connection.

11.21.2.1.3 Parallel Conductors. Verify uniform resistance of parallel conductors. Deviations in resistance between parallel conductors should be investigated.

11.21.2.2 3-Phase 4-Wire Neutral-Current Testing.

11.21.2.2.1 Situations exist where it is possible for the neutral current of 3-phase systems to exceed the ampacity of the neutral conductor in normal operation. This is usually due to unbalanced phase loading, nonsinusoidal load currents (harmonics), or a combination of the two.

11.21.2.2.2 There are certain conditions where even perfectly balanced loads result in significant neutral currents. Nonlinear loads, such as rectifiers, computers, variable-speed drives, electrical discharge lighting fixtures, and switching mode power supplies, cause phase currents that are not sinusoidal.

11.21.2.2.3 Symptoms of a nonsinusoidal condition might be overheating of the neutral conductor, deterioration of conductor insulation, carbonized insulation, and measurable voltage between the neutral and ground conductors (common-mode noise). This condition can cause a fire or malfunction of microprocessor-based equipment.

11.21.2.2.4 The neutral current problem can be detected using a true rms ammeter to measure the current flowing in the neutral conductor. The use of an average responding ammeter calibrated to read the rms value of a sine wave should not be used, because it will not yield valid results when used on nonsinusoidal waveforms. If the neutral current is found to be excessive, the current in each phase should be measured to determine if an abnormal condition exists. If excessive neutral current exists and the phase currents are not excessive, harmonic content is the most likely cause. A means of analyzing neutral current containing harmonic components is through the use of a wave or spectrum analyzer. Most analyzers on the market today have the ability to provide a direct readout of the harmonic's magnitude.

11.21.2.2.5 Verification should be made that the neutral is bonded to the grounding electrode conductor only at the service and at each separately derived source, where used.

11.21.3 Medium- and High-Voltage Cables (2.3 kV–138 kV).

11.21.3.1 Insulation Resistance. Measure the insulation resistance individually on each conductor with all other conductors and shields grounded. Apply dc voltage in accordance with the manufacturer's published data. Insulation resistance values should be in accordance with the manufacturer's published data. In the absence of manufacturer's published data, use Table 11.21.3.1(a). See Table 11.21.3.1(b) for temperature correction factors.

11.21.3.2 Shield Continuity. Measure the shield continuity on each power cable shield. Shielding should exhibit continuity. Investigate resistance values in excess of 10 ohms per 1000 ft of cable.

Table 11.21.3.1(a) Insulation Resistance Test Values Electrical Apparatus and Systems

Nominal Rating of Equipment (Volts)	Minimum Test Voltage (dc)	Recommended Minimum Insulation Resistance (Megohms)
250	500	25
600	1,000	100
1,000	1,000	100
2,500	1,000	500
5,000	2,500	1,000
8,000	2,500	2,000
15,000	2,500	5,000
25,000	5,000	20,000
34,500 and above	15,000	100,000

Notes: (1) Test results are dependent on the temperature of the insulating material and the humidity of the surrounding environment at the time of the test.

(2) Insulation resistance test data can be used to establish a trending pattern. Deviations from the baseline information permit evaluation of the insulation.

Source: ANSI/NETA MTS-2011, *Maintenance Testing Specifications for Electrical Power Distribution Equipment and Systems*, Table 100.1.

Table 11.21.3.1(b) Insulation Resistance Conversion Factors (20°C)

Temperature		Multiplier for Apparatus Containing Solid Insulation
°C	°F	
-10	14	0.25
-5	23	0.32
0	32	0.40
5	41	0.50
10	50	0.63
15	59	0.81
20	68	1.00
25	77	1.25
30	86	1.58
35	95	2.00
40	104	2.50
45	113	3.15
50	122	3.98
55	131	5.00
60	140	6.30
65	149	7.90
70	158	10.00
75	167	12.60
80	176	15.80
85	185	20.00
90	194	25.20
95	203	31.60
100	212	40.00
105	221	50.40
110	230	63.20

Source: ANSI/NETA MTS-2011, *Maintenance Testing Specifications for Electrical Power Distribution Equipment and Systems*, Table 100.14.

11.21.3.3 Medium-Voltage Cable Testing Methods. Medium-voltage cable systems are typically very reliable and provide years of service; however, like all electrical equipment, they begin to deteriorate from the first day of installation. There are many testing options for the field assessment of medium voltage cables. In addition to the many factors affecting test results there is not a consensus within industry as to the most reliable and repeatable testing method to employ, which in turn further complicates one's analysis of which field testing procedure should be utilized. The decision process as to which technology to utilize for field testing can be achieved by taking into account several factors. Considerations should be made as to the following:

(1) Cable performance and failure history
(2) Cable system components and composition
(3) System reliability requirements
(4) Availability of historical testing data and previous test results
(5) Impact to operations of a cable failure while under test

11.21.3.3.1 Cable Withstand Testing at Elevated Voltages.

11.21.3.3.1.1 Withstand Voltage Testing Direct Current (dc). A dc withstand voltage test, also commonly referred to as a dc high-potential or dc hipot test, consists of applying dc voltage across a cable at or above the dc equivalent of the 60 Hz operating crest voltage. This test can be applied either as a dielectric absorption test or a step-voltage test. A dc high-potential test is an appropriate method for an acceptance test. The dc withstand voltage test should not be applied to cables over 5 years old.

11.21.3.3.1.2 Withstand Voltage Testing Alternating Current (ac). An ac withstand voltage test at power line (60 Hz) frequencies stresses the cable insulation system by applying an ac voltage waveform similar to what the cable would experience in normal operation. It is also a similar test to the factory test on new cables.

11.21.3.3.1.3 Withstand Voltage Testing Very Low Frequency (VLF). A very low frequency (VLF) test is a withstand voltage test of cable system insulation. The test is similar to a dc withstand voltage (dc hipot) test except the VLF test unit provides an ac voltage in the frequency range from 0.01 Hz to 1.0 Hz. To successfully pass the test, the cable system must withstand the test voltage (usually 3 times rated voltage or less) for a specified time (usually 1 hour or less).

11.21.3.3.2 Cable Diagnostic Testing.

11.21.3.3.2.1 Dissipation Factor (Tan Delta) Testing. Also called loss angle or insulation power factor testing, dissipation factor (tan delta) testing is a diagnostic method of testing cables to determine the loss factor of the insulation material. Because the loss factor increases during the aging process of the cable, the dissipation factor measurement can be used as a diagnostic method to predict overall cable health. For cables, the dissipation factor of each conductor with respect to ground should be obtained, and a hot-collar test should be performed on each pothead or porcelain termination assembly.

11.21.3.3.2.2 Partial Discharge (PD) Testing in General. The insulation system of a medium-voltage distribution system has partial discharges into air, across surfaces, and through the insulating material. These discharges emit energy in various parts of the electromagnetic spectrum. Partial discharges appear as individual events of very short duration and are always accompanied by emissions of light, sound, and heat, as well as electromagnetic pulses, and often result in chemical reactions. The severity of these discharges is an indication of the health of the insulation system. PD testing measures these discharges off line or on line and compares them to a database of discharge signatures to determine the severity.

(A) Different Methods of PD Measuring. Some of the different methods of measuring PD are as follows:

(1) Radio frequency interference (RFI) detection: uses an RF sensor to measure PD pulses occurring in an insulation system.
(2) Electromagnetic detection: can be made with oscilloscopes combined with other detectors.
(3) Acoustical detection: uses an ultrasonic sensor to detect PD. With power cables this method is usually applied to terminations, joints, and cable sections that are accessible so that direct contact with the device under test can be achieved.
(4) Ultraviolet detection: uses a camera that "sees" discharges into air or across surfaces.

(B) PD Testing: Off Line. Some advantages of off-line PD testing are as follows:

(1) There is no direct interface with electrical components while under normal source power.
(2) PD sites can be more accurately located in an insulation system.
(3) Partial discharge inception voltage (PDIV) and partial discharge extinction voltage (PDEV) can be measured if a variable voltage source is used.
(4) PD characteristics can be obtained at different voltages, which can aid in the identification of certain types of defects.

(C) PD Testing: On Line. Some advantages of on-line PD testing are as follows:

(1) Cables are not disconnected.
(2) PD characteristics can be obtained while the equipment is under normal load connection, operation, and loading patterns, which can help identify certain types of defects.
(3) A system outage interrupting normal operations is not required.

11.22 Adjustable-Speed Drive Testing. Detailed test procedures should be obtained from the manufacturer. Adjustable-speed drives (ASDs) are frequently referred to by other names and acronyms, such as variable-frequency drives (VFDs) and adjustable-frequency drives (AFDs). The following are, at a minimum, routine tests that can be performed on an ASD:

(1) Measuring currents and voltages and checking for balance and proper levels
(2) Using an oscilloscope, checking firing signals for proper waveform
(3) Testing for proper output of printed circuit board power supplies
(4) Testing manual and automatic reference signals

> **CAUTION:** If an adjustable-speed drive has been deenergized for more than a year, the output voltage and frequency should be brought up very slowly (typically 10 percent of rated output voltage per 15 minutes), to avoid capacitor failure.

11.23 Switchgear and Switchboard Assemblies.

11.23.1 Field Testing in General.

11.23.1.1 Where field testing is required, it is recommended that a qualified field service team be employed and that instructions be followed as recommended by appropriate standard and manufacturer's instructions. If evaluation of the switchgear components indicates results that differ significantly from the recommended values, the switchgear component should be removed and repaired.

11.23.1.2 The applicable industry field evaluation standards include the following:

(1) ANSI/NETA ATS, *Standard for Acceptance Testing Specifications for Electrical Power Distribution Equipment and Systems*
(2) ANSI/NETA *Standard for Maintenance Testing Specifications for Electrical Power Distribution Equipment and Systems*
(3) ANSI/NEMA PB 2.1 *General Instructions for Proper Handling, Installation, Operation, and Maintenance of Dead Front Distribution Switchboards Rated 600 Volts or Less*

11.23.1.3 For further information on switchgear assemblies, see Section 15.2, Switchgear Assemblies.

11.23.2 Field Tests.

11.23.2.1 Insulation Resistance.

11.23.2.1.1 Measure insulation resistance on each bus section, phase-to-phase and phase-to-ground for one minute. Apply voltage in accordance with manufacturer's published data. For further information on insulation resistance testing, see 11.9.2.3, Insulation Resistance Testing.

11.23.2.1.2 Insulation resistance values of each bus section should be in accordance with manufacturer's published data.

11.23.2.2 Direct-Current (dc) Overpotential Test.

11.23.2.2.1 Measure leakage current on each bus section, phase to phase to ground and phase to phase with phases not under test grounded. Test voltage should be in accordance with manufacturer's published data. For further information on dc overpotential testing, see 11.9.2.6, Direct-Current (dc) Overpotential Testing.

11.23.2.2.2 If no evidence of distress or insulation failure is observed by the end of the total time of voltage application during the dc overpotential test, the bus section is considered to have passed the test.

11.23.2.3 Ground Resistance. Measure switchgear ground resistance. For further information on ground resistance, see Section 11.13.

11.23.3 System Function Tests.

11.23.3.1 It is the purpose of system function tests to prove the correct interaction of all sensing, processing, and action devices.

11.23.3.2 Perform system function tests upon completion of the acceptance or maintenance tests on specified equipment.

11.23.3.3 Develop test parameters and perform tests for the purpose of evaluating performance of all integral components and their functioning as a complete unit within design requirements and manufacturer's published data.

11.23.3.4 Verify the correct operation of all interlock safety devices for fail-safe functions in addition to design function.

11.23.3.5 Verify the correct operation of all sensing devices, alarms, and indicating devices.

11.24 Surge Arresters.

11.24.1 The indicated testing and protocol is to verify system integrity and that when the surge protective device is triggered the discharge path to the electric supply electrode system has minimal resistance.

11.24.1.1 Visual and electrical testing of the discharge path from the surge protective device to the electrical supply electrode system verifies the low-resistance path had been maintained.

11.24.1.2 Commissioning of the surge protective device will establish a baseline for future comparison with follow-up testing for subsequent evaluation of the test results over the life of the arrester-discharge path assembly.

11.24.1.3 This testing is not intended to evaluate whether the arrester will function during a surge energy event, only that if triggered an adequate arrester-electrode system path exists.

11.24.2 Low-Voltage Surge Arresters (Surge Protection Devices).

11.24.2.1 Field Testing in General.

11.24.2.1.1 Where field testing is required, it is recommended that a qualified field service team be employed and that instructions be followed as recommended by appropriate standard and manufacturer's instructions. If evaluation of the arrester indicates results that differ significantly from the recommended values, the arrester should be removed.

11.24.2.1.2 The applicable industry field evaluation standards include the following:

(1) ANSI/NETA ATS, *Standard for Acceptance Testing Specifications for Electrical Power Distribution Equipment and Systems*
(2) ANSI/NETA MTS, *Standard for Maintenance Testing Specifications for Electrical Power Distribution Equipment and Systems*

11.24.2.1.3 For further information on surge arresters, see 15.9.2, Surge Arresters.

11.24.2.2 Field Tests.

11.24.2.2.1 Insulation Resistance.

11.24.2.2.1.1 Measure insulation resistance on each arrester, from the phase terminal to ground. Apply voltage in accordance with the manufacturer's published data. For further information on insulation resistance testing, see 11.9.2.3, Insulation Resistance Testing.

11.24.2.2.1.2 Insulation resistance values should be in accordance with manufacturer's published data.

11.24.2.2.2 Grounding Connection.

11.24.2.2.2.1 Measure grounding connection. For further information on grounding, see Section 11.13.

11.24.2.2.2.2 Resistance between the arrester ground terminal and the ground system should be less than 0.5 ohm.

11.24.3 Medium- and High-Voltage Surge Arresters (Surge Protection Devices).

11.24.3.1 Field Testing in General.

11.24.3.1.1 Where field testing is required, it is recommended that a qualified field service team be employed and that instructions be followed as recommended by appropriate standard and manufacturer's instructions. If evaluation of the arrester indicates results that differ significantly from the recommended values, the arrester should be removed.

11.24.3.1.2 The applicable industry field evaluation standards include the following:

(1) ANSI/NETA ATS, *Standard for Acceptance Testing Specifications for Electrical Power Distribution Equipment and Systems*
(2) ANSI/NETA MTS, *Standard for Maintenance Testing Specifications for Electrical Power Distribution Equipment and Systems*

11.24.3.1.3 For further information on surge arresters, see 15.9.2, Surge Arresters.

11.24.3.2 Field Tests.

11.24.3.2.1 Insulation Resistance.

11.24.3.2.1.1 Measure insulation resistance on each arrester, from the phase terminal to ground. Apply voltage in accordance with manufacturer's published data. For further information on insulation resistance testing, see 11.9.2.3, Insulation Resistance Testing.

11.24.3.2.1.2 Insulation resistance values should be in accordance with manufacturer's published data.

11.24.3.2.2 Watts-Loss/Milliwatts-Loss Test.

11.24.3.2.2.1 Measure arrester watts-loss or milliwatts-loss.

11.24.3.2.2.2 Watts-loss or milliwatts-loss values are evaluated on a comparison basis with similar units and test equipment manufacturer's published data.

11.24.3.2.3 Grounding Connection.

11.24.3.2.3.1 Measure grounding connection. For further information on grounding, see Section 11.13.

11.24.3.2.3.2 Resistance between the arrester ground terminal and the ground system should be less than 0.5 ohm.

11.25 Power Factor Correction Capacitors.

11.25.1 Field Testing in General.

11.25.1.1 The applicable industry field evaluation standards include the following:

(1) ANSI/NETA ATS, *Standard for Acceptance Testing Specifications for Electrical Power Distribution Equipment and Systems*
(2) ANSI/NETA MTS, *Standard for Maintenance Testing Specifications for Electrical Power Distribution Equipment and Systems*
(3) *NFPA 70, National Electrical Code,* Article 460

11.25.1.2 For further information on capacitors, see 15.9.3, Capacitors.

11.25.2 Field Tests.

11.25.2.1 Insulation Resistance.

11.25.2.1.1 Measure insulation resistance on each capacitor, from the phase terminal to ground. Apply voltage in accordance with manufacturer's published data. For further information on insulation resistance testing, see 11.9.2.3, Insulation Resistance Testing.

11.25.2.1.2 Insulation resistance values should be in accordance with manufacturer's published data.

11.25.2.2 Capacitance. Measure capacitance of all terminal combinations. Investigate capacitance values differing from manufacturer's published data.

11.25.2.3 Capacitance Discharge Resistor.

11.25.2.3.1 Measure the internal capacitor discharge resistor resistance. Investigate discharge resistance values that differ from manufacturer's published data.

11.25.2.3.2 In accordance with *NFPA 70, National Electrical Code,* Article 460, residual voltage of a capacitor should be reduced to 50 volts in 1 minute for capacitors rated 600 volts and less and 5 minutes for capacitors rated greater than 600 volts after being disconnected from the source of supply.

11.26 Emergency Systems.

11.26.1 Automatic Transfer Switches.

11.26.1.1 Field Testing in General.

11.26.1.1.1 Where field testing is required, it is recommended that a qualified field service team be employed and that instructions be followed as recommended by appropriate standard and manufacturer's instructions. If evaluation of the arrester indicates results that differ significantly from the recommended values, the arrester should be removed.

11.26.1.1.2 The applicable industry field evaluation standards include the following:

(1) ANSI/NETA ATS, *Standard for Acceptance Testing Specifications for Electrical Power Distribution Equipment and Systems*
(2) ANSI/NETA MTS, *Standard for Maintenance Testing Specifications for Electrical Power Distribution Equipment and Systems*

11.26.2 Field Tests.

11.26.2.1 Contact/Pole Resistance or Millivolt Drop Tests. Measure contact/pole resistance or millivolt drop. These tests are used to test the quality of the contacts. The contact resistance or millivolt drop should be kept as low as possible to reduce power losses at the contacts with the resultant localized heating, which will shorten the life of both the contacts and nearby insulation. Microhm or millivolt drop values should not exceed the high levels of the normal range as indicated in the manufacturer's published data.

11.26.2.2 Automatic Transfer Tests. Perform the following tests:

(1) Simulate loss of normal power.
(2) Return to normal power.
(3) Simulate loss of emergency power.
(4) Simulate all forms of single-phase conditions.

Automatic transfers should operate in accordance with manufacturer's and/or system design requirements.

11.26.2.3 Timers and Relays. Verify correct operation and timing of the following functions:

(1) Normal source voltage-sensing relay
(2) Engine start sequence
(3) Time delay upon transfer
(4) Alternate source voltage-sensing relay
(5) Interlocks and limit switch functions
(6) Time delay and retransfer upon normal power restoration
(7) Engine cool down and shutdown feature

Operation and timing should be in accordance with manufacturer's and/or system design requirements.

11.27 Test or Calibration Decal System.

11.27.1 General. After equipment testing, device testing, or calibration, a decal on equipment, in conjunction with test records, can communicate the condition of electrical equipment to maintenance and service personnel. This can be important for assessing the hazard identification and risk assessment for electrical safety procedures as well as the condition of electrical equipment.

11.27.2 Decal. After a piece of electrical equipment or device is tested and/or calibrated, a color-coded decal should be attached on the exterior enclosure to that particular equipment. The decal should include the following:

(1) Date of test or calibration
(2) Person or outside company who performed the testing or calibration
(3) Color coding indicating the service classification as described in 11.27.3

11.27.3 Service Classifications and Related Decal Color Codes. The test or calibration decal system has a color code that communicates one of three service classifications in 11.27.3.1 through 11.27.3.3. *(See Figure 11.27.3.)*

11.27.3.1 White Decal: Serviceable. If a device passes all tests satisfactorily and has met the requirements of the testing specifications, then a white decal should be attached to the device. This indicates that the device is electrically and mechanically sound and acceptable for return to service. There could be some minor deficiencies with the equipment but none that affect the equipment electrically or mechanically to any large degree. Examples of deficiencies include evidence of slight corrosion, incorrect circuit ID, and nameplate missing.

11.27.3.2 Yellow Decal: Limited Service. If the device under test has a minor problem that is not detrimental to the protective operation or major design characteristics of that particular device, then a yellow "Limited Service" decal should be attached to the device. Examples of limited service classifications include indicating trip targets that don't function properly, slightly lower than acceptable insulation resistance readings, and chipped arc chute.

11.27.3.3 Red Decal: Nonserviceable. If the device under test has a problem that is detrimental to the proper electrical or mechanical operation of that device, then a red "Nonserviceable" decal should be attached to the device. The nonserviceable decal would be attached to the device after attempts at field repair were made. Examples of nonserviceable classifications include no trip on one or more phases, low insulation resistance readings, mechanical trip problems, and high contact resistance readings. In addition, management or the owner should be advised of this condition.

Chapter 12 Maintenance of Electrical Equipment Subject to Long Intervals Between Shutdowns

12.1 Introduction.

12.1.1 Due to the more extensive and costly damage possible from electrical failures in continuous-process operations, plus the longer intervals between shutdowns, more thorough and comprehensive maintenance procedures are recommended. The need for and frequency of inspection and maintenance are determined by the effect on safety, plant operation, and severity of service.

12.1.1.1 The primary effects of electrical failure or malfunction are those directly associated with the failure and usually involve damage to electrical equipment. The secondary effects are those associated with the process or product. Damages resulting from secondary effects can be much more extensive and, in some cases, catastrophic.

12.1.2 In addition to more intensive maintenance procedures, this chapter covers system design considerations insofar as they relate to safety and maintainability as well as first and future costs.

12.2 General Aspects of Maintaining Medium- and Low-Voltage Distribution Systems.

12.2.1 Unless an electrical distribution system is adequately engineered, designed, and constructed, it will not provide reliable service, no matter how good the maintenance program. Therefore, the recommendations of 12.2.1.1 through 12.2.1.5 are much more essential for electrical distribution systems that supply production equipment that can operate for long periods between shutdowns.

12.2.1.1 Careful planning in the engineering and design stages permits maintenance work without load interruptions. Alternative electrical equipment and circuits should be provided to permit routine or emergency maintenance on one while the other supplies the load that cannot be shut down. For instance, automatic or manual transfer equipment permits the load to be switched, with minimal interruption, from a source of a circuit that fails to one that is operating.

12.2.1.2 High-quality equipment has sufficient capacity and features that permit reasonable inspection of the energized parts while in operation without hazard to an inspector using proper precautions. Viewing windows or expanded metal guards inside hinged doors provide a safe means for inspecting energized components inside enclosures. Complete barriers between adjacent switch and breaker sections, and so on, allow personnel to work safely inside a deenergized compartment while adjacent ones are energized. Close inspection of the equipment before shipment is the best way to certify compliance with specifications.

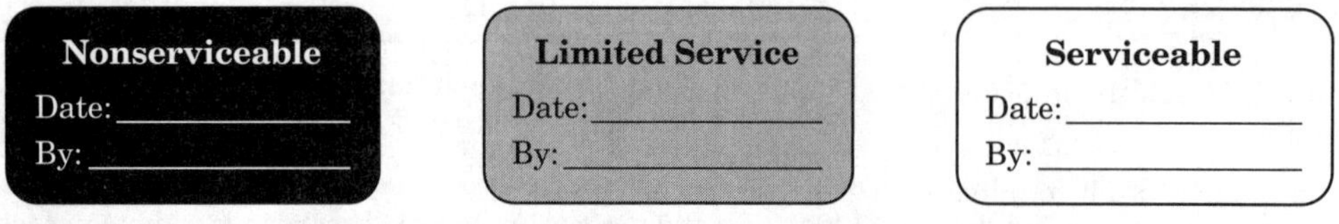

FIGURE 11.27.3 Examples of Service Classification Color Decals.

12.2.1.3 Construction specifications should be strictly adhered to, complete with detailed drawings and installation procedures.

12.2.1.4 Close scrutiny during all phases of construction is essential to ensure adequate quality workmanship and that cables, insulating materials, and other components are not damaged by poor practices.

12.2.1.5 Acceptance testing (in accordance with applicable recognized standards), including functional testing and inspection, is invaluable in detecting equipment that is defective, badly damaged, or installed in an inferior manner. In addition, reinspection and retesting within 1 or 2 years after energization might reveal conditions that can lead to in-service failures.

12.2.2 After the prerequisites in 12.2.1 are satisfied, an adequate electrical preventive maintenance (EPM) program helps keep the system in good condition and provides the necessary reliability over a long period.

12.2.3 Maintenance, inspection, and test methods for equipment that can operate for long periods are essentially the same as for equipment that might be shut down frequently. However, the recommended work should be performed with more care and diligence to obtain the desired reliability for service to loads that can operate continuously for months or years.

12.2.4 The following should effect an adequate EPM program for reliable long-term operation of an electrical power system:

(1) Good knowledge of the entire power system by all associated personnel, with posted or readily available diagrams, procedures, and precautions as highly beneficial aids in keeping personnel up to date
(2) General understanding of the loads served and their electrical quality and continuity of service requirements
(3) Length of time between scheduled maintenance shutdowns for utilization equipment, process changes, and so on, that will influence the length of intervals between electrical power system maintenance shutdowns
(4) A complete list of all the electrical system equipment associated with a given process or manufacturing system to ensure that all of it is maintained during one shutdown instead of being done piecemeal, which would require additional shutdowns
(5) The amount of time during the utilization equipment shutdown when the electrical power system can be deenergized for EPM
(6) Knowledge of electrical power system components, including operating and maintenance data, which are often included in the manufacturers' maintenance instructions
(7) Knowledge of ambient conditions, such as heat, moisture, and vibration, that can affect the equipment
(8) Ability to recognize abnormal conditions and early evidence of potential problems, such as overheating and surface tracking on insulating materials, that can cause failure if not corrected in sufficient time
(9) Standardized maintenance procedures shown in other portions of the text, modified by the information in 12.2.4(1) through 12.2.4(8) and knowledge gained through experience
(10) Knowledge of services available from local, area, and national electrical maintenance contractors that have specialized test equipment and highly qualified personnel who routinely perform this work, such as relay calibration and testing, circuit-breaker overcurrent trip-device calibration and testing, high-potential testing, power-factor testing, insulating-liquid testing and reconditioning, switchgear maintenance and testing, maintenance and testing of solid-state devices, and infrared inspection

12.2.4.1 Unless the amount of specialized work is sufficient to keep plant electrical maintenance personnel adept in the performance of such work, the use of specialized electrical maintenance contractors should be considered.

12.2.4.2 Plant maintenance supervision should have sufficient electrical knowledge to decide with the contractor on the recommended work to be done and to closely follow the contractor's performance to ensure full compliance. Merely telling a contractor to maintain or test the equipment usually creates a false sense of security that can be shattered by a serious failure caused by inadequate or incorrect maintenance procedures. The result is often the same when plant supervision does not sufficiently instruct plant maintenance personnel.

12.2.5 When a piece of equipment or a component fails, merely making repairs or replacement is not sufficient. A complete analysis should be made to determine the cause and to formulate corrective action to prevent recurrence in the same and similar equipment.

12.2.5.1 Following is a list of equipment for which maintenance, inspection, and testing guide tables are located in Annex K. The material contained in those tables is of a general nature and might have to be revised to conform more closely to the equipment being maintained to ensure the coverage necessary for the required reliability. Experience has indicated that the frequencies of maintenance, and so on, shown in the tables are sufficient for most installations. They might have to be tailored to suit installations where the ambient conditions are more or less severe.

(1) Medium-voltage equipment (over 1000 volts)
  (a) Cables, terminations, and connections
  (b) Liquid-filled transformers
  (c) Dry-type transformers
  (d) Metal-clad switchgear
  (e) Circuit breakers
  (f) Metal-enclosed switches
  (g) Buses and bus ducts
  (h) Protective relays
  (i) Automatic transfer control equipment
  (j) Fuses
  (k) Lightning arresters
(2) Medium- and low-voltage equipment: outside overhead electric lines
(3) Low-voltage equipment (below 1000 volts)
  (a) Low-voltage cables and connections
  (b) Dry-type transformers
  (c) Switchgear
  (d) Drawout-type circuit breakers
  (e) Buses and bus ducts
  (f) Panelboards
  (g) Protective relays
  (h) Automatic transfer control equipment
  (i) Circuit-breaker overcurrent trip devices
  (j) Fuses
  (k) Lightning arresters

12.3 Utilization.

12.3.1 General.

12.3.1.1 The utilization of electrical energy in industry is the conversion of electrical energy into useful work such as mechanical operations, lighting, and heating. Of primary concern is the maintenance of the many kinds of utilization equipment used with processes that operate for long intervals between shutdowns. Utilization equipment as covered here is considered to operate at 480 volts and less.

12.3.1.2 Chapters 6 and 8 make reference to the need for planning and developing an EPM program and describe some of the fundamentals. Utilization equipment that serves equipment that operates for long intervals between shutdowns should receive special consideration. The serviceability and safety of the equipment should be thoroughly studied. During the initial design stages, thought should be given to EPM, with ease of maintenance and accessibility being of extreme importance in the design considerations and emphasis on access for adequate visual and infrared inspection of all bus bars and joints.

12.3.1.3 Maintenance personnel who are going to service the equipment should be consulted during the design phases.

12.3.2 Records and Inspection Tours.

12.3.2.1 Keeping records on utilization equipment that operates over long intervals is more important than for short-interval equipment. Wiring changes, parts replacement, and other modifications should all be accurately recorded.

12.3.2.2 Schedules should be laid out for periodic inspection tours of utilization equipment. Records of findings on these inspection tours help indicate trends. Another important reason for good recordkeeping is that personnel often change, and it is necessary for current personnel to know what has been done prior to their involvement.

12.3.2.3 Power and lighting panel directories should be kept up to date and accurate.

12.3.3 Power-Distribution Panels.

12.3.3.1 Power-distribution panels are either fuse or circuit-breaker type. Where critical circuits are involved, power-distribution panels should be appropriately identified by tags, labels, or color coding.

12.3.3.2 Seldom are power panels deenergized, and then only for circuit changes; it is for those times that EPM can be scheduled. Although procedures can be developed for working on them live, it is not recommended because of the safety hazards involved. There is always the possibility of an error or accidental tripping of a main breaker causing an unscheduled shutdown. During operating periods, the panels can be checked only for hot spots or excessive heat. This EPM should be done at reasonable intervals in accordance with the importance of the circuit. A record should be made of areas that have given trouble; memory should not be relied on.

12.3.3.3 During a shutdown and while the panel is dead, all bolted connections should be checked for tightness and visually inspected for discoloration. Should there be discoloration, further investigation should be made and possibly the parts affected replaced. *(For further information, refer to Chapters 17 and 18.)*

12.3.4 Lighting Panels. Lighting panels generally have the same problems as power panels. However, experience indicates an increased probability of circuit overloading and thus protective-device overheating. Since such panels applied in long-term maintenance areas usually feed important circuits, overheating problems should be corrected immediately.

12.3.5 Plug-in-Type Bus Duct. Since plug-in bus duct is seldom used in long-term areas, maintenance of this equipment is not covered here. *(See Chapters 16, 17, and 18 for related information.)*

12.3.6 Wiring to Utilization Equipment. Maintenance procedures outlined in Chapter 19 are recommended. Visual inspection intervals should be based on the importance of the circuits and on previous experience. In addition, more extensive insulation testing might be warranted during shutdown periods to ensure higher reliability.

12.3.7 Rotating Equipment.

12.3.7.1 Maintenance. Proper maintenance of electric motors and rotating equipment is essential to prevent unscheduled downtime. Their most trouble-prone parts are bearings. The quantity of lubricant, the frequency of lubrication, the method of application, and the type of lubricant are of prime concern. Although lubrication of rotating equipment is discussed in Chapter 25, it is important enough with equipment that operates for long periods between shutdowns, especially motors, that further mention is made here. Suggestions for both oil and grease lubrication systems are listed in 12.3.7.2 through 12.3.7.8.

12.3.7.2 Grease Lubrication Systems. Grease is the most common lubricant used for electric motor bearings. It provides a good seal against the entrance of dirt into the bearing, has good stability, is easy to apply, and is easy to contain without elaborate seals. For extended service intervals, an extremely stable grease is required. Grease should be selected on the basis of the expected temperature range of service. The motor manufacturer can provide advice on exactly which grease to use. A grease that is compatible with the grease already in the bearing should be used.

12.3.7.3 Regreasing.

12.3.7.3.1 The correct quantity of lubricant in a rolling contact bearing is vital to its proper operation. Both insufficient and excessive lubrication will result in failure. Excessive lubrication can cause motor failure due to migration of grease into the motor winding. Table 12.3.7.3.1 is a guide to determining regreasing intervals by the type, size, and service of the motor, to obtain the most efficient operation and the longest bearing life. Where a variety of motor sizes, speeds, and types of service are involved in a single plant, a uniform relubrication period is sometimes selected. A yearly basis is common, for instance, and such a yearly regreasing might conveniently be carried out on a plantwide basis during a vacation shutdown.

12.3.7.3.2 Motors equipped with grease fittings and relief plugs should be relubricated by a low-pressure grease gun using the following procedure:

(1) The pressure-gun fitting and the regions around the motor grease fittings should be wiped clean.
(2) The relief plug should be removed and the relief hole freed of any hardened grease.
(3) Grease should be added with the motor at standstill until new grease is expelled through the relief hole. In a great majority of cases, it is not necessary to stop the motor during relubrication, but regreasing at standstill will minimize the possibility for grease leakage along the shaft seals.
(4) The motor should be run for about 10 minutes with the relief plug removed to expel excess grease.
(5) The relief plug should be cleaned and replaced.

12.3.7.4 Regreasing of Totally Enclosed, Fan-Cooled (TEFC) Motors. For TEFC motors, the instructions in 12.3.7.3.2 apply for greasing the drive-end bearing. The fan-end housing frequently is equipped with a removable grease relief pipe that extends to the outside of the fan casing. First, the pipe should be removed,

Table 12.3.7.3.1 Guide for Maximum Regreasing Intervals

	Motor Regreasing Intervals			
Type of Service	**Up to 7½ hp**	**10–40 hp**	**50–150 hp**	**Over 150 hp**
Easy: infrequent operation (1 hr per day), valves, door openers, portable floor sanders	10 years	7 years	4 years	1 year
Standard: 1- or 2-shift operation, machine tools, air-conditioning apparatus, conveyors, garage compressors, refrigeration apparatus, laundry machinery, textile machinery, wood-working machines, water pumping	7 years	4 years	1½ years	6 months
Severe: motors, fans, pumps, motor generator sets running 24 hr per day, 365 days per year; coal and mining machinery; motors subject to severe vibration; steel-mill service	4 years	1½ years	9 months	3 months
Very severe: dirty, vibrating applications, where end of shaft is hot (pumps and fans), high ambient	9 months	4 months	3 months	2 months

cleaned, and replaced. Next, during the addition of new grease from a grease gun, the relief pipe should be removed several times until grease is observed in the pipe. After grease is observed to have been pushed out into this pipe, no more should be added. The pipe, after again being cleaned and replaced, will then act as a sump to catch excess grease when expansion takes place during subsequent operation of the motor.

12.3.7.4.1 In many vertical motors, the ball-bearing housing itself is relatively inaccessible. In such cases, a grease relief pipe is frequently used in a manner similar to that in the TEFC motors. The same regreasing procedures should be used as described in 12.3.7.4 for TEFC motors.

12.3.7.4.2 Motors with sealed bearings cannot be relubricated.

12.3.7.5 Regreasing of Small Motors. In many small motors, no grease fittings are used. Such motors should be relubricated by removing the end shields, cleaning the grease cavity, and refilling three-quarters of the circumference of the cavity with the proper grade of grease. In the end shields of some small motors, threaded plugs are provided that are replaceable with grease fittings for regreasing without disassembly.

12.3.7.5.1 Because regreasing of motor bearings tends to purge the old grease, a more extensive removal of all the used grease is seldom necessary. Whenever a motor is disassembled for general cleaning, however, the bearings and housing should be cleaned by washing with a grease-dissolving solvent. To minimize the chance of damaging the bearings, they normally should not be removed from their shaft for such a washing. After thorough drying, each bearing and its housing cavity should be filled approximately one-half to three-fourths full with new grease before reassembly. Spinning the bearing with an air hose during cleaning should be avoided. Any bearing that has been removed from the shaft by pulling on the outer ring should not be reused.

12.3.7.6 Oil Lubrication Systems. Oil lubrication is recommended when a motor is equipped with sleeve bearings. It is sometimes used for roller contact bearings under certain conditions.

12.3.7.6.1 Oils for lubricating electrical motors should be high-quality circulating oils with rust and oxidation inhibitors.

12.3.7.6.2 The oil viscosity required for optimum operation of motor bearings is determined by the motor speed and the operating temperature.

12.3.7.6.3 In general, it is recommended that 150 S.U.S. oil be used for motor speeds above 1500 rpm, and 300 S.U.S. oil be used for motor speeds below 1500 rpm. These recommendations might vary with specific application and, in particular, with the ambient temperature to which the motor or generator is exposed. The motor manufacturer's recommendations relative to oil viscosity should be followed.

12.3.7.7 Methods and Quantity.

12.3.7.7.1 Wick Oiling. Fractional horsepower motors that can be relubricated generally use felt, waste, or yarn packing to feed sleeve bearings. The packing should be saturated at each lubrication interval.

12.3.7.7.2 Ring Oiling. Integral horsepower motors can have ring-lubricated sleeve bearings. The rings are located in a slot in the upper half of the bearing and ride loosely on the shaft. Normally there are no more than two rings for each bearing. Free turning of the rings should be checked on starting a new motor, at each inspection period, and after maintenance work. The oil level should be such that a 60-degree segment of the oil ring on the inside diameter is immersed while the motor shaft is at rest, as shown in Figure 12.3.7.7.2. A sight glass, constant level oiler, or some other unit should be provided to mark and observe the oil level. Levels should be marked for the at-rest condition and the operating condition.

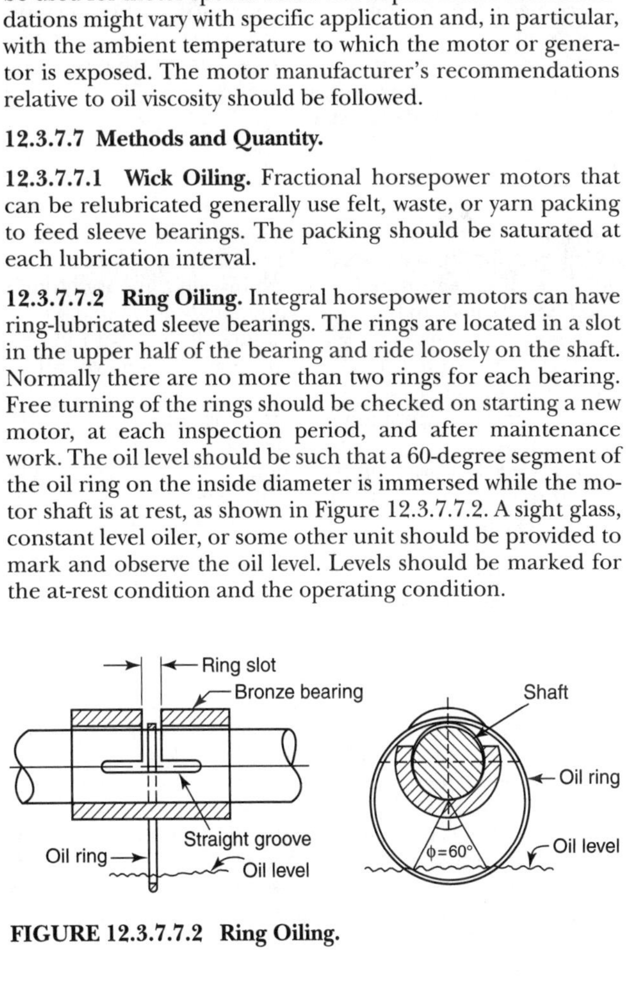

FIGURE 12.3.7.7.2 Ring Oiling.

12.3.7.7.3 Bath Oiling. Large, vertical motors frequently have a surrounding oil bath for lubrication of either rolling-element bearings or plate-thrust bearings. Horizontal units equipped with ball and roller bearings might also have an oil bath. The proper oil level is determined by the manufacturer and depends on the bearing system. A sight glass or some other unit should be provided to mark and observe the oil level. This level can change depending on whether the motor is operating or at rest. It should be marked for both situations.

12.3.7.7.4 Oil-Mist Lubrication. Pressurized oil-mist systems are being increasingly used in refinery applications. These applications normally involve interlocked controls such that the source of mist pressure should be in operation to permit energization of the lubricated motor. Often a single centralized mist source supplies a number of motors. Maintenance should include checking of drain/discharge openings at each bearing to see that pressure can be discharged freely to the atmosphere and that the mist pressure–regulation equipment is functioning properly.

12.3.7.8 Frequency. In oil-lubricating systems, it is required that the oil level be maintained. The oil level is observed by means of a sight glass, constant level oiler, and so on, and oil added as needed. Normally, these systems should be drained and refilled on an annual basis. Wick-oil systems require addition of oil quarterly, and the wick should be saturated.

12.3.7.9 Motor Inspections. Visual inspections should be performed on a periodic basis. These inspections are necessary to detect mechanical or lubrication deficiencies before they become serious. The inspection should include a check for increase in temperature, excessive bearing noise, excessive vibration, and lubricant leakage. If any of these conditions exists, the cause should be located and corrected.

12.3.8 Vibration Tests and Analysis.

12.3.8.1 The life of a ball bearing or roller bearing is defined as the number of revolutions or hours of operation at constant speed that the bearing is capable of running before fatigue develops. If a bearing is properly lubricated, mounted, and handled, all causes of failure are eliminated except one, which is fatigue of the material. These failures initiate with the removal of metal from the races or rolling elements. Vibration-analyzing equipment can be used to predict these failures when it monitors vibration velocity or is able to distinguish vibration displacement as a function of frequency. Such equipment is useful in isolating the source of vibration that might appear to be the result of other malfunctions within a motor. It is also useful for ensuring proper installation of critical production equipment. Antifriction bearings fail due to a loss of oil film resulting from wear, leakage, and so on. These failures are sudden, and without constant vibration-monitoring equipment, they cannot be predicted.

12.3.8.2 Vibration analyzers are handy tools for detecting trouble and preventing downtime. A formal vibration analysis program can reduce costly machine failures. The program can range from the use of simple hand-held analyzers to sophisticated multichannel recorders with permanently mounted sensors to provide data for comparison. Such a program makes it possible to keep track of the condition of rotating equipment, particularly high-speed types. Trend charts assist in establishing maintenance needs. The degree of sophistication depends on the application, but even a hand-held vibrograph is a useful tool in EPM. *(See Chapter 26.)*

12.3.9 Dirt. Where rotating equipment is exposed to dirt, regular inspection is recommended to detect when cleaning is needed. A major cause of burned-out motors is clogged air passages. On motors in dirty atmospheres, filters (where used) frequently become clogged; therefore, filter cleaning or changing should be scheduled. The external surface of motors should be kept cleaned because a pileup of dirt restricts heat dissipation. This is particularly important with T-frame motors. *(See Chapter 25 for cleaning methods.)* In dirty locations and critical applications, more extensive insulation testing might be warranted, as described in Chapter 11. Excessive leakage current might well indicate that a motor failure is imminent.

12.3.10 Control for Rotating Equipment.

12.3.10.1 This involves the motor starters, contactors, and other devices that are directly involved with the control of equipment operating over long periods between shutdowns. The maintenance recommendations in Chapter 16 are pertinent to equipment operating for long periods between shutdowns.

12.3.10.2 While the equipment is in operation, EPM procedures should be modified. Where control panels can be opened while energized, any terminals with a voltage greater than 150 volts to ground should be covered with a transparent protective covering to permit visual inspection. Essentially, EPM is limited to visual inspection. Maintenance of adequate ventilation should be ensured within enclosures. Gaskets should be kept in good repair where used and where the atmosphere is dirty. Contact wear should be observed where possible.

12.3.11 Redundancy. Although it is expensive, redundant circuits and equipment often are necessary to ensure continuity of operation. During initial design stages and even at later times, consideration should be given to what is needed to prevent unscheduled shutdowns and high maintenance costs. Frequently, redundancy of critical circuits provides the solution.

12.3.12 Heating Equipment.

12.3.12.1 In general, this equipment cannot be maintained while it is in operation. Perhaps rotating parts are not involved, but certainly there is heat, and the potential for serious burns therefore exists.

12.3.12.2 In most process-heating systems, continuous cycling or on-off operation is carried out. Cycling causes a certain amount of temperature change. As a result, particular attention should be paid to all connections and joints. The use of Belleville washers has been successful in maintaining tight connections. During the time the equipment is in operation, all joints and terminations should be visually inspected and looked at for signs of heating or arcing, which would indicate loose joints. The cycling frequently causes some movement of the wiring; therefore, check the insulation on the wiring where it passes through nipples, access holes, and other openings.

12.3.13 Electrostatic Discharge (ESD) Grounding.

12.3.13.1 General. The purpose of ESD grounding is to remove the accumulation of static electricity that can build up during machine operation on equipment, on materials being handled or processed, or on operating personnel. On equipment that is in continuous operation, regular inspection and repair procedures should be developed and maintained to retain the integrity of the grounding path.

12.3.13.1.1 Because the static charge can build up to several thousand volts, consideration should be given during the initial construction of equipment to reduce the buildup. Equipment is made up of conductors (metal machine frame) and insulators (conveyor belts, plastic parts, and so on). Usually some part of a machine is grounded either electrically or by virtue of construction. Machine parts can be grounded directly or by bonding them to other machine parts that are grounded. Clean, unpainted metal nuts and bolts holding together clean, unpainted metal parts provide adequate continuity. Bonding and grounding can be accomplished by permanently attached jumper wires. Where such wires are attached by lugs or placed under bolt heads or nuts, all parts should be clean and unpainted before installation. Any painting of parts used for ESD grounding should be done only after such parts have been properly installed and the adequacy of the ground verified. Slowly rotating parts normally are adequately bonded or grounded through the bearings. However, parts rotating at high RPMs, such as baskets or centrifuges, should be bonded or grounded by wipers, carbon brushes, or other devices. Portable equipment can be temporarily grounded by clamping an ESD ground to the equipment.

12.3.13.2 Adequate ESD Grounding. It might be necessary to obtain the recommendations of experts in a particular ESD grounding problem. However, some guidelines that will provide adequate ESD grounding are provided in 12.3.13.2.1 through 12.3.13.2.5.

12.3.13.2.1 Static electrical charging currents rarely exceed 1 microampere and often are smaller. Thus, leakage currents of the order of microamperes provide protection against the accumulation of static electricity to dangerously high potentials.

12.3.13.2.2 A leakage resistance between a conductor and ground as high as 10,000 megohms provides adequate ESD grounding in many cases. However, when charges are generated rapidly, a leakage resistance as low as 1 megohm might be necessary.

12.3.13.2.3 The leakage resistance necessary for adequate ESD grounding varies among different operations and should be established by a qualified authority. In the absence of any specifications, the leakage resistance from any conductor to ground should not exceed 1 megohm.

12.3.13.2.4 There is no electrical restriction in conductor size for ESD ground wires and jumpers, but larger size conductors might be necessary to limit physical damage. However, where the equipment-grounding conductor for a power circuit is also used for ESD grounding, the conductor should be sized in accordance with Table 250.122 of *NFPA 70, National Electrical Code.* Any equipment-grounding conductor that is adequate for power circuits is more than adequate for static grounding.

12.3.13.2.5 An ESD ground wire need not be insulated.

12.3.13.3 Inspection and Maintenance. An inspection and maintenance program is essential in ensuring that the integrity of ESD grounding systems is retained. Inspections should consist of both resistance measurements and a visual check.

12.3.13.3.1 The resistance from all conductive parts to ground should be measured with a suitable megohmmeter (*see 12.3.13.5*). Corrective measures should be made to bring all resistance values within specifications.

12.3.13.3.2 A visual inspection should be made for frayed wires, wires with broken strands, and other physical damage. Such damage should be repaired regardless of measured resistance values.

12.3.13.4 Installations and Alterations. Inspections should be made of all new installations and wherever alterations are made to or parts replaced in an installation. Inspections should be made at regular intervals. The frequency of regular periodic inspections can be determined from experience. Inspections should be most frequent in areas where corrosion is a problem and in areas classified as hazardous.

12.3.13.5 Megohmmeters. A suitably calibrated resistance-measuring device having a nominal open-circuit output voltage of 500 volts dc and a short-circuit current not exceeding 5 mA should be used to check static grounding systems.

12.3.13.6 Hazardous Locations. If the inspections are made in hazardous (classified) locations, the area should be verified nonhazardous if the megohmmeter is not of an intrinsically safe type. The area should be verified as nonhazardous during the testing period when a megohmmeter is used.

12.3.13.7 Recordkeeping. Precise records should be made and retained of the results of all inspections and of the corrective actions taken. Precise records will aid in determining the necessary inspection frequency and point out weak spots in the static grounding system that might need modification.

12.3.13.8 Precautions During Inspections. If inspections and corrective measures have to be made when flammable vapors are apt to be present, certain precautions should be taken by the inspector and maintenance personnel, as follows:

(1) Care should be taken that personnel are adequately grounded to prevent a dangerous accumulation of static electricity on their bodies.
(2) Care should be taken that no spark discharge occurs between improperly grounded conductors and personnel, instrumentation, or tools.
(3) Only nonferrous, nonsparking tools should be used in the area.

12.3.13.9 Typical Checkpoints for Inspection. All conductors in a hazardous area should be inspected for adequate static grounding.

12.3.13.9.1 Since machines and operations differ considerably, a checklist should be prepared of all points to be checked.

12.3.13.9.2 The following are typical for many machines and operations:

(1) Permanently installed jumper wires
(2) Static ground wires and clamps used for the temporary grounding of portable and mobile equipment
(3) Metal hose couplings
(4) Metal hose clamps
(5) Metal bolts and nuts used to connect sections of either conductive or nonconductive pipes and ducts
(6) All sections of metal pipes and ducts
(7) Rotating parts and shafts
(8) Rotating baskets of centrifuges
(9) Handles and stems of ball valves and plug valves

12.3.13.9.3 All rotating parts should be checked for the accumulation of electrical charge while in motion.

12.4 Process Instrumentation and Control.

12.4.1 Introduction. The following systems and equipment are covered in this section: power supplies; interlock and logic systems; safety and shutdown systems; sensing, control, and indicating systems; and alarm systems.

12.4.2 Design to Accommodate Maintenance.

12.4.2.1 Section 8.1 of this recommended practice stated that, except for limited visual inspection such as observing operating temperatures, examination for contamination, recording load readings, and so on, the apparatus should be taken out of service for efficient and effective maintenance. Further, unless flexibility is built into the system in the way of duplication or alternative transfer schemes, maintenance of vital electrical apparatus should be scheduled with planned production outage.

12.4.2.2 The importance of identifying and designing for the vital elements of the process control system cannot be overstressed. The elements of the process instrumentation and control system that should be inspected, tested, or maintained while the plant or process remains in operation should be identified in the design stage. The necessary duplication of facilities and provision for test and inspection should be provided.

12.4.2.3 Examples of such provisions are alternative power sources to permit shutdown and inspection of normal power sources, bypass switches for inverters, provisions for on-stream function testing of shutdown circuits, provision of dual sensing components for critical controls, test circuits to permit simulation of alarm conditions, and monitoring devices for important interlock and logic systems. Selection of quality equipment was also mentioned in Section 8.1 as a means of reducing maintenance requirements. Again, the importance of long-run facilities cannot be overemphasized.

12.4.2.4 Whenever possible, control modules should be the plug-in type, replaceable with normal precautions and procedures. Test and adjustment of major components should be possible without disconnection or removal from enclosures and with use of standard instruments such as volt-ohm-milliammeter and oscilloscope.

12.4.2.5 Cabinets should be fully compartmented to allow maintenance access to sections not in service without risk to personnel or continuity of service. For instance, the inverter, standby transformer/voltage regulator, and transfer-switch power supply should be in physically separate compartments. Removal or replacement of components in one cabinet section should not require access to other sections.

12.4.3 Power Supplies.

12.4.3.1 Power supplies can be divided into two categories: power supplies normally in service and alternate (emergency or standby) power supplies. NFPA 110, *Standard for Emergency and Standby Power Systems*, and ANSI/IEEE 446, *Recommended Practice for Emergency and Standby Power systems for Industrial and Commercial Applications (IEEE Orange Book)* should be consulted because they cover installation, maintenance, operation, and testing requirements as they pertain to the performance of the emergency power supply system.

12.4.3.2 Power supplies that are normally in service should be inspected on a regular basis. This inspection should include the following typical checks and inspections:

(1) Reading of meters to detect changes in or abnormal load or voltage conditions
(2) Check of ground detection equipment for presence of grounds
(3) Integrity of trip and transfer circuits where monitoring lights are provided
(4) State of charge on batteries
(5) Battery charger supply and output load and voltages
(6) Visual inspection of accessible current-carrying parts for signs of overheating
(7) Check on equipment environment for heat, moisture, or dust that exceeds the conditions for which the equipment is designed

12.4.3.3 The inspection interval can be daily, weekly, or monthly, depending on equipment environment and operating conditions. Tasks such as reading of meters and checks on monitoring lights can be incorporated as part of a daily walk-through inspection.

12.4.3.4 Where redundancy in facilities is provided, equipment components should be taken out of service for a thorough inspection and testing and for any recommended maintenance at intervals dictated by service and operating conditions. The initial interval should be in line with manufacturers' recommendations and later shutdowns scheduled in line with the as-found condition of the equipment.

12.4.3.5 Where power supply components are in standby or emergency service, periodic testing should be carried out to ensure that the standby equipment is ready to function and can assume the supply function. This requires periodic start-up of emergency generators, operation of autotransfer switches, and so on. Testing should simulate actual operating conditions as closely as possible. For critical facilities, testing intervals such as once a week are suggested.

12.4.3.6 Where it is possible to put critical standby facilities in operation to supply the normal load without disturbing plant operations, the standby facilities should be switched in at regular intervals and operated for a sufficient period to ensure that they are functioning properly. An interval of once a month is suggested for operating standby facilities. Where standby facilities are fully rated, they are permitted to share operating time on an equal basis with the normal supply.

12.4.4 Interlock and Logic Systems.

12.4.4.1 Maintenance procedures on interlock and logic systems are limited to visual inspections of components and wiring and checks on monitoring devices unless design features permit onstream functional testing. Also, in some plants, the process operation or equipment arrangement permits periodic function testing.

12.4.4.2 Where functional testing can be done and where the system does not function during normal operations, once-a-week function testing is suggested for systems whose failure can result in hazard to personnel, fire, damage to equipment, or serious degradation or loss of product. Systems of lesser importance should be tested initially on a once-per-month basis with subsequent testing intervals determined by experience and assessment of operating environment.

12.4.5 Sensing, Indicating, and Control Systems.

12.4.5.1 The need for and frequency of inspection and maintenance are determined by the effect on safety, plant operations, and the severity of service. Also, some components can be readily isolated, while others can be inspected only during plant or process shutdowns.

12.4.5.2 Visual inspection either by plant operators during normal operations or as part of a scheduled inspection can assist in detection of deficiencies such as loose connections, overheating, and excessive vibration.

12.4.5.3 Sensing, indicating, and control devices can be divided into two categories: primary elements and secondary elements.

(1) Primary elements are elements in contact with the process medium directly or indirectly and that might or might not be isolated from the process medium.
(2) Secondary elements are transmitting, recording, or controlling devices. Some are normally in use and thus receive an automatic day-to-day check. Some are remotely located or infrequently used and require a check at regular intervals.

12.4.6 Level Devices.

12.4.6.1 Primary devices installed within process vessels can be checked only with the vessel out of service. Visual inspection should indicate need for maintenance.

12.4.6.2 Where the device can be isolated from the process, visual inspection should be made at least once a year and more frequently if extreme accuracy is needed or if the service is severe or critical.

12.4.7 Temperature Devices.

12.4.7.1 Primary devices are generally installed in wells and can be checked at any time the device appears to be malfunctioning. The well should be visually inspected at each plant shutdown and necessary maintenance carried out.

12.4.7.2 The secondary device or instrument usually can be checked at any time without seriously affecting normal operations.

12.4.8 Pressure Devices.

12.4.8.1 Primary devices usually have block valves to permit isolation from the process and checking any time malfunction is indicated.

12.4.8.2 Secondary devices usually can be isolated from the primary device and checked at any time.

12.4.8.3 Process impulse connections should be checked during equipment shutdown.

12.4.9 Indicating, Recording, and Controlling Signal Receivers. Checks are limited to day-to-day observation of performance by plant operators. Receiver construction usually permits substitution of spare units for faulty units.

12.4.10 Safety and Shutdown Systems.

12.4.10.1 On-line testing facilities for safety and shutdown systems should be provided in all designs. Where practical, the facilities should include multiple sensors and safe bypass systems around the final control element. This permits testing of the entire shutdown circuit.

12.4.10.2 Safety and shutdown circuits should be tested in the range of once-per-shift to once-per-week unless the circuit functions regularly in normal operation. This might be the case for some shutdown circuits.

12.4.10.3 Because of the frequency of testing, these functional tests might be part of the plant operators' normal duties with maintenance personnel involved only if problems are indicated.

12.4.11 Alarm Systems.

12.4.11.1 Alarm systems are usually equipped with lamp test switches that permit checking lamp and alarm circuit integrity at any time during normal operation. These tests should be made on a once-per-shift to once-per-day basis to detect lamp burnout or circuit defects in alarms that operate infrequently. This can be done as part of the plant operators' normal duties with maintenance personnel involved only if further attention is needed.

12.4.11.2 Alarms for critical conditions that can result in hazard to personnel, fire, equipment damage, or serious degradation or loss of product should be function tested at regular intervals. A once-per-week to once-per-month interval is suggested depending on the importance and vulnerability of the alarm devices to hostile environments. Function testing requires that either provision be made in the system design for the testing facilities or that it be possible to test by manipulating the process variable or otherwise simulating the alarm conditions.

12.4.12 Wiring Systems. These systems can be visually checked for loose connections, proper grounding and shielding, and signs of deterioration or corrosion. Usually maintenance during plant operation is limited to circuits that malfunction or show evidence of possible malfunction.

Chapter 13 Ground-Fault Protection

13.1 Introduction. Ground-fault protective devices intended to protect personnel or systems from ground-faults are of two distinct types: ground-fault circuit interrupters and ground-fault protection of equipment. IT IS EXTREMELY IMPORTANT TO UNDERSTAND THE DIFFERENCE BETWEEN THE TWO TYPES.

13.1.1 Ground-Fault Circuit Interrupter (GFCI). A GFCI is designed to protect a person from electrocution when contact between a live part of the protected circuit and ground causes current to flow through a person's body. A GFCI disconnects the circuit when a current equal to or higher than the calibration point (4 mA to 6 mA) flows through the protected circuit to ground. It does not eliminate the shock sensation, since normal perception level is approximately 0.5 mA. It does not protect from electrocution on line-to-line contact, because the nature of line-to-line loads cannot be distinguished.

13.1.2 Ground-Fault Protection of Equipment. There are two applications where ground-fault protection of equipment is intended to be used: where there might be excessive ground-fault leakage current from equipment and where equipment and conductors are to be protected from damage in the event of a higher-level ground fault (either solid or arcing). These types of protective equipment are for use only on ac, grounded circuits and cause the circuit to be disconnected when a current equal to or higher than its pickup setting or rating flows to ground. They are not designed to protect personnel from electrocution.

13.1.2.1 Equipment ground-fault protective devices are intended to operate on a condition of excessive ground-fault leakage current from equipment. The ground current pickup level of these devices is from above 6 mA to 50 mA.

13.1.2.2 Circuit breakers with equipment ground-fault protection are combination circuit breaker and equipment ground-fault protective devices designed to serve the dual function of providing overcurrent protection and ground-fault protection for equipment. The ground current pickup level of these breakers is typically 30 mA. They are intended to be used in accordance with Articles 426 and 427 of *NFPA 70, National Electrical Code.*

13.1.2.3 Ground-fault sensing and relaying equipment is intended to provide ground-fault protection of equipment at services and feeders. They are rated for ground current pickup levels from 4 amperes to 1200 amperes.

13.1.3 Definitions. The following definitions are unique to this chapter.

13.1.3.1 Ground-Fault Circuit Interrupter (GFCI). A device intended for the protection of personnel that functions to deenergize a circuit or portion thereof within an established period of time when a current to ground exceeds the values established for a Class A device. Note: Class A GFCIs trip when the current to ground has a value in the range of 4 mA to 6 mA. *(For further information, see UL 943, Standard for Ground-Fault Circuit Interrupters.)* [**70,** 2008]

13.1.3.2 Ground-Fault Protection of Equipment (GFP). A system intended to provide protection of equipment from damaging line-to-ground fault currents by operating to cause a disconnecting means to open all ungrounded conductors of the faulted circuit. This protection is provided at current levels less than those required to protect conductors from damage through the operation of a supply circuit overcurrent device. [**70,** 2008]

13.2 Ground-Fault Protective Equipment for Excessive Leakage Currents.

13.2.1 Equipment Ground-Fault Protective Devices. These are typically cord and plug-connected devices. Recommended maintenance is that specified in Section 24.2 and Section 29.4.

13.2.2 Circuit Breakers with Equipment Ground-Fault Protection. Recommended maintenance is the same as that specified in Chapter 17 for molded-case circuit breakers.

13.2.3 Maintenance.

13.2.3.1 The devices are sealed at the factory, and maintenance should be limited to that described in 13.2.3.2 through 13.2.3.6 or as recommended by the manufacturer.

13.2.3.2 In addition to the maintenance specified for the individual types of GFCIs, tripping tests should be performed with the test button on the unit in accordance with the frequency recommended by the manufacturer. Results and dates of tests should be recorded on the test record label or card supplied with each permanently installed GFCI unit.

13.2.3.3 GFCIs are equipped with an integral test means for checking the tripping operation.

13.2.3.4 Separate test instruments are available that can be used for testing and troubleshooting GFCIs. Such testers should be listed by a nationally recognized testing laboratory to UL 1436, *Outlet Circuit Testers and Similar Indicating Devices.* Separate GFCI test instruments should not be used to test GFCIs protecting 2-wire circuits — doing so can result in electric shock.

13.2.3.5 When a separate GFCI test instrument is used, if the tester indicates "No Trip" and the GFCI integral test button indicates "Trip," the following miswiring scenarios should be investigated:

(1) Line and load wires transposed
(2) Reverse polarity
(3) Open ground

13.2.3.6 Only after it has been ascertained that the GFCI is properly wired should the test result be considered indicative of an improperly functioning GFCI.

13.2.4 GFCI Types. The following are the four types of GFCIs:

(1) Circuit-breaker
(2) Receptacle
(3) Portable
(4) Permanently mounted

13.2.5 Circuit-Breaker-Type GFCI.

13.2.5.1 A circuit-breaker-type GFCI is designed in the form of a small circuit breaker and is completely self-contained within the unit housing. The circuit-breaker-type GFCI provides overload and short-circuit protection for the circuit conductors in addition to ground-fault protection for personnel. It is intended to be mounted in a panelboard or other enclosure.

13.2.5.2 Recommended maintenance is the same as that specified in Chapter 17 for molded-case circuit breakers.

13.2.6 Receptacle-Type GFCI.

13.2.6.1 A receptacle-type GFCI is designed in the form of a standard receptacle, is completely self-contained within the unit housing, and does not provide overload or short-circuit protection. It is intended for permanent installation in conventional-device outlet boxes or other suitable enclosures.

13.2.6.2 Maintenance required is the same as that specified in Section 24.3 for standard receptacle outlets.

13.2.7 Portable-Type GFCI.

13.2.7.1 A portable-type GFCI is a unit intended to be easily transported and plugged into a receptacle outlet. Cords, tools, or other devices to be provided with ground-fault protection for personnel are then plugged into receptacles mounted in the unit.

13.2.7.2 Recommended maintenance is that specified in Section 24.3 for receptacles and in Section 29.4 for connecting cords.

13.2.8 Permanently Mounted-Type GFCI.

13.2.8.1 A permanently mounted-type GFCI is a self-contained, enclosed unit designed to be wall- or pole-mounted and permanently wired into the circuit to be protected.

13.2.8.2 Maintenance beyond tightness of connections and cleanliness should not be attempted. Any repairs needed should be referred to the manufacturer.

13.3 Ground-Fault Protective Equipment to Prevent Damage.

13.3.1 Ground-Fault Sensing and Relaying Equipment. Ground-fault sensing and relaying equipment is used to prevent damage to conductors and equipment. The protective equipment consists of three main components: (1) sensors, (2) relay or control unit, and (3) a tripping means for the disconnect device controlling the protected circuit. Refer to Section 230.95 of *NFPA 70, National Electrical Code,* for performance testing and recordkeeping when this equipment is first installed at a site.

13.3.2 Sensing Methods. Detection of ground-fault current is done by either of two basic methods. With one method, ground current flow is detected by sensing current in the grounding conductor. With the other method, all conductor currents are monitored by either a single large sensor or several smaller ones.

13.3.3 Sensors. Sensors are generally a type of current transformer and are installed on the circuit conductors. The relay or control unit can be mounted remote from the sensors or can be integral with the sensor assembly.

13.3.4 Combination Units. Circuit breakers with electronic trip units might have a combination ground-fault sensing and relaying system integral with the circuit breaker. Any maintenance work performed on the electronic circuitry should adhere to manufacturers' instructions. Maintenance on the mechanical operating mechanism components should be done as indicated in Chapter 17.

13.3.5 Maintenance.

13.3.5.1 Maintenance recommendations for the sensors are as specified in 15.9.5.2 for indoor-type instrument transformers. Careful inspection for tight terminal connections and cleanliness should be made.

13.3.5.2 If interconnections between components are disconnected, they should be marked and replaced to maintain the proper phasing and circuitry.

13.3.5.3 A formal program of periodic testing should be established. The manufacturer or ANSI/NETA MTS, *Standard for Maintenance Testing Specifications for Electrical Power Distribution Equipment and Systems* should be consulted for sample specifications.

Chapter 14 Grounding

14.1 Introduction.

14.1.1 *Grounding* is a term that has many different facets, depending on the application. For example, certain current-carrying electrical system conductors (or common of a 3-phase wye electrical system) are intentionally grounded. This intentional connection stabilizes the voltage under normal operating conditions and maintains the voltage at one level relative to earth or something that serves in place of the earth.

14.1.2 Electrically conductive surfaces are also normally grounded for safety purposes. Grounding is necessary to keep the metal enclosures, metal housings, or non-current-carrying parts of the electrical equipment at earth potential and to avoid hazardous voltages between the equipment and earth.

14.1.3 During maintenance or construction, deenergized, ungrounded conductors are also temporarily grounded for personnel protection against the energizing of circuit conductors. Therefore, grounding is also a temporary protective measure involving connecting the deenergized lines and equipment to earth through conductors.

14.1.4 Common reasons for grounding both electrical systems and equipment are to limit the voltage imposed by lightning, line surges (transients) or unintentional contact with higher voltages; to stabilize the voltage to earth under normal operation; and to establish an effective path for fault current. This fault current path should be capable of safely carrying the maximum fault likely to be imposed on it, and should have sufficiently low impedance to facilitate the operation of overcurrent devices under fault conditions. This path should also be designed and installed to limit touch and step potentials to safe values.

14.1.5 Grounding is one of the most important and essential aspects of an electrical system. However, it is often misunderstood because of its many different interpretations and misuse of definitions.

14.1.6 Special Terms. The following special terms are used in this chapter.

14.1.6.1 Bonding (Bonded). The permanent joining of metallic parts to form an electrically conductive path that will ensure electrical continuity and the capacity to conduct safely any current likely to be imposed. The "permanent joining" can be accomplished by the normal devices used to fasten clean, noncorroded parts together. Machine screws, bolts, brackets, or retainers necessary to allow equipment to function properly are items typically employed for this purpose. While welding and brazing can also be utilized, these preclude easy disassembly, and welding can increase rather than decrease resistance across joints. Metallic parts that are permanently joined to form an electrically conductive path that will ensure electrical continuity and the capacity to conduct safely any current likely to be imposed are bonded.

14.1.6.2 Bonding Jumper. A reliable conductor to ensure the required electrical conductivity between metal parts required to be electrically connected. This conductor can be solid or stranded or braided, and connected by compatible fittings to separate parts to provide this electrically conductive path. The bonding jumper can also be a screw or a bolt. This bonding jumper can be used alone or in conjunction with other electrically conductive paths. It generally is associated with the equipment-grounding path, but might or might not be electrically linked for a lowest impedance path.

14.1.6.3 Case (Enclosure) Ground. See 14.1.6.29, Grounding Terminal.

14.1.6.4 Central Grounding Point. The location where the interconnected parts of the grounding system are connected in a common enclosure. The central grounding point provides a common connection point for termination of the feeder or branch-circuit equipment-grounding conductors.

14.1.6.5 Counterpoise. A conductor or system of conductors arranged beneath the transmission/distribution supply line; located on, above, or most frequently below the surface of the earth; and connected to the grounding system of the towers or poles supporting the line. (This conductor(s) might or might not be the continuous length of the supply path. It is often used to provide a lower surge impedance path to earth for lightning protection when there is a transition from overhead supply conductors to underground insulated cable.) Counterpoise is also used in communication systems, where it is a system of conductors, physically elevated above and insulated from the ground, forming a lower system of conductors of an antenna. Note that the purpose of a counterpoise is to provide a relatively high capacitance and thus a relatively low impedance path to earth. The counterpoise is sometimes used in medium- and low-frequency applications where it would be more difficult to provide an effective ground connection. Sometimes counterpoise is confused with equipotential plane. *(See also 14.1.6.17, Equipotential Plane.)*

14.1.6.6 Down Conductor. A conductor from a lightning protection system to earth ground designed to provide a low impedance path for the current from a lightning strike and/or dissipate the charge buildup that precedes a lightning strike. This conductor typically goes from the air terminals to earth. Due to the very high currents at very high frequencies, the impedance of the entire system is very critical. Normal wiring conductors are not suitable for the down conductor. Typically, they are braided conductors. There might be certain instances where additional investigation about the interconnection between the lightning and the grounding electrode system is warranted.

14.1.6.7 Earth Grounding. The intentional connection to earth through a grounding electrode of sufficiently low impedance to minimize damage to electrical components and prevent an electric shock that can occur from a superimposed voltage from lightning and voltage transients. In addition, earth grounding helps prevent the buildup of static charges on equipment and material. It also establishes a common voltage reference point to enable the proper performance of sensitive electronic and communications equipment.

14.1.6.8 Earthing. An IEC term for *ground. (See 14.1.6.18, Ground.)*

14.1.6.9 Effective Grounding Path. The path to ground from circuits, equipment, and metal enclosures for conductors should (1) be permanent and electrically continuous, (2) have capacity to conduct safely any fault current likely to be imposed on it, and (3) have sufficiently low impedance to limit the voltage to ground and to facilitate the operation of the circuit protection devices. The earth should not be used as the sole equipment-grounding conductor.

14.1.6.10 Effectively Grounded (as applied to equipment or structures). Intentionally connected to earth (or some conducting body in place of earth) through a ground connection or connections of sufficiently low impedance and having sufficient current-carrying capacity to prevent the buildup of voltages that might result in undue hazards to connected equipment or to persons.

14.1.6.11 Effectively Grounded (as applied to systems). This is defined by ratios of impedance values that must be within prescribed limits.

14.1.6.12 Electrostatic Discharge (ESD) Grounding. The conductive path created to reduce or dissipate the electrostatic charge where it builds up as a result of equipment operation or induced from an electrostatically charged person or material coming in contact with the equipment. Also referred to as *static grounding*.

14.1.6.13 Equipment Bonding Jumper. The connection between two or more portions of the equipment-grounding conductor.

14.1.6.14 Equipment Ground. An ambiguous term that can mean either case ground, equipment-grounding conductor, or equipment bonding jumper; hence, use of this term should be avoided.

14.1.6.15 Equipment-Grounding Conductor. The conductor used to connect the non-current-carrying metal parts of equipment, raceways, and other enclosures to the system grounded conductor, the grounding electrode conductor, or both, at the service equipment or at the source of a separately derived system.

14.1.6.16 Equipotential Bonding. Electrical connection putting various exposed conductive parts and extraneous conductive parts at a substantially equal potential.

14.1.6.17 Equipotential Plane. (1) (as applied to livestock) An area accessible to livestock where a wire mesh or other conductive elements are embedded in concrete, are bonded to all metal structures and fixed nonelectrical metal equipment that might become energized, and are connected to the electrical grounding system to prevent a difference in voltage from developing within the plane. (2) (as applied to equipment) A mass or masses of conducting material that, when bonded together, provide a uniformly low impedance to current flow over a large range of frequencies. Sometimes the *equipotential plane* is confused with *counterpoise.*

14.1.6.18 Ground. The earth. [**70,** 2011]

14.1.6.19 Ground Fault. Unintentional contact between an ungrounded conductor and earth or conductive body that serves in place of earth. Within a facility, this is typically a fault between a current-carrying conductor and the equipment-grounding path that results in the operation of the overcurrent protection.

14.1.6.20 Ground Leakage Current. Current that is introduced into the grounding conductor by normal equipment operation, such as capacitive coupling. Many RFI/EMI filters in electronic equipment have capacitors from current-carrying conductors to the equipment-grounding conductor to shunt noise emitted from or injected into their power supplies. While there are relatively low current level limits imposed by regulatory agencies (e.g., UL specifies maximum 3.5 mA, hospital equipment 0.5 mA), not all equipment is listed. Even with listed equipment, the sum of the current from a large quantity of such equipment in a facility can result in significant ground currents.

14.1.6.21 Ground Loop. Multiple intentional or unintentional connections from a conductive path to ground or the conductive body that serves in place of earth. Current will flow in the ground loop if there is voltage difference between the connection nodes. Regrounding of the grounded circuit conductor (neutral) beyond the service point will result in ground loops. This might or might not be harmful depending on the application.

14.1.6.22 Ground Resistance/Impedance Measurement. The use of special test equipment to measure the grounding electrode resistance or impedance to earth at a single frequency at or near power line frequency.

14.1.6.23 Ground Well. See 14.1.6.28, Grounding Electrode System.

14.1.6.24 Grounded (Grounding). Connected (connecting) to ground or to a conductive body that extends the ground connection. [**70,** 2011]

14.1.6.25 Grounded Conductor. A system or circuit conductor that is intentionally grounded. This intentional grounding to earth or some conducting body that serves in place of earth takes place at the premises service location or at a separately derived source. Control circuit transformers are permitted to have a secondary conductor bonded to a metallic surface that is in turn bonded to the supply equipment-grounding conductor. Examples of grounded system conductors would be a grounded system neutral conductor (3 phase or split phase) or a grounded phase conductor of a 3-phase, 3-wire, delta system.

14.1.6.26 Grounding Conductor. A conductor used to connect equipment or the grounded circuit conductor of a wiring system to a grounding electrode or electrodes. This ensures the electrical continuity of equipment bonding and grounding into the system, and provides for an equipotential bonding to minimize voltage differences between individual units and between these units and the grounding electrode conductor.

14.1.6.27 Grounding Electrode. A conductive body deliberately inserted into earth to make electrical connection to earth. Typical grounding electrodes include the following:

(1) The nearest effectively grounded metal member of the building structure

(2) The nearest effectively grounded metal water pipe, but only if the connection to the grounding electrode conductor is within 5 ft of the point of entrance of the water pipe to the building
(3) Any metal underground structure that is effectively grounded
(4) Concrete encased electrode in the foundation or footing (e.g., Ufer ground)
(5) Ground ring completely encircling the building or structure
(6) Made electrodes (e.g., ground rods or ground wells)
(7) Conductive grid or mat used in substations

14.1.6.28 Grounding Electrode System. The interconnection of grounding electrodes.

14.1.6.29 Grounding Terminal. A terminal, lug, or other provision provided on some equipment cases (enclosures) to connect the conductive portion of the enclosure to the equipment-grounding conductor.

14.1.6.30 Grounding-Type Receptacle. A receptacle with a dedicated terminal that is to be connected to the equipment grounding conductor.

14.1.6.31 High-Impedance Grounded. High-impedance grounded means that the grounded conductor is grounded by inserting a resistance or reactance device that limits the ground-fault current to a low value.

14.1.6.32 Insulated Ground. See 14.1.6.33, Isolated Equipment-Grounding Conductor.

14.1.6.33 Isolated Equipment-Grounding Conductor. An insulated equipment-grounding conductor that has one intentional connection to the equipment-grounding system. The isolated equipment-grounding conductor is typically connected to an equipment-grounding terminal either in the facility's service enclosure or in the first applicable enclosure of a separately derived system. The isolated equipment-grounding conductor should be connected to the equipment-grounding system within the circuits' derived system.

14.1.6.34 Lightning Ground. See 14.1.6.28, Grounding Electrode System.

14.1.6.35 Multipoint Grounding. Multipoint grounding consists of interconnecting primary and secondary neutrals of the transformer. The secondary and primary neutral are common, and they both utilize the same grounding electrode that connects the system to earth.

14.1.6.36 Noise(less) Ground. The supplemental equipment-grounding electrode installed at machines, or the isolated equipment-grounding conductor, intended to reduce electrical noise.

14.1.6.37 Personnel Protective Ground. Bonding jumper that is intentionally installed to ground deenergized, normally ungrounded circuit conductors when personnel are working on them, to minimize voltage differences between different parts of the equipment and personnel, so as to protect against shock hazard and/or equipment damage.

14.1.6.38 Protective Bonding Circuit. See 14.1.6.16, Equipotential Bonding.

14.1.6.39 Protective Conductor. See 14.1.6.15, Equipment-Grounding Conductor.

14.1.6.40 Protective Ground. See 14.1.6.16, Equipotential Bonding.

14.1.6.41 RFI/EMI Grounding. See 14.1.6.20, Ground Leakage Current.

14.1.6.42 Safety Ground. See 14.1.6.37, Personnel Protective Ground.

14.1.6.43 Separately Derived System. A premises wiring system whose power is derived from a battery, a solar photovoltaic system, or from a generator, transformer, or converter windings, and that has no direct electrical connection, including a solidly connected grounded circuit conductor, to supply conductors originating in another system. Equipment-grounding conductors are not supply conductors and are to be interconnected.

14.1.6.44 Shield Ground. Intentional grounding of one or both ends of the shield of a cable.

14.1.6.44.1 Data Communications Cables. The shield of data communication cables can be connected to the equipment-grounding conductor at either one end of the cable (single end) or at both ends (double ended). When both ends of a shield are grounded, another shield should be provided inside the outer shield and that one single end grounded.

14.1.6.44.1.1 There are advantages and disadvantages to both types, single- or double-ended. Single-ended grounding minimizes the ground loop potential but can result in the shield voltage at the ungrounded end rising above safe levels for equipment or personnel. Single-end grounded shields can have the ungrounded end grounded through a high-frequency drain, such as a surge device, to help control this.

14.1.6.44.1.2 Double-ended grounding can minimize the potential voltage rise but can result in a ground loop that exceeds the current-carrying capacity of the outer shield.

14.1.6.44.2 Shield Ground, Power Cables. The shield of power cables can be connected to the equipment-grounding conductor at either one end of the cable (single ended) or at both ends (double ended). Shielding will ensure uniform dielectric stress along the length of the cable. When grounded at both ends, cable derating might be necessary because of heat due to ground loop current.

14.1.6.45 Single-Point Grounding. The single-point grounding of a transformer means connecting the secondary side of the transformer to earth ground through one or more grounding electrodes. This connection should be made at any point on the separately derived system from the source to the first system-disconnecting means or overcurrent device.

14.1.6.46 Solid Grounded. Solidly grounded means that the grounded conductor is grounded without inserting any resistor or impedance device.

14.1.6.47 Substation Ground. Grounding electrode system (grid) in a substation. See 14.1.6.28, Grounding Electrode System.

14.1.6.48 System Grounding. The intentional connection of an electrical supply system to its associated grounding electrode(s).

14.1.6.49 Luminaire Grounding.

14.1.6.49.1 Luminaires (fixtures) shall be inspected to verify that they are properly grounded.

14.1.6.49.1.1 If the wiring method utilizes a metallic armored cable wiring method or nonmetallic-sheathed cable with ground, proper connection of the wiring provides an acceptable equipment ground. Grounding of a surface-mounted luminaire is accomplished by securing the luminaire to a properly grounded metal outlet box. Metal outlet boxes have a location to place a grounding screw. The bare copper equipment grounding conductor in the nonmetallic sheathed cable is usually terminated under this screw.

14.1.6.49.1.2 If the outlet box is nonmetallic, the small bare equipment grounding conductor from the luminaire is connected to the equipment grounding conductor in the outlet box. For suspended ceiling luminaires, grounding of the luminaire is accomplished by using metallic fixtures whips or nonmetallic sheathed cable with ground between the outlet box and the luminaire.

14.1.6.49.2 For a more complete list of possible wiring methods or if there is no equipment grounding means found, refer to *NFPA 70, National Electrical Code,* for proper grounding.

14.2 Symptoms and Causes of Inadequate Grounding.

14.2.1 Common mode noise voltages can develop when the equipment-grounding conductor and the grounded conductor are not effectively bonded.

14.2.2 Common mode noise can be produced in wiring without an equipment-grounding conductor and without electrically continuous raceway.

14.2.3 Ground loops can be undesirable because they create a path for noise currents to flow.

14.2.4 Undesirable touch potentials can result from contacting metallic surfaces that are improperly grounded.

14.2.5 Equipment misoperation due to unequal ground potentials results in improper data communication or improper readings of transducers.

14.2.6 Shutdown or damage of electronic equipment can be due to electrostatic discharge (ESD).

14.2.7 Nonoperation or malfunction of protective circuit devices or voltage sag can be due to high-impedance ground fault paths.

14.2.8 Damage, nonoperation, or misoperation of electronic components can be caused by poor connections in the grounding path.

14.2.9 Damage or destruction of the neutral conductor or cable shields can result from improper sizing of a high-impedance neutral grounding device.

14.2.10 Voltage can be present on deenergized circuits during testing of these conductors.

14.2.11 Destruction of equipment and surge protection devices can follow a voltage transient, such as a lightning strike.

14.3 Grounding System Inspection, Testing, and Monitoring.

14.3.1 A visual and physical inspection should be made to verify the integrity of the grounding and bonding conductors and associated connections.

14.3.2 The integrity of the grounding electrode system and substation grids should be checked on a periodic basis. The electrical connection to earth can be measured using one of several available methods and technologies. *(See 11.13.3.)* Also refer to ANSI/IEEE 142, *Recommended Practice for Grounding of Industrial and Commercial Power Systems (IEEE Green Book),* and ANSI/IEEE 80, *Guide for Safety in AC Substation Grounding.*

14.3.3 A ground loop impedance test should be performed on the equipment-grounding path with a four-lead, low-resistance ohmmeter such as a wheatstone bridge, a kelvin bridge, or a digital low-resistance ohmmeter. Impedances should be appropriate for the type, length, and size of the path.

14.3.4 Measure the voltage between the equipment-grounding conductor and the grounded conductor at multiple locations throughout the system, as applicable.

14.3.4.1 At the bonding jumper, the voltage normally should be less than 0.1 volt ac.

14.3.4.2 It is normal to find voltage downstream from the main bonding jumper in energized circuits, due to current flow in the grounded conductor. Readings in excess of 3 volts ac or less than 0.5 volt ac at locations remote from the bonding jumper should be investigated to determine if this represents a problem for the system.

14.3.5 The current on the equipment-grounding conductor should be measured for objectionable levels, which will depend on the location and type of the facility. The source of currents on equipment-grounding conductors should be determined and corrected. Use of a true rms ammeter is recommended.

14.3.6 The voltage from the chassis of equipment and an external ground point should be measured. Differences should be less than 2 volts.

14.3.7 Continuous monitoring of ground and neutral currents in information technology areas is recommended.

14.3.8 In the absence of any specifications, when ESD systems are being examined, the leakage resistance should not exceed 1 megohm from any conductor to ground. *(See 12.3.13.2.)*

14.3.9 Testing of the ground integrity of data communication cable shields might require special instrumentation and expertise.

14.3.10 If a result of testing indicates that changes to a substation grounding system are necessary or required, reference should be made to ANSI/IEEE 80, *Guide for Safety in AC Substation Grounding,* for appropriate design requirements.

14.4 Solutions to Inadequate Grounding.

14.4.1 To minimize the resistance between the grounding electrode system and the earth, the following should be done:

(1) Clean and tighten and test connections as needed, using appropriate safety precautions.
(2) Replace or repair damaged or corroded components.
(3) Size the grounding electrode conductor in accordance with Article 250 of *NFPA 70, National Electrical Code.*
(4) Use soil enhancement material as necessary.

14.4.2 The grounded conductor should be connected to the equipment-grounding conductor only as permitted by Article 250 of *NFPA 70, National Electrical Code.* The grounded conductor and the equipment-grounding conductor should be sized in accordance with *NFPA 70.*

14.4.3 Many of the grounding electrode corrosion problems are caused by galvanic action. This problem can be minimized by using a system of cathodic protection (active or passive). The use of dissimilar metals should be avoided.

14.4.4 Isolation Transformer.

14.4.4.1 An isolation transformer has separate primary and secondary windings. The bonding jumper between the equipment-grounding conductor and the secondary grounded conductor provides protection from common mode electrical noise.

14.4.4.2 It is recommended that a shielded isolation transformer be used. It contains an electrostatic shield between the primary and secondary windings that is connected to the equipment-grounding terminal.

14.4.5 Signal Circuit Isolation. See 10.7.3.4. *[See ANSI/IEEE 1100, Recommended Practice for Powering and Grounding Sensitive Electronic Equipment (IEEE Emerald Book).]*

14.4.6 Isolated Ground Receptacles. One solution is to install an isolated ground receptacle (identified by orange color and/or orange triangle) in which the equipment-grounding terminal is insulated from the mounting strap. An insulated equipment-grounding conductor is then connected from *NFPA 70, National Electrical Code.* The insulated equipment-grounding conductor is connected to the applicable derived system or service grounding terminal.

Chapter 15 Substations and Switchgear Assemblies

15.1 Substations.

15.1.1 Introduction.

15.1.1.1 Substations in an electrical system perform functions such as voltage transformation, system protection, power factor correction, and circuit switching. They are comprised of electrical power products, such as transformers, regulators, air switches, circuit breakers, capacitors, and lightning arresters.

15.1.1.2 Maintenance of the individual power system components is discussed under the appropriate headings.

15.1.1.3 The recommended frequency of maintenance depends on the environment in which the substation is operating and the criticality of the circuit that it controls. In many cases, it is an outdoor installation and exposed to atmospheric contaminations. In areas of industrial contamination or in coastal areas where ocean vapors are prevalent, inspections are recommended at intervals of 6 weeks to 2 months. Less frequent inspections are recommended in areas of relatively clean atmosphere.

15.1.2 Insulators.

15.1.2.1 Insulators should be inspected for evidence of contaminated surfaces or physical damage, such as cracked or broken segments. Contaminated insulator surfaces should be cleaned and damaged insulators replaced.

15.1.2.2 Evidence of corona when the substation is energized should be reported and corrected. Mild corona might be normal and is more pronounced when humidity is high. *(See 3.3.13, Corona.)* Ultrasonic detection and light amplification (night vision) equipment are useful for detecting corona.

15.1.3 Conductors. All exposed conductors should be visually inspected for evidence of overheating at bolted joints. Extreme overheating can discolor copper conductors, deteriorate the insulation, and may require additional maintenance. When the substation is deenergized, these bolted connections should be checked. Bolts should be loosened and retorqued to proper values. There are infrared detectors that can be used on energized systems to check for overheating by scanning from a distance. Where aluminum-to-copper joints exist, they should be inspected carefully for evidence of corrosion, overheating, or looseness. In all cases manufacturer's specifications should be followed. See Section 15.10 for torque tables.

15.1.4 Air-Disconnecting Switches.

15.1.4.1 Air-disconnecting switches are operated infrequently in service and usually are energized during routine substation maintenance. Maintenance of the switch is limited to those areas that can be approached safely. The insulators and conducting parts should be examined as described earlier in 15.1.2 and 15.1.3. Interphase linkages and operating rods should be inspected to make sure that the linkage has not been bent or distorted and that all fastenings are secure. The position of the toggle latch of the switch-operating linkage should be observed on all closed switches to verify that the switch is mechanically locked in a closed position.

15.1.4.2 Power-operated switches should be operated periodically to ensure that the switches, their mechanism, and control features are functioning properly. When the circuit condition does not permit operating the switch while energized and the circuit cannot be deenergized for routine maintenance, the operating mechanism should be disengaged from the linkage to allow the control circuits and mechanism to be checked, provided that this method does not adversely affect the overall adjustment.

15.1.4.2.1 The maintenance instructions of the particular manufacturer of each mechanism should be followed.

15.1.4.2.2 The following features should be checked:

(1) Limit switch adjustment
(2) Control relays for damaged contacts, defective coils, and inadequate supply voltage
(3) Any other condition that might inhibit proper functioning of the switch assembly
(4) Lubrication

15.1.4.3 A scheduled outage should be planned and thorough maintenance performed as follows:

(1) A switch should be operated several times and checked for simultaneous closing of all blades and for complete contact closing, and the blade lock or latch should be checked in the fully closed position.
(2) Contacts should be inspected for alignment, pressure, pitting and arcing, or corrosion.
    (a) Badly pitted or burned contacts should be replaced.
    (b) If pitting is of a minor nature, the surface should be left as is.
    (c) Arcing horns should be inspected for signs of excessive burning and should be replaced if necessary.
(3) Insulation should be inspected for breaks, cracks, or burns and cleaned. Where abnormal conditions prevail, such as salt deposits, cement dust, acid fumes, or other contaminants, more frequent cleaning may be required.
(4) Gear boxes, linkages, and contact pivots should be checked for proper lubricants per the manufacturer's recommendation, and for evidence of moisture that could cause corrosion or difficulty in switch operation due to ice formation.
(5) Flexible braids or slip ring contacts commonly used for grounding the operating handle should be inspected, and braids showing signs of corrosion, wear, or broken strands should be replaced.
(6) All safety interlocks should be inspected, checked, and tested for proper operation.

15.1.4.4 If it is known that a switch has been subjected to a short-circuit condition, special effort should be made to inspect contacts at the earliest possible time.

15.1.4.5 For further information on air switch testing see Section 11.15, Switches.

15.1.5 Grounding Equipment. All grounding connections should be inspected for tightness and absence of corrosion. The substation, enclosure, and apparatus grounds should be inspected and tested when possible. For further information on grounding systems and testing, see Section 11.13, Grounding Systems.

15.1.6 Enclosures. The security of fences or other enclosures should be checked to ensure against entry of animals or unauthorized personnel. The gates or doors, especially where equipped with panic hardware, should be checked for security and proper operation. The enclosed area should not be used for storage of anything other than the most frequently used spare parts directly associated with the enclosed equipment, which should not be stored within the required working space.

15.1.7 Miscellaneous Equipment.

15.1.7.1 The availability and condition of rack-out devices, hoisting or handling apparatus, grounding equipment, live line tools, rubber gloves, voltage detectors, and other test equipment should be inspected and, if defective, replaced or repaired according to manufacturer's specifications.

15.1.7.2 The proper operation of floodlights and other auxiliary apparatus, such as cooling fans on transformers, should be inspected and, if defective, replaced or repaired according to manufacturer's recommendations.

15.1.7.3 Any activated warning lights or flags on temperature, pressure, or liquid level gauges should be reported and corrective action taken.

15.2 Switchgear Assemblies.

15.2.1 Introduction.

15.2.1.1 A switchgear assembly might be an open type or an enclosed type. These switchgear assemblies are normally constructed in modules or cubicles, each of which contains either one or more interrupting devices or auxiliary equipment, such as metering transformers, auxiliary power transformers, control relaying, and battery chargers. Power is supplied throughout the assembly by the main power bus.

15.2.1.2 Low-voltage metal-enclosed switchgear assemblies have a maximum nominal voltage rating of 600 volts. Over 600 volts metal-enclosed power switchgear assemblies have nominal voltage ratings from 5,000 volts through 69,000 volts.

15.2.1.3 Metal-enclosed switchgear assemblies are normally connected to one or more supply transformers, either directly connected to the transformer or remotely connected by cable or bus. They might be found outdoors as a part of a substation or indoors as a power distribution center.

15.2.1.4 Circuit interrupters in switchgear assemblies are either circuit breakers or interrupter switches or fuses. Fuses are covered separately in Chapter 18.

15.2.2 Grounding. Grounding of the switchgear assembly should be verified by visual inspection where possible.

15.2.3 Frequency of Maintenance.

15.2.3.1 Recommended frequency of maintenance depends on environmental and operating conditions, along with the criticality of equipment, so no fixed rule can govern all applications.

15.2.3.1.1 An annual inspection of the entire switchgear assembly, including withdrawable elements during the first 3 years of service, is suggested as a minimum when no other criteria can be identified.

15.2.3.1.2 Inspection frequency can be increased or decreased depending on observations, level of importance of the system, personnel safety, trending, and experience. Specific manufacturers' recommendations regarding inspection and maintenance should be considered. It is recommended that frequent inspections be made initially; the interval can then be gradually extended as conditions warrant taking criticality of equipment into consideration. In cases where low frequency of maintenance is assigned to low reliability equipment, consideration should be made regarding hazards to personnel maintaining the equipment. For further information on switchgear assembly testing, see Section 11.23, Switchgear and Switchboard Assemblies.

15.2.3.2 The following factors affect the decision on when to inspect:

(1) Scheduled shutdowns
(2) Emergency shutdowns
(3) Periods of sustained unusual or abnormal operating conditions (e.g., switching or sustained overloads)
(4) Feeder, bus, or system fault occurrence
(5) Extremes in atmospheric conditions, such as heat, cold, heavy dust, high winds, rain, snow, fog, smog, fumes of many kinds, fly ash, salt spray, high humidity, unusual temperature changes, and lightning
(6) Maintenance requirements and schedules for related equipment, either component parts of the switchgear assembly or items apart from but connected to the switchgear circuits [Time is the most universal criterion, but other indicators, such as reliability centered maintenance (RCM), mean time to failure, or number of operations, can be used as a guide. See Chapter 30 for additional information on RCM.]
(7) Criticality of the equipment. (The more critical the equipment, the more frequent the inspections.)

15.2.3.3 Partial inspections can be made even when the entire switchgear assembly cannot be deenergized.

15.2.3.4 Specific circuits can be taken out of service even though the main bus is not deenergized. This permits an insulation inspection of bus risers and supports in the load side or disconnected side of the switchgear unit.

15.2.3.5 Where a full shutdown is impractical, partial inspections can indicate whether a full shutdown is required to avoid an unscheduled outage. Partial inspection cannot be guaranteed to be indicative of conditions in inaccessible areas.

15.2.4 Enclosure. A good maintenance program will ensure the continuation of the two functions of an enclosure:

(1) To prevent exposure to live parts and operating mechanisms
(2) To protect the equipment from exposure to moisture and contaminants outside the enclosure

15.2.5 Security. All doors and access panels should be inspected to ensure that all hardware is in place and in good condition. Hinges, locks, and latches should be lubricated. Screens covering ventilation openings should be in place to prevent entry of rodents or small animals.

15.2.6 Leakage. On outdoor assemblies, roof or wall seams should be checked for evidence of leakage, and any leaking seams should be repaired. The base should be checked for openings that could permit water to drain into the interior, and any such openings should be repaired, such as by caulking or grouting.

15.2.7 Moisture.

15.2.7.1 Moisture accumulation might occur on internal surfaces of enclosures even though they are weathertight. The source of this moisture could be condensation. All floor openings, other than those specifically provided for drainage or ventilation purposes, should be effectively sealed. All unused conduits or openings around cables at entrance ducts should be sealed with an appropriate grade of caulking compound. The source of water pooling should be determined and eliminated.

15.2.7.2 Conditions causing condensation could be intermittent and might not be prevalent at the time of inspection. All internal surfaces should be examined for signs of previous moisture such as the following:

(1) Droplet depressions or craters on dust-laden surfaces
(2) Excessive rust anywhere on the metal housing

15.2.7.3 Condensation is prevented by heat and air circulation. It is important, therefore, to make sure the heating and ventilating systems are functioning properly.

15.2.8 Heating. Heat losses in operating switchgear assemblies may not be adequate to prevent condensation. Where space heaters are provided to supply supplementary heat, they should be checked to ensure that they are in good condition and are operating properly. If they are thermostatically controlled, the thermostat should be checked for proper operation and setting. A thermostat set too low will not control the heaters properly under all climatic conditions.

15.2.9 Ventilation. Where ventilators are supplied on enclosures, including metal-enclosed bus enclosures, they should be checked to ensure that they are clear of obstructions and that the air filters are clean and in good condition. Base foundations should be examined to ensure that structural members have not blocked floor ventilation.

15.2.10 Lighting and Housekeeping. All interior and exterior lighting should be checked for proper operation. Availability of spare equipment and handling devices should be checked. They should be stored in such a manner as to be readily available yet not hamper normal operation, block ventilation passages, or interfere with the required working space.

15.2.11 Insulation.

15.2.11.1 With proper maintenance, the insulation of metal-enclosed switchgear assemblies is designed and expected to withstand operating voltages for decades. During this time, the insulation will be subject to an accumulation of deteriorating conditions that detract from its insulating capability.

15.2.11.2 Moisture combined with dirt is one of the greatest deteriorating factors for insulation. Dirt and/or dust could also result in electrical leakage that leads to tracking and eventual flashover if allowed to continue to accumulate. It is important in the maintenance of switchgear to know the condition of the insulation.

15.2.12 Dielectric Stress.

15.2.12.1 The following specific areas in which insulation failure is more likely to occur should be given special attention:

(1) Boundaries between two adjoining insulators
(2) Boundaries between an insulating member and the grounded metal structure
(3) Taped or compounded splices or junctions
(4) Bridging paths across insulating surfaces, either phase-to-phase or phase-to-ground
(5) Hidden surfaces such as the adjacent edges between the upper and lower members of split-type bus supports or the edges of a slot through which a bus bar protrudes
(6) Edges of insulation surrounding mounting hardware either grounded to the metal structure or floating within the insulating member

15.2.12.2 Damage caused by dielectric stress could be evident on the surface of insulating members in the form of corona erosion or markings or tracking paths.

15.2.13 Corona.

15.2.13.1 If corona occurs in switchgear assemblies, it is usually localized in thin air gaps that exist between a high-voltage bus bar and its adjacent insulation or between two adjacent insulating members. It might form around bolt heads or other sharp projections that are not properly insulated or shielded.

15.2.13.2 Organic insulating materials, when exposed to corona discharge, initially develop white powdery deposits on their surface. These deposits can be wiped off with solvent. If the surface has not eroded, further maintenance is not required. Prolonged exposure to corona discharge will result in erosion of the surface of the insulating material. In some materials, corona deterioration has the appearance of worm-eaten wood. If the corrosion paths have not progressed to significant depths, surface repair probably can be accomplished. Manufacturers' recommendations should be followed for such repair.

15.2.14 Tracking.

15.2.14.1 Tracking is an electrical discharge phenomenon caused by electrical stress on insulation. This stress can occur phase-to-phase or phase-to-ground. Tracking, when it occurs in switchgear assemblies, typically is found on insulation surfaces.

15.2.14.2 Tracking develops in the form of streamers or sputter arcs on the surface of insulation, usually adjacent to electrodes. One or more irregular carbon lines in the shape of tree branches are the most common sign of tracking.

15.2.14.3 Surface tracking can occur on the surfaces of organic insulation or on contaminated surfaces of inorganic insulation. The signs of tracking on organic materials are eroded surfaces with carbon lines. On track-resistant organic materials, these erosion patterns are essentially free of carbon.

15.2.14.3.1 Tracking can propagate from either the voltage terminals or the ground terminal. It does not necessarily progress in a regular pattern or by the shortest possible path.

15.2.14.4 Tracking conditions on surfaces of inorganic material can be completely removed by cleaning the surfaces, because no actual damage to the material occurs. In the case of organic material, the surface is damaged in varying degrees, depending on the intensity of the electric discharge and the duration of exposure. If the damage is not too severe, it can be repaired by sanding and application of track-resistant varnish in accordance with the manufacturers' instructions. Organic material that has been damaged should be replaced or repaired in accordance with manufacturer's instructions.

15.2.15 Thermal Damage.

15.2.15.1 Temperatures over design levels for prolonged periods can reduce the electrical life of organic insulating materials. Prolonged exposure to higher than rated temperatures can also cause physical deterioration of the materials, resulting in lowered mechanical strength.

15.2.15.2 Localized heating (hot spots) can sometimes occur and can be masked because the overall temperature of the surroundings is not raised appreciably. Loosely bolted connections in a bus bar splice or void spaces (dead air) in a taped assembly are examples of this problem.

15.2.15.3 Infrared thermography inspections can be used to detect potentially damaging heat *(see Section 11.17)*. However, infrared inspection should not be utilized as the only method of inspection. External conditions that provide evidence of heat damage include:

(1) Discoloration, usually a darkening, of materials or finishes
(2) Crazing, cracking, and flaking of varnish coatings
(3) Embrittlement of tapes and cable insulation
(4) Delamination of materials or finishes
(5) Generalized carbonization of materials or finishes
(6) Melting, oozing, or exuding of substances from within an insulating assembly

15.2.15.3.1 Insulating materials that have been physically damaged should be replaced. Mild discoloration is permissible if the cause of overheating is corrected.

15.2.16 Summary. In summary, there are three important things to remember in maintenance of switchgear: KEEP IT CLEAN, KEEP IT DRY, AND KEEP IT TIGHT.

15.3 Circuit Interrupters. Circuit interrupters in switchgear assemblies are either circuit breakers or interrupter switches or fuses. Fuses are covered separately in Chapter 18. For further information on switchgear assembly testing, see Section 11.23. For further information on interrupter switch testing, see Section 11.15.

15.4 Air Circuit Breakers.

15.4.1 Introduction.

15.4.1.1 Before any maintenance work is performed, manufacturers' instruction manuals should be obtained and read carefully. If the breaker is a drawout type, it should be removed from its cubicle and placed in a secure, convenient location for maintenance. A stored-energy-type circuit breaker or its mechanism never should be serviced while its closing spring is charged.

15.4.1.2 Maintenance on fixed- or bolted-type circuit breakers normally should be performed with the breaker in place inside its cubicle. Special precautions should be exercised to ensure that the equipment is deenergized and the circuit in which it is connected is properly secured from a safety standpoint. All control circuits should be deenergized. Stored-energy closing mechanisms should be discharged. Follow all applicable lockout/tagout procedures in accordance with Chapter 7.

15.4.2 Insulation. Interphase barriers should be removed and cleaned, along with all other insulating surfaces, with a vacuum cleaner or clean lint-free rags and solvents as recommended by the manufacturer, if needed, to remove hardened or encrusted contamination. An inspection should be made for signs of corona, tracking, or thermal damage as described in 15.2.13 through 15.2.15. The maintenance theme here is KEEP IT CLEAN and KEEP IT DRY.

15.4.3 Contacts.

15.4.3.1 The major function of the air circuit breaker depends on, among other things, correct operation of its contacts. Air circuit breakers normally have at least two distinct sets of contacts on each pole — main and arcing. Some have an intermediate pair of contacts that open after the main current-carrying contacts and before the arcing contacts. When closed, practically the entire load current passes through the main contacts. Also, high-overload or short-circuit current passes through them during opening or closing faulted lines. If the resistance of these contacts becomes high, they will overheat. Increased contact resistance can be caused by pitted contact surfaces, foreign material embedded on contact surfaces, or weakened contact spring pressure. This resistance will cause excessive current to be diverted through the arcing contacts, with consequent overheating and burning. The pressure should be kept normal, which is usually described in the manufacturer's instructions.

15.4.3.2 Arcing contacts are the first to close and the last to open; any arcing usually originates on them. In circuit interruption, they carry current only momentarily, but that current might be equal to the interrupting rating of the breaker. In closing against a short circuit, they can momentarily carry considerably more than the short-circuit interrupting rating. Therefore, they must maintain sufficient contact when they are touching. If not, the main contacts can be badly burned when interrupting heavy faults. Failure to properly interrupt the fault might also result.

15.4.3.3 On magnetic blow-out air breakers, the arc is quickly removed from the arcing contacts by a magnetic blow-out field and travels to arcing horns, or runners, in the arc interrupter, also known as an arc chute. The arcing contacts are expendable and can eventually burn enough to require replacement.

15.4.3.4 The general rules for maintaining contacts on all types of breakers are as follows:

(1) They should be kept clean, smooth, and in good alignment.
(2) The pressure should be as prescribed in the manufacturers' literature.

15.4.3.5 The main contact surfaces should be clean and bright. Discoloration of the silvered surfaces, however, is not usually harmful unless it is caused by insulating deposits. Insulating deposits should be removed with alcohol or a silver cleaner. Slight impressions on the stationary contacts are caused by the pressure and wiping action of the movable contacts. Minor burrs or pitting are allowed, and projecting burrs can be removed by dressing. Nothing more abrasive than crocus cloth should be used on the silvered contact surfaces. Where serious overheating is indicated by discoloration of metal and surrounding insulation, the contacts and spring assemblies should be replaced in accordance with the manufacturers' instructions.

15.4.3.6 The circuit breaker should be closed manually to check for proper contact wipe, pressure, and contact alignment and to ensure that all contact surfaces are made simultaneously. The spacing between stationary and movable contacts should be checked in the fully open position. Adjustments should be made in accordance with the manufacturers' recommendations.

15.4.3.7 Laminated copper or brush-style contacts found on older circuit breakers should be replaced when they are badly burned. Repairs are impractical because the laminations tend to weld together when burning occurs, and contact pressure and wipe are greatly reduced. They can be dressed with a file to remove burrs or to restore their original shape. They should be replaced when they are burned sufficiently to prevent adequate circuit-breaker operation or when half of the contact surface is burned away. Carbon contacts, used on older circuit breakers, require little maintenance. However, inadequate contact pressure caused by erosion or repeated dressing might cause overheating or interfere with their function as arcing contacts.

15.4.3.8 The drawout primary disconnect contacts on the circuit breaker and the stationary contacts in the cubicle should be cleaned and inspected for overheating, proper alignment, and broken or weak springs. The drawout primary disconnect contact surfaces should be lightly coated with a contact lubricant to facilitate ease of the mating operation.

15.4.4 Arc Interrupters.

15.4.4.1 Modern arc interrupters of medium-voltage air circuit breakers are built with only inorganic materials exposed to the arc. Such materials line the throats of the interrupter and constitute the interrupter plates or fins, which act to cool and disperse the arc. The insulation parts of the interrupter remain in the circuit across contacts at all times. During the time that the contacts are open, these insulating parts are subject to full potential across the breaker. The ability to withstand this potential depends on the care given the insulation.

15.4.4.2 Particular care should be taken at all times to keep the interrupter assembly clean and dry.

15.4.4.3 The interrupters should be inspected each time the contacts are inspected. Any residue, dirt, or arc products should be removed with a cloth or by a light sanding. A wire brush or emery cloth should not be used for this purpose because of the possibility of embedding conducting particles in the ceramic material.

15.4.4.4 An interrupter should be inspected for broken or cracked ceramic parts, erosion of ceramics, and dirt.

15.4.4.4.1 Broken or Cracked Ceramic Parts. Small pieces broken from the ceramics or small cracks may not indicate a problem. Breaks or expansive cracks can interfere with reliable performance of the interrupter. If broken or badly cracked plates are apparent, replacement of the ceramic stack is recommended.

15.4.4.4.2 Erosion of Ceramics. When an arc strikes a ceramic part in the interrupter, the surface of the ceramic will be melted slightly. When solidified again, the surface will have a glazed, whitish appearance. At low and medium currents, the effect is slight. However, large-current arcs repeated many times can boil away appreciable amounts of the ceramic. When that happens, the ceramic stack assembly should be replaced.

15.4.4.4.3 Dirt in Interrupter. While in service, the arc chute assembly can become dirty. Dust or loose soot deposited on the inside surface of the arc chute can be removed by vacuuming or by wiping with cloths that are free of grease or metallic particles. Deposits can accumulate on ceramic arc shields from the arcing process. These deposits, from the metal vapors boiled out of the contacts and arc horns, can accumulate to a harmful amount in breakers that receive many operations at low- or medium-interrupting currents. Particular attention should be paid to any dirt on the plastic surfaces below the ceramic arc shield. These surfaces should be wiped clean, if possible, especially if the dirt contains carbon or metallic deposits. On breakers that operate thousands of times at low and medium currents, dirt can accumulate on the ceramic arc shields to impair proper interrupting performance. These arc chutes are of a very hard material, and a hard nonconducting abrasive is necessary for cleaning. The ceramic arc shields might appear dirty and yet have sufficient dielectric strength. The following insulation test can be used as a guide in determining when a complete or major cleaning operation is required. The arc chutes of medium-voltage circuit breakers should withstand the 60 Hz–rated maximum voltage for 1 minute between the front and rear arc horns. In some applications, circuit breakers can be exposed to overvoltages, in which case such circuit breakers should have an appropriate overpotential test applied across the open contacts. Some manufacturers also recommend a surface dielectric test of the ceramic surfaces near the contacts to verify adequate dielectric strength of these surfaces.

15.4.4.5 Air-puffer devices used to blow the arc up into the interrupter should be checked for proper operation. One accepted method is as follows. With the interrupter mounted on the breaker in its normal position, a piece of tissue paper is placed over the discharge area of the interrupter and observed for movement when the breaker is opened. Any perceptible movement of the paper indicates that the puffer is functioning properly.

15.4.4.6 Low-voltage air circuit–breaker arc chutes are of relatively simple construction, consisting primarily of a wedge-shaped vertical stack of splitter plates enclosed in an insulating jacket. An arc chute is mounted on each pole unit directly above the main contacts. Arc interruptions produce erosion of the splitter plates. The lower inside surfaces of the insulating jackets will also experience some erosion and sooty discoloration.

15.4.4.7 The arc chutes should be removed and examined as part of routine maintenance. If the splitter plates are seriously eroded, they should be replaced. If the interior surfaces of the enclosing jackets are discolored or contaminated with arc products, they should be cleaned or replaced. Occasionally, the entire arc chute could need replacing, depending on the severity of the duty and extent of the damage.

15.4.5 Operating Mechanism.

15.4.5.1 The purpose of the operating mechanism is to open and close the contacts. This usually is done by linkages connected, for most power breakers, to a power-operating device such as a solenoid or closing spring for closing, and that contains one or more small solenoids or other types of electromagnets for tripping. Tripping is accomplished mechanically, independently from the closing device, so that the breaker contacts will open even though the closing device still might be in the closed position. This combination is called a mechanical trip-free mechanism. After closing, the primary function of the operating mechanism is to open the breaker when it is desired, which is whenever a trip coil is energized at above its rated minimum operating voltage.

15.4.5.2 The operating mechanism should be inspected for loose or broken parts, missing cotter pins or retaining keepers, missing nuts and bolts, and binding or excessive wear. All moving parts are subject to wear. Long-wearing and corrosion-resistant materials are used by manufacturers, and some wear can be tolerated before improper operation occurs.

15.4.5.2.1 Excessive wear usually results in the loss of travel of the breaker contacts. It can affect operation of latches; they could stick or slip off and prematurely trip the breaker. Adjustments for wear are provided for certain parts. In others, replacement is necessary.

15.4.5.2.2 The closing and tripping action should be quick and positive. Any binding, slow action, delay in operation, or failure to trip or latch must be corrected prior to returning to service.

15.4.5.3 The two essentials to apply in maintenance of the operating mechanism are KEEP IT SNUG and KEEP IT FRICTION FREE.

15.4.6 Breaker Auxiliary Devices.

15.4.6.1 The closing motor or solenoid, shunt trip, auxiliary switches, and bell alarm switch should be inspected for correct operation, insulation condition, and tightness of connections.

15.4.6.2 On/off indicators, spring-charge indicators, mechanical and electrical interlocks, key interlocks, and padlocking fixtures should be checked for proper operation and should be lubricated where required. In particular, the positive interlock feature that prevents the insertion and withdrawal of the circuit breaker should be tested while it is in the closed position.

15.4.6.3 The protective relay circuits should be checked by closing the breaker in the test position and manually closing the contacts of each protective relay to trip the circuit breaker. Test procedures are given in 11.12.2.

15.4.6.4 Electromechanical Series Trip Devices. Trip devices on low-voltage breakers could be the electromechanical series overcurrent type with an air or fluid dashpot for time delay. These devices should be tested periodically for proper calibration and operation with low-voltage/high-current (primary injection) test devices. Calibration tests should be made to verify that the performance of the breaker is within the manufacturer's published curves. It is important that manufacturers' calibration curves for each specific breaker rating be used. Time-current curves are plotted as a band of values rather than a single line curve, and this factor should be taken into account. If the trip devices do not operate properly, the calibration and timing components should be repaired or replaced in accordance with the manufacturer's recommendations. For further information on testing, see Section 11.10.

15.4.6.5 Solid State (Static) Trip Devices. If the breakers are equipped with static-tripping devices, they should be checked for proper operation and timing in accordance with the manufacturer's recommendations. Some manufacturers recommend replacement of electromagnetic devices with static devices in the interest of realizing more precision and a higher degree of reliability with the latter devices. Secondary injection test devices are available for verifying the logic, timing, and trip circuits of static trip devices; however, these devices do not test the power supply, the current sensor, or interconnecting wiring. For further information on testing, see Section 11.10.

15.5 Vacuum Circuit Breakers.

15.5.1 The principal difference between vacuum circuit breakers and air circuit breakers is in the main contact and interrupter equipment. In the vacuum circuit breaker, these components are in an enclosed vacuum bottle assembly and are not available for cleaning, repair, or adjustment. Contact-wear indicators are available for measuring contact wear within the bottle assembly.

15.5.2 Vacuum integrity tests are performed by application of test voltage across the open contacts of the bottle. These tests should be performed strictly in accordance with the manufacturer's instructions. Ac or dc overpotential tests can be performed. However, some dc test sets may not provide appropriate filtering of the output voltage, resulting in potential damage to the vacuum bottle. For further information on testing, see 11.16.3.

CAUTION: Application of high dc voltage across an open gap in vacuum can produce x-ray emission.

15.5.3 The level of x-ray emission from a vacuum breaker with proper contact spacing and subjected to standard test voltages is extremely small and well below the maximum level permitted by standards. In view of the possibility that the contacts are out of adjustment or that the applied voltage is greater than prescribed, it is advisable that during the overvoltage test all personnel remain behind the front steel barrier and remain farther from the breaker than would otherwise be necessary for reasons of electrical safety. During the high-voltage test, the vapor shield inside the interrupter can acquire an electrostatic charge. This should be discharged to ground immediately after the test.

15.5.4 All other maintenance on vacuum circuit breakers should be performed in accordance with the previous recommendations on air circuit breakers.

15.6 Oil Circuit Breakers.

15.6.1 Introduction.

15.6.1.1 Oil circuit breakers are seldom found in modern metal-enclosed switchgear assemblies. They are prevalent in older metal-enclosed switchgear assemblies and in open-type outdoor substations.

15.6.1.2 Although oil circuit breakers perform the same function in switchgear assemblies as air circuit breakers, they are different in appearance and mechanical construction. The principal insulating medium is mineral oil rather than air.

15.6.2 Insulation.

15.6.2.1 External insulation is provided by insulating bushings. Outdoor oil circuit breakers typically have porcelain bushings, whereas indoor breakers can have either porcelain or epoxy bushings or organic tubing. The bushings should be examined for evidence of damage or surface contamination. If they are damaged to the extent that the electrical creepage path has been reduced or the glazed surface on porcelain bushings is damaged, they should be replaced. Otherwise they should be cleaned thoroughly as required to remove all surface contamination.

15.6.2.2 The oil, in addition to providing insulation, acts as an arc-extinguishing medium in current interrupters. In this process, it absorbs arc products and experiences some decomposition in the process. For that reason, maintenance of the oil is important. Oil maintenance involves detection and correction of any condition that would reduce its overall quality and electrical insulating properties. The principal contaminants are moisture and carbon. Moisture will appear as droplets on horizontal members, while free water will accumulate in the bottom of the tank. Moisture may appear as a milky translucent substance. Carbon initially appears as a black trace. It eventually will disperse and go into suspension, causing the oil to darken.

15.6.2.3 A dielectric breakdown test is a positive method of determining the insulating value of the oil. Oil should be tested periodically or following a fault interruption. For further information on testing, see 11.16.2.

15.6.2.4 In replacing the oil, only the oil recommended by the manufacturer should be used, and it should have been stored in sealed containers. In addition, the oil should have a dielectric breakdown test performed immediately prior to use. Avoid air entrapment when oil is added to the breaker by using an oil pump. For further information, see Section 11.19 and 11.16.2.

15.6.2.5 In the event entrapment of air cannot be avoided, the equipment should be allowed to stand for 8 to 12 hours prior to being energized.

15.6.3 Contacts. The main contacts of an oil circuit breaker are typically not readily accessible for routine inspection. Contact resistance should be measured. Contact engagement can be measured by measuring the travel of the lift rod from the start of contact opening to the point where contacts separate, as indicated by an ohmmeter.

15.6.3.1 More extensive maintenance on main contacts could require removal of the oil and lowering the tank and should therefore be performed less frequently than routine maintenance. The frequency should be determined by the severity of the breaker duty, for example, the number of operations and operating current levels. Any time the breaker has interrupted a fault current at or near its maximum rating, this type of maintenance should be performed. The contacts should be inspected for erosion or pitting. Contact pressures and alignment should be checked. All bolted connections and contact springs should be inspected for looseness.

15.6.4 Arc-Quenching Assemblies.

15.6.4.1 Arc-quenching assemblies should be inspected for carbon deposits or other surface contamination in the areas of arc interruption.

15.6.4.2 If cleaning of these surfaces is necessary, manufacturers' instructions should be followed.

15.6.5 Operating Mechanism. Maintenance of the operating mechanism should follow the same procedure as recommended for air circuit breakers. *(See 15.4.5.)*

15.6.6 Breaker Auxiliary Devices. Breaker-auxiliary-device maintenance should follow the same procedure as recommended for air circuit breakers *(see 15.4.6)* when applicable. Other accessories, such as oil level gauges, sight glasses, valves, gaskets, breathers, oil lines, and tank lifters should be inspected. The breaker should be taken out of service immediately if the oil level is below the level gauge or sight glass.

15.7 Interrupter Switches.

15.7.1 A medium-voltage interrupter switch is an air switch equipped with an interrupter for making or breaking specified currents, or both. It can be either the fixed-mounted type or the draw-out type and can be either manually or electrically operated. If fixed-mounted, it will be interlocked with access doors or panels to prevent access to closed switches.

15.7.2 Maintenance procedures should correspond to those recommended for air circuit breakers except for the interrupter device. This device, on most interrupter switches, is of a simple open-type construction and can be inspected and cleaned easily without being removed from the switch. Enclosed interrupters should be removed from the switch and disassembled for maintenance in accordance with the manufacturer's recommendation. Dielectric tests are not usually required as a part of maintenance. Air puffers are not employed in this type of interrupter. For further information, see Section 11.15.

15.8 Gas-Insulated Substations and Gas-Insulated Equipment.

15.8.1 Introduction. A gas-insulated substation (GIS) is a manufactured assembly of gas-insulated equipment (GIE) typically installed on electric systems rated 72.5 kV and above. While some GISs are operated by industrials, most are operated by utilities. A GIS might include such GIE as circuit breakers, disconnect switches, ground switches, voltage transformers, current transformers, capacitors, gas-to-air bushings, gas-to-cable terminations, buses, associated enclosures, and control and monitoring equipment.

15.8.1.1 In addition to the general guidelines provided in this document, manufacturers' maintenance procedures for GIE should be followed.

15.8.1.2 Some circuit-interrupting devices are hermetically sealed inside an enclosure containing a prescribed amount of sulfur hexafluoride (SF_6) gas and are not serviceable. These devices typically are found in some medium-voltage metal-enclosed switchgear and medium-voltage outdoor power circuit breakers, and they also are used in some medium- and high-voltage interrupter switches. Generally, this equipment utilizes devices that provide a visual indication of loss of gas. Replacement of the entire sealed unit is required when this occurs.

15.8.1.3 Other GIE units contain gas density gauges to monitor temperature and pressure, providing indication of sufficient gas to maintain the equipment ratings.

15.8.2 Sulfur Hexafluoride (SF_6) Gas.

15.8.2.1 SF_6 gas, under pressure, is used as the dielectric and interrupting medium in circuit breakers and switches. SF_6 has been identified as a "greenhouse gas," whose release to the environment could be detrimental. Therefore, SF_6 gas should be reused and recycled whenever possible and should never be released into the atmosphere unnecessarily. SF_6 in its pure state is odorless, colorless, tasteless, nonflammable, noncorrosive, and nontoxic. It is five times heavier than air and will settle to the bottom of an enclosed vessel, displacing any breathable air to the top of the vessel. Although nontoxic, SF_6 gas does not support life by itself and will result in asphyxiation. For that reason, any vessel previously containing SF_6 gas should not be entered unless thorough ventilation has been achieved and the oxygen content verified. See OSHA 29 CFR 1910.146, "Occupational Safety and Health Standards," for practices and procedures to protect employees from the hazards of entry into permit-required confined spaces and Part 7 of 29 CFR 1910.269, "Electric Power Generation, Transmission, and Distribution," paragraph (e), for enclosed space entry.

15.8.2.2 SF_6 gas used in GIE can be tested and should conform to ASTM D 2472, *Standard Specification for Sulfur Hexafluoride.*

15.8.3 Causes of SF_6 Decomposition. SF_6 decomposes as a result of excessive heating, electric sparks, power arcs, and partial discharges. The rate of decomposition of the gas during operation is determined by the equipment design and the inclusion of desiccants and adsorbents within the equipment. A power arc associated with a fault in the equipment results in decomposition of SF_6 within a compartment and the generation of gaseous and solid by-products.

15.8.4 Decomposition By-Products.

15.8.4.1 Gaseous By-Products. The major gaseous by-products include S_2F_2, SOF_2, SO_2F_2, SF_4, SO_2, and HF. Some of these gases are highly toxic. The reaction of some of these gases with available moisture produces additional quantities of toxic gases. Arcing causes SF_6 to decompose into other sulfur fluorides and, in the presence of moisture, hydrogen fluoride. These decomposition products are toxic and harmful to the eyes, nose, and lungs. When arcing has occurred, breathing of any SF_6 by-products should be avoided by using appropriate personal protective equipment. A rotten-egg odor in the vicinity of the equipment is indicative of contaminated SF_6.

15.8.4.2 DO NOT vent gas from the equipment or attempt to sniff for it.

15.8.4.3 Solid Arcing By-Products. Solid decomposition by-products are produced in the form of a fine, talcum-like powder. This powder is a metal fluoride and is white or tan in color. The danger from solid arcing by-products comes more from the gases adsorbed on the surface area of the powder than from the toxicity of the base material. The aluminum fluoride (AlF_3) powder that normally dominates solid arcing by-products is so fine that the lungs do not easily expel it. Powder by-products are hazardous waste and should be disposed of in accordance with manufacturer instructions and government regulations.

15.8.5 Maintenance and Repair of the GIS and GIE.

15.8.5.1 General.

15.8.5.1.1 Instruction books or equipment manuals furnished with the GIS or GIE are necessary for operating and maintenance personnel. This instructional literature should include information pertaining to safe operating and maintenance procedures.

15.8.5.1.2 Safety during maintenance and repair requires that the components on which work is to be performed are electrically isolated, deenergized, grounded, and locked/tagged out.

15.8.5.1.3 Equipment should never be depressurized until it is deenergized and grounded.

15.8.5.1.4 Cleanliness, in accordance with manufacturer's instructions, should be observed at all times. The area around the access point to be opened, including supporting steel and other parts from which dirt or contaminants could fall or be blown into the enclosure, should be vacuumed and wiped with lint-free cloths.

15.8.5.1.5 Do not stand or step on small piping or connections.

15.8.5.1.6 Gas is handled through commercially available gas-processing trailers (carts) that contain vacuum-pumping equipment, gas storage tanks, compressors, filters, and dryers. Suitable evacuating equipment and a heat source to counteract the chilling effect of the expanding gas can permit filling directly from gas cylinders or gas-handling equipment. The specific requirements for gas purity, handling, processing, filling, and refilling that the equipment manufacturer provides should be followed to ensure proper equipment operation.

15.8.5.1.7 Evacuate to 133 Pa all gas from the compartment by a closed evacuation system and pass it through a filter capable of removing arc decomposition by-products. Allow dry air to enter and refill the compartment to atmospheric pressure (101 kPa) before opening the access port.

15.8.5.1.8 Upon opening of the compartment, ensure proper ventilation and oxygen content of enclosure prior to personnel entry.

15.8.5.1.9 Maintenance workers who conduct the initial opening of the faulted gas compartment and removal of the solid arcing by-products should employ air respirators and wear disposable protective clothing covering all garments, boots, hair, and hands. Avoid direct contact with arc decomposition by-products (fine powder).

15.8.5.1.10 Work quickly, because the exposure of the solid arcing by-products to moist air will result in toxic fumes with a strong rotten-egg odor. Immediate removal of the solid by-products should be the first priority, as they can become sticky and more difficult to remove with continued exposure to moist air.

15.8.5.1.11 A commercial-type vacuum cleaner with high-efficiency particulate air (HEPA) filters and nonmetallic accessories should be used to remove the arc solid by-products. Precautions should be taken to avoid breathing the exhaust air from the vacuum cleaner, since dust particles might go through the collection system. Following vacuuming, the affected area should be wiped down with a safety solvent by workers continuing to wear respirators, appropriate personal protective equipment, and disposable clothing.

15.8.5.1.12 All work should be completed safely and as quickly as possible. When delays are encountered, any open sections should be covered immediately with suitable seals. It might also be necessary to add heat to prevent condensation. When any section is left overnight or longer, it should be pressurized with dry air to a pressure of approximately 136 kPa to avoid condensation or entrance of moist air.

15.8.5.1.13 To achieve the required fill density, it is important that the gas pressure and temperature curve from the manufacturer's instruction book be used. Sufficient time should be allowed for equalization of the gas temperature.

15.8.5.1.14 After recharging, several measurements of the moisture content of the gas in the equipment should be taken to ensure that the moisture content of the SF_6 remains within acceptable limits. If the moisture content of the gas rises to an unacceptable level, recirculation through the dehydration portion of the gas-processing trailer (cart) is required to remove the excess moisture. For further information on testing, see 11.16.4.

15.8.5.2 References. The following should be considered for specific maintenance, repair considerations, and procedures:

(1) The equipment manufacturer's instructions
(2) ANSI/IEEE 1125, *IEEE Guide for Moisture Measurement and Control in SF6 Gas-Insulated Equipment*
(3) ANSI/NETA MTS, *Standard for Maintenance Testing Specifications for Electrical Power Distribution Equipment and Systems*, Section 7.5.4, SF_6 Switches, and Section 7.6.4, SF_6 HV Circuit Breakers
(4) Sections of IEEE C37.122.1, *IEEE Guide for Gas-Insulated Substations*, as follows:
   (a) 4.2, Installation and equipment handling
   (b) 4.4, Gas handling-SF_6 and GIS
   (c) 4.5, Safe operating procedures
   (d) 4.8, Partial discharge (PD) testing

(e) 4.10, Field dielectric testing
(f) 4.11, Maintenance and repair

15.9 Auxiliary Equipment.

15.9.1 Fuses. Fuse maintenance is covered in Chapter 18. For further information on fuse testing, see Section 11.18.

15.9.2 Surge Arresters.

15.9.2.1 Surge arresters should be inspected periodically for evidence of damage to the porcelain housing or surface contamination. If the porcelain is damaged to the extent that the creepage path over its surface is reduced or the porcelain glazed surface is seriously damaged, the arrester should be replaced. Otherwise, the porcelain surface should be cleaned thoroughly as required to remove all surface contamination.

15.9.2.2 There are certain tests that can be performed that will give sufficient information to determine whether the arrester can be relied on to be an insulator under normal conditions. For further information, see Sections 11.24 and 11.13.

15.9.3 Power-Factor Correction Capacitors.

15.9.3.1 Capacitors should always be discharged before handling or making connections by closing the ground devices that are usually installed with large capacitor banks. A grounding stick that has an internal resistor should be used for dissipating the charge; however, it should be applied only with full knowledge of the circuit and with the use of appropriate protective equipment.

CAUTION: Capacitors, even though they have discharge resistors, could possess a stored charge that is capable of injuring a person coming into contact with the terminals.

15.9.3.2 The capacitor case, the insulating bushings, and any connections that are dirty or corroded should be cleaned. Each capacitor case should be inspected for leaks, bulges, or discoloration. Any liquid-filled capacitor found to be bulging or leaking should be replaced.

15.9.3.3 Power capacitors are generally provided with individual fuses to protect the system in case of a short circuit within the capacitor. In addition to a faulty capacitor, a fuse can also be opened by an abnormal voltage surge. A check should be made for open fuses, which should be replaced with the type recommended by the manufacturer. Fuses should not be removed until the capacitor has been discharged completely.

15.9.3.4 Adequate ventilation is necessary to remove the heat generated by continuous full-load duty. Any obstructions of ventilation openings in capacitor housings should be cleared, and adequate ventilation must be provided and maintained.

15.9.3.5 For further information on capacitor testing, see Section 11.25.

15.9.4 Stationary Batteries and Battery Chargers.

15.9.4.1 General. Stationary batteries are a power source for critical systems, ac power generation equipment, switchgear, and control circuits. Stationary batteries are most often used as the reserve power source for critical equipment during power outages. These batteries are typically vented lead–acid (VLA) batteries consisting of either lead calcium or lead antimony grids. Other technologies can be utilized in these applications but are not specifically addressed in the following paragraphs. Due to the reliability requirements for these applications, specific safety guidelines as well as well-defined maintenance practices should be utilized to ensure a reliable system. The following sections of 15.9.4 offer guidance on the safety and maintenance requirements. In addition, the manufacturers' recommendations should be followed and the applicable IEEE standards should be referred to for additional information.

15.9.4.1.1 Battery chargers play a critical role in battery maintenance because they supply normal dc requirements and maintain batteries at appropriate levels of charge. Chargers should be set and maintained according to manufacturers' instructions.

15.9.4.1.2 References. The following references should be considered for specific maintenance, repair considerations, and procedures:

(1) IEEE 450 — *IEEE Recommended Practice for Maintenance, Testing and Replacement of Vented Lead-Acid Batteries for Stationary Applications*
(2) IEEE 1188 — *IEEE Recommended Practice for Maintenance, Testing and Replacement of Valve-Regulated Lead-Acid (VRLA) Batteries for Stationary Applications*
(3) IEEE 1106 — *IEEE Recommended Practice for Installation, Maintenance, Testing and Replacement of Vented Nickel-Cadmium Batteries for Stationary Applications*
(4) IEEE 1145 — *IEEE Recommended Practice for Installation and Maintenance of Nickel-Cadmium Batteries for Photovoltaic (PV) Systems*
(5) IEEE Std 1657 — *IEEE Recommended Practice for Personnel Qualifications for Installation and Maintenance of Stationary Batteries*

15.9.4.2 Maintenance Program. Battery maintenance normally consists of periodic inspections and tests. As a minimum, a maintenance program should be established based on manufacturers' installation and operation manuals. For further information on testing, see Section 11.14.

15.9.4.3 Safety Guidelines. Personnel should be aware of the hazards associated with stationary batteries. A vented battery emits a mixture of hydrogen and oxygen gas. Under abnormal conditions, such as a lack of space ventilation, it is possible for the mixture to reach a flammable level. Voltages present can cause injury and death. Exposing skin and eyes to electrolyte can cause severe burns and blindness. Vented lead-acid batteries can emit a mixture of hydrogen and oxygen gas. Under abnormal conditions, such as a lack of space ventilation, it is possible for the mixture to reach a flammable level. As a minimum, the safety precautions in 15.9.4.3.1 through 15.9.4.3.6 should be observed.

15.9.4.3.1 Maintenance personnel should be trained to perform the tasks properly. Training should include using personal protective equipment, handling the electrolyte safely, using the proper tools, and following the battery manufacturer's service and maintenance instructions.

15.9.4.3.2 The technician should verify that the ventilation system in the room or compartment in which operating lead–acid batteries are located is operating as required.

15.9.4.3.3 Appropriate safety equipment should be used in accordance with *NFPA 70E, Standard for Electrical Safety in the Workplace*, and manufacturer's instructions.

15.9.4.3.4 Open flames, sparks, and other ignition sources should be kept away from storage batteries, gas ventilation paths, and places where hydrogen can accumulate.

15.9.4.3.5 Conductive objects should not be used near battery cells. Insulated tools should be used to protect against shorting of cells.

15.9.4.3.6 Unauthorized access to exposed batteries should be prohibited.

15.9.4.4 Guide for Visual Inspections of Electrical and Mechanical Equipment.

15.9.4.4.1 Containers and covers should be checked for cracks and structural damage. Damaged units and damaged or missing removable vent caps should be replaced.

15.9.4.4.2 Plates and internal parts within clear containers should be checked for damage such as buckling, warping, scaling, swelling, cracking, hydration rings, excessive sedimentation, mossing, copper contamination, internal post seal cracks, and changes in color. Cells exhibiting any of these characteristics should be evaluated for repair or replacement.

15.9.4.4.3 The charger should be checked for proper operation. Interconnection cables, cell connectors, and other conductors should be examined for wear, contamination, corrosion, and discoloration. Racks should be checked for corrosion, cleanliness, proper grounding, and structural integrity.

15.9.4.4.4 A check should be made for spilled electrolyte. A solution of water and bicarbonate of soda (baking soda) should be used to neutralize lead–acid battery spills, and a solution of boric acid and water should be used for Ni-Cad spills. The battery manufacturer's instructions should be consulted for proper proportions.

15.9.4.4.5 The electrolyte level before water addition should be checked, and corrective measures should be noted in accordance with the owner's maintenance program and the manufacturer's recommendations. Excessive water consumption can be a sign of overcharging or cell damage. For lead–antimony batteries, water consumption increases gradually with age. Distilled or deionized water should be used unless otherwise recommended by the battery manufacturer.

CAUTION: Never add anything but water to a battery unless recommended to do so by the manufacturer.

15.9.4.4.6 Ventilation and the suitability and condition of electrical equipment in the area should be checked for its possible effect on the battery. Local sources of heating and cooling can create cell temperature differentials that cause battery damage. Battery room ventilation openings should be checked to be sure they are clear of obstructions.

15.9.4.4.7 Ambient temperature should be checked to be within the manufacturer's recommended range. High ambient temperatures reduce cell life. Lower ambient cell temperatures reduce cell capacity.

15.9.4.4.8 Area heating, air conditioning, seismic protection, dc circuit overcurrent protection, distilled water supply, grounding connections, cable clamps, and all other installed protective systems and devices should be checked.

15.9.4.4.9 Voltage potential between the battery's most positive and most negative terminals should be verified to be within the manufacturer's recommended float voltage range for the observed ambient temperature.

15.9.4.4.10 Terminal connectors, battery posts, and cable ends should be checked and all corrosion and dirt removed. Battery posts should be cleaned according to manufacturers' recommendations.

15.9.4.4.11 Lead–acid battery surfaces should be cleaned with a solution of water and sodium bicarbonate to avoid leakage currents caused by electrolyte on the battery. Ni-Cad battery surfaces should be cleaned with a solution of boric acid and water. Cleaners, soaps, or solvents should not be used to clean battery containers and covers since damage can result. Consult the battery manufacturer for the proper solution and dilution.

15.9.4.4.12* All battery connections should be checked on a routine basis for high connection resistance. When a connection resistance is high, then the connection should be cleaned and torqued in accordance with the manufacturer's procedures.

15.9.4.4.13 Alarm relays, lights, and horns should be checked for proper operation. The battery room emergency lighting should be checked.

15.9.4.5 General Observations.

15.9.4.5.1 Vibration reduces battery life. Excessive vibration can be detected by observing vibration of plates and sediment in the jar. If this condition is observed, then steps should be taken to reduce the vibration source, isolate the batteries from the vibration, and/or plan for an earlier than normal scheduled replacement of the batteries.

15.9.4.5.2 Some lead-acid batteries support as few as 50 full discharges while others can tolerate over 1000 discharges, depending upon the battery construction and the depth of discharges. Excessive discharges can shorten the life of a battery. Fully discharging a stationary battery more than twice in one year is not recommended.

15.9.4.6 Battery Test and Measurement Guidelines. For further information on battery testing, see Section 11.14.

15.9.5 Instrument Transformers and Auxiliary Transformers.

15.9.5.1 Instrument transformers and auxiliary transformers might be the outdoor type, although in some cases they can be mounted inside metal-enclosed switchgear assemblies. These transformers are similar to other outdoor transformers in that they are liquid filled and equipped with outdoor bushings. All recommendations for maintenance of outdoor transformers apply.

15.9.5.2 Indoor-type instrument and auxiliary transformers are normally of a dry type construction and should be kept clean and dry. All of the transformers above are of the completely molded type, with only the terminals exposed. The maintenance recommendations in 15.2.12 can be applied to indoor transformer types.

15.9.6 Alarm and Indicators.

15.9.6.1 Alarms. Alarms associated with transformer overtemperature, high or low pressure, circuit-breaker trip, accidental ground on an ungrounded system, cooling waterflow or overtemperature, or other system conditions should be tested periodically to ensure proper operation.

15.9.6.2 Indicators. Circuit-breaker open/close indicators can be checked during their regular maintenance.

15.9.6.2.1 Flow, overtemperature, and excess pressure indicators, and so forth, should be checked or operated periodically to ensure proper operation.

15.9.7 Protective Relays and Metering Devices.

15.9.7.1 The current elements of protective relays and metering devices are usually connected in the secondary circuit of current transformers.

CAUTION: Opening the secondary circuit of an energized current transformer will produce a very high voltage that can be lethal.

15.9.7.1.1 The secondary terminals of an energized current transformer are required to be short-circuited before the secondary circuit is opened. Some protective relays and metering devices have special test terminals or test switches that make a closed circuit in the current transformer secondaries during test. Upon completion of tests, it is necessary to remove the short-circuit jumper to permit the current transformer to function. For further information on relay testing, see Section 11.12.

15.9.7.2 Protective relays and metering devices play an important role in the prevention of hazard to personnel and facility equipment. Typically, they only operate during an abnormal electric power system condition, and the only way to ensure correct operation is by a comprehensive inspection, maintenance, and testing program.

15.9.7.3 Protective relays should be examined to ensure that all moving parts are free of friction or binding. Metering devices should be calibrated or verified for proper indication. Wiring should be checked for loose connections. Contacts should be inspected for pitting or erosion. Solenoid coils, armatures and circuit boards should be inspected for evidence of overheating. Cracked glass or damaged covers or cases should be replaced. For further information on relay testing, see Section 11.12.

15.9.8 Interlocks and Safety Devices. Interlocks and safety devices are employed for the protection of personnel and equipment and should never be made inoperative or bypassed. Proper functioning of these devices should be ensured by the following procedures:

(1) The adjustments and operation of the devices should be checked as follows:
    (a) Mechanical interlocks on drawout mechanisms should prevent withdrawal or insertion of electrical devices in the closed position.
    (b) Safety shutters, where provided, should automatically cover the energized components.
    (c) Limit switches should prevent overtravel of motorized lifting devices.
(2) Key interlock systems should be operated in proper sequence, and suitable operation ensured by the following:
    (a) Adjustments should be made and the system lubricated in accordance with manufacturer's instructions.
    (b) For complicated systems, interlock operational procedures should be available to authorized personnel where the interlocks might be operated only annually or in emergencies.
(3) Spare keys should be identified and controlled to prevent unauthorized use.
(4) Grounding and test devices used in medium-voltage switchgear should be maintained to the same degree as the circuit breaker itself.
    (a) If stored indoors, they should be covered to prevent dust accumulation.
    (b) If stored outdoors, they should be stored in a weatherproof covering.

15.9.9 Grounding.

15.9.9.1 Equipment-grounding circuits are not inherently self-monitoring. Equipment-grounding conductors should be checked periodically.

15.9.9.2 Checking a system to determine the adequacy of the equipment ground involves inspection of connections that can be supplemented by an impedance test to enable an evaluation of those parts of the system not accessible for inspection. For further information on ground testing, see 11.13.3.

15.9.9.3 Terminal connections of all equipment-grounding conductors and bonding jumpers should be checked to see that they are tight and free of corrosion. Bonding jumpers should also be examined for physical abuse, and those with broken strands should be replaced. Where metal raceway is used as the equipment-grounding path, couplings, bushings, set-screws, and locknuts should be checked to see that they are tight and properly seated. Any metal raceway used as the equipment-grounding path should be examined carefully for rigid mounting and secure joints; screws and bolts should be checked to confirm that they are properly torqued.

15.9.10 Ground-Fault Indicators.

15.9.10.1 Ground-fault indicators can be installed on all ungrounded or resistance-grounded low-voltage systems. The indicator can consist of a simple set of lamps wired phase-to-ground. A ground on one phase will cause the lamp on that phase to be dark, while the other two lamps will have increased brilliance. Ground indicator lamps for ungrounded electric systems should be checked daily or weekly for proper operation.

15.9.10.2 A more elaborate system provides audible as well as visual indication so the ground is more readily detected.

15.9.10.3 Once a ground has been detected, prompt location and correction are important, since the system will be highly vulnerable in the event of a ground on another phase. Through the process of elimination, searching for the ground requires circuit or system interruptions and isolation of the circuit(s) until the ground fault is located and eliminated. The use of an instrument that permits location of such ground faults without power interruptions is recommended.

15.9.10.4 Maintenance of ground fault detectors should include a complete inspection of the signal elements such as lamps, horns, or buzzers. Audible devices should be operated to ensure that they are in operable condition. Wiring should be checked for loose connections or damaged wiring.

15.9.10.5 A complete, effective maintenance program for substations and assembled switchgear will result if the four "keepers" are observed:

(1) If it is insulation, KEEP IT CLEAN and KEEP IT DRY.
(2) If it is mechanical, KEEP IT SNUG and KEEP IT FRICTION FREE.

15.9.11 Network Protectors.

15.9.11.1 A network protector is an air circuit breaker equipped with specialized relays that sense network circuit conditions and command the circuit breaker to either open or close. There is no separate power source for control. All control power is taken from the system.

15.9.11.1.1 A routine maintenance schedule for network protectors should be observed. Frequency of inspection will vary to a great extent depending on the location and the environment in which a protector is installed.

15.9.11.1.2 Maintenance should include the cleaning of any accumulated dust from the unit, a thorough visual inspection, and overall operational test. Should any part appear damaged, defective, or out of adjustment, the manufacturer's instructions for operation, adjustment, and replacement of these parts should be consulted. If relays are out of calibration, they should be recalibrated by competent personnel. For further information on protective relay testing, see Section 11.12.

15.9.11.2 Safety. The first procedure in performing maintenance is to trip the protector to the open position and where possible remove the protector from the system. See Chapter 7 for further safety guidelines.

15.9.11.2.1 Extensive use of barriers is an important feature in the design of network protectors. The barriers should be kept in place, and any that have been broken should be replaced immediately. Only skilled maintenance personnel who are thoroughly familiar with the construction and operation of network protectors should be permitted to perform maintenance on an energized unit.

15.9.11.3 Maintenance. The circuit-breaker mechanism and relay panel assembly are usually constructed as an integral drawout unit that should be withdrawn from the housing for maintenance. Removal of the fuses at the top and the disconnecting links at the bottom (some modern protectors have bolt-actuated disconnecting fingers at the bottom) isolates the unit electrically from the system. Although this procedure provides comparative safety, work should be done cautiously, since it should be assumed that normally there is voltage on the transformer and the network leads. With the drawout unit outside the enclosure on the extension rails, the following inspection and maintenance operations should be performed on the drawout unit.

15.9.11.3.1 The complete unit should be cleaned. Use of a vacuum cleaner is preferred, or cloth rags free of oil or grease should be used to remove dirt.

15.9.11.3.2 Arc chutes should be inspected, and any damaged splitter plates should be replaced.

15.9.11.3.3 Main contacts should be inspected.

15.9.11.3.3.1 Any damaged area should be smoothed with a fine file, stone, crocus cloth, or other suitable abrasive that does not shed particles.

15.9.11.3.3.2 The hinge joint should be protected from falling particles during dressing.

15.9.11.3.4 Arcing contacts might become rough and high projections of metal should be smoothed.

15.9.11.3.5 All electrical connections should be checked and torqued to manufacturer's specifications. Bolts should be loosened and re-torqued to proper values.

15.9.11.3.6 Any abrasion of wire insulation should be repaired.

15.9.11.3.7 Control wiring and current-carrying parts should be inspected for overheating.

15.9.11.3.8 Springs should be in place and unbroken.

15.9.11.3.9 Nuts, pins, snap rings, and screws should be inspected to see that they are in place and tight.

15.9.11.3.10 Any broken barriers should be replaced.

15.9.11.3.11 With the rollout unit removed, the following maintenance operations should be performed inside the enclosure.

CAUTION: Both network and transformer connections should be treated as though they are energized. When working in housing or on frame, use only insulated tools and wear safety protective equipment. Do not remove any barriers from the enclosure.

15.9.11.3.11.1 Hardware on the floor or beneath the frame should be traced to its source and reinstalled.

15.9.11.3.11.2 Stand-off bus insulators should be cleaned.

15.9.11.3.11.3 Any oxide film should be removed from terminal contacts if necessary.

15.9.11.3.12 The protector should be closed manually in accordance with the manufacturer's instructions.

15.9.11.3.12.1 It should close with a definite snap action. Sluggish closing indicates excessive friction.

15.9.11.3.12.2 The trip level should be moved to the "tripped" position.

15.9.11.3.12.3 The breaker should snap open.

15.9.11.3.13 An operational test is best performed using a network protector test kit.

15.9.11.3.14 An insulation resistance test, a dielectric test, and electrical operating tests should be performed strictly in accordance with the manufacturers' recommendations. For further information on insulation resistance testing, see 11.9.2.3.

15.10 Torque Tables. Table 15.10(a) through Table 15.10(d) are derived from Square D Company, Anderson Products Division, General Catalog: Class 3910 Distribution Technical Data, Class 3930 Reference Data Substation Connector Products, and Penn-Union Catalog data.

Table 15.10(a) Bolt-Torque Values for Electrical Connections U.S. Standard Fasteners[a] Heat-Treated Steel — Cadmium or Zinc Plated[b]

Grade	SAE 1&2	SAE 5	SAE 7	SAE 8
Head Marking				
Minimum Tensile (Strength) (lbf/in.^{2})	64K	105K	133K	150K
Bolt Diameter (in.)	Torque (Pound-Feet)			
¼	4	6	8	8
5⁄16	7	11	15	18
⅜	12	20	27	30
7⁄16	19	32	44	48
½	30	48	68	74
9⁄16	42	70	96	105
⅝	59	96	135	145
¾	96	160	225	235
⅞	150	240	350	380
1.0	225	370	530	570

[a] Consult manufacturer for equipment supplied with metric fasteners.
[b] Table is based on national coarse thread pitch.

Table 15.10(b) U.S. Standard Fasteners[a] Silicon Bronze Fasteners[b, c] Torque (Pound-Feet)

Bolt Diameter (in.)	Nonlubricated	Lubricated
5⁄16	15	10
3⁄8	20	15
1⁄2	40	25
5⁄8	55	40
3⁄4	70	60

[a] Consult manufacturer for equipment supplied with metric fasteners.
[b] Table is based on national coarse thread pitch.
[c] This table is based on bronze alloy bolts having a minimum tensile strength of 70,000 lb/in.^{2}

Table 15.10(c) U.S. Standard Fasteners[a] Stainless Steel Fasteners[b, c] Torque (Pound-Feet)

Bolt Diameter (in.)	Lubricated
5⁄16	10
3⁄8	14
1⁄2	25
5⁄8	40
3⁄4	60

[a] Consult manufacturer for equipment supplied with metric fasteners.
[b] Table is based on national coarse thread pitch.
[c] This table is based on aluminum alloy bolts having a minimum tensile strength of 55,000 lb/in.^{2}

Table 15.10(d) U.S. Standard Fasteners[a] Stainless Steel Fasteners[b, c] Torque (Pound-Feet)

Bolt Diameter (in.)	Uncoated
5⁄16	15
3⁄8	20
1⁄2	40
5⁄8	55
3⁄4	70

[a] Consult manufacturer for equipment supplied with metric fasteners.
[b] Table is based on national coarse thread pitch.
[c] This table is to be used for the following hardware types: bolts, cap screws, nuts, flat washers, locknuts (18-8 alloy) Belleville washers (302 alloy).

Chapter 16 Motor Control Equipment

16.1 Introduction.

16.1.1 There are many varieties of motor controllers, motor control centers, switchboards, and power panels. Following are some of the more common motor starters:

(1) Manual across-the-line starters
(2) Magnetic across-the-line starters
(3) Combination starters
  (a) Breaker-protected starters
  (b) Fuse-protected starters
  (c) Fused breaker–protected starters
(4) Reduced-voltage starters
  (a) Autotransformer starters
  (b) Resistance starters
  (c) Part-winding starters
  (d) Wye-delta starters
  (e) Solid-state starters
(5) Two-speed starters
(6) Starters and speed regulators for ac wound rotor and dc motors
(7) Adjustable-speed/frequency starters
(8) Miscellaneous types
  (a) Reversing starter
  (b) Motor control center

16.1.2 The maintenance recommendations in this chapter are general in nature and can be adapted to a wide variety of product types.

16.2 Components and Maintenance of Motor Controls. Motor control equipment should be inspected and serviced at the same time as the motors. As a general rule, overhaul procedures for control equipment are less involved than motor overhauling. Most repairs can be made on-site. Motor starters represent one area in which manufacturers have emphasized simplicity of construction and wiring. Improvements have resulted in starters that are simple to install, maintain, and operate. Connections are readily accessible. Some parts are the plug-in type and can be replaced easily. Coils are often encapsulated in epoxy compounds and are less likely to burn out. Practically all newer starters have provisions for adding several auxiliary contacts with very little effort. Spare parts for starters are usually available from local suppliers. Spare starters, as well as spare parts, for the most used types and sizes should be stocked in the regular shop supply channels.

16.2.1 See Table 16.2.1 for guidance on motor control equipment troubleshooting.

Table 16.2.1 Motor Control Equipment Troubleshooting Chart

Problem	Possible Causes	Remedies
Contactor or relay not closing	No supply voltage	Check fuses and disconnect switches.
	Low voltage	Check power supply; wire might be too small.
	Coil open or shorted	Replace.
	Wrong coil	Check coil number.
	Mechanical obstruction	With power off, check for free movement of contact and armature assembly.
	Pushbutton contacts not making contact	Clean or replace if badly worn.

(continues)

Table 16.2.1 ***Continued***

Problem	Possible Causes	Remedies
	Interlock or relay contact not making contact	Adjust or replace if badly worn.
	Loose connection	Turn power off first, then check the circuit visually with a flashlight.
	Overload relay contact open	Reset.
Contactor or relay not opening	Pushbutton not connected correctly	Check connections against wiring diagram.
	Shim in magnetic circuit (dc only) worn, allowing residual magnetism to hold armature closed	Replace.
	Interlock or relay contact not opening circuit	Adjust contact travel.
	"Sneak"circuit	Check control wiring for insulation failure.
	Gummy substance on pole faces	Clean with solvent.
	Worn or rusted parts causing burning	Replace parts.
	Contacts welded shut	See next item.
Contacts welding shut or freezing	Insufficient contact spring pressure causing contacts to burn and draw arc on closing	Adjust, increasing pressure. Replace if necessary.
	Very rough contact surface causing current to be carried by too small an area	Smooth surface or replace if badly worn.
	Abnormal inrush of current	Use larger contactor or check for grounds, shorts, or excessive motor load current.
	Rapid jogging	Install larger device rated for jogging service or caution operator.
	Low voltage preventing magnet from sealing	Correct voltage condition. Check momentary voltage dip during starting.
	Foreign matter preventing contacts from closing	Clean contacts with approved solvent.
	Short circuit	Remove short-circuit fault and check that fuse or breaker size is correct.
Contact chatter	Broken pole shader	Replace.
	Poor contact in control circuit	Improve contact or use holding circuit interlock (3-wire control).
	Low voltage	Correct voltage condition. Check voltage condition. Check momentary voltage dip during starting.
Arc lingering across contacts	If blowout is series, possible short	Check wiring diagram to see kind of blowout.
	If blowout is shunt, possible open circuit	Check wiring diagram through blowout.
	Arc box left off or not in correct place	See that arc box is on contactor as it should be.
	No blowout, but increased travel of contacts, which increases rupturing capacity	Note travel of contacts.
Excessive corrosion of contacts	Chattering of contacts as result of vibration outside the control cabinet	Check control spring pressure and replace spring if it does not give rated pressure. If that does not help, move control so vibrations are decreased.
	High contact resistance because of insufficient contact spring pressure	Replace contact spring.
Abnormally short coil life	High voltage	Check supply voltage and rating of controller.
	Gap in magnetic circuit (ac only)	Check travel of armature. Adjust so magnetic circuit is completed.
	Ambient temperature too high	Check rating of contact. Get coil of higher ambient rating from manufacturer, if necessary.
	Filing or dressing	Do not file silver-faced contacts. Rough spots or discoloration will not harm contacts.
	Interrupting excessively high currents	Install larger device or check for grounds, shorts, or excessive motor currents. Use silver-faced contacts.

Table 16.2.1 ***Continued***

Problem	Possible Causes	Remedies
	Excessive jogging	Install larger device rated for jogging or cautionoperator.
	Weak contact pressure	Adjust or replace contact springs.
	Dirt on contact surface	Clean contact surface.
	Short-circuits	Remove short-circuit fault and check for proper fuse or breaker size.
	Loose connections	Clean and tighten.
	Sustained overload	Install larger device or check for excessive load current.
Panel and apparatus burned by heat from resistor	Motor being started frequently	Use resistor of higher rating.
Coil overheating	Overheating or high ambient temperature	Check application and circuit.
	Incorrect coil	Check rating; if incorrect, replace with proper coil.
	Shorted turns caused by mechanical damage or corrosion	Replace coil.
	Undervoltage, failure of magnet to seal in	Correct pole faces.
	Dirt or rust on pole faces increasing air gap	Clean pole faces.
Overload relays tripping	Sustained overload	Check for grounds, shorts, or excessive motor currents.
	Loose connection on load wires	Clean and tighten.
	Incorrect heater	Replace relay with correct size heater unit.
Overload relay failing to trip	Mechanical binding, dirt, corrosion, etc.	Clean or replace.
	Wrong heater or heaters omitted and jumper wires used	Check ratings. Apply proper heaters.
	Motor and relay in different temperatures	Adjust relay rating accordingly or make temperature the same for both.
Noisy magnet (humming)	Broken shading coil	Replace shading coil.
	Magnet faces not mating	Replace magnet assembly or realign.
	Dirt or rust on magnet faces	Clean and realign.
	Low voltage	Check system voltage and voltage dips during starting.

16.2.2 Table 16.2.2 is a guide for motor control preventive maintenance.

16.3 Enclosures.

16.3.1 External Care.

16.3.1.1 An enclosure located in a clean, dry, and noncorrosive atmosphere and where it is not likely to incur physical damage does not require scheduled maintenance. However, internal components should be inspected and serviced as necessary.

16.3.1.2 Enclosures in a marginal atmosphere should be inspected periodically for excessive dust and dirt accumulation as well as for corrosive conditions. The more contaminated the atmosphere, the more frequently the inspections should be conducted. Any accumulation should be removed with a vacuum cleaner or manually during equipment maintenance shutdown periods.

16.3.1.3 Badly corroded enclosures should be properly cleaned and refinished, or replaced.

16.3.2 Opening Enclosures. Compliance with Section 8.3 is essential before opening the door or cover of a cabinet or enclosure. Foreign material, dirt, hardware, and debris should be removed from the outside top surfaces to avoid the risk of anything falling into the equipment.

16.3.3 Internal Inspection. Upon opening of the cabinet or enclosure, equipment should be inspected for any dust, dirt, moisture or evidence of moisture, or other contamination. If any is found, the cause should be eliminated. Internal contamination could be an indication of an incorrectly selected, deteriorated, or damaged enclosure; unsealed enclosure openings; internal condensation; condensate from an unsealed conduit; or improper operating procedures (e.g., operating with enclosure door or cover open).

16.3.3.1 Ventilation passages should be checked for obstructions.

16.3.3.2 If equipment depends on auxiliary cooling or heating, the temperature control system should be checked and repaired if necessary to ensure proper functioning.

Table 16.2.2 Motor Control Preventive Maintenance Guide

What to Inspect	What to Inspect For
Exterior and surroundings	Look for dust, grease, oil, high temperature, rust, corrosion, mechanical damage; check condition of gaskets, if any.
Interior of enclosure, nuts, and bolts	Check same items as for exterior and surroundings plus excess vibration, which might have loosened nuts, bolts, or other mechanical connections.
Contactors, relays, solenoids	
General	Check control circuit voltage; inspect for excess heating of parts evidenced by discoloration of metal, charred insulation or odor; freedom of moving parts; dust, grease, and corrosion; loose connections.
Contact tips	Check for excessive pitting, roughness, copper oxide; do not file silver contacts.
Springs	Check contact pressure; is pressure same on all tips?
Flexible leads	Look for frayed or broken strands; be sure lead is flexible — not brittle.
Arc chutes	Check for breaks or burning.
Bearings	Check for freedom of movement; do not oil.
Coils	Look for overheating, charred insulation, or mechanical injury.
Magnets	Clean faces; check shading coil; inspect for misalignment, bonding.
Fuses and fuse clips	Check for proper rating, snug fit; if copper, polish ferrules; check fuseclip pressure.
Overload relays	Check for proper heater size; trip by hand; check heater coil and connection; inspect for dirt, corrosion.
Pushbutton station and pilot devices	Check contacts; inspect for grease and corrosion.
Dashpot-type timers and overload relays	Check for freedom of movement; check oil level.
Resistors	Check for signs of overheating, loose connections; tighten sliders.
Connections	Tighten main line and control conductor connections; look for discoloration of current-carrying parts.
Control operation	Check sequence of operation of control relays; check relay contacts for sparking on operation; check contacts for flash when closing and, if necessary, adjust to eliminate contact bounce; check light switches, pressure switches, temperature switches, and other sensing devices.
Circuit breakers	Inspect for case damage, exercise by opening and closing while de-energized. Reference Chapter 17.
Disconnects and switching devices	Check the line and load side for excess heating of parts evidenced by discoloration of metal, charred insulation or odor; freedom of moving parts; dust, grease, and corrosion; loose connections; and excessive pitting and roughness on switch blades. Exercise by opening and closing while de-energized.

16.3.4 Internal Environment. If a cooling, heating, or air-conditioning system is installed to maintain a safe environment inside an enclosure, it should be verified that the system functions as designed. As appropriate, air temperatures, air pressures, air quality, heat exchanges, fans, pumps, filters, and power supplies should be checked. For instance, a compressed-air cooling system might be installed on a cabinet to provide a positive-pressure enclosed environment, component cooling, and continual fresh air purging. A simple, periodic check of the filter, exhaust port opening, and the air supply helps determine that heat from normal operations is satisfactorily and reliably dissipated and that a clean, dry atmosphere is being maintained in the enclosure.

16.3.5 Cleaning. Cleaning should be done in accordance with the appropriate recommendations of Section 8.7, Equipment Cleaning.

16.4 Bus Bar, Wiring, and Terminal Connections.

16.4.1 Introduction. Any loose bus bar or terminal connection will cause overheating that will lead to equipment malfunction or failure. Loose bonding or grounding can compromise safety and function. Overheating in a bus or terminal connection will cause a discoloration in the bus bar, which can easily be spotted where connections are visible, often too late to avoid replacement. An overheating bus bar condition will feed on itself and eventually lead to deterioration of the bus system as well as the equipment connected to the bus, such as protective devices, bus stabs, and insulated leads. Aluminum connections usually utilize plated parts that should not be cleaned with abrasives.

16.4.2 Loose Connections. Bus bar and terminal connections should be inspected periodically to ensure that all joints are properly tightened. Proper torque is a function of bolt size, bolt type, terminal material, washer type, and type of bus bar. Proper bolt torque values for all types of joints involved normally are available in manufacturers' maintenance and instructional literature. It should not be assumed that bus bar and terminal hardware, once tightened to proper torque values, remains tight indefinitely. *(See Section 11.17, which describes one method of detecting loose connections during the periods between shutdowns.)*

16.4.3 Special Operating Environments. Special attention should be given to bus bars and terminal connections in equipment rooms where excessive vibration or heating/cooling cycles can cause more than normal loosening of bolted bus and terminal connections.

16.4.4 Bus Bar Support Insulators. Bus bar support insulators and barriers should be inspected to ensure that they are free of contamination. Insulators should be checked periodically for cracks and signs of arc tracking. Defective units should be replaced. Loose mounting hardware should be tightened.

16.4.5 Power and Control Wiring. Insulation on conductors should be examined for overheating or chafing that could progress into an insulation failure. Damaged conductors should be replaced. Replacement conductors should be rerouted, braced, or shielded as needed to avoid similar damage in future operation. Temporary wiring should be removed or replaced by permanent wiring.

16.5 Disconnects.

16.5.1 Introduction. Disconnects should be examined on both the line side and the load side for proper maintenance evaluation. Prior to initiating such an evaluation, an electrically safe work condition should be established. See Chapter 7.

16.5.1.1 Switches used in drawout units normally supplied in motor control centers can be opened, safely withdrawn, and examined on a workbench, thus avoiding a potential hazard.

16.5.2 Safety. It should never be assumed that a disconnect is in the open position because the handle mechanism is in the open position. For safety, always double check. Compliance with Section 8.3 is essential.

16.5.3 Inspection and Cleaning. Routine maintenance should include a procedure for inspecting and removing excessive dust accumulations. *(See Section 8.7.)*

16.5.4 Loose Connections. Loose connections are the major source of excessive heat, which can lead to deterioration of the insulation and eventual failure of the device. Terminal and bus bar connections as well as cable connections should be examined and tightened as required using the manufacturer's torque recommendations. Any device that has evidence of overheated conductors and carbonized insulation should be repaired or replaced. Disconnects showing any evidence of damage and contacts showing evidence of welding or excessive pitting should be repaired or replaced.

16.5.5 Mechanical Operation. Mechanisms should be operated manually to ensure proper working condition. Factory-lubricated mechanisms sometimes dry out after a period of time in the dry, heated atmosphere of a motor control center enclosure. Manufacturers' maintenance literature should be followed for proper lubrication instructions.

16.6 Molded Case Breakers. A wide variety of circuit breakers are used with motor control equipment. Molded case breaker maintenance is covered in Chapter 17.

16.7 Fuses. Fuses normally are used in conjunction with disconnect switches. A dummy fuse, copper slug, or length of wire should never be used as a fuse substitute. Fuse and fuseholder maintenance is covered in Chapter 18.

16.8 Contactors.

16.8.1 Introduction. Because contactors are the working portion of a motor controller, normal wear can be expected.

16.8.2 Contacts and Arc Chutes.

16.8.2.1 Inspection. Contacts and arc chutes of electromechanical contactors should be checked for excessive burning, beads of molten material, and unusual erosion of the contact faces.

16.8.2.2 Servicing. Excessively worn or pitted contacts should be replaced with manufacturer-recommended renewal parts, or the contactor should be replaced. All contacts of multipole devices should be replaced simultaneously to avoid misalignment and uneven contact pressure. Contacts should not be filed or dressed unless recommended by the manufacturer.

16.8.2.2.1 Arc chutes and arc hoods should be replaced if they are broken or deeply eroded.

16.8.2.2.2 Easily dislodged dust or granules should be removed by vacuuming, wiping, or light brushing. Insulating surfaces should not be scraped, sandpapered, or filed.

16.8.3 Alternating Current (ac) Magnet Solenoids. A noisy solenoid in a relay or contactor indicates failure to seat properly or a broken or loose shading coil. The cause should be determined and corrected to avoid overheating and coil damage. If a coil exhibits evidence of overheating (cracked, melted, or burned insulation), it should be replaced, after the cause of overheating has been detected and corrected. This could include the preceding ac magnet symptoms, binding that keeps the magnet from seating properly, and overvoltage and undervoltage conditions. If melted coil insulation has flowed onto other parts, they should be cleaned or replaced.

16.9 Motor Overload Relays — Thermal Types.

16.9.1 Introduction. Motor overload relays perform the vital supervisory function of monitoring the overload current conditions of the associated motor. The most commonly used overload relays employ a thermal element designed to interpret the overheating condition in the motor windings by converting the current in the motor leads to heat in the overload relay element. As the heat in the thermal element reaches a predetermined amount, the control circuit to the magnetic contactor holding coil is interrupted and the motor branch circuit is opened. The two most common types of thermal elements in overload relays employ either a bimetal or a melting alloy joint to initiate the opening action of the contactor.

16.9.1.1 Overload relays that trip during operation are usually resettable. The cause of the trip should be identified before resetting. Some overload relays are adjustable.

16.9.1.2 Failure of the thermal element can occur when the element is subjected to short-circuit conditions. The cause should be identified and corrected.

16.9.1.3 Replacement or adjustment of the heater element to a higher rating should not be done without full consideration of the ambient temperatures in which the motor and controller operate, as well as all the factors in 16.9.3.

16.9.2 Other Types. The manufacturers' literature should be consulted for maintenance of other types of overload devices.

16.9.3 Motor Data. Overload thermal elements are applied on the basis of motor full load current and the motor service factor found on the motor rating nameplate. Complete records on all motors, including motor full-load amps together with proper manufacturer's heater selection and application charts, should be included as a part of any maintenance file on motor starters. General heater application charts usually are secured inside the starter enclosure.

16.9.4 Inspection and Replacement. Routine maintenance should include a check for loose terminal or heater connections and signs of overheating. Overheating can cause carbonization of the molding material, creating potential dielectric breakdowns as well as possibly altering the calibration of the overload relay. Overload elements can be tested with primary injection current and compared to the manufacturer's curve for performance. Overload elements operating outside the manufacturer's curve or showing signs of excessive heating should be replaced.

16.10 Pilot and Miscellaneous Control Devices.

16.10.1 Introduction. Pilot and other control devices consist of the control accessories normally employed with motor starters, such as push buttons, selector switches, indicating lights, timers, and auxiliary relays.

16.10.2 Inspection. Routine maintenance checks on these types of devices generally should include the following:

(1) Check for loose connections
(2) Check for proper mechanical operation of operators and contact blocks
(3) Inspection of exposed contacts
(4) Check for signs of overheating
(5) Replacement of pilot lamps, if necessary

16.11 Interlocks.

16.11.1 Electrical Interlocks.

16.11.1.1 Auxiliary Contacts. A contactor or starter could be provided with auxiliary contacts that permit interlocking with other devices.

16.11.1.2 Inspection. Proper maintenance of the electrical auxiliary contacts should include the following:

(1) Check for loose connections.
(2) Check for proper mechanical operation and alignment with the contactor.
(3) Inspect exposed contacts.

16.11.2 Mechanical Interlocks. Mechanical interlocks can be classified into two categories according to their application: safety and functional performance. Safety interlocks are designed to protect operating personnel by preventing accidental contact with energized conductors and the hazards of electrical shock. Functional interlocks, such as those found on reversing contactors, are designed to prevent the inadvertent closing of parallel contactors wired to provide alternate motor operating conditions. A mechanical interlock should be examined to ensure that the interlock is free to operate and that bearing surfaces are free to perform their intended function. Interlocks showing signs of excessive wear and deformation should be replaced. Several types of locking or interlocking features are used, including those described in 16.11.2.1 through 16.11.2.4.

16.11.2.1 Primary Disconnect Mechanism. This device usually is mounted directly on the disconnect device. It is mechanically interlocked with the door to ensure that the door is held closed with the disconnect in the on position. A maintenance check should be made to ensure that the adjustment is correct and that the interlock is providing proper engagement.

16.11.2.2 Padlock Mechanism. Disconnect operating mechanisms usually are provided with padlocking means whereby the mechanism can be padlocked in the off position. During maintenance checks of the equipment and the motor, these mechanisms should be padlocked in the off position for personnel safety.

16.11.2.3 Defeat Mechanisms. Most disconnects are equipped with defeater mechanisms that can be operated to release door interlock mechanisms with the disconnect device in the on position. The use of this release mechanism should be limited to qualified maintenance and operating personnel.

16.11.2.4 Unit Lock. Motor control centers can be provided with plug-in starters for ease of inspection and interchangeability. Plug-in motor-starter units normally are held locked in their connected positions by a unit latch assembly. Although maintenance on this assembly normally is not required, it should be understood by maintenance personnel.

Chapter 17 Insulated-Case/Molded-Case Circuit-Breakers

17.1 Introduction.

17.1.1 An insulated-case/molded-case circuit breaker consists of two basic parts. One part consists of the current-carrying conductors, contacts, and appropriate operating mechanism necessary to perform the circuit-switching functions. The second part consists of the protective element, including the tripping mechanism associated therewith.

17.1.2 Insulated-case/molded-case circuit breakers undergo extensive production testing and calibration at the manufacturers' plants. These tests are based on ANSI/UL 489, *Molded-Case Circuit Breakers, Molded-Case Switches and Circuit Breaker Enclosures.* Circuit breakers carrying the UL label have factory-sealed, calibrated elements; an unbroken seal ensures that the mechanism has not been subjected to alteration or tampering and that the breaker can be expected to perform according to UL specifications. A broken seal voids the UL label and jeopardizes the manufacturer's warranty.

17.2 Application Considerations. Insulated-case/molded-case circuit breakers trip from exposure to continuous currents beyond their ratings, and many trip from unduly high ambient temperatures, poor or improper connections, damaged plug-in members, and other conditions that transfer undue heat to the breaker mechanism. Some of these conditions violate application specifications. An insulated-case/molded-case circuit breaker applied in a panelboard should not be loaded in excess of 80 percent of its continuous current rating, where in normal operation the load will continue for 3 hours or more.

17.3 Phase-Fault Current Conditions. A typical insulated-case/molded-case circuit breaker is equipped with both time-delay and instantaneous tripping devices. Time-delay tripping has inverse time characteristics that provide a shorter tripping time for higher overloads. Under moderate, short-duration overloads, the circuit breaker allows sufficient time for applications such as motor starting. Under severe overloads, the circuit breaker trips quickly, providing adequate protection for conductors and insulation. For high-fault currents, the magnetic tripping device responds to open the circuit breaker immediately.

17.4 Ground-Fault Tripping. It should be recognized that standard thermal magnetic insulated-case/molded-case circuit breakers generally are not equipped with ground-fault sensing and protection devices and, therefore, will not normally trip and clear low-level ground faults, which can do immense damage. Special ground-fault sensing and protective devices should be specified to achieve this type of equipment protection where necessary. *(See Section 13.2.)*

17.5 Types of Insulated-Case/Molded-Case Circuit Breakers.

17.5.1 Insulated-case/molded-case circuit breakers can be divided into the following three major categories depending on the type of trip unit employed:

(1) Factory-sealed, noninterchangeable trip units
(2) Interchangeable trip units
(3) Solid-state units

17.5.2 The most common type of trip unit under 17.5.1(1) and 17.5.1(2) is the thermal-magnetic trip unit. This type of trip unit employs a thermal element to provide inverse characteristics giving overload protection and a magnetic circuit to provide short-circuit protection. Solid-state units also accomplish other functions, including ground-fault protection, not normally available as an integral part of thermal-magnetic breakers.

17.6 Special-Purpose Breakers. A special design of an instantaneous-only circuit breaker having an adjustable instantaneous pickup is utilized in motor-circuit protection schemes.

17.7 Types of Maintenance. Maintenance of insulated-case/molded-case circuit breakers generally can be divided into two categories: mechanical and electrical. Mechanical maintenance consists of inspection involving good housekeeping, maintenance of proper mechanical mounting and electrical connections, and manual operation as outlined in Sections 17.8 through 17.10. Electrical testing under field test conditions is covered in 11.10.2.

17.8 Inspection and Cleaning. Insulated-case/molded-case circuit breakers should be kept clean of external contamination so that internal heat can be dissipated normally. Further, a clean case reduces potential arcing conditions between live conductors and between live conductors and ground. The structural strength of the case is important in withstanding the stresses imposed during fault-current interruptions. Therefore, the case should be inspected for cracks and replaced if necessary.

17.9 Loose Connections. Excessive heat in a circuit breaker can cause a malfunction in the form of nuisance tripping and possibly an eventual failure. Loose connections are the most common cause of excessive heat. Periodic maintenance checks should involve checking for loose connections or evidence of overheating. Loose connections should be tightened as required, using manufacturers' recommended torque values. Insulated-case/molded-case circuit breakers having noninterchangeable trip units are properly adjusted, tightened, and sealed at the factory. Those having interchangeable trip units installed away from the factory could overheat if not tightened properly during installation. All connections should be maintained in accordance with manufacturers' recommendations.

17.10 Mechanical Mechanism Exercise. Devices with moving parts require periodic checkups, and an insulated-case/molded-case circuit breaker is no exception. Manual operation of the circuit breaker will help keep the contacts clean and will help the lubrication perform properly. Although manual operations will exercise the breaker mechanism, none of the mechanical linkages in the tripping mechanisms will be moved with this exercise. Some circuit breakers have push-to-trip buttons that should be manually operated to exercise the tripping mechanism linkages. *(Refer to Annex K, Long-Term Maintenance Guidelines; Annex L, Maintenance Intervals; and ANSI/NETA MTS, Standard for Maintenance Testing Specifications for Electrical Power Distribution Equipment and Systems, for guidance on frequency of exercise as appropriate.*

17.11 Electrical Testing. For further information on testing see 11.10.1, Insulated-Case/Molded-Case Circuit Breaker Testing.

Chapter 18 Fuses

18.1 Fuses Rated 1000 Volts or Less.

18.1.1 Installing and Removing Fuses. Fuseholders should be deenergized before installing or removing fuses. Where it is not feasible or would result in a greater hazard to deenergize fuseholders, installation or removal of fuses should be performed in accordance with appropriate safety-related work practices for the task.

18.1.2 Inspection. Fuse terminals and fuseclips should be examined for discoloration caused by heat from poor contact or corrosion. Early detection of overheating is possible through the use of infrared examination. If evidence of overheating exists, the cause should be determined.

18.1.3 Cleaning and Servicing. The power source to fuseholders should be disconnected before servicing. All fuseholder connections should be tightened. All connections to specifications should be torqued where available. Fuseclips should be checked to ascertain that they exert sufficient pressure to maintain good contact. Clips making poor contact should be replaced or clip clamps used. Contact surfaces of fuse terminals and clips that have become corroded or oxidized should be cleaned. Silver-plated surfaces should not be abraded. Contact surfaces should be wiped with a noncorrosive cleaning agent. Fuses showing signs of deterioration, such as discolored or damaged casings or loose terminals, should be replaced.

18.1.4 Replacement. Many different types of fuses are used in power distribution systems and utilization equipment. Fuses differ by performance, characteristics, and physical size. It should be verified that fuses, whether new or replacement, are the proper type and rating. When fuses are replaced, fuseholders should never be altered or forced to accept fuses that do not readily fit. An adequate supply of spare fuses with proper ratings, especially those that are uncommon, minimizes replacement problems.

18.1.4.1 Type. The most common fuse classes for 0 ampere through 600 ampere applications on power systems are Class H, K, R, J, T, G, and CC. Class H, K, and R are the same physical size and are interchangeable in standard nonrejection style fuseholders. Special rejection-style fuseholders accept only Class R fuses. Note that Class R fuses are manufactured in two types, Class RK1 and Class RK5. Class RK1 fuses are more current limiting than Class RK5 fuses and are generally recommended to upgrade older distribution systems. Class L fuses are available in the range of 601 amperes through 6000 amperes. Class J, T, G, CC, and L are size rejection fuses. One type of fuse should never arbitrarily be replaced with a different type simply because it fits into the fuseholder.

18.1.4.2 Ratings. Five characteristics should be considered when fuses need to be replaced: interrupting rating, voltage rating, current rating, degree of time delay, and degree of current limitation.

18.1.4.2.1 Interrupting Rating. Fuses should have an interrupting rating equal to or greater than the maximum fault current available at their point of application. Fuses have interrupting ratings from 10,000 amperes to 300,000 amperes. *(See Section 9.2.)*

18.1.4.2.2 Voltage. The voltage rating of the fuse should be at least equal to or greater than the system voltage.

18.1.4.2.3 Current. Fuse ampere ratings should be adequate for the applications. Ratings are determined by the service, feeder, and branch-circuit conductors, and the loads served. Consult *NFPA 70, National Electrical Code,* and the electrical system single-line diagram for proper fuse sizing. Fuse manufacturers can be contacted for application information.

18.1.4.2.4 Time Delay. Most fuse classes are manufactured in time-delay and non-time-delay versions. Time-delay fuses are especially useful on inductive circuits such as motor and transformer circuits with inrush currents. Time-delay fuses are the most commonly used fuses on power distribution and motor circuits.

18.1.4.2.5 Current Limitation. Fuses are designated as either current limiting or non-current limiting based on their speed of response during short-circuit conditions. Non-current-limiting fuses can be replaced with current-limiting fuses, but current-limiting fuses should not be replaced with non-current-limiting fuses unless a review of the specific application is undertaken.

18.1.4.3 Listing. It is important that the fuses bear the listing mark of a nationally recognized testing laboratory. Testing laboratories test fuses for both ac and dc performance characteristics, and the ratings are marked on the fuse label. Be sure to select the proper fuse for the specific application.

18.1.4.4 Special Purpose. Special-purpose fuses are used for supplementary protection of power systems and for utilization equipment such as power rectifiers, variable speed drives, and solid-state controllers. High-speed or semiconductor-type fuses are most commonly used in these applications. These fuses have unique performance characteristics and physical size. They should be matched to the utilization equipment.

18.2 Fuses Rated over 1000 Volts.

18.2.1 Introduction. Fuses rated over 1000 volts consist of many parts, some current carrying and some non-current carrying, all subject to atmospheric conditions. These fuses can be current limiting or non-current limiting, sand or liquid filled, or vented expulsion type. The frequency of inspection is necessarily a function of the conditions at a given fuse location and should be determined by the user.

18.2.2 Installing and Removing Fuses. Manufacturers' instructions regarding installing and removing fuses should be followed. If the fuse does not have a loadbreak rating, the system should be deenergized before the fuse is removed.

18.2.3 Inspection and Cleaning.

18.2.3.1 The fuse should be disconnected and the mounting deenergized from all power sources before servicing, and an electrically safe work condition should be established. *(See Chapter 7.)* Insulators should be inspected for breaks, cracks, and burns. The insulators should be cleaned, particularly where abnormal conditions such as salt deposits, cement dust, or acid fumes prevail, to avoid flashover as a result of the accumulation of foreign substances on their surfaces.

18.2.3.2 Contact surfaces should be inspected for pitting, burning, alignment, and pressure. Badly pitted or burned contacts should be replaced, and the following criteria should be met:

(1) The fuse unit or fuse tube and renewable element should be examined for corrosion of the fuse element or connecting conductors, excessive erosion of the inside of the fuse tube, discharge (tracking) and dirt on the outside of the fuse tube, and improper assembly that might prevent proper operation.
(2) Fuse tubes or units showing signs of deterioration should be replaced.

18.2.3.3 Bolts, nuts, washers, pins, and terminal connectors should be in place and in good condition, and the following criteria should be met:

(1) The lock or latch should be checked.
(2) Fuse tubes made of organic (Class A) material should be refinished as required and specified by the manufacturer.

18.2.3.4 Vented expulsion fuses might be equipped with condensers or mufflers to restrict expulsion of gases during operation. They might have a dropout feature that automatically disengages the fuse when it operates. The lower, or discharge end, of the expulsion fuse might have a sealing disc over the expulsion chamber to prevent entrance of moisture if the fuse is left in an inverted, disconnected position in service. These seals should be inspected to ensure that moisture has not entered the interrupting chamber. If the seals are damaged or show evidence of leakage, the fuses should be replaced.

Chapter 19 Power Cables

19.1 Introduction. Preventive maintenance is the one of best ways to ensure continued reliable service from electrical cable installations. Visual inspection and electrical testing of the insulation are the major maintenance procedures. However, it should be stressed that no amount of maintenance can correct improper application or physical damage done during installation.

19.2 Visual and Mechanical Inspection.

19.2.1 If, in addition to the visual inspection, cables are to be touched or moved, they should be deenergized.

19.2.2 Cables in manholes should be inspected for sharp bends, physical damage, excessive tension, oil leaks, pits, cable movement, insulation swelling, soft spots, cracked jackets in nonlead cables, damaged fireproofing, poor ground connections, deterioration of metallic sheath bonding, as well as corroded and weakened cable supports and the continuity of any main grounding system. Terminations and splices of nonlead cables should be inspected for tracking or signs of corona. The ground braid should be inspected for corrosion and tight connections. The bottom surface of the cable should be inspected for wear or scraping, due to movement, at the point of entrance into the manhole and also where it rests on the cable supports.

19.2.3 The manhole should be inspected for deterioration of the concrete, both internal and above ground. In some instances, the manhole can be equipped with drains that might require cleaning. In some instances, it might be necessary to pump water from the manhole prior to entrance. A manhole should not be entered unless a test for dangerous gas has been made and adequate ventilation is provided. The inspection crew should always consist of two or more persons with at least one remaining outside the manhole, and the rules and regulations for confined space entry should be followed. *[See OSHA requirements in 29 CFR 1910.146, "Permit-Required Confined Spaces," for practices and procedures to protect employees from the hazards of entry into permit-required confined spaces, and 29 CFR 1910.269(e), "Electric Power Generation, Transmission, and Distribution," Enclosed Spaces, for enclosed space entry.]*

19.2.4 Potheads, a type of insulator with a bell or pot-like shape typically used to connect underground electrical cables to overhead lines, should be inspected for oil or compound leaks and cracked or chipped porcelain. The porcelain surfaces should be cleaned, and if the connections are exposed, their tightness should be checked.

19.2.5 Cable identification tags or markings should be checked.

19.3 Aerial Installations. Aerial cable installations should be inspected for mechanical damage due to vibration, deteriorating supports, or suspension systems. Special attention should be given to the dead-end supports to ensure that the cable insulation is not abraded, pinched, or bent too sharply. Terminations should be inspected as covered in 19.2.2. Aerial cable installations should be inspected for animal and bird infestation.

19.4 Raceway Installations. Because the raceway is the primary mechanical support for the cable, it should be inspected for signs of deterioration or mechanical damage or if the cable jacket is being abraded or mechanically damaged. In many installations, the raceway serves as a part of the ground-fault current circuit. Joints should be inspected for signs of looseness or corrosion that could result in a high resistance. The recommendations for splices and terminations covered in 19.2.2 also apply in this section.

19.5 Electrical Testing. *(See Chapter 11.)* When performing electrical testing of cables, there are many factors that need to be considered before applying a specific test methodology. The two most commonly used tests for cable insulation are insulation resistance testing and dc over-potential testing. Other tests are listed in ANSI/IEEE 400, *Guide for Making High-Direct-Voltage Tests on Power Cable Systems in the Field.* In many instances it can be desired to achieve a more comprehensive analysis of cable condition, doing so with techniques and methods other than insulation resistance. The various cable testing methods are covered in Section 11.21.

19.6 Inspection and Testing Records. Because inspection intervals normally are one year or more, comprehensive records are an important part of any maintenance program. Comprehensive records should be arranged to facilitate comparison from year to year.

Chapter 20 Cable Tray and Busway

20.1 Introduction.

20.1.1 A cable tray system is a unit or assembly of units or sections and associated fittings made of metal or other noncombustible materials forming a rigid structural system used to support cables. Cable tray systems include ladders, troughs, channels, solid-bottom trays, and other similar structures.

20.1.2 The frequency of maintenance depends on the environment in which the cable tray is installed. In areas of heavy industrial contamination or coastal areas, frequent inspections might be necessary.

20.2 Cable.

20.2.1 Cable insulation should be visually inspected for damage. Among the factors that might cause insulation damage are sharp corners, protuberances in cable tray, vibration, and thermal expansion and contraction.

20.2.2 Cable insulation should be tested in accordance with Chapter 19.

20.2.3 The number, size, and voltage of cables in the cable tray should not exceed that permitted by *NFPA 70, National Electrical Code,* Article 392. Communication or data-processing circuits are susceptible to interference problems when mixed with power circuits.

20.3 Cable Tray.

20.3.1 The cable tray should be inspected for intrusion of such items as pipe, hangers, or other equipment that could damage cables.

20.3.2 Deposits of dust, industrial process materials, and trash of any description should be checked and evaluated in terms of reduced ventilation and potential fire hazard.

20.3.3 Bolted connections between sections should be visually checked for corrosion and a sample retorquing done in suspect areas.

20.3.4 Certain atmospheric conditions might create fastener failure; therefore, a visual inspection should check for missing or damaged bolts, bolt heads, or nuts. Where necessary, they should be replaced with suitable hardware.

20.3.5 A visual and mechanical check should be made for adequacy of cable tray grounding, and all takeoff raceways should be bonded to the cable tray.

20.3.6 Covers should be inspected to ensure that physical damage does not reduce spacings or damage cables.

20.4 Low-Voltage (600-Volt) Busway.

20.4.1 General. For the purpose of this section, a busway is considered to be a grounded metal enclosure containing factory-mounted, bare, or insulated conductors that are usually copper or aluminum bars, rods, or tubes.

20.4.1.1 A feeder busway is a busway that has no plug-in openings and that is intended primarily for conducting electric power from sources of supply to centers of distribution. It can have provisions for bolt-on devices.

20.4.1.2 A plug-in busway is a busway that has plug-in openings on one or both sides at spaced intervals, offering means for electrical connection of plug-in or bolt-on devices to the bus bars.

20.4.1.3 Metal-Enclosed Busway (5 kV to 15 kV). Busway over 600 volts is referred to as metal-enclosed busway. Rated 5 kV and 15 kV, it consists of three types: isolated phase, segregated phase, and nonsegregated phase. Isolated phase and segregated phase are utility-type busways used in power-generation stations; industrial plants use nonsegregated phase for connection of transformers and switchgear and interconnection of switchgear lineups.

20.4.2 Electrical Joints.

20.4.2.1 Infrared inspection of busway joints can reveal loose connections and should be performed in accordance with Section 11.17.

20.4.2.2 Belleville spring washers are designed to help maintain proper tightness at the joints of bus bars and cable connections as the bus material expands and contracts under load. A flat or discolored Belleville spring washer could be a sign that the bolt has been overtorqued or that it has been overheated and lost its temper. If either of these situations exists, the washer should be replaced according to the manufacturers' specifications with regard to components and installation procedures.

20.4.3 Housing.

20.4.3.1 A visual check should be made to ensure that all joint covers and plug-in covers are in place and tight, to prevent accidental contact with energized conductors.

20.4.3.2 A visual check should be made for bonding of a bus and the equipment to which it is connected.

20.4.3.3 Trash, combustible material, and other debris should be removed from a busway. Ventilation openings should be clear.

20.4.3.4 On an indoor busway, a visual check should be made for evidence of exposure to liquids and the source eliminated or necessary protection provided.

20.4.3.5 On an outdoor busway, a visual check should be made to ascertain if weep hole screws have been removed in accordance with the manufacturer's instructions.

20.4.4 Plugs.

20.4.4.1 Circuit breaker and fusible plugs should be checked for proper operation.

20.4.4.2 Plug hangers should be checked for tightness to ensure proper grounding.

20.4.4.3 If plug installation requires hook sticks for operation, hook sticks should be checked for ready accessibility.

20.4.5 Conduit and Raceways. Cable and raceways should be visibly checked for proper bonding to fittings (plugs, tap boxes).

20.4.6 Insulators. Bus supports should be visually inspected for dirt or tracking. Dirty insulators should be cleaned, and insulators that are cracked or show evidence of tracking should be replaced.

20.4.7 Heaters for Metal-Enclosed Busway. A check should be made for proper operation of space heaters. Ammeters in heater supply circuits provide means for quick and frequent observation for proper heater loads to determine if one or more heater units is defective.

20.4.8 Testing.

20.4.8.1 Insulation resistance testing should be performed in accordance with 11.9.2.3.

20.4.8.2 If there is uncertainty concerning the adequacy of the insulation after insulation resistance testing, a high-potential test should be conducted. *(See 11.9.3.1.)* Normal high-potential voltages are twice rated voltage plus 1000 volts for 1 minute.

20.4.8.3 High-Potential Testing for Metal-Enclosed Busway. High-potential tests in accordance with IEEE C37.20.1, *Standard for Metal-Enclosed Low-Voltage Power Circuit Breaker Switchgear*, and IEEE C37.23, *Standard for Metal-Enclosed Bus and Calculating Losses in Isolated-Phase Bus*, should be conducted at 75 percent of the rated insulation withstand levels shown in Table 20.4.8.3. Because this might be above the corona starting voltage of some busways, frequent testing is undesirable.

Table 20.4.8.3 Metal-Enclosed Bus Dielectric Withstand Test Voltages

Metal-Enclosed Bus Nominal Voltage (kV, rms)	Insulation Withstand Level (kV, rms)[a]	High-Potential Field Test (kV, rms)[b]
4.16	19.0	14
13.8	36.0	27
23.0	60.0	45
34.5	80.0	60

[a] 1 minute.

[b] 75 percent of insulation withstand level.

Chapter 21 Power and Distribution Transformers

21.1 Introduction.

21.1.1 Transformers are usually used to step down a distribution level voltage to a utilization level. Because a transformer failure is usually of a very serious nature, requiring extensive repair and long downtime, regular maintenance procedures are the best assurance of continued high reliability.

21.1.2 The extent and frequency of maintenance is based on factors such as replacement lead time and relative importance of the transformer in the system. The failure of a small distribution transformer serving a critical load can have more impact on an operation than the failure of a larger or higher-voltage unit. Also, on some smaller systems, the failure of a distribution transformer can result in an outage of the complete system.

21.1.3 Transformers can be grouped into two general categories, according to their insulating medium and construction: liquid filled and dry type. For further information on transformer testing, see Section 11.11.

21.2 Liquid-Filled Transformers.

21.2.1 Introduction.

21.2.1.1 The core and coils of liquid-filled transformers are immersed in a liquid. The liquid serves two purposes. It is an important part of the insulating medium, and it serves to transfer heat away from the windings to be dissipated by the heat exchanger, cooling fins, tank surface, or radiator.

21.2.1.2 There are several types of insulating liquid in use, with mineral oil being the most common. Each liquid has definite characteristics, and they should only be mixed after consulting with both manufacturers. Manufacturers' instructions should be carefully followed with all insulating liquids.

21.2.1.3 Askarel is an older insulating liquid and is identified by various brand names and consists largely of polychlorinated biphenyls (PCBs). It is subject to strict government regulation as a toxic substance. Knowledge of government regulations is necessary because any liquid-filled transformer might contain some level of PCBs. One reference is 40 CFR 761, "Protection of Environment — Polychlorinated Biphenyls (PCBs) Manufacturing, Processing, Distribution in Commerce, and Use Prohibitions."

21.2.1.4 There are several types of transformer construction and these affect the preservation of the liquid. *Preservation* means minimizing exposure of the insulating liquid to the atmosphere. The types are as follows:

(1) Free breathing (open to the atmosphere)
(2) Restricted breathing (open to the atmosphere through dehydrating compounds)
(3) Conservator or expansion tank (exposure to air limited to the liquid in the conservator tank)
(4) Sealed tanks (an inert gas space above the liquid serves as a cushion for internal pressure)
(5) Gas–oil seal (exposure to air limited to the oil in the auxiliary tank)
(6) Automatic gas pressure (gas space above liquid maintained under positive pressure by gas supplied from a nitrogen cylinder)

21.2.1.5 Common cooling methods identified on older transformer nameplates are as follows:

(1) Self-cooled [OA (oil to air cooled) or ONAN (oil natural convection air natural convection)] — heat is dissipated by the tank surface and cooling fins or tubes
(2) Forced-air cooled [FA or ONAF (oil natural convection air fan) — fans are employed to force air over the cooling surfaces to augment the self-cooled rating
(3) Forced-air cooled/forced-oil cooled [FA/FOA or OFAF (oil forced air forced)] — an oil pump circulates oil through a fan-blown oil-to-air heat exchanger
(4) Water cooled [FOW or OFWA (oil forced water forced)] — heat exchange by means of water pumped through a pipe coil installed inside or outside the transformer tank

Refer to Annex M for additional information.

21.2.2 Regular Inspections.

21.2.2.1 Inspections of transformers should be made at regular intervals based on the importance of the transformer, the operating environment, and the severity of the loading conditions. *(See Table L.1 for guidelines.)* Regular inspection data include load current, voltage, liquid level, top-oil temperature, winding hot-spot temperature, ambient temperature, leaks, and general condition.

21.2.2.2 The current, voltage, and temperature readings should be taken at the time of peak load. The liquid level reading should be taken at the end of a low-load period. Permanent records should be kept of the readings, ensuring that readings will be made and providing a means of ready comparison with previous conditions. Further explanations are covered in 21.2.3 through 21.2.5.2.

21.2.3 Current and Voltage Readings.

21.2.3.1 Load current readings are a very important part of the recommended regular inspections. If current in any phase exceeds the rated full-load value, and the rated maximum temperature rise is exceeded, steps should be taken to reduce the load.

21.2.3.2 Overvoltages and undervoltages can be detrimental to the transformer and the load it serves. The cause should be investigated immediately and corrective action taken to bring the voltage within acceptable limits.

21.2.4 Temperature Readings.

21.2.4.1 Transformers are rated to carry their nameplate load in kVA with a given temperature rise when the ambient temperature is at the specified maximum. Exact values are stated on the nameplate. For instance, a liquid-filled transformer might be rated to deliver nameplate capacity with a 65°C (149°F) temperature rise above a 30°C (86°F) ambient temperature.

21.2.4.2 Readings on transformers having temperature gauges should be recorded at regular intervals. If the gauge is also equipped with a maximum-temperature indicator, present-temperature and maximum-temperature readings should be recorded, and the maximum-temperature indicator should be reset. Excessive temperature indicates an overload or some interference with the normal means of cooling. Prolonged operation at overtemperature significantly reduces the service life of the transformer.

21.2.5 Liquid-Level Indicator and Pressure/Vacuum Gauges.

21.2.5.1 The liquid level should be checked regularly, especially after a long period of low load at low ambient temperature, when the level should be at its lowest point. It is important that liquid be added before the level falls below the sight glass or bottom reading of the indicator. It is necessary to deenergize the transformer prior to performing any of the following procedures. If a transformer is not equipped with a liquid-level indicator, the liquid level can be checked by removing the inspection plate on the top of the transformer or by removing the top if no inspection plate is available. *(See 21.2.7 for precautions relative to de-energizing the transformer and for the recommended procedures for adding liquid.)*

21.2.5.2 Pressure/vacuum gauges are often found on sealed tank and automatic pressure transformers and indicate the integrity of the sealed construction. Most transformers have provisions for adding a pressure/vacuum gauge, and, if feasible, the gauge should be added. The readings should be compared to the recommendations of the manufacturer for normal operating ranges. High pressures indicate an overload or internal defect and should be investigated immediately. A sustained zero pressure reading indicates a gas leak or a defective gauge.

21.2.6 Miscellaneous. Regular inspections for specific types of transformer construction should include the following:

(1) The water-in and water-out temperatures of FOW or OFWF transformers.
(2) The oil-in and oil-out temperatures of ONAF or OFWF transformers with oil-to-air or oil-to-water heat exchangers.
(3) The pressure in the nitrogen cylinder for a transformer equipped with an automatic gas–pressure system — if the nitrogen cylinder pressure drops below the desired value, the cylinder should be replaced and leaks repaired.
(4) Dehydrating breathers should be checked to ensure that they are free from restriction and have not absorbed excessive moisture.

21.2.7 Special Inspections and Repairs.

21.2.7.1 Because of the wide variety of liquid-filled transformer types, sizes, and uses, the special inspection and repair recommendations given in 21.2.7.2 through 21.2.7.8 are general in nature. For specific directions, the manufacturer's recommendations should be followed.

21.2.7.2 The case of the transformer should be regarded as energized until the tank ground connection is inspected and found to be adequate. A transformer should be deenergized for any procedure more extensive than an external visual inspection and establish an electrically safe work condition *[see 7.1.3.2(1)]* except for oil sampling as noted in 11.11.8.1.3. Then verify the integrity of the tank ground before proceeding. Before grounding of the terminals or any other work is performed, the terminals of the transformer should be tested and found to be absent of nominal voltage. The transformer terminals should be grounded prior to the start of any work. *(See Section 7.2, Grounding of Equipment to Provide Protection for Electrical Maintenance Personnel.)*

21.2.7.3 An inspection should be made for the following problems:

(1) Cracked, contaminated or chipped bushings or insulators.
(2) Loose, overheated or corroded electrical connections.
(3) Bushing or insulator tracking.
(4) Gasketed bushing or insulator bases should be checked for leaks and repaired.
(5) Leaking gaskets.
(6) Cracked or leaking pressure-relief device diagrams. Cracked or leaking diaphragm should be replaced immediately.

21.2.7.4 The tank, heat exchanger, cooling fins, tubes, radiators, no-load tap changer, and all gasketed or other openings should be inspected for leaks, deposits of dirt, or corrosion. Leak repair, cleaning, and painting should be done as required. Infrared inspection can be used to detect fluid levels as well as flow restrictions in cooling tubes.

21.2.7.5 The tank ground should be inspected for corrosion or a loose connection. A grounding-electrode resistance test should be performed. *(For further information on ground testing, see 11.13.3.)*

21.2.7.6 Auxiliary equipment such as a load tap changer, cooling fans, circulating oil pumps, and protective relays (e.g., Bucholtz relays and sudden-gas relays) should be inspected regularly in accordance with manufacturers' recommended practices.

21.2.7.7 The conservator tank, inert gas atmosphere, and dehydrating breather equipment should be inspected and tested according to the manufacturer's instructions. Large, liquid-filled power transformers have features to minimize exposure of the liquid to air. Opening of this type of transformer for internal inspection is recommended only when the need is positively indicated, and the manufacturer's instructions should be followed or technical assistance employed.

21.2.7.7.1 Contamination of the insulating liquid should be avoided. If the humidity is high, exposure should be avoided entirely unless the work is absolutely necessary and cannot be postponed.

21.2.7.8 If insulating liquid is to be added, a dielectric-breakdown test of the liquid should be performed before being added. For further information on liquid maintenance and testing see 11.11.8, Liquid Maintenance and Analysis. If a large amount of liquid is added, the transformer should remain deenergized for a period of time sufficient to permit the escape of entrapped air bubbles. A desirable method is to add the liquid with the transformer tank under a vacuum. *(Check the manufacturer's instructions for further information.)*

21.2.8 Liquid Maintenance and Analysis. For further information on liquid maintenance and testing see 11.11.8, Liquid Maintenance and Analysis.

21.2.9 Other Tests. For further information on transformer testing see Section 11.11, Transformer Tests.

21.3 Dry-Type Transformers.

21.3.1 Introduction.

21.3.1.1 Dry-type transformers operate in air or gas rather than being liquid filled. The two general types of construction are the open or ventilated dry-type and the sealed or closed-tank type. Dry transformer windings are usually varnish impregnated or cast coil construction. Sealed transformers are cooled and insulated by a high-dielectric inert gas, such as nitrogen, sulfur hexafluoride, or perfluoropropane.

21.3.1.2 The air or gas serves as an insulating medium and also to dissipate heat from the windings. Standard insulation classes are 80°C (176°F) rise, 115°C (239°F) rise, and 150°C (302°F) rise.

21.3.2 Regular Inspections. The recommendations in 21.2.2 regarding regular inspections of liquid-filled transformers also apply to dry-type transformers, with the exception of those that pertain strictly to liquid-filled construction.

21.3.3 Current and Voltage Readings. The recommendations in 21.2.3 regarding current and voltage readings also apply to dry-type transformers.

21.3.4 Temperature Readings. The recommendations in 21.2.4 regarding temperature readings also apply to dry-type transformers. However, dry-type transformers typically have high-temperature insulation and operate at higher temperatures than liquid-filled units.

21.3.5 Inspections and Repairs.

21.3.5.1 Ventilation. The louvers in the enclosures of ventilated dry-type transformers should be inspected to verify that they are not clogged with dirt or otherwise obstructed. The operation of integral ventilating fans should be checked. When dry-type transformers are installed indoors or in a vault, the temperature of the vault or room should be measured regularly and recorded. Any material or obstruction that might prevent the free circulation of air around a transformer should be removed. If the room or vault has power-driven ventilating fans, their correct operation should be determined and overtemperature alarms, if provided, should be tested.

21.3.5.2 Enclosure Integrity. Corrosion of the transformer enclosure, the intrusion of dirt, and evidence of water leaks into the room or vault should also be carefully checked and corrective measures taken as required.

21.3.5.3 Noise. A high noise level or change in level could indicate improper installation or loose windings or barriers.

21.3.5.4 Safety Procedures. When a dry-type transformer is given an external visual examination, the transformer case should be regarded as energized until the case-ground connection is inspected and found to be connected and secure. If any procedure more extensive than an external visual examination is to be performed, the first precaution that should always be taken is to deenergize the transformer. Deenergization should be accompanied by approved lockout procedures to ensure against an unplanned reenergization and resulting hazard to personnel or equipment. Deenergization should be followed immediately by a test to ensure that the equipment is deenergized. The equipment should be grounded prior to the start of any work. *(See Section 7.2.)* A grounding-electrode resistance test should be made as covered in 11.13.3.

21.3.5.5 Visual Inspection. Enclosure covers of ventilated dry-type transformers should be removed carefully. An inspection should be made for the following problems:

(1) Accumulations of dirt on windings, insulators, and where cooling airflow might be restricted
(2) Discoloration caused by overheating
(3) Tracking and carbonization
(4) Cracked or chipped insulators
(5) Loose insulators, clamps, or coil spacers
(6) Deterioration of barriers
(7) Corroded or loose electrical connections.
(8) Support or restraint of electrical control wiring that could have become loose

21.3.5.5.1 In addition, the equipment ground should be inspected for corrosion or loose connections. A grounding-electrode resistance test should be made, as covered in 11.13.3.

21.3.5.6 Cleaning. Dirt and dust should be cleaned from the windings with a vacuum cleaner. If vacuum cleaning is insufficient, compressed air should be used only if it is clean, dry, and applied at a low pressure to avoid damage to windings. The use of compressed air should comply with OSHA regulations in 29 CFR 1910.242(b), "Hand and Portable Powered Tools and Other Hand Held Equipment," including limiting air pressure for such cleaning to less than a gauge pressure of 208.85 kPa (30 psi) and the provision of effective chip guarding and appropriate personal protective equipment. In particular, ventilating ducts and the top and bottom of the windings should be cleaned. Liquid cleaners should be used only when it is known that they will not have a deteriorating effect on the insulation.

21.3.6 Out of Service Precautions. Best service life results if the windings are maintained above the ambient-temperature level. For that reason, transformers operating in high humidity should be kept energized, if feasible. If a transformer is to be deenergized long enough for it to cool to ambient temperature, special drying procedures might be necessary before the transformer is reenergized. Refer to the manufacturer's recommendations for the drying procedures to be followed.

21.3.7 Sealing severe leaks or opening and resealing the tanks of sealed dry-type transformers requires special procedures and equipment. The manufacturer of the transformer, an experienced transformer repair facility, or a qualified electrical maintenance contractor should perform this work. In addition, special procedures for drying out the windings, plus purging and refilling the tank, might be necessary.

21.3.8 Insulation Tests. For further information on transformer insulation resistance testing, see 11.13.3, Insulation Resistance Tests.

Chapter 22 Electronic Equipment

22.1 Introduction. This chapter describes the maintenance of electronic equipment in general terms. Specific maintenance procedures normally are available from the equipment manufacturer or are contained in the instruction book supplied with the apparatus. In some cases, these procedures require the services of trained specialists.

22.2 Reasons for Maintenance.

22.2.1 Maintenance procedures are designed to do the following:

(1) Protect the equipment from adverse effects of heat, dust, moisture, and other contaminants
(2) Maintain top reliability and minimize costly downtime
(3) Prolong the useful life of the equipment
(4) Recognize incipient problems and take corrective action

22.2.2 The importance of maintenance cannot be overemphasized. Equipment should be kept operating efficiently to contribute to the success of the process or operation in which the equipment is used. Apparatus that is improperly maintained can become unreliable.

22.2.3 Persons charged with maintenance responsibility should have a keen appreciation as to why the work is required and the importance of even routine aspects of maintenance to the overall performance of the equipment.

22.3 Special Precautions.

22.3.1 Special safety precautions should be observed before and during the preventive maintenance operation. Extreme care should be taken to ensure that all power is removed from the apparatus before it is serviced. To prevent accidental shock from stored energy, capacitors should be discharged in accordance with the equipment manufacturer's instructions. Capacitors having high stored energy can be lethal or could be damaged by the application of a direct short circuit. Discharging the capacitor through a resistor followed by a direct short circuit might be required. Connecting charged capacitors to earth ground will not discharge capacitors that are used in circuits that normally are isolated from ground. After power has been removed, parts, such as tubes, resistors, and heat sinks, can remain extremely hot and cause painful burns.

22.3.2 Occasionally, some equipment requires troubleshooting while the circuits are energized. If so, it should be ensured that the insulation on test equipment leads is fully rated for the operating voltage under test and in good mechanical condition. Special care should be observed in the use or servicing of equipment that employs the chassis as one side of the circuit. Such equipment can be hazardous in the presence of grounded or some ungrounded 3-phase circuits.

22.3.3 In the absence of other instructions, it should be assumed that all electronic equipment is electrostatic discharge (ESD) sensitive. Industry standard ESD procedures should be followed.

22.4 Preventive Maintenance Operations. Actual work performed during maintenance of electronic equipment should include the following operations:

(1) Inspection
(2) Cleaning
(3) Adjustments
(4) Testing
(5) Servicing

22.4.1 Inspection. Inspection is most important in the maintenance program. Slight abnormalities might not immediately interfere with the equipment performances, but deviations from normal should be discovered early. Time and effort can be saved if defects are corrected before they lead to major breakdowns. Inspections consist of careful observation of all parts of equipment, noticing their color, placement, state of cleanliness, and so on. Inspection should be made for conditions such as the following:

(1) Overheating as indicated by discoloration or other visual characteristics. Infrared inspection can reveal abnormal temperatures and possible problem areas and should be performed in accordance with Section 11.17.
(2) Placement. Leads and cable clearances, rub points, and so on, should be observed.
(3) Cleanliness. Recesses should be examined for accumulation of dust, especially between connecting terminals. Parts, connections, and joints should be free of dust, corrosion, and other foreign material.
(4) Tightness. Soldered or screw terminal connections and mountings should be tested by slightly pulling on the wire or feeling the lug or terminal screw. Printed circuit boards should be inspected to determine that they are fully inserted into the edge board connectors. Board locking tabs also should be engaged. Unless connector malfunctions are suspected, routine unplugging and replugging of connectors to verify seating is not recommended; doing so can shorten the useful life of the connectors.
(5) Moisture. Look for evidence of moisture or corrosion. Consider a space heater if the surrounding air is repeatedly or continuously high in humidity. Verify the operation of any space heaters.
(6) Blockages. Keep air passages, fans, and ducting clear and clean to prevent overheating. Check fans for proper direction of rotation.

22.4.2 Cleaning. Cleaning the apparatus, both inside and out, is essential for good operation. Dust and the like increase chances of current leakage or flashover with resultant malfunction or damage to critical parts. Any accumulation of dust should be removed with a vacuum cleaner, if possible, or manually cleaned during maintenance shutdown periods. Enclosure filters should be cleaned at regular intervals and replaced if they are damaged or clogged. Solvents should not be used on printed circuit boards.

22.4.3 Adjustments. Adjustments should be made only when performance indicates that they are necessary to maintain normal operating conditions. Specific adjustments vary with each type of equipment and are described in the instruction booklets supplied with the apparatus. Equipment calibrations should be scheduled on a routine basis, with the frequency depending on the operating conditions particular to the process or equipment.

22.4.4 Testing. Reference to manufacturers' instructions is recommended.

22.4.5 Servicing. Necessary replacements should be made only at the printed circuit board or plug-in component level unless otherwise recommended by the equipment manufacturer. Manufacturers' recommendations should be followed for removal, handling, packaging, shipping, and replacement of such components or modules. Unnecessary strains on wires, cables, and connections should be avoided.

Chapter 23 Lighting

23.1 Introduction. A planned maintenance program is an essential part of any initial lighting design and recommendation. The maintenance of lighting systems is aimed at preserving the light-producing capability at the original design level. Dirt and lamp aging are the two major factors that reduce the light output.

23.2 Cleaning.

23.2.1 Lighting equipment — lamps, reflectors, and lenses — should be cleaned periodically. The cleaning interval depends on the amount and type of dirt in the air, although the design of the luminaire affects the rate at which dust collects. Periodic light meter readings can be taken and cleaning intervals established when the lighting level falls 15 percent to 20 percent, corrected for lamp lumen depreciation (aging).

23.2.1.1 Cleaning can economically be combined with group relamping, although in dirty environments cleaning should also be done between relampings. During spot relamping, the luminaires should be cleaned, and a separate planned cleaning program should be considered.

23.2.2 Washing is generally better than wiping. The cleaning procedure should be in accordance with the instructions of the luminaire manufacturer. Strong alkaline or abrasive cleaners should be avoided.

23.3 Relamping.

23.3.1 The longer a lamp remains in service, the less light it produces. The different types of lamps — filament, fluorescent, and high-intensity discharge — depreciate at different rates. Since the life expectancies also differ, replacement intervals will vary.

23.3.2 The two general relamping procedures are spot relamping and group relamping. Spot relamping is the replacement of individual lamps as they fail. Group relamping is the replacement of all lamps at a time, typically at 70 to 80 percent of their rated average life or when the light output falls below the desired level. It is economical to clean the luminaires at the time of replacement. It is also advantageous to inspect the sockets, hangers, reflectors, and lenses at the time of lamp replacement. General replacement recommendations and study results are available from the major lamp manufacturers.

23.3.3 Normally, replacement lamps should be of the same type, color, wattage, and voltage as those being replaced. However, where energy conservation is being considered, replacements might warrant appropriate substitutes. Such substitute lamps should conform to the luminaire relamp label instructions, and the luminaire manufacturer should be consulted. Appropriate lumen levels should be maintained.

23.3.4 With group relamping, it is appropriate to consider conversion to more energy-efficient lighting. Operating costs can be reduced by a planned conversion to energy-saving lamps or more efficient ballasts. The Energy Policy Act of 1992 eliminated the availability of many full-wattage fluorescent and incandescent reflector lamps. The lamp manufacturer should be consulted for compliant energy-efficient replacements.

23.4 Voltage.

23.4.1 Lamps and ballasts are designed to provide rated-average life expectancy and light output at the rated operating voltage.

23.4.2 A filament lamp operating at 5 percent overvoltage will have its life expectancy reduced almost 50 percent, while the light output will be increased by about 18 percent. Five percent undervoltage operation will increase lamp life to about 195 percent, and light output will be reduced to about 84 percent.

23.4.3 Fluorescent lamp ballasts are designed for operation at nominal rated voltages of 120, 208, 240, 277, or 480 volts. The ranges of permissible variations are 110–126, 191–218, 220–252, 254–291, and 440–504 volts. Higher voltage shortens lamp and ballast life, while lower voltage can shorten lamp life and might cause uncertain starting.

23.4.4 High-intensity discharge lamp ballasts are designed for operation at 120, 208, 240, 277, and 480 volts. Line voltage higher than rated voltage shortens ballast and lamp life, while lower voltages reduce light output and might cause uncertain starting. If a multiple-voltage primary winding is used, the connected tap typically should match the line voltage. Some high-intensity discharge lamp ballasts are provided with taps to accommodate variations from rated voltage.

23.5 Lamps and Ballasts.

23.5.1 Efficiency requirements influence the need to replace entire light fixtures or selected components (e.g., ballasts, lamps) during maintenance procedures. The U.S. Government Energy Star program provides identification of products meeting these energy efficiency requirements.

23.5.2 Fluorescent Lamps. Frequent starting might shorten lamp life and can damage a ballast. When blinking occurs, the lamps should be replaced. If that does not solve the problem, the ballast should be replaced.

23.5.2.1 Except as identified in Section 410.130 of *NFPA 70, National Electrical Code,* all fluorescent lamp ballasts for luminaires installed indoors are required to contain thermal protection and to be marked "Class P."

23.5.2.2 If inline fuseholders or fuses are used for individual luminaires, the luminaire fuseholder or fuse marking for replacement should be followed.

23.5.3 High-Intensity Discharge Lamps. High-intensity discharge (HID) lamps include metal halide, mercury vapor, and high-pressure sodium lamps. These lamps typically are constructed of an outer bulb with an internal arc tube. Metal halide arc tubes operate at higher pressures and temperatures (approximately 1100°C). Metal halide arc tubes and outer bulbs can rupture, particularly if the lamp is misapplied. Metal halide lamp types are as follows:

(1) O-type lamps are designed for open fixtures. They contain a shrouded arc tube strong enough to prevent lamp shattering.
(2) E-type lamps are intended for use in enclosed fixtures. Such fixtures include integral containment barriers that enclose the lamp.
(3) S-type lamps can be used in either open or enclosed fixtures. These lamps have no shroud. The design is limited to certain lamps between 350 watts and 1000 watts. The lamps must be operated vertically if they are used in open fixtures. S-type lamps in open fixtures offer the least protection in the event of a rupture.

23.5.3.1 Luminaires should be listed for the location and purpose. Lamp type and rating should be appropriate for the luminaire and should meet the manufacturer's specification.

23.5.3.2 Replacement lamps should conform to the luminaire relamp label instructions. Covers on enclosed luminaires should be replaced properly.

23.5.3.3 Metal halide systems should be turned off at least once a week for a minimum of 15 minutes. Failure to do this increases the risk of rupture.

23.5.3.4 To further reduce risk of rupture, metal halide fixtures should be group re-lamped at 70 percent of rated life.

23.5.4 Light Emitting Diode (LED) Lamps. An LED lamp consists of an array of LEDs as the light source, which can include semiconductor, organic, or polymer LEDs.

23.5.4.1 Relamping Fixtures with LED Lamps.

23.5.4.1.1 Two common approaches to relamping with LEDs are as follows:

(1) A direct lamp replacement that does not involve any rewiring or changes in the luminaire other than replacement of the existing lamp, such as replacing an incandescent lamp with a more energy efficient LED lamp. It is important to look for the listing mark and follow the use markings for the correct application of the replacement lamp.
(2) A luminaire conversion using a retrofit luminaire conversion kit. These kits generally involve replacing the lamp, rewiring the luminaire, and, in some cases, replacing parts such as a ballast with an LED power supply or directly connecting the lamp to the supply circuit.

23.5.4.1.2 Disconnect the electrical power to the entire fixture when servicing a fixture for any reason.

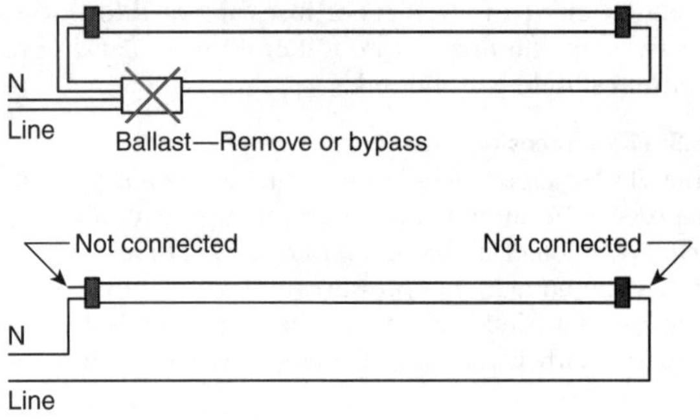

FIGURE 23.5.4.1.3(a) Example of Removal of Ballast and the Wiring Reconnection for Retrofit of Luminaire for LED Lamps.

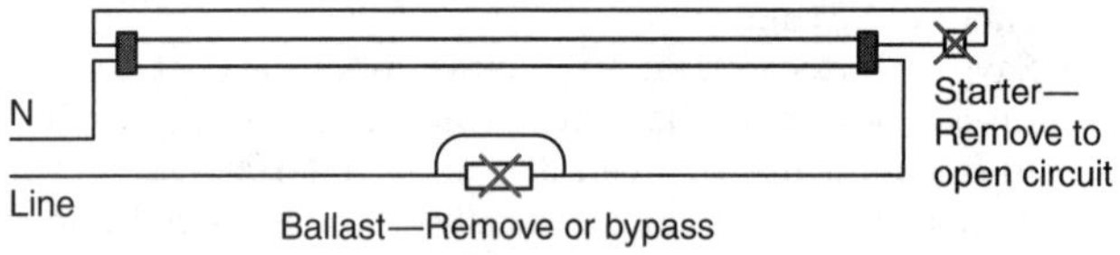

FIGURE 23.5.4.1.3(b) Example of Removal of Ballast and Starter with the Wiring Reconnection for Retrofit of Luminaire for LED Lamps.

23.5.4.1.3 In some luminaires, electronic ballasts are to be removed and the wiring reconnected directly to the end sockets. Refer to Figure 23.5.4.1.3(a). For older fixtures with a magnetic ballast and starter that require removal or opening of the starter circuit and removal or bypassing the ballast, refer to Figure 23.5.4.1.3(b).

23.5.4.1.4 Retrofit luminaire conversion kits selected for use should be evaluated by a testing organization to confirm that the kits, when installed into the identified luminaires in accordance with the installation instructions included with the kit, will result in a luminaire that continues to comply with safety requirements. It is important for installers and luminaire maintenance personnel when installing a retrofit luminaire conversion kit to note the following essential safety elements:

(1) Before installation, a thorough review of the luminaire and the kit installation instructions is necessary to verify that the retrofit kit is appropriate for the luminaire.
(2) Review and follow the instructions for preparation of the existing luminaire prior to installation of the kit and note all warnings and other instructions with respect to kit limitations.

23.5.4.1.5 Use only the exact replacement parts provided with the kit or reuse existing luminaire parts if specifically directed to by the kit installation instructions.

23.5.4.1.6 Lamp replacement directions must be followed to avoid risk of fire or electric shock hazard.

23.5.4.1.6.1 Where replaceable LED lamps are used in a converted luminaire, it is important to follow the markings added to the luminaire describing the exact replacement lamps that must be used for proper operation.

23.5.4.1.6.2 Use of anything other than the exact replacement lamp specified for the converted luminaire could pose a risk of shock or fire when installed in the luminaire.

23.5.4.1.6.3 Follow the instructions for insertion of replacement lamps since orientation and specific polarity could be required for the luminaire to properly operate.

23.5.4.2 Retrofitting Fluorescent Luminaries with LED Lamps.

23.5.4.2.1 Special considerations must be given when retrofitting fluorescent luminaries that have bi-pin lampholders that are being reused.

23.5.4.2.2 A retrofit kit cannot be installed in luminaires employing shunted bi-pin lampholders when the retrofit kit instructions specify that line and neutral power are to be connected to the same lampholder, since this can result in a potential fire or electric shock hazard. Contact the LED luminaire retrofit kit manufacturer when in doubt about the compatibility of the kit and the lamp fixture.

23.5.4.2.3 Push-in terminals of existing fluorescent lampholders are not intended to have wires removed and reinstalled. If rewiring and/or connections to existing fluorescent lampholders are indicated by the installation instructions, the connections can be made by splicing the lampholder leads using an appropriate splicing means using listed connectors.

23.6 Disposal. Certain lamps and ballasts require special disposal considerations. Contact the lamp and ballast manufacturer for more information. *(See 8.8.3.)*

Chapter 24 Wiring Devices

24.1 Introduction. This section covers the maintenance of attachment plugs, cord connectors, and receptacles rated not more than 200 amperes nor more than 600 volts.

24.1.1 The connection of equipment to supplies of different electrical ratings of current, voltage, phase, or frequency can be hazardous or can cause damage to equipment. Therefore, attachment plugs, cord connectors, and equipment are provided with different ratings and configurations to prevent hazardous interconnection. See Annex I for configuration charts from NEMA WD 6, *Wiring Devices — Dimensional Specifications.*

24.1.2 The use of these devices for the connection of equipment provides for rapid removal and replacement and facilitates relocation.

24.1.3 Up to 60 amperes, all devices are tested for the capability of being connected or disconnected under full load. Devices rated above 60 amperes are marked as to whether they are listed for this mode of operation.

24.1.4 Use of these devices to disconnect some equipment under some load conditions, such as welders, and running or stalled motors can be hazardous. Other load-interrupting means intended for this purpose should be used.

24.1.5 If the plug or connector housing or interior is cracked or distorted, if pieces are missing or damaged, or if the pins or contacts are bent, missing, or discolored, the complete interior should be replaced. For particularly adverse environments, such as highly corrosive environments, high-temperature locations, or hazardous (classified) locations, devices specifically intended for the purpose should be used.

24.1.5.1 If the receptacle or plug insulation is cracked, broken, or discolored, the defective parts should be replaced.

24.1.5.2 Receptacle contacts should retain inserted plugs firmly. Corroded, deformed, or mechanically damaged contacts should be replaced. A check should be made for proper wire connections on receptacles and proper polarity of power connection, including the integrity of the equipment-grounding conductor.

24.1.5.3 If there is abnormal heating of the receptacle, plug, or connector insulation, a check should be made for loose terminations or insufficient pressure between contacts, and they should be corrected or replaced. If there is arc tracking or evidence of burning of the insulation or other damage, the insulation should be replaced.

24.2 Connector and Receptacle.

24.2.1 Plugs should fit firmly when inserted into the mating connector or receptacle. Insufficient mating force can result in contact erosion caused by arcing of the contacts or accidental disengagement. The connector or receptacle should be checked to ensure that adequate contact pressure is present. The complete interior should be replaced if there is discoloration of the housing or severe erosion of the contact.

24.2.1.1 When continuity of service is essential, consideration should be given to the installation of a mechanically held or interlocked assembly.

24.2.2 The equipment-grounding conductor (green insulation) of the cord must be attached to the grounding terminal of the device, thereby ensuring grounding continuity.

24.2.3 The face of the receptacle, plug, or connector should occasionally be thoroughly cleaned.

24.2.4 Cracked, bent, or broken spring doors or covers should be replaced.

24.2.5 All mounting and assembly screws should be present and checked to ensure that they are tight because they can provide grounding, prevent the entrance of adverse environmental products, and provide cable retention.

24.2.6 All gaskets, if used, should be inspected to determine if they are present and maintain the integrity of the enclosure.

24.2.7 To ensure proper selection of replacement parts, the nameplates should be kept clean and legible, and the instructions supplied with the product should be maintained on file, together with a list of the manufacturer's replacement parts.

24.2.8 Because the grounding circuit path for the equipment can include the external shell, pin, and sleeve devices, these surfaces should not be painted.

24.2.9 Control contacts are occasionally used in conjunction with power pins. These control contacts should be inspected to ensure that they make last and break first.

24.2.10 Devices used in hazardous (classified) locations require some additional inspections. All mechanically and electrically interlocked plugs and receptacles should be inspected for proper operation and for excessively worn or broken parts; they should be replaced as required. All parts and surfaces of these devices should be clean and free of foreign material or corrosion. Flame paths should be inspected to ensure that safe gaps are not exceeded and that no scratches are on the ground joints. All screws holding the receptacle to the body should be installed and tight. Covers and threaded openings should be properly tightened. These devices should be checked to make sure that the plug and receptacle marking agree with the present classification of the area in regard to the class, group, and division.

24.3 Receptacles.

24.3.1 If the receptacle is badly worn, cracked, or broken, or if contacts are exposed, the receptacle should be replaced.

24.3.2 Receptacle contacts should hold and retain inserted plugs firmly. If accidental disengagement of the plug from the receptacle is a recurring problem, the receptacle should be replaced. When continuity of service is essential, consideration should be given to the installation of a locking-type device.

24.3.3 A check should be made to ensure proper wire connections on receptacles and proper polarity of power connections, including the integrity of the equipment ground.

24.3.4 When replacing 15- and 20-ampere nongrounding-type receptacles, refer to Section 406.3(D) of *NFPA 70, National Electrical Code.*

24.3.5 If there is abnormal heating on the receptacle face, a check should be made for loose terminations, and if found, they should be corrected or replaced. If there is arc tracking or evidence of burning of the device or other damage, the receptacle should be replaced.

24.4 Adapters. Adapters between locking and nonlocking configurations provide flexibility in obtaining power for maintenance functions. However, adapters should not be used to bypass the equipment ground, nor should adapters with pigtails be used.

24.5 General-Use Snap Switches.

24.5.1 Switches of the ac-dc (T-rated) type should not be used to control inductive loads such as fluorescent lighting or motors where the load exceeds 50 percent of the switch rating. Switches that are rated ac-only are permitted to control up to 100 percent of their rating for inductive loads or 80 percent of their rating for motor loads.

24.5.2 If the switch is broken or the mechanism does not function in a normal manner, the switch should be replaced. Where repeated abuse is incurred, consideration should be given to relocating the switch or replacing it with a switch having a guarded operating means or a switch with a low profile.

24.5.3 The switch should be firmly fastened to the box to ensure electrical and mechanical integrity.

24.5.4 If there is evidence of abnormal heating, the switch should be checked for loose terminals or switch malfunction and corrected or replaced.

24.6 Cover Plates.

24.6.1 All switches and receptacles should be installed with wall plates or covers suitable for the environment and location.

24.6.2 Cracked, bent, or broken wall plates or spring doors or covers should be replaced.

24.7 Boxes. Boxes used for the containment of receptacles and switches should be rigidly secured in place. Locknuts and conduit fittings should be made up tight, and proper box-fill of conductors should be observed. Closures should be placed in unused knockout holes. Where boxes, particularly the surface-mounted type, sustain repeated abuse, consideration should be given to flush mounting or additional guarding means.

24.8 Pin and Sleeve Devices.

24.8.1 Heavy-Duty Industrial-Type Plugs, Cord Connectors, and Receptacles.

24.8.1.1 Introduction. This section covers the maintenance of heavy-duty industrial-type plugs, cord connectors, and receptacles rated not more than 400 amperes nor more than 600 volts.

24.8.1.2 General. Plugs, cord connectors, and receptacles of this type are provided with different ratings and polarizations to prevent hazardous interconnection of different current ratings, voltages, or frequencies.

24.8.1.2.1 Devices connected to circuits having different voltages, frequencies, or types of current on the same premises should not be interchangeable.

24.8.1.2.2 Noninterchangeability is accomplished in these products by at least two methods. The first is the size and location of the contacts. The second is by keying arrangements of the plug sleeve and receptacle housing. By varying these parameters, sufficient variations can be obtained to accomplish noninterchangeability.

24.8.1.2.3 A detailed plan should be prepared specifying the devices, based first on performance requirements and then defining the specific configuration for each voltage, amperage, and frequency of use on the premises.

24.8.1.2.4 The use of these devices for the connection of equipment provides for rapid removal and replacement and facilitates relocation of electrical equipment.

24.8.1.2.5 Most of these assemblies are designed and listed to disconnect the equipment under full-load or locked-rotor currents. If they are not suitable, other load-interrupting means, such as interlocked receptacles, should be used.

24.8.1.3 Plugs. Cord clamps and strain-relief fittings should be checked to ensure that they are tight and that the outer cord jacket is completely within the clamping area.

24.8.1.4 Abnormal heating on the plug surface might be caused by loose terminations, overloading, high ambients, or equipment malfunction. Insulators and contacts should be inspected visually for discoloration of the insulator or pitting of contacts. Inspection of other parts should be initiated if discoloration or pitting is observed. The assembly of individual conductors to terminals should be periodically checked. Individual conductor strands should be properly confined and terminations made tight. Conductor strands should not be soldered when used with binding head screws because that can cause overheating.

24.8.1.5 If the plug or connector housing or interior is cracked or distorted, if pieces are missing or damaged, or if the pins or contacts are bent, missing, or discolored, the complete interior should be replaced. For particularly adverse environments, such as highly corrosive environments, high-temperature locations, or hazardous (classified) locations, devices specifically intended for the purpose should be used.

24.8.1.5.1 If the receptacle or plug insulation is cracked, broken, or discolored, the defective parts should be replaced.

24.8.1.5.2 Receptacle contacts should retain inserted plugs firmly. Corroded, deformed, or mechanically damaged contacts should be replaced. A check should be made for proper wire connections on receptacles and proper polarity of power connections, including the integrity of the equipment-grounding conductor.

24.8.1.5.3 If there is abnormal heating of the receptacle, plug, or connector insulation, a check should be made for loose terminations or insufficient pressure between contacts, and they should be corrected or replaced. If there is arc tracking or evidence of burning of the insulation or other damage, the insulation should be replaced.

24.8.2 Connector and Receptacle.

24.8.2.1 Plugs should fit firmly when inserted into the mating connector or receptacle. Insufficient mating force can result in contact erosion caused by arcing of the contacts or accidental disengagement. The connector or receptacle should be checked to ensure that adequate contact pressure is present. The complete interior should be replaced if there is discoloration of the housing or severe erosion of the contact. When continuity of service is essential, consideration should be given to the installation of a mechanically held or interlocked assembly.

24.8.2.2 The equipment-grounding conductor (green insulation) of the cord must be attached to the grounding terminal of the device, thereby ensuring grounding continuity.

24.8.2.3 Occasionally, the face of the receptacle, plug, or connector should be cleaned thoroughly.

24.8.2.4 Cracked, bent, or broken spring doors or covers should be replaced.

24.8.2.5 All mounting and assembly screws should be present and checked to ensure that they are tight, because they can provide grounding, prevent the entrance of adverse environmental products, and provide cable retention.

24.8.2.6 All gaskets, if used, should be inspected to determine if they are present and to maintain the integrity of the enclosure.

24.8.2.7 To ensure proper selection of replacement parts, the nameplates should be kept clean and legible, and the instructions supplied with the product should be maintained on file, together with a list of the manufacturer's replacement parts.

24.8.2.8 Because the grounding circuit path for the equipment can include the external shell, pin and sleeve devices, these surfaces should not be painted.

24.8.2.9 Control contacts are occasionally used in conjunction with power pins. These control contacts should be inspected to ensure that they make last and break first.

24.8.2.10 Devices used in hazardous (classified) locations require some additional inspections. All mechanically and electrically interlocked plugs and receptacles should be inspected for proper operation and for excessively worn or broken parts, which should be replaced as required. All parts and surfaces of these devices should be clean and free of foreign material or corrosion. Flame paths should be inspected to ensure that safe gaps are not exceeded and that no scratches are on the ground joints. All screws holding the receptacle to the body should be installed and tight. Covers and threaded openings should be properly tightened. These devices should be checked to make sure that the plug and receptacle marking agree with the present classification of the area in regard to the class, group, and division.

Chapter 25 Rotating Equipment

25.1 Introduction.

25.1.1 The various classes of rotating equipment have many common features in routine maintenance, both electrical and mechanical. The recommendations in this chapter are of a general nature and are not intended to cover in detail large or special applications, such as gear pump motors, or those designed for hazardous (classified) locations.

25.1.2 A complete list of the machines in operation, the functions they perform, and the past history of operation form the basis for a schedule of routine maintenance. Frequency of inspection depends on the nature of the service, the hours of operation, and the environment under which the equipment operates. Periodic inspection and appropriate maintenance assist in making continuous operation of the equipment possible. In some instances, disassembly is necessary for a complete inspection and necessary repairs.

25.2 Safety Precautions. The following safety precautions should be observed:

(1) A machine should be locked out/tagged out before work begins and properly protected against unintentional reenergization.
(2) Workers should wear personal protective equipment such as goggles, gloves, aprons, and respirators when working with solvents.
(3) Great care should be exercised in the selection of cleaning agents for any particular task. Be sure to follow all applicable environmental regulations.
(4) Where cleaning agents are used, adequate ventilation should be provided to avoid fire, explosion, and health hazards.
(5) A metal nozzle used for spraying flammable cleaning agents should be bonded to the supply drum and to the equipment being sprayed.
(6) Rubber insulating gloves should be used in connecting and operating high-voltage test instruments.
(7) After tests have been made, stored energy should be discharged from windings before test leads are handled.

25.3 Stator and Rotor Windings. The life of a winding depends on keeping it near to its original condition as long as possible. Insulation failure causes immediate outage time. The following points should be carefully examined and corrective action taken during scheduled inspections to prevent operational failures.

25.3.1 Dust and dirt are almost always present in windings that have been in operation under average conditions. Some forms of dust are highly conductive and contribute materially to insulation breakdown as well as restrict ventilation. *(See 25.6.2 for recommended cleaning methods.)*

25.3.2 Evidence of moisture, oil, or grease on the winding should be noted, and, if necessary, the winding should be cleaned thoroughly with a solvent solution. Generally, after a major cleaning, a drying process is necessary to restore the insulation to a safe level for operation. *(See 25.6.3 for drying methods.)*

25.3.3 Winding tightness in the slots or on the pole pieces should be checked. One condition that hastens winding failure is movement of the coils due to vibration during operation. The effects of varnishing and oven treatment serve to fill the air spaces caused by insulation drying and shrinking and will maintain a solid winding.

25.3.4 Insulation surfaces should be checked for cracks, crazing, flaking, powdering, or other evidence of the need to renew insulation. Usually, under these conditions, when the winding is still tight in the slots, a coat or two of air-drying varnish can restore the insulation to a safe value.

25.3.5 The winding mechanical supports should be checked for insulation quality and tightness. The ring binding on stator windings and the glass or wire-wound bands on rotating windings also should be checked.

25.3.6 Squirrel-cage rotors should be examined for excessive heating, for discolored or cracked rotor bars, or for cracked end rings, which can indicate open circuits or high-resistance points between the end rings and the rotor bars. The symptoms of such conditions are a slowing down under load and reduced starting torque. Brazing or welding broken bars or replacing bars should be done only by a qualified person or repair shop.

25.4 Brushes, Collector Rings, and Commutators. In general, the machine should be observed while in operation, if possible, and any evidence of maloperation, such as sparking, chatter of brushes in the holder, cleanliness, should be noted as an aid to inspection repairs later.

25.4.1 Brushes. Successful brush operation depends on the proper selection and maintenance of the brush most suitable for the service requirements.

25.4.1.1 Brushes in holders should be checked for fit and free play, and those that are worn down almost to the brush rivet should be replaced.

25.4.1.2 Brush studs that might have become loose from the drying and shrinking of insulating washers should be tightened.

25.4.1.3 Brush faces should be examined for chipped toes or heels and for heat cracks. Any that are damaged should be replaced.

25.4.1.4 A check of brush spring pressure should be made using the spring balance method. The spring pressure should be readjusted in accordance with the manufacturers' instructions.

25.4.1.5 The brush shunts should be checked to ensure that they are properly secured to the brushes and holders.

25.4.1.6 In some instances, if changes have occurred in the operation of equipment since installation, it might be necessary to check the following points that ordinarily would not be disturbed:

(1) Brushes should be reset at the correct angle.
(2) Brushes should be reset in the neutral plane.
(3) Brushes should be properly spaced on the commutator.
(4) The brush holders should be correctly staggered.
(5) Brush holders should be properly spaced from the commutator.
(6) A check should be made to ensure that the correct grade of brush as recommended by the manufacturer is being used.

25.4.2 Collector Rings. The surest means of securing satisfactory operation is maintaining the slip-ring surface in a smooth and concentric condition.

25.4.2.1 Insulation resistance should be checked between ring and shaft to detect cracked or defective bushings and collars.

25.4.2.2 A thorough cleaning is usually recommended, using a solvent cleaner and stiff brush.

25.4.2.3 Brush holder end play and staggering should be checked to prevent grooving of the rings during operation.

25.4.2.4 When the rings have worn eccentric with the shaft, the ring face should be machined.

25.4.3 Commutators. In general, sources of unsatisfactory commutation are due to either improper assembly of current-collecting parts or faulty operating conditions.

25.4.3.1 Commutator concentricity should be checked with a dial gauge if sufficient evidence indicates that the commutator is out of round. A dial indicator reading of 0.001 in. on high-speed machines to several thousandths of an inch on low-speed machines can be considered normal.

25.4.3.2 The commutator surface should be examined for high bars, grooving, evidence of scratches, or roughness. In light cases, the commutator can be hand stoned, but for extreme roughness, turning of the commutator in the lathe is recommended.

25.4.3.3 A check should be made for high or pitted mica, and it should be undercut where deemed advisable.

25.4.3.4 After conditioning a commutator, it should be completely clean, with every trace of copper, carbon, or other dust removed. *(See ANSI/EASA AR100, Recommended Practice for the Repair of Rotating Electrical Apparatus.)*

25.5 Bearings and Lubrication.

25.5.1 General. The bearings of all electrical equipment should be carefully inspected at scheduled periodic intervals to ensure maximum life. The frequency of inspection is best determined by a study of the particular operating conditions.

25.5.2 Sleeve Bearings.

25.5.2.1 In the older types, the oil should be drained, the bearing flushed, and new oil added at least every year.

25.5.2.2 The new type of sealed sleeve bearings requires very little attention, since oil level is frequently the only check needed for years of service.

25.5.2.3 Bearing currents on larger machines are usually eliminated by installing insulation under the pedestals or brackets or by insulating the shell of the bearing from its support housing. Elimination of this circulating current prevents pitting of the bearing and shaft. From a maintenance standpoint, a check should be made to ensure that the bearing insulation is not short-circuited by bearing temperature detectors or by lubricating-oil piping. This type of check might require uncoupling the machine or lifting the noninsulated end (after disassembling the bearing) of the shaft if both bearings are not insulated.

25.5.3 Ball Bearings and Roller Bearings.

25.5.3.1 External inspection at the time of greasing will determine whether the bearings are operating quietly and without undue heating.

25.5.3.2 The bearing housings can be opened to check the condition of the bearings and grease. The bearing and housing parts should be thoroughly cleaned and new grease added.

25.5.3.3 Where special instructions regarding the type or quantity of lubricant are recommended by the manufacturer, they should be followed. In all cases, standard greasing practices should be strictly adhered to.

25.5.4 Hydrodynamic Thrust Bearings. Established lubrication practice for sleeve bearings applies in general for thrust bearings.

25.6 Cleaning and Drying Insulation Structures.

25.6.1 General. Refer to Section 8.7, Equipment Cleaning, for basic recommendations.

25.6.2 Cleaning. The recommended methods for cleaning electrical equipment are given in 25.6.2.1 and 25.6.2.2.

25.6.2.1 Apparatus that has been clogged with mud from dust storms, floods, or other unusual conditions require a thorough water washing, usually with a hose with pressure not exceeding 1.72 kPa (25 psi). Initial cleaning should be made with hot non-saline water plus detergent, followed by a rinse with hot nonsaline water (no detergent). Chemical tests should be made to verify that the water is nonsaline. The machine should be completely dismantled, terminal boxes opened, and all corroded parts identified for repair or replacement. All components that are to be reused should be washed in a tank of hot, fresh, nonsaline water for at least four hours. The water tank should have a water inlet and outlet such that the water is constantly changing at a minimum rate of 10 gpm. All washed components should be dried at 85°C, or less, for 2 hours with continuous air circulation. Electrical insulation should be dried at 85°C for an additional 4 hours, followed by 105°C to 120°C for at least four more hours. The winding insulation resistance should be measured with a 500 volt insulation test instrument every 2 hours until the insulation resistance has stabilized. Allow the insulation to cool in a dry environment to avoid moisture absorption. Before being placed in service, the minimum recommended insulation resistance levels should be in accordance with ANSI/IEEE 43, *Recommended Practice for Testing Insulation Resistance of Rotating Machinery.* Sleeve bearings and housing should be cleaned, and antifriction bearings should be replaced with the same type as originally supplied.

25.6.2.2 Silicone-treated windings require special treatment, and the manufacturer should be contacted for advice.

25.6.3 Drying. After being cleaned, stored, or shipped, apparatus should be dried before being placed in operation if tests indicate that the insulation resistance is below a safe minimum level. Two general methods are commonly used: external heat or internal heat. External heat is preferred because it is the safer application.

25.6.3.1 Where available, low-pressure steam can be used through radiators or steam pipes placed below the end windings with a temporary built-in enclosure to hold the heat.

25.6.3.2 Forced hot air can be heated electrically, by steam, or by open fire. This method is usually inefficient and costly unless built into the original installation.

25.6.3.3 Electric space heaters or infrared lamps can be used. They should be distributed so as not to overheat the insulation.

25.6.3.4 Coil insulation can be dried by circulating current through the winding. There is some hazard involved with this method because the heat generated in the inner parts is not readily dissipated. This method should be followed only under competent supervision.

25.6.3.5 For synchronous motors, the short-circuit method is sometimes used by shorting the armature windings and driving the rotor, applying sufficient field excitation to give somewhat less than full-load armature current.

25.7 General Overhaul. When indicated by visual inspection or tests, the equipment should be disassembled, and the winding should be cleaned, dried, and re-insulated or dipped and baked, and the bearings checked and relubricated. Rewinding or other repair decisions should be made at this time. *(Refer to Section 26.5 for information on methods of balance.)*

25.8 Records. Sample record forms are shown in Annex H.

25.9 Testing. See Chapter 11 for recommended tests.

25.10 Energy Efficiency of Motors. When replacing existing motors or installing new motors, energy efficiency ratings should be considered. *(Refer to Annex O for details.)*

Chapter 26 Vibration

26.1 Introduction. Many rotating machinery failures occur for mechanical reasons, such as poor alignment, bearing failure, dynamic unbalance, or improper mounting.

26.2 Machine Vibration. All equipment vibrates when it is running. Excessive vibration indicates a problem. The cause might be in the mechanical integrity of the machine, for example, dynamic unbalance, misalignment, loose parts, or faulty bearings. It might be in the electrical integrity of the machine, for example, an open rotor bar or cracked end ring in a squirrel cage motor, or a faulty power supply to a dc motor. Frequently, it is a combination of factors that causes vibration.

26.2.1 The most common methods of measuring vibration are in units of velocity. When measured as displacement, the units are microns peak-to-peak or mils peak-to-peak. Velocity measurements are in millimeters per second or inches per second. Acceleration measurements are expressed in grams peak. Vibration is usually measured at the bearing housing.

26.2.2 Displacement is generally used as an indicator of vibration severity for both low-speed equipment operating at less than 1200 rpm and low-frequency vibration. Examples include dynamic unbalance, belt vibration, and shaft seal rub. The acceptable value of displacement for machine vibration decreases with increasing speed. For example, a machine rotating at 900 rpm might have an acceptable vibration displacement limit of 2.5 mils (1 mil = 0.001 in.). Running at 3600 rpm, the acceptable vibration displacement limit might be 1 mil.

26.2.3 For higher-frequency problems, either vibrational velocity or acceleration measurements are generally used for bearing housing or support vibration. Velocity is independent of machine speed and therefore a better general indicator of overall vibration severity. *(See Table 26.2.3.)* Acceleration is used to evaluate high-frequency problems such as those related to bearings and gears.

26.2.4 Unfiltered Vibration Limits. Suggested vibration limits for larger machines are specified in Table 26.2.4.

Table 26.2.3 Vibration Severity Chart

Velocity rms					
mm/sec	**in./sec**	**Class 1**	**Class 2**	**Class 3**	**Class 4**
0.71	0.028	A	A	A	A
1.12	0.044	B	A	A	A
1.8	0.071	B	B	A	A
2.8	0.110	C	B	B	A
4.5	0.177	C	C	B	B
7.1	0.279	D	C	C	B
11.2	0.440	D	D	C	C
18.0	0.708	D	D	D	C
28.0	1.10	D	D	D	D

Notes:

(1) Class 1: up to 20 hp on fabricated steel foundation; Class 2: 25 hp–100 hp on fabricated steel foundation, 100 hp–400 hp on heavy solid foundation; Class 3: above 400 hp on heavy solid foundation; Class 4: above 100 hp on fabricated steel foundation.

(2) Grade A: good; Grade B: usable; Grade C: just acceptable; Grade D: not acceptable.

Table 26.2.4 Unfiltered Vibration Limits

Speed (rpm)	Rotation Frequency (Hz)	Peak Velocity mm/sec	Peak Velocity in./sec
3600	60	3.8	0.15
1800	30	3.8	0.15
1200	20	3.8	0.15
900	15	3.0	0.12
720	12	2.3	0.09
600	10	2.0	0.08

Note: These levels pertain to bearing housing monitoring in the vertical, horizontal, and axial directions. Test conditions are uncoupled and without load.

26.2.5 Large machines can also use noncontacting shaft vibration probes and instrumentation to measure the rotor shaft vibration relative to the bearing housing.

26.3 Types of Instruments. Analog and digital instruments are available to measure displacement, velocity, and acceleration. In addition, there are computerized data collecting analyzers that store vibration spectrums, using Fast Fourier Transform (FFT) methodology. In addition to detecting unbalanced vibration, FFT analysis of the instruments can identify faults in state or windings, rotate bars and end rings, and bearings.

26.4 Resonance. All machines have certain natural frequencies of vibration. When vibration occurs at a frequency equal to one of the machine's natural frequencies (critical frequencies), the machine or component will exhibit a large amplitude of vibration. When this happens, the machine is said to have a resonant vibration. It is suggested that machine speed be at least 15 percent removed from any critical frequency. Where a machine should pass through one or more critical frequencies in coming up to running speed, it should pass through these quickly.

26.5 Methods of Balance. Static unbalance is an unbalance on one side (plane) of a rotating device, and dynamic unbalance is an unbalance in two planes of a rotating device. The solution is either to remove the excess weight or to add an equal amount at the opposite side. Narrow (length less than ½ diameter) slow speed (less than 1000 rpm) rotors can be statically balanced in a test stand with loose bearings or a knife edge. Dynamic balance, also known as two-plane balancing, is typical of cylindrical rotating devices, such as a roller or an electric motor rotor. Dynamic balance requires specialized equipment, such as a balancing machine. Modern dynamic balance machines are equipped with computers to calculate and display the amount and location of balance correction weights. Balancing might be possible with machinery in place, using portable balancing instruments. Modern portable balancing equipment is equipped with computers that calculate the exact location and amount of weight to add or remove to correct the unbalance.

26.6 Assembly and Installation Guidelines.

26.6.1 Installation of Accessories. Where possible, bearings, gears, and couplings should be uniformly preheated before installation to minimize damage. All rotating bodies should be dynamically balanced to within standard tolerance, as established by the manufacturer. Equipment with accessories should be balanced without these extra items. Accessories should then be installed individually and the equipment rebalanced if necessary.

26.6.2 Alignment. All rotating equipment should be properly aligned when installed. The human eye, straight edges, feeler gauges, and bubble gauges do not provide the precision required. Rim and face or reverse indicator methodology should be used. Dial indicators on all laser alignment equipment should be used for alignment. Laser alignment provides the advantages of accuracy, speed, and minimum chance for operator error. If the operating temperature of equipment changes significantly, thermal expansion should be considered.

26.7 Baseline Data. Together with other tests, it is important to keep data on vibration levels. With time, vibration tends to increase. Scheduled maintenance can reduce such problems, usually at a fraction of the cost of a breakdown caused by insufficient maintenance. Readings should be made at 3- to 6-month intervals or more often as required. Computerized data collectors, using TFF technology vibration analysis, can be used to measure vibration and trend results. Causes of a substantial change in vibration should be investigated promptly.

26.8 Noise. All machines produce some sound when running. Changes in the sound level might indicate problems and should be investigated. Manufacturers often can supply equipment with low noise levels when necessary. Excess noise can be caused by many factors, such as using rigid conduit connections instead of flexible connections, locating a machine in the corner of a room with hard, sound-reflecting side walls, and designing an installation with inadequate vibration isolators. Totally enclosed fan-cooled motors can be expected to produce more noise than open drip-proof or weather-protected motors of equivalent rating.

26.8.1 Adjustable-speed drives (ASDs), which operate ac motors at varying speeds, sometimes have switching frequencies that will cause noise in the load motor. Methods to correct this problem include a higher drive switching frequency, isolation transformer, line reactor, or a motor with a skewed rotor design.

Chapter 27 Hazardous (Classified) Location Electrical Equipment

27.1 Types of Equipment. Hazardous location electrical equipment is used in areas that commonly or infrequently contain ignitible vapors or dusts. Designs of hazardous location electrical equipment include explosionproof, dust-ignition-proof, dust-tight, purged pressurized, intrinsically safe, nonincendive, oil immersion, hermetically sealed, and other types. Maintenance of each type of equipment requires attention to specific items.

27.2 Maintenance of Electrical Equipment for Use in Hazardous (Classified) Locations.

27.2.1 Electrical equipment designed for use in hazardous (classified) locations should be maintained through periodic inspections, tests, and servicing as recommended by the manufacturer. Electrical preventive maintenance (EPM) documentation should define the classified area (the class, group, and division specification and the extent of the classified area) and the equipment maintenance required. EPM documentation should identify who is authorized to work on this equipment, where the maintenance is to be performed, and what precautions are necessary. Although repairs to certain equipment should be done by the manufacturer or authorized representatives, inspection and servicing that can be performed in-house should be clearly identified.

27.2.2 Maintenance should be performed only by qualified personnel who are trained in safe maintenance practices and the special considerations necessary to maintain electrical equipment for use in hazardous (classified) locations. These individuals should be familiar with requirements for obtaining safe electrical installations. They should be trained to evaluate and eliminate ignition sources, including high surface temperatures, stored electrical energy, and the buildup of static charges, and to identify the need for special tools, equipment, tests, and protective clothing.

27.2.3 Where possible, repairs and maintenance should be performed outside the hazardous (classified) area. For maintenance involving permanent electrical installations, an acceptable method of compliance can include deenergizing the electrical equipment and removing the hazardous atmosphere for the duration of the maintenance period. All sources of hazardous vapors, gases, and dusts should be removed, and enclosed, trapped atmospheres should be cleared.

27.2.4 Electrical power should be disconnected and all other ignition sources abated before any electrical equipment is disassembled in a hazardous (classified) location. Time should be allowed for parts to cool and electrical charges to dissipate, and other electrical maintenance precautions followed.

27.2.5 Electrical equipment designed for use in hazardous (classified) locations should be fully reassembled with original components or approved replacement before the hazardous atmosphere is reintroduced and before power is restored. Special attention should be given to joints and other openings in the enclosure. Cover(s) should not be interchanged unless identified for the purpose. Foreign objects, including burrs, pinched gaskets, pieces of insulation, and wiring, prevent the proper closure of mating joints designed to prevent the propagation of flame upon explosion.

27.2.6 An approved system of conduit and equipment seals conforming to the requirements of *NFPA 70, National Electrical Code,* and manufacturer's specifications should be maintained. Corrective action should be taken on maintenance actions that damage or discover damage to a seal. Damage to factory-installed seals within equipment can necessitate replacing the equipment.

27.2.7 Wherever electrical equipment cover bolts or screws require torquing to meet operating specifications, the bolts or screws should be maintained with the proper torque as specified by the manufacturer. Electrical equipment should not be energized when any such bolts or screws are missing. All bolts and screws should be replaced with original components or approved replacements.

27.2.8 Special care should be used in handling electrical devices and components approved for use in hazardous (classified) locations. Rough handling and the use of tools that pry, impact, or abrade components can dent, scratch, nick, or otherwise mar close-tolerance, precision-machined joints and make them unsafe.

27.2.8.1 Grease, paint, and dirt should be cleaned from machined joints with a bristle (not wire) brush, an acceptable noncorrosive solvent, or other methods recommended by the manufacturer.

27.2.8.2 Prior to replacing a cover on an enclosure designed to prevent flame propagation upon an explosion, mating surfaces should be cleaned and lubricated in accordance with the manufacturer's instructions.

27.2.9 Field modifications of equipment and parts replacement should be limited to those changes acceptable to the manufacturer and approved by the authority having jurisdiction. Normally, modifications to equipment void any listing by nationally recognized testing laboratories.

27.2.10 The requirements of *NFPA 70, National Electrical Code,* should be followed.

27.2.10.1 Explosionproof enclosures, dust-ignition-proof enclosures, dusttight enclosures, raceway seals, vents, barriers, and other protective features are required for electrical equipment in certain occupancies. Equipment and facilities should be maintained in a way that does not compromise equipment performance or safety.

27.2.10.2 Intrinsically safe equipment and wiring is permitted in locations for which specific systems are approved. Such wiring should be separate from the wiring of other circuits. *NFPA 70, National Electrical Code,* Article 504, Intrinsically Safe Systems, describes control drawings, grounding, and other features involved in maintenance programs.

27.2.10.3 Purged and pressurized enclosures can be used in hazardous (classified) areas. NFPA 496, *Standard for Purged and Pressurized Enclosures for Electrical Equipment,* provides guidance useful to maintenance personnel.

Chapter 28 Uninterruptible Power Supply (UPS) Systems

28.1 Introduction. The basic function of uninterruptible power supply (UPS) systems is to preserve power to electrical or electronic equipment. Most UPS systems are intended to provide regulated power to prevent power supply fluctuations or aberrations that can damage or cause malfunction of sensitive electrical/electronic equipment, such as computers and process controllers. A UPS system represents a sizable investment in equipment specifically installed to provide reliable regulated power to equipment. Therefore, it is essential that the UPS system be maintained in a manner that the UPS itself will not fail.

28.1.1 The general recommendations in this chapter can be applied to all UPS systems; however, it should be noted that UPS systems are equipment-specific. As a result, manufacturers' instructions should be followed carefully in the performance of any maintenance on UPS equipment.

28.1.2 The maintenance program should be planned at the time the UPS system is put into service, to provide early attention to ensuring the continuing reliability of the system. The development of an EPM program should not be deferred until the end of the warranty period.

28.1.3 Maintenance should be scheduled at times that will least affect operations. Actual maintenance procedures should not be started until the users have been notified.

28.1.4 Only fully trained and qualified persons with proper test equipment should perform UPS maintenance.

28.2 Types of UPS Systems.

28.2.1 There are two basic types of UPS systems: static and rotary. Some systems are hybrid versions that incorporate some features of both. A basic rotary system is essentially a motor-generator set that provides isolation between the incoming power supply and the load and buffers out power supply aberrations by flywheel mechanical inertia effect.

28.2.2 A static unit rectifies incoming ac power to dc and then inverts the dc into ac of the proper voltage and frequency as input power to the load. A battery bank connected between the rectifier and inverter sections ensures an uninterrupted supply of dc power to the inverter section.

28.2.3 In the UPS industry, the term *module* refers to a single self-contained enclosure containing the power and control elements needed to achieve uninterrupted operation. These components include transformers, rectifier, inverter, and protective devices.

28.2.4 UPS systems can consist of one or more UPS modules connected in parallel either to increase the capacity of the system power rating or to provide redundancy in the event of a module malfunction or failure. Figure 28.2.4 illustrates a typical single-module static 3-phase UPS configuration. Note that in this configuration the solid-state switch (SSS) is internal to the UPS module.

28.2.4.1 Figure 28.2.4.1 illustrates a typical multimodule static 3-phase UPS configuration. Note that in this configuration the SSS is located in the stand-alone static transfer switch (STC) control cabinet.

28.2.5 Almost all UPS systems comprise these common elements: disconnecting means, bypass and transfer switches, protective devices and power switchgear, molded-case circuit breakers, and fuses. Depending on the type of UPS (static, rotary, or

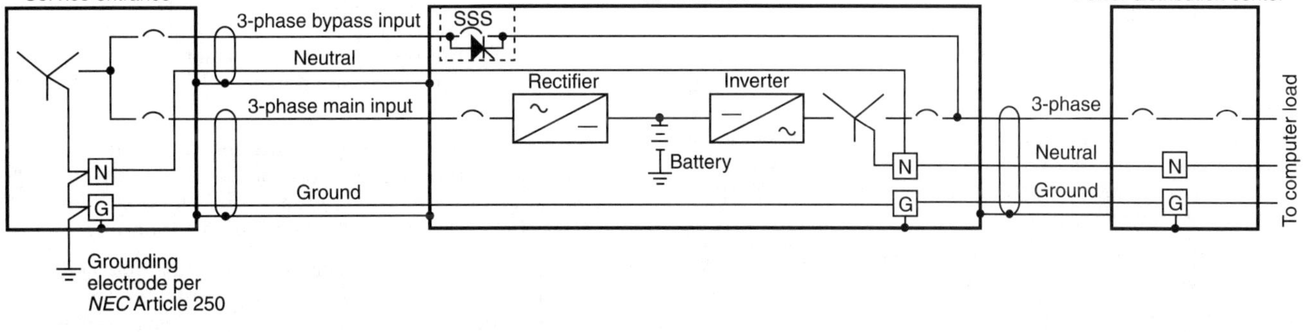

FIGURE 28.2.4 Typical Single-Module Static 3-Phase UPS Configuration.

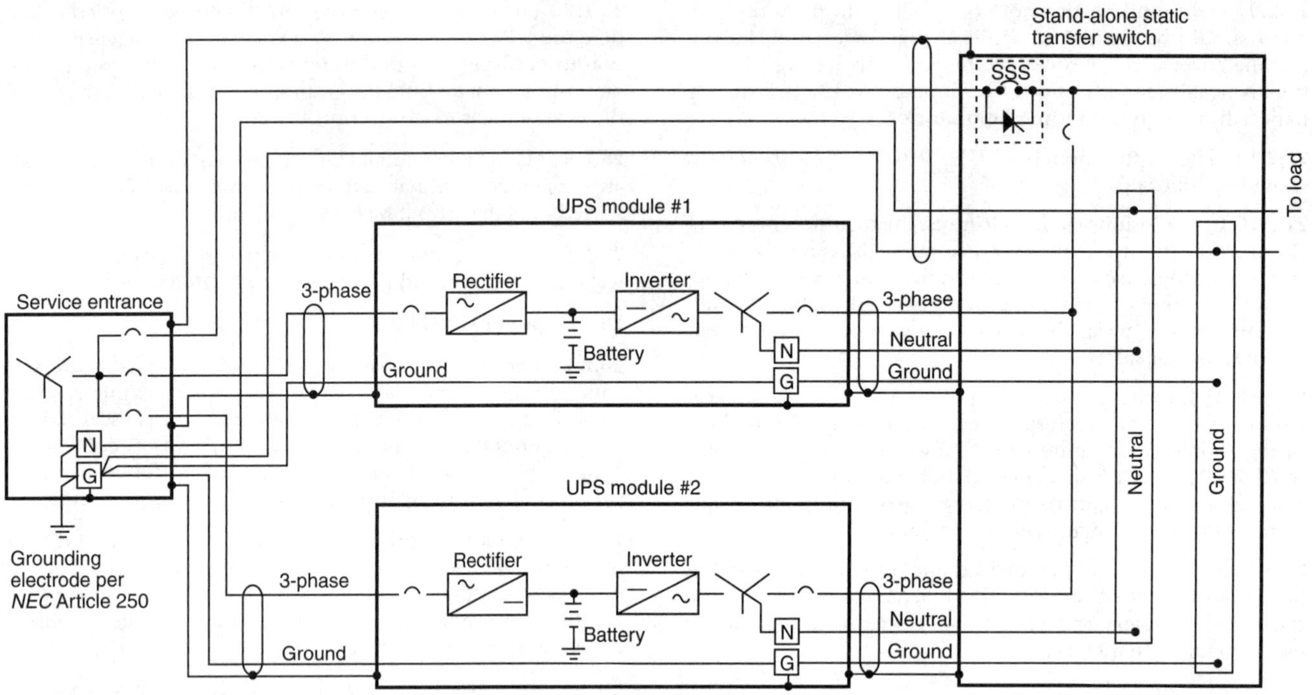

FIGURE 28.2.4.1 Typical Multimodule Static 3-Phase UPS Configuration.

hybrid), the system might also include transformers, batteries, a battery charger, a rectifier/inverter unit (static system), and a motor-generator set (rotary system). The system might also be supported by a standby generating unit to permit operations to continue during sustained power outages.

28.3 UPS System Maintenance Procedures — General. The routine maintenance procedures for components of UPS systems are covered in the particular equipment sections of this publication (i.e., switches, transfer switches, motor controllers, protective devices, batteries and battery chargers, transformers, rotating equipment). However, to aid in an organized preventive maintenance program, the following procedures are recommended.

CAUTION: It is important to avoid interruption of the power output of the UPS system. Extreme caution should be used in the servicing of the system to prevent unscheduled outages.

28.3.1 Disconnecting Means and Bypass Switches. These elements of the system should be maintained in accordance with the general maintenance procedures prescribed for the particular device in this document or the manufacturer's instructions as applicable.

28.3.2 Transfer Switches. Transfer switches in UPS systems can be of either the manually operated or automatic type. Switching devices should be maintained in accordance with the appropriate sections of this document. If of the static type, they should be maintained in accordance with the general procedures for maintaining electronic equipment in Chapter 22 and the specific procedures provided by the manufacturer.

28.3.2.1 Transfer switches also should be maintained in accordance with the manufacturer's guidelines.

28.3.3 Circuit Protective Devices. Molded-case circuit breakers should be maintained in accordance with Chapter 17, fuses in accordance with Section 18.1, and other protective devices in accordance with Chapter 15. It is especially important to keep an ample supply of the proper types of spare fuses on hand. UPS systems are generally protected with special fuses. Installing an improper fuse on a UPS can result in severe damage to the UPS and the load equipment.

28.3.4 Batteries and Chargers. Batteries and chargers should be maintained in accordance with manufacturers' instructions. See Chapter 15 for lead–acid batteries and chargers.

28.3.5 UPS Support Standby Generator. If the UPS is supported by a standby generating unit, the generator should be maintained in accordance with the general procedures for maintaining rotating equipment in Chapter 25. It is important that a program be in effect to ensure that the generating unit is test-run on a regular basis and also subjected to a full-load test at least monthly for a minimum of 2 hours. In addition, generator start-up, transfer, restoration of power, retransfer, and auxiliary generator shutdown operation should be checked at least twice a year.

28.3.6 UPS Ventilation. Ventilation air filters should be inspected on a regular basis. The frequency of cleaning or replacement depends on the amount of dust or dirt in the air at the installation and could range from as little as a week to as much as 6 months.

28.3.7 UPS Record Keeping. It is strongly recommended that a complete and thorough logbook be maintained for the UPS in a suitable location. The logbook should be used to record all items concerning the UPS, including the following:

(1) System operation — normal settings and adjustments
(2) Meter readings such as voltmeter, ammeter, and frequency meter at input and output, taken on a weekly basis (more frequently as necessary)

(3) Record of abnormal operations, failures, and corrective action taken
(4) Maintenance history

28.3.7.1 This log should be used for comparison to detect changes and degradation of the UPS circuitry, need for adjustment of controls, or other maintenance and testing.

28.3.7.2 Schematics, diagrams, operating procedures, record drawings, spare parts lists, troubleshooting techniques, maintenance procedures, and so on, should be kept in the same suitable location as the logbook.

28.3.8 Routine Maintenance. On a semiannual basis, the insides of cabinets should be vacuumed and the tightness of all electrical connections verified. On an annual basis, tightness of electrical connections should be checked using infrared scanning techniques or testing with a digital low-resistance ohmmeter *(see 11.10.5.1.5)*. Possible loose or corroded connections should be identified and cleaned and retightened as necessary.

28.3.8.1 All system alarms and indicating lights should be checked periodically for proper operation. On a quarterly basis, a visual inspection should be made for signs of overheating and corrosion. Wherever additional loads are connected to the UPS, the protective-device coordination, calibration, and proper operation of the modified system should be checked.

28.3.8.2 All heating, ventilating, air-conditioning, and humidity-control systems should be checked for proper operation and to ensure that the flow of cooling air is not blocked by obstructions in front of the vents. A check should be made for unusual sounds and odors because these signs might be the first indication of a potential malfunction.

28.3.8.3 The integrity of the grounding system should be maintained as required by Article 250 of *NFPA 70, National Electrical Code*. For separately derived systems, it should be ascertained that the neutral is properly grounded.

28.3.8.4 The neutral output current should be measured during peak loads every 3 months or when new equipment is added to the system. Measurements should be taken using a true rms-type ammeter to verify that the neutral conductor ampacity is not exceeded. Excessive current readings could indicate the presence of harmonics.

28.3.9 Rectifier and Inverter (Static Systems). This equipment should be maintained in the manner prescribed for electronic equipment in Chapter 22. In many cases, a common enclosure houses the rectifier, inverter, and support battery charger sections of the UPS system.

28.3.9.1 On a semiannual basis, the inverter should be inspected visually for signs of leaking fluid from wave-forming capacitors, and the capacitors should be checked for swelling or discoloration. *(See 15.9.3.)*

28.3.9.2 The transformers and the heat sinks should be inspected visually for signs of overheating.

28.3.9.3 Replacement of electrolytic capacitors should be considered at regular intervals not to exceed 5 years.

28.3.10 Motor and Generator (Rotary Systems). The motor and generator should be maintained in accordance with the general procedures for maintaining rotating electrical equipment in Chapter 25.

28.3.11 UPS Modifications. It is extremely important that all modifications be reflected in the record drawings and other pertinent documentation *(see 28.3.7)*. Modifications to procedures should be recorded. Component failures and corrective action that affect the documentation, such as a change in components, should be indicated.

28.3.11.1 The manufacturer should be contacted periodically (2-year interval, maximum) for information on equipment upgrades and recommended revisions.

28.4 UPS Testing.

28.4.1 Introduction.

28.4.1.1 UPS systems require periodic testing to determine if the system is functioning as designed. Each manufacturer provides, with the equipment, specifications delineating the stated equipment performance (i.e., voltage variation, balance, regulation, and harmonic distortion). Batteries can weaken, which will shorten the backup time of the particular manufacturer's specifications. Transfer operations might be generating transients or momentary outages that can create havoc in a computer system. The recommendations in 28.4.2 through 28.5.2 are intended to identify problems and apprise maintenance personnel of the actual capabilities of the UPS system.

28.4.1.2 Testing should not be attempted unless those performing this work are completely familiar with the manufacturer's recommendations, specifications, tolerances, and safety precautions.

28.4.2 Preliminary Testing.

28.4.2.1 Prior to testing, all operating parameters, such as frequency, voltage, and current, at the bypass switch, UPS input, UPS output, batteries, and modules should be recorded where applicable.

28.4.2.2 Tests should be performed with the unit under load to ascertain the condition and reserve capability of the batteries. Refer to 15.9.4 for preparation of batteries prior to load testing of the system.

28.4.2.3 An infrared scan of the batteries and UPS equipment should be performed. The scan should look specifically at the battery connections with ac input power disconnected and the battery supplying power to the load. The unit should not be operated under load for long periods of time with covers removed, because cooling might be inhibited, and damage to the unit might result.

28.4.2.4 Any abnormalities that have been detected should be corrected prior to proceeding with further testing.

28.5 System Tests.

28.5.1 Introduction.

28.5.1.1 Certain system tests might be necessary to fully determine the operating condition of a UPS system. These tests should be performed when warranted by special circumstances, such as repeated failure of a system to pass routine maintenance checks. The tests also should be conducted on a 2-year cycle or other periodic basis when the desired degree of reliability justifies the procedure. It might be necessary for an independent testing company or the equipment manufacturer to conduct these tests, because of the complexity and the sophisticated test instruments recommended. The units should be placed under load by the use of external load banks during such tests.

28.5.1.2 All UPS tests should require that the batteries be fully charged. (Some systems do not utilize battery backup.) Critical loads should be placed on isolation bypass, if available, or connected to another source.

28.5.1.3 It should be verified that all alarm and emergency shutdown functions are operating. It should be ascertained that the load transfers manually and automatically from UPS to bypass. It should be verified that all modules, when applicable, are functioning by load-testing each module individually prior to parallel load testing.

28.5.2 Special Tests. Simultaneous input and output readings of voltage, current, and frequency should be recorded. The external power source should be removed and reapplied to verify output stability.

28.5.2.1 Voltage and frequency recordings of UPS operation during transient response voltage tests should be provided; a high-speed recording device such as an oscillograph should be used to document the load tests described in 28.5.2.2 through 28.5.2.5.

28.5.2.2 The load should be stepped from 0 percent to 50 percent to 0 percent; 25 percent to 75 percent to 25 percent; 50 percent to 100 percent to 50 percent; 0 percent to 100 percent to 0 percent of UPS system rating.

28.5.2.3 It should be verified that the voltage regulation and frequency stability are within the manufacturer's specifications. In accordance with the manufacturer's specifications, the load bank should be increased to greater than 100 percent system load to ascertain that the system is within the manufacturer's ratings for input and output current overload rating.

28.5.2.4 Where applicable, UPS ac input power should be removed while the system is supplying 100 percent power to a load bank. The elapsed time until low battery voltage shutdown occurs should be recorded and compared with specifications. Voltage, current, and frequency should be read and recorded during tests. On restoration of UPS input power, it should be verified that the battery is recharging properly.

28.5.2.5 Any abnormalities should be corrected, and a check should be made to ensure that the battery is fully recharged prior to returning the system to service.

Chapter 29 Portable Electrical Tools and Equipment

29.1 Introduction.

29.1.1 Dependable performance and long service life of power tools is becoming more important as the need for mechanization and the use of power tools increase. A plant's entire inventory of portable tools can be kept in top operating condition for maximum production quality and cost efficiency with planned routine and periodic inspection.

29.1.2 There are many and varied types of portable power tools and many and varied causes of power tool failure. Therefore, the procedures for their maintenance can be general recommendations only. Variations exist and depend on the type of tool and the particular conditions of its use. Information on proper use and maintenance given in the tool manufacturer's use and care manual, supplied with each tool, should be carefully followed.

29.1.3 Periodic electrical testing will uncover many operating defects, and their immediate correction will ensure safe operation and prevent breakdown and more costly repairs. This testing and the related maintenance should be systematic. Tools should be visually inspected for damage and defects before and after each use.

29.1.4 Use tools only for their intended purpose.

29.1.5 Use the personal protective equipment necessary when using tools to protect from the hazards of falling, flying, abrasive, and splashing objects and from harmful dusts, fumes, mists, vapors, or gases.

29.2 Employee Training.

29.2.1 Employee training in the proper care and use of portable power tools is an important part of preventive maintenance. Employees should be given instructions in selecting the proper tool for the job and the limitations of the tool. Using an underpowered tool for the work load can cause overloading.

29.2.2 Employees should be trained to recognize obvious defects such as cut, frayed, spliced, or broken cords; cracked or broken attachment plugs; and missing or deformed grounding prongs. Such defects should be reported immediately.

29.2.3 Employees should be instructed to report all shocks immediately, no matter how minor, and to cease using the tool. Tools that cause shocks should be examined and repaired before further use.

29.3 Tool Maintenance. The maintenance procedure in 29.3.1 through 29.3.3.2 are general recommendations. The best source for maintenance information is the original manufacturer.

29.3.1 Periodic Inspection of Crucial Wear Points. Brushes and commutators should be inspected periodically. This is easily accomplished by removal of brush-holder plugs or inspection plates, depending on the construction of the tool. Brushes worn down to 50 percent of their original size should be replaced. When a brush is replaced, always be sure to use the manufacturer's original equipment.

29.3.2 Excessive Dirt Accumulation. All universal motors are fan ventilated to prevent excessive heat. Even though many tools have filters and deflectors to prevent destructive material from damaging the motor, a small amount of it will pass through. Excessive buildup affects the brush operation and reduces the air volume necessary to cool the motor. When necessary, a tool used in a normal environment should be blown out with low-pressure, dry-compressed air. More frequent specialized maintenance should be considered if the atmosphere is heavy in abrasives or conducting dusts.

29.3.3 Insufficient or Improper Lubrication. Lubricant inspection is recommended at frequent intervals to ensure sufficient lubrication to prevent wear to mechanical parts. Dirty lubricants should be removed and replaced. Because lubricant varies from tool to tool, it is recommended that proper lubricant be obtained from the manufacturer or the manufacturer's distribution outlet.

29.3.3.1 Lubricants. Manufacturers carefully match lubricants to be compatible with speeds, heat, seals, bearings, and pressure to ensure long gear and mechanism life. Substitutions can damage the tool and invalidate the warranty.

29.3.3.2 The wrong amount of lubricant can cause serious problems. Too little means that surfaces are not adequately covered, and excess wear will result. Too much lubricant can cause excess pressure in the gear case and eventually ruin seals.

29.4 Cord and Attachment Plug Care.

29.4.1 The cord of an electric power tool is its lifeline. It should be kept free of oil, grease, and other material that might ruin the rubber cover. Tangling knots or dragging across sharp surfaces should be avoided. The cord should not be used as a towline to carry or drag the tool.

29.4.2 All power tools, unless they are double insulated and so marked, are required to be grounded through an additional grounding conductor in the cord and the grounding prong of the attachment plug. The integrity of this grounding circuit is necessary for the protection of life and should be inspected visually before each use. Experience has shown that the grounding prongs of attachment caps are frequently cut off for use in ungrounded receptacles. This practice should not be permitted.

29.4.3 If a cord is cut, broken, spliced, or frayed, the attachment plug is damaged, or the grounding prong is removed, it should be immediately withdrawn from service until it can be repaired. Cords can be replaced in their entirety, or a damaged cord can be repaired by cutting out the damaged portion and applying a plug and connector to rejoin the two sections. Replacement cords should be of the same type and conductor size and suitable for use.

29.4.4 To avoid accidents, the green insulated conductor is to be used only for connecting the frame of the tool to the equipment-grounding terminal of the attachment plug meeting the conditions of *NFPA 70, National Electrical Code*, Section 400.24. It should not be used for any other purpose.

29.4.5 Flexible cords and cables should be protected from damage. Contact with sharp corners and projections should be avoided.

29.5 Extension Cords.

29.5.1 Before an extension cord is placed into service, the plug and connector should be checked for proper polarity, and the grounding conductor should be tested for continuity and integrity. Extension cords of the proper conductor size should be used to avoid excessive voltage drop, which can result in poor operation and possible damage to the tool. Table 29.5.1 lists the recommended sizes of extension cords.

29.5.2 Only extension cords with an equipment grounding conductor should be used.

29.5.3 Extension cords should be protected from damage, and not run through doorways or windows where the doors or windows may close, causing damage to the cord.

29.6 Major Overhauls. Major overhauls and repairs should be performed by the manufacturer. Large companies that use power tools and that prefer to do their own repairs and overhaul should obtain the necessary parts, schematics, connection diagrams, lubricant charts, and other technical information from the manufacturer.

29.7 Leakage Current Testing. Portable and cord-connected equipment should be tested periodically for the amount of leakage current present to help ensure against shock hazards.

Chapter 30 Reliability-Centered Maintenance (RCM)

30.1 Reliability-Centered Maintenance (RCM). Operational data should be retained through the trending and data storage features of the SCADA system for use in automated performance monitoring electro-mechanical systems that are supporting an RCM program.

30.1.1 General. RCM is the process of developing preventive maintenance (PM) programs for electrical and mechanical systems used in facilities based on the reliability characteristics of those systems and economic considerations, while ensuring that safety is not compromised.

30.1.2 Reliability-Centered Maintenance Concept.

30.1.2.1 The RCM approach provides a logical way of determining if PM makes sense for a given item. This might require contracting the work to firms that specialize in providing such

Table 29.5.1 Recommended Extension Cord Sizes for Portable Electric Tools

	Nameplate Ampere Rating											
Extension Cord	**0–2.0**		**2.1–3.4**		**3.5–5.0**		**5.1–7.0**		**7.1–12.0**		**12.1–16.0**	
Length (ft)	**115 V**	**230 V**	**115 V**	**230 V**	**115 V**	**230 V**	**115 V**	**230 V**	**115 V**	**230 V**	**115 V**	**230 V**
25	18	18	18	18	18	18	18	18	16	18	14	16
50	18	18	18	18	18	18	16	18	14	16	12	14
75	18	18	18	18	16	18	14	16	12	14	10	12
100	18	18	16	18	14	16	12	14	10	12	8	10
200	16	18	14	16	12	14	10	12	8	10	6	8
300	14	16	12	14	10	14	8	12	6	10	4	6
400	12	16	10	14	8	12	6	10	4	8	4	6
500	12	14	10	12	8	12	6	10	4	6	2	4
600	10	14	8	12	6	10	4	8	2	6	2	4
800	10	12	8	10	6	8	4	6	2	4	1	2
1000	8	12	6	10	4	8	2	6	1	4	0	2

Notes:

(1) Size is based on current equivalent to 150 percent of full load of tool and a loss in voltage of not over 5 volts.

(2) If voltage is already low at the source (outlet), voltage should be increased to standard, or a larger cord than listed should be used to minimize the total voltage drop.

services. The approach is based on the following precepts *(see Annex N.5)*:

(1) The objective of maintenance is to preserve an item's function(s). The RCM program should preserve system or equipment functionality. Redundancy improves functional reliability but increases cost in terms of procurement and life cycle cost.
(2) RCM examines the entire system and should be focused on maintaining system function rather than individual component function.
(3) Reliability is the basis for decisions. The failure characteristics of the item in question must be understood to determine the efficacy of preventive maintenance. The conditional probability of failure at specific ages (i.e., the probability that failure will occur in each given operating age bracket) should be determined.
(4) RCM acknowledges design limitations. Maintenance cannot improve the inherent reliability as it is dictated by design. Maintenance should sustain the design level of reliability over the life of an item.
(5) RCM is an ongoing process. Throughout the life of the equipment, the difference between the expected life and the actual design life and failure characteristics should be addressed.

30.1.2.2 The RCM concept will completely change the way in which PM is viewed. It is widely accepted that not all items benefit from PM. It might be less expensive to allow an item to "run to failure," provided safety is not compromised, rather than to do PM. Although RCM should focus on identifying PM actions, corrective actions are identified by default. RCM should focus on optimizing readiness, availability, and sustainability through effective and economical maintenance. Pertinent information can be found in Annex N.

30.1.2.3 Part of an effective RCM program is to determine the failure modes effects and conduct criticality analysis of all systems (FMECA), determine the risk priority based on the product of the severity level of a component, failure occurrence level, and detection level. Detailed information on how to compute this can be found in Annex K.

30.2 Operations and Maintenance (O&M) Documentation. Before installing or employing a SCADA system, an O&M analysis should be performed to provide the maintenance parameter data.

30.3 Technical Support. Ongoing maintenance of a SCADA system might require specialized technical support from the system vendor. This might require the vendor to have remote modem access to the system, though this provision represents a vulnerability to infiltration of the system by unauthorized personnel, and appropriate precautions need to be observed. Such access should be monitored while in use and physically disconnected when not in use.

30.4* Availability Metric. If availability is to be compared, the availability metric should be used as the assessment method for comparing different systems.

Chapter 31 EPM from Commissioning (Acceptance Testing) Through Maintenance

31.1 Introduction.

31.1.1 Commissioning, also referred to as acceptance testing, integrated system testing, operational tune-up, and start-up testing, is the process by which baseline test results verify the proper operation and sequence of operation of electrical equipment, in addition to developing baseline criteria by which future trend analysis can help to identify equipment deterioration. This process begins at the project planning stage and sequentially follows design, construction, acceptance testing, commissioning, and operation and maintenance.

31.1.2 It is not unusual for electrical systems to have problems during startup and installation. Sometimes it takes experienced engineers and technicians to identify operational problems and provide solutions to fine-tune the system to operate as it was designed. When implemented correctly, a realistic commissioning plan minimizes startup and long-term problems, reduces operational costs, and minimizes future maintenance requirements.

31.2 Purpose. The purpose of electrical testing on system and components is twofold. Electrical-system commissioning on new projects is critical to ascertain that the system is installed correctly and that it will remain in service for its projected life cycle.

31.3 Requirements. The first requirement is to check the installation of the equipment and to perform component and system tests to ensure that, when energized, the system will function properly. The second requirement is to develop a set of baseline test results for comparison in future maintenance testing to identify equipment deterioration. This process or set of tests usually is performed by independent contractors, installation contractors, or the manufacturer and is usually called *commissioning*.

31.4 Commissioning Planning Stages. The commissioning planning stages should be followed in the establishment of a commissioning testing program as shown in Figure 31.4.

31.4.1 Initial Commissioning (Kick-Off) Meeting. The first commissioning meeting might start as early as before a contract is awarded or as late as after the system has been installed (before it is turned over to the customer). The participants in the meeting should include the customer, government contractor personnel, participating engineers, commissioning personnel (might be an independent commissioning contractor or government engineer), general contractor, and electrical and mechanical subcontractors. Topics of the meeting should include areas of responsibility, expectations, overall presentation of system, methodology, potential problem areas, and so on. All the participants should be considered part of the team.

31.4.2 Review Initial Statement of Work (SOW). All participants should review the SOW that describes the requirements of the commissioning process. The customer (or designated authority) should have the ultimate control where decisions are required.

31.4.3 Review Drawing Submittals. After the systems are installed, drawings should be submitted for review and comment. Sometimes the review is before systems are installed. Care must be taken to verify that the drawings submitted reflect the actual installed system.

31.4.4 Approval for SOW and Design Intent. After the SOW and drawing submittals have been submitted, reviewed, comments made, and comments incorporated, approval of documents should be provided.

31.4.5 Obtain or Develop System Operating Documents (SOD) and System Operation Maintenance Manuals (SOMM). The prime contractor should provide a SOD or a SOMM. These documents are required to develop specific commissioning tests.

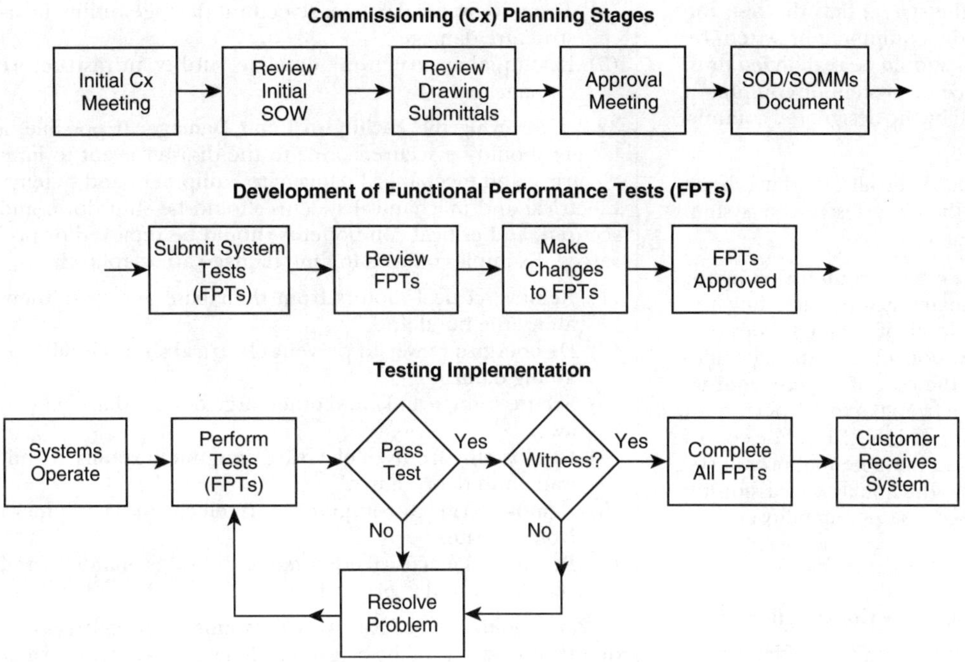

FIGURE 31.4 Sample of a Commissioning Plan. *(Source: TM 5-694, Commissioning of Electrical Systems for Command, Control, Communications, Computer, Intelligence, Surveillance, and Reconnaissance (C4ISR) Facilities, 2006 Edition.)*

31.5 Developing of Functional Performance Tests (FPTs). System/component tests are commonly known as functional performance tests (FPTs). FPTs are critical in establishing a baseline reference for future maintenance testing requirement criteria. Examples of sample documents that can be used in the development of FPTs can be found in the ANSI/NETA ATS, *Standard for Acceptance Testing Specifications for Electrical Power Distribution Equipment and Systems*, Square D Services' *Procedures for Startup and Commissioning of Electrical Equipment*, or manufacturers' materials. The following steps should be taken during the development of FPTs.

31.5.1 Submit Functional Performance Tests (FPTs).

31.5.1.1 System/component tests, or FPTs, should be developed from submitted drawings, SODs, and SOMMs. The tests should include large-component testing (i.e., transformers, cable, generators, UPS) and how components operate as part of the total system. The commissioning authority should develop the test. The commissioning authority should not be the installation contractor (or subcontractor).

31.5.1.2 As the equipment/components/systems are installed, quality assurance procedures should be administered to verify that components are installed in accordance with minimum manufacturers' recommendations, safety codes, and acceptable installation practices. Quality assurance discrepancies should be identified and added to a commissioning action list, which must be rectified as part of the commissioning program. These items usually are discussed during commissioning meetings and discrepancies identified initially by visual inspection.

31.5.2 Review FPTs. The tests should be reviewed by the customer, electrical contractors, quality assurance personnel, maintenance personnel, and the rest of the commissioning team. Areas of concern should include the following:

(1) Are all functions of the system being tested?
(2) Are all major components included?
(3) Do the tests reflect the system operating documents?
(4) Do the tests make sense? etc.

31.5.3 Make Changes to FPTs as Required. This is where corrections, answers to questions, and additions will be implemented. The commissioning authority will do this.

31.5.4 Approval of FPTs. After the changes have been made to the FPTs, they should again be submitted to the commissioning team. When they are acceptable, the customer or the designated approval authority should approve the FPTs. It should be noted that even though the FPT has been approved, problems that arise during the test or in areas not covered should be addressed.

31.6 Testing Implementation. The third and final step in the successful commissioning plan is testing and proper execution of system-integrated tests. *(See the pertinent checklists in Annex H: Figure H.21, Figure H.23, Figure H.24, Figure H.25, and Figure H.35.)*

31.6.1 Checklists. For consistency purposes, it is recommended that forms and checklists similar to those shown in Annex H be utilized. These forms will ensure that the results of the FPTs are accurately recorded, witnessed, and approved. In addition, these forms will ensure that baseline data exist to compare with future measurements, thus establishing needed maintenance requirements.

31.6.2 Systems Ready to Operate. The FPTs can be implemented as various systems become operative (i.e., test the generator system) or when the entire system is installed. However, the final "pull-the-plug" test is performed after all systems have been completely installed. If the electrical contractor (or subcontractor) implements the FPTs, a witness should initial each step of the test. The witness should not be employed by the electrical contractor either directly or indirectly.

31.6.3 Perform Tests (FPTs). If the system fails the test, the problem should be resolved and the equipment or system retested, or the testing requirements should be reanalyzed until successful tests are witnessed. Once the system or equipment passes the test, it should be verified by the designated commissioning official.

31.6.4 Customer Receives System. After all tests have been completed (including the pull-the-plug test), the system should be turned over to the customer.

31.7 Costs of Commissioning. The costs of commissioning for an electrical system depend on many factors, including the system size, complexity, and the level of reliability desired. New-building construction, renovation of an existing building, or modernization also affect the cost of commissioning, which for a new building can range from 0.5 to 1.5 percent of the total construction cost, as shown in Table 31.7. The cost of commissioning is small compared to the potential overall system cost. Experience has shown that the initial commissioning cost is more than offset by increased system reliability and reduced operating costs.

Table 31.7 Costs of Commissioning, New Construction

Commissioning Scope	Cost
Entire building (HVAC, controls, electrical, mechanical) commissioning	0.5%–1.5% of total construction cost
HVAC and automated control system commissioning	1.5%–2.5% of mechanical system cost
Electrical systems commissioning	1.0%–1.5% of electrical system cost

(Source: U.S. General Services Administration and U.S. Department of Energy, *Building Commissioning Guide*, July 30, 1998.)

Chapter 32 Electrical Disaster Recovery

32.1 Introduction. When industrial facilities are faced with a catastrophic event such as a fire, flooding, hurricane, tornado, earthquake, or similar disaster, a very specific and detailed sequence of events should occur to return the facility to productive operation in a safe and expeditious manner. To accomplish this after a disaster event, it is especially critical to analyze and repair the electrical power system in a safe and logical sequence. This chapter describes the recovery steps for an electrical power system and related equipment that should be followed after an electrical disaster event occurs.

32.2 Catastrophic Event Categories. The events surrounding a disaster can be detailed into specific event phases.

32.2.1 The Initial Event. Disaster recovery efforts can be a result of fire, flooding, hurricane, tornados, and earthquakes. Disaster scenarios including, but not limited to, the following inflict damage of varying degrees to facilities:

(1) Fire: soot, material and equipment damage, water damage, structural damage
(2) Flooding: water damage, structural damage
(3) Hurricane: water damage, structural damage, utility infrastructure damage
(4) Tornado: water damage, structural damage, utility infrastructure damage
(5) Earthquake: structural damage, utility infrastructure damage

32.2.2 Securing the Facility to Limit Damage. If possible, a facility should be secured prior to the disaster event to limit electrical and mechanical damage to equipment and systems. Electrical and mechanical systems should be shut down and secured, and critical components should be removed or preserved. Examples of tasks to limit damage are as follows:

(1) Remove critical motors from their base and raise them above the flood line.
(2) De-energize power to prevent electrical short circuit and arcing damage.
(3) Secure storage tanks and other large devices that can float away.
(4) Sandbag the fronts of electrical equipment rooms to limit water and debris entry.
(5) Remove critical computer and electronic equipment from the site.
(6) Remove all electrical equipment, drawings, manuals, and supplies stored at ground level.

32.2.3 Mobilization of Recovery Personnel. During large-scale disaster events, one of the biggest challenges for a commercial or industrial facility is providing enough qualified contractors and disaster recovery specialists to perform required remediation to the facility. Prior to a disaster event occurring, a preplan should be developed for the mobilization of recovery personnel.

32.2.3.1 In-House Personnel. Before a disaster event occurs, personnel responsible for the disaster recovery operations and facility repair should be designated and have possession of any applicable action plans. Depending on the magnitude of the event, the recovery effort can be done solely with in-house personnel or with the assistance of professional restoration companies.

32.2.3.2 Outsourced (Contract) Personnel. Prior to a disaster event, facilities should consider establishing master service agreements (MSAs) with multiple qualified vendors who specialize in electrical disaster recovery services. Doing so prevents confusion and delays in the recovery efforts. Decide who will perform the cleanup (debris removal and electrical equipment restoration) and supply support equipment (e.g., flood pumps, heavy equipment and operators, emergency power equipment, and temporary electrical services). Qualified repair facilities should be identified prior to a disaster recovery event.

32.2.3.3 Notification to Insurance Carrier. As soon as feasible the site insurance carrier's claims representative should be notified of the event.

32.2.4 Developing a Safety Plan. A site-specific safety plan should be developed before a disaster occurs. When performing recovery of electrical equipment, safety, environmental, and health are paramount. Lockout/tagout, test before touch, and the application of safety grounds are typically covered in site electrical safety plans. While these are key safety aspects of placing equipment into an electrically safe condition, there are other items of safety that need to be addressed and integrated into the safety plan, such as the following:

(1) Air quality
(2) Structural issues
(3) Chemical and biological hazard spill exposure
(4) Site-specific hazards
(5) Site-specific PPE requirements

32.2.5 Temporary and Emergency Power Generation. When disaster events occur, often times there is a loss of normal utility power. This creates a unique safety and logistical challenge to provide the required electrical power in a facility for critical systems and lighting. The temporary power portion of the project should be managed to reduce the risk of shock and arc-flash hazards. There should be dedicated personnel responsible for temporary power, and they should develop all written standards and procedures to be followed. Typical emergency power procedures should identify elements such as the following:

(1) Backfeeding of equipment
(2) Individual motor starters for pumps
(3) Temporary signage and barricades
(4) Site generator location maps.
(5) Fueling schedules
(6) Written form for the addition of electrical power
(7) Access and exhaust flow

32.2.6 Initial Damage Assessment. One of the first tasks in assessing equipment and system damage to electrical equipment involved in a disaster event is to gather all pertinent drawings and documentation available and perform a walkthrough and initial assessment of the entire electrical infrastructure.

32.2.6.1 Drawings, Schematics, Equipment Documentation. In some instances drawings and documentation are not available due to destruction from the disaster event. All equipment instruction books, operation and maintenance (O&M) manuals, and documentation should be identified and centrally located.

32.2.6.2 Priority Assessment. Equipment repair priorities should be assessed with a focus on the highest priority equipment. Examples of typical equipment categories are as follows:

(1) Category 1: medium-voltage equipment including distribution transformers
(2) Category 2: low-voltage distribution equipment
(3) Category 3: electric motors
(4) Category 4: balance of the plant

32.2.7 Documentation. All electrical components or equipment should be properly documented prior to removal to ensure the equipment is reinstalled properly as found. The documentation process includes the following:

(1) Tag each piece of equipment.
(2) Label all control and power wires.
(3) Take a digital picture of each piece of equipment.
(4) Sketch an accurate diagram of each piece of equipment on the electrical equipment drawing sheet.
(5) Fill out the electrical equipment tracking form.
(6) Save all pictures on a local database.
(7) File the electrical equipment drawing sheet.
(8) Create a master electrical equipment tracking document.
(9) Ship documents of all electrical equipment.

32.2.7.1 Service Shop Activities. If equipment is to be removed from the affected facility for repair at an offsite service center, the equipment should be tagged, identified, and tracked and the status updated on a master equipment repair database.

32.2.7.2 Equipment Tag. Information on each tag should include a unique sequence number, plant identification number, plant description, date, power center, or room number. The tag should be filled out with a medium point permanent marker so the information is legible. The tag should be attached to the equipment with a secure plastic wire tie.

32.2.7.3 Labeling of Wires. All control wires should be labeled with wire numbers and the power wires with colored phasing tape. Make sure that each side of the termination, both wire and connected device, is identified. This will ensure the wiring will be re-connected as it was originally installed.

32.2.7.4 Photographs of Equipment. After the equipment is tagged and the wires are labeled, a minimum of three photographs should be taken of each piece of equipment. The first photo should include the equipment tag in the picture, making sure the tag is legible and the picture is clear. The second photo is an overall view for the sole purpose of wire clarification/documentation during the reinstallation process and should include all wiring associated with the applicable device. The third photograph should be of the equipment nameplate. Additional photographs should be taken as deemed appropriate.

32.2.7.5 Field Sketch. An accurate field sketch of the electrical equipment should be generated. The sketch should be recorded on a site-specific electrical equipment drawing sheet template. This drawing sheet should include the job name, job number, power center, sequence number, plant equipment number, plant description, technician name, date, and enough room to sketch the piece of equipment.

32.2.7.6 Equipment Tracking Sheet. After a sketch is made of the piece of equipment, the equipment should be added to an electrical equipment tracking sheet. The electrical equipment tracking sheet should be customized and detailed. The tracking sheet should include general information such as overall condition, item number, sequence number, priority, area of the plant, power center or room number, transformer, substation, cell position, equipment type, circuit identification, plant identification number, manufacturer, percent water level, model number, frame size, and voltage. Field tracking information should also include date documented, date pulled, date shipped, date returned, date installed, and any dates that quality assurance procedures were performed.

32.2.7.7 Repair or Replace. During the documentation process, initial decisions should be made pertaining to each piece of equipment that is damaged. Seeking the services of qualified equipment assessment personnel, whether manufacturer representatives or subject matter experts, is important in the decision making process.

32.2.7.7.1 Repair or Replace Decisions. Many factors can affect the repair/replace decision. Some of the likely decisions are as follows:

(1) Can the equipment be repaired or does the equipment need to be replaced?
(2) Can the repairs take place on site or does the equipment need to be sent to a repair facility?

32.2.7.7.2 Repair or Replace Factors. Some of the factors that can affect the repair or replace decision are as follows:

(1) Is the equipment currently manufactured?
(2) Are there long lead times to replace with new?
(3) Will equipment performance be compromised if repaired?
(4) What is the age of the equipment?
(5) What is the reliability requirement?
(6) Can it be effectively repaired?
(7) Is the manufacturer still in business?
(8) Is the repair contractor qualified for the task?
(9) Will the authority having jurisdiction allow repair or replacement?
(10) What is the financial impact?
(11) What is the total outage time required?

32.2.8 Industry Standards and Guidelines. Industry standards and guidelines should be referred to for information. Information is available from the following:

(1) Electrical Apparatus Service Association (EASA), ANSI/EASA AR100, *Recommended Practice for the Repair of Rotating Electrical Apparatus*
(2) Federal Emergency Management Agency (FEMA), FEMA P-348, *Protecting Building Utilities From Flood Damage: Principles and Practices for the Design and Construction of Flood Resistant Building Utility Systems*
(3) Institute of Electrical and Electronic Engineers (IEEE), IEEE 3007.2, *Recommended Practice for the Maintenance of Industrial and Commercial Power Systems*
(4) National Electrical Manufacturers Association (NEMA) NEMA, *Guidelines for Handling Water-Damaged Electrical Equipment*
(5) InterNational Electrical Testing Association (NETA) ANSI/NETA ATS, *Standard for Acceptance Testing Specification*
(6) ANSI/NETA MTS, *Standard for Maintenance Testing Specifications for Electrical Power Distribution Equipment and Systems*
(7) National Fire Protection Association (NFPA), *NFPA 70, National Electrical Code*, and *NFPA 70E, Standard for Electrical Safety in the Workplace*
(8) PowerTest Annual Technical Conference, 2009, *Flood Repair of Electrical Equipment; March 12, 2009, Pat Beisert, Shermco Industries*

32.2.9 Medium-Voltage Equipment. Medium-voltage equipment typically serves as the backbone to the electrical power system and should be the primary focus of the initial recovery activities.

32.2.10 Low-Voltage Distribution Equipment. Affected components of low-voltage equipment should be removed to facilitate cleaning and drying of the structures. During the removal of the equipment, care should be taken to keep all wiring for each component well marked and together.

32.2.11 Electric Motors. When a disaster event involves water, electric motor repair is a major component of a flood recovery project. The documentation process is very similar to other electrical equipment but there are additional items that should be documented. The documentation process should include the following:

(1) Record nameplate data and location of the motor.
(2) Tag the motor base and the motor with a unique sequence number.
(3) Mark and record electrical connections.
(4) Record coupling information and condition of coupling.
(5) Mark and record shim information.
(6) Collect all mounting hardware, couplings, and shims and store in its own labeled container. This equipment stays on site and should be stored in a central location.

32.2.12 Power and Control Wiring. Power and control wiring should be tested to determine serviceability. *(See 11.21, Cables.)*

32.2.13 Balance of Plant Electrical Repair. The balance of plant consists of all equipment other than medium-voltage equipment, low-voltage distribution equipment, and motors. This includes items such as receptacles, light switches, lighting transformers, panelboards, start-stop stations, fire alarm panels, instrumentation and process control components, and metering. These devices are typically repaired by replacement.

32.2.14 Re-Energization of the Facility. Initial re-energization to utility power of a facility damaged by a disaster event should be carefully planned and methodically implemented. To reduce the possibility of accidental energization of equipment it might be prudent to forego any utility energization until all affected equipment has been repaired or replaced.

32.2.15 System Commissioning During. Re-energization of the facility, the equipment operation, and performance should be verified. *[See Chapter 31, EPM from Commissioning (Acceptance Testing) Through Maintenance and Figure H.35 in Annex H.]* A period of monitoring should be established to verify and document proper operation has been restored.

32.2.16 Project Summary. After a disaster recovery event there is information gathered that should be available for future reference. The final project report should contain this data and should include information such as the following:

(1) As-found conditions of the electrical infrastructure
(2) Listing of equipment repaired or replaced
(3) Test results of all equipment tested
(4) Assessment of individual equipment condition
(5) Long-term equipment replacement plan

Chapter 33 Photovoltaic Systems

33.1 Introduction.

33.1.1 A solar photovoltaic electrical energy system is a renewable source of energy. The major electrical system components include the array circuit(s), inverter(s), and controller(s). The arrays are generally found mounted either on a building roof or on supports in a ground mounted array. Photovoltaic systems can be interactive with other electrical power production sources or standalone, with or without electrical energy storage such as batteries.

33.1.2 The maintenance program should be planned at the time the system is installed in order to ensure the greatest level of safety to the maintenance worker and that the highest level of efficiency can be obtained from the operation of the system.

33.1.3 The elements of the weather and climate changes can have a significant impact on the photovoltaic system, and attention should be given to the system, after significant weather or environmental events.

33.1.4 Photovoltaic systems typically generate voltages in the 400 V dc to 1000 V dc range. Only qualified persons should perform maintenance on photovoltaic systems due to the unique hazards associated with the arrays always producing electrical energy.

33.2 Maintenance of the Photovoltaic System.

33.2.1 A newly installed photovoltaic system should include supporting documentation that should include specifications, electrical schematics, mechanical drawings, and a material list.

33.2.2 Development of a comprehensive maintenance plan should include the following:

(1) Energy monitoring
(2) Visual inspection
(3) Array cleaning
(4) Emergency response

33.2.3 The visual inspection of the array should include the following:

(1) Wiring and terminations and grounding
(2) Wiring harness secureness
(3) Array cleanliness, absences of damage, and structural integrity
(4) Roof penetrations and weather sealing
(5) Inverters, switches, and combiner boxes
(6) Batteries (if applicable)

33.3 Markings and Labeling. Proper signage should be installed to identify the location of rooftop panels on the building prior to completion of installation. Marking is needed to provide appropriate warning and guidance with respect to working around and isolating the solar electric system. This can facilitate identifying energized electrical lines that connect the solar modules to the inverter.

Chapter 34 Electric Vehicle Charging Systems

34.1 Introduction.

34.1.1 Electric vehicles require the installation and maintenance of electrical infrastructure, which include the electrical conductors and equipment external to an electric vehicle that connect an electric vehicle to a supply of electricity by conductive or inductive means, and the equipment and devices related to electric vehicle charging.

34.1.2 Businesses, building owners, and municipalities should include a maintenance program when an electric vehicle charging system is installed. The maintenance program should be planned at the time the system is installed in order to ensure the greatest level of safety to the maintenance worker and the highest level of reliability and safety to the user and operator of the vehicle charging station.

34.1.3 The elements of the weather, climate change, or physical impact on system components can have a significant impact on the electric vehicle charging station, and attention should be given to the system after a significant weather or environmental event or physical impact.

34.1.4 Electric vehicle charging systems typically are supplied by 240 V or 480 V systems. Only qualified persons should perform maintenance on electric vehicle charging systems due to the unique hazards associated with the unique controls and potential hazardous level of energy provided at this point on the system.

34.2 Maintenance of the Electric Vehicle Charging Stations.

34.2.1 An installed electric vehicle charging system should include supporting documentation, specifications, electrical schematics, mechanical drawings, and a material list.

34.2.2 Electric vehicle charging system cords should be inspected on a periodic basis to identify damage. Cord damage can result from physical impact or environmental degradation. A damaged cord should be replaced immediately.

34.2.3 Electric vehicle charging system connectors should be inspected on a periodic basis to identify damage. The connection to the car serves a significant safety role by ensuring a good electrical connection. A damaged connector should be replaced immediately.

34.2.4 Electric vehicle supply equipment (EVSE) should be inspected on a periodic basis to identify damage. The cord connection to the EVSE should be inspected on a periodic basis to verify that the strain relief is intact and stress is not placed on the cord terminations in the EVSE. Damage should be addressed immediately.

34.2.5 The mounting of the EVSE should be inspected on a periodic basis to ensure the integrity of the mounting means.

Chapter 35 Wind Power Electric Systems and Associated Equipment

35.1 Introduction.

35.1.1 The electrical equipment associated with wind power electric systems requires maintenance attention that might be more frequent than other installations due to the nature of these installations. Environmental conditions are often very severe; therefore, vibration and temperature variations can affect the safety of these installations. Many different designs are used, and the technology is constantly changing.

35.1.2 This service should be undertaken by trained and qualified workers, and all electrical safety practices outlined in *NFPA 70E, Standard for Electrical Safety in the Workplace,* need to be referenced and used.

35.2 Towers and Foundations.

35.2.1 Check tower grounding connections, cables, and grounding electrode resistance. *(See Section 14.3.)*

35.2.2 Check tower top electrical equipment, including navigational warning lighting and weather measurement devices.

35.2.3 Special attention should be paid to lightning protection components and their associated wiring systems. *(See* NFPA 780, *Standard for the Installation of Lightning Protection Systems, for guidance in testing and maintenance of these systems.)*

35.3 Yaw Systems. Electrical components and wiring should be checked for signs of damage or overheating.

35.4 Generators. See Chapter 25, Sections 25.1 through 25.9.

35.5 Pitch Systems. Electrical components and wiring should be checked for signs of damage or overheating.

35.6 Instrumentation and Controls. Operational checks should be performed to ensure systems for emergency and safety shutdown are functional.

35.7 Supervisory Control and Data Acquisition System (SCADA). Connection and data transfer tests should be performed to ensure operation.

35.8 Transformers and Converters. Electrical inspections should be performed as specified in Chapter 21 for transformers and Chapter 22 for converters.

35.9 Circuit Breakers. Electrical inspections should be performed as specified in Chapter 17 for circuit breakers.

35.10 Cable Support Systems, Cables, and Terminations. Inspect cable support systems, terminations, and cables for damage due to vibration. *(See Section 19.2 for visual inspection.)* Terminations should be inspected as covered in 19.2.2 and tested in accordance with the recommendations in Section 19.5.

35.11 Collector Substations and Switchgear. Electrical inspections should be performed as specified in Chapter 15 for collector substations and switchgear.

35.12 Associated Electrical Equipment.

35.12.1 Station power distribution panels should be checked for loose connections.

35.12.2 Fire alarm and emergency lighting systems should be checked for correct operation and loose connections.

35.12.3 Motor control centers should be checked. *(See Sections 16.3, 16.4, and 16.7.)*

Annex A Explanatory Material

Annex A is not a part of the recommendations of this NFPA document but is included for informational purposes only. This annex contains explanatory material, numbered to correspond with the applicable text paragraphs.

A.3.2.1 Approved. The National Fire Protection Association does not approve, inspect, or certify any installations, procedures, equipment, or materials; nor does it approve or evaluate testing laboratories. In determining the acceptability of installations, procedures, equipment, or materials, the authority having jurisdiction may base acceptance on compliance with NFPA or other appropriate standards. In the absence of such standards, said authority may require evidence of proper installation, procedure, or use. The authority having jurisdiction may also refer to the listings or labeling practices of an organization that is concerned with product evaluations and is thus in a position to determine compliance with appropriate standards for the current production of listed items.

A.3.2.2 Authority Having Jurisdiction (AHJ). The phrase "authority having jurisdiction," or its acronym AHJ, is used in NFPA documents in a broad manner, since jurisdictions and approval agencies vary, as do their responsibilities. Where public safety is primary, the authority having jurisdiction may be a federal, state, local, or other regional department or individual such as a fire chief; fire marshal; chief of a fire prevention bureau, labor department, or health department; building official; electrical inspector; or others having statutory authority. For insurance purposes, an insurance inspection department, rating bureau, or other insurance company representative may be the authority having jurisdiction. In many circumstances, the property owner or his or her designated agent assumes the role of the authority having jurisdiction; at government installations, the commanding officer or departmental official may be the authority having jurisdiction.

A.3.2.3 Listed. The means for identifying listed equipment may vary for each organization concerned with product evaluation; some organizations do not recognize equipment as listed unless it is also labeled. The authority having jurisdiction should utilize the system employed by the listing organization to identify a listed product.

A.3.3.13 Corona. High electrical gradients exceeding the breakdown level of air lead to corona discharges. Mild corona has a low sizzling sound and might not be audible above ambient noise in the substation. As the corona increases in activity, the sizzling sound becomes louder and is accompanied by popping, spitting, or crackling as flashover level nears. Corona ionizes the air, converting the oxygen to ozone, which has a distinctive, penetrating odor.

A.3.3.23 Electrical Preventive Maintenance (EPM). Electrical preventive maintenance relies on knowing the electrical systems and equipment being maintained and on knowing the operating experience, loss exposures, potential for injury, and maintenance resources.

A.3.3.38 Ground-Fault Circuit Interrupter (GFCI). A GFCI does not eliminate the electric shock sensation since normal perception level is approximately 0.5 mA; nor does it protect from electric shock hazard from line-to-line contact.

A.3.3.39 Ground-Fault Protection of Equipment (GFP). There are two applications where ground-fault protection of equipment is intended to be used: where there may be excessive ground-fault leakage current from equipment and where equipment and conductors are to be protected from damage in the event of a higher-level ground fault (either solid or arcing). These types of protective equipment are for use only on ac, grounded circuits; they cause the circuit to be disconnected when a current equal to or higher than its pickup setting or rating flows to ground. They are not designed to protect personnel from electrocution. Equipment ground-fault protective devices are intended to operate on a condition of excessive ground-fault leakage current from equipment. The ground current pickup level of these devices is from above 6 mA to 50 mA. Circuit breakers with equipment ground-fault protection are combination circuit breaker and equipment ground-fault protective devices designed to serve the dual function of providing overcurrent protection and ground-fault protection for equipment. The ground current pickup level of these breakers is typically 30 mA. They are intended to be used in accordance with *NFPA 70, National Electrical Code*, Articles 426 and 427. Ground-fault sensing and relaying equipment is intended to provide ground-fault protection of equipment at services and feeders. They are rated for ground current pickup levels from 4 amperes to 1200 amperes.

A.4.2.8 Table A.4.2.8 represents the results of a study performed by only one of the major insurance groups (Factory

Table A.4.2.8 1987–1991 Losses Associated with Electrical Failures, Including Electrical and Fire Damage

Class of Equipment	Number of Losses from All Causes, Including Unknown	Gross Dollar Loss from All Causes, Including Unknown ($1000)	Number of Losses from Causes Unknown	Gross Dollar Loss from Causes Unknown ($1000)	Number of Losses from Known Causes Due to Inadequate Maintenance	Gross Dollar Loss from Known Causes Due to Inadequate Maintenance ($1000)
Transformers	529	185,874	229	27,949	71	47,973
Generators	110	110,951	31	39,156	14	40,491
Cables	230	99,213	68	59,881	23	7,756
Motors	390	57,004	199	17,027	34	15,343
Circuit breakers	104	24,058	32	6,874	10	5,054
Controllers, switches, switchgear, and switchboards	108	17,786	36	5,537	17	2,308
Total	1,471	494,886	595	156,424	169	118,925

Notes:

(1) Statistics compiled by only one of the major insurance groups (Factory Mutual) that specialize in industrial fire and machinery insurance.

(2) Gross dollar losses are indexed to 1992 values.

Mutual) that specialize in industrial fire and machinery insurance. The table indicates that in a 2-year period, half of the losses associated with electrical equipment failures might have been prevented by an effective EPM program.

A.6.4.4.1 In the absence of manufacturer's recommendations, refer to Annex K, Long-Term Maintenance Guidelines, and Annex L, Maintenance Intervals, for guidance on maintenance frequency and tests and ANSI/NETA MTS, *Standard for Maintenance Testing Specifications for Electrical Power Distribution Equipment and Systems,* for guidance on maintenance frequency, methods, and tests.

A.10.2.2.2.3 More often, typical problems with harmonics are overheated motors, transformer, and neutral conductors, audible distortion on telephone circuits, timing errors for multiple zero crossing, inadequate ride-through due to flat-topping of waveforms, and overheating of capacitors due to harmonic equivalent surge resistance (ESR) losses. In addition, the problem with the resonance is actually the excessive voltage generated by the resonance. The harmonics are the stimulus of the resonance.

A.11.11.11 Detection of metals in low concentration in transformer oil can provide information as to the location of incipient internal faults where a gas-in-oil analysis has indicated that there is an incipient fault, or, to detect excessive wear in oil circulation pumps. Typically these metals include iron, copper, aluminum, lead, silver, tin and zinc. Different types of faults result in the presence of metals associated with a specific fault. Knowledge obtained from the manufacturer as to where a particular metal is used in the transformer can lead to a determination of the fault location. Typical use of various metals in transformers is shown in Table A.11.11.11. At present, there are no guidelines as to acceptable concentrations of metals in oil so it is important to establish a baseline of concentrations for a particular transformer. Subsequent changes from the baseline concentrations observed along with the dissolved gas-in-oil analysis described in 11.11.10 aid in the determination of the nature and location of an incipient fault.

Table A.11.11.11 Typical Use of Various Metals in Transformers

Aluminum	Windings, corona shields, ceramic bushings
Copper	Windings, bronze and brass components
Iron	Core and tank
Lead	Solder joints
Lead, tin, silver, zinc	Connectors, lugs, bolts, and peripheral components

A.15.9.4.4.12 The battery connection resistance value can vary depending upon the type of battery and the application. See IEEE 1188, *Recommended Practice for Maintenance, Testing, and Replacement of Valve-Regulated Lead-Acid (VRLA) Batteries for Stationary Applications,* for valve-regulated lead–acid (VRLA) batteries and IEEE 450, *Recommended Practice for Maintenance, Testing, and Replacement of Vented Lead-Acid Batteries for Stationary Applications,* for vented lead–acid batteries.

Table A.30.4(a) "Nines," Applications, and Expected Downtime

Number of 9s	Downtime/Year	Typical Application
3 Nines (99.9%)	~9 Hours	Typical Desktop Computer **Baseline Power Delivery** General Home Use
4 Nines (99.99%)	~1 Hour	Enterprise Server Desktop Software General Factory Use
5 Nines (99.999%)	~5 Minutes	**Carrier Class Server** Airports
6 Nines (99.9999%)	~32 Seconds	**Carrier Switching Equip.**
9 Nines (99.9999999%)	30 Milliseconds	Online Markets

A.30.4 Availability Metric — the "Nines." A metric used in the industry to compare levels of availability is often referred to as "the nines." This reflects the number of 9s in the percentage of availability, as shown below in Table A.30.4(a).

The number of nines is often correlated to the cost of down time. The higher the cost of downtime, the greater the number of "nines" that is typically required for the availability. *[See Table A.30.4(b).]*

When parts of the system are in series, the availability is the product of the two numbers, whereas in paralleled systems, the availability is the lower of the two numbers, as shown in Figure A.30.4(a) through Figure A.30.4(e).

Table A.30.4(b) The Costs of Outage for Selected Commercial Customers

Industry	Average Cost of Downtime
Cellular communications	$41,000 per hour
Telephone ticket sales	$72,000 per hour
Airline reservations	$90,000 per hour
Credit card operations	$41,000 per hour
Telephone ticket sales	$41,000 per hour

Source: Leiter, David, "Distributed Energy Resources."

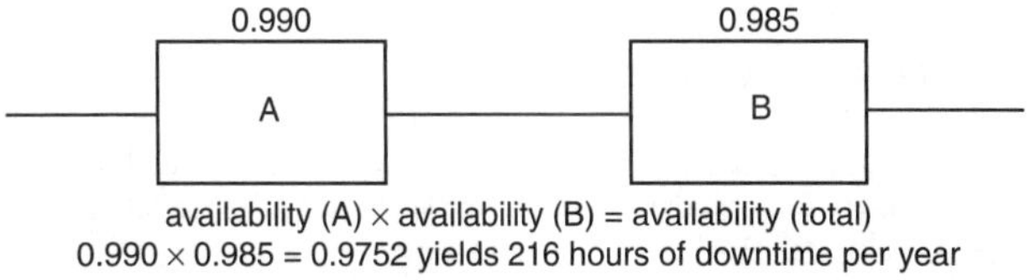

FIGURE A.30.4(a) Calculation of Availability and Downtime Single Point Failure.

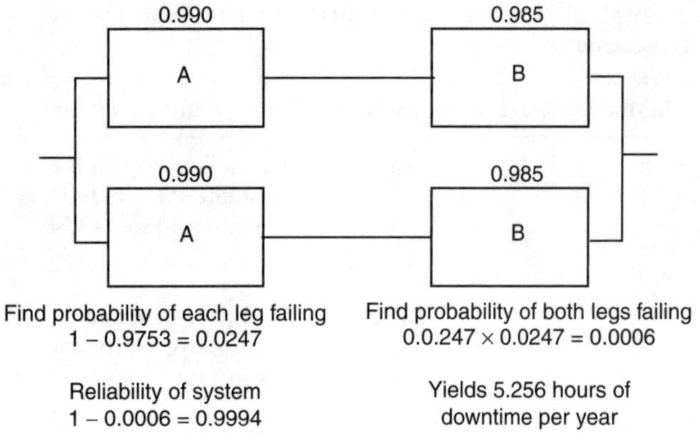

FIGURE A.30.4(b) Calculation of Availability and Downtime with Parallel Redundancy.

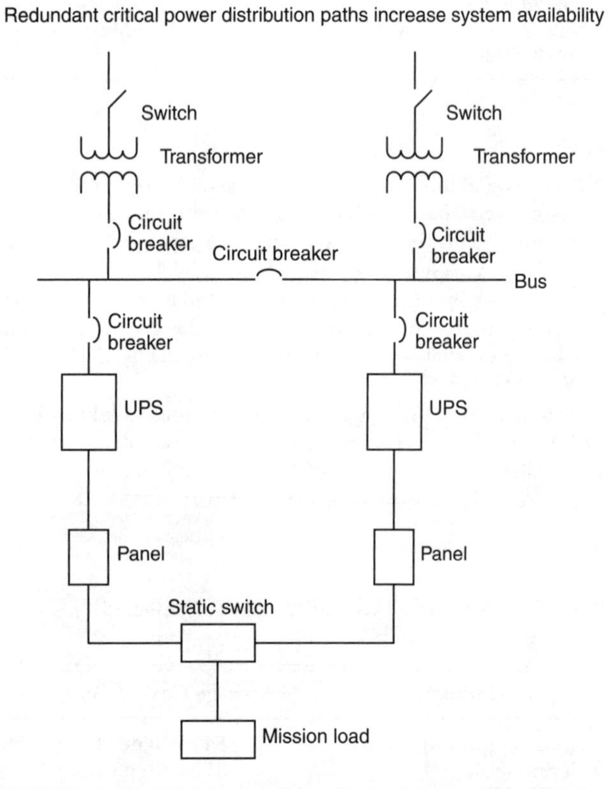

FIGURE A.30.4(d) Redundant Critical Power Distribution Paths Increase System Reliability.

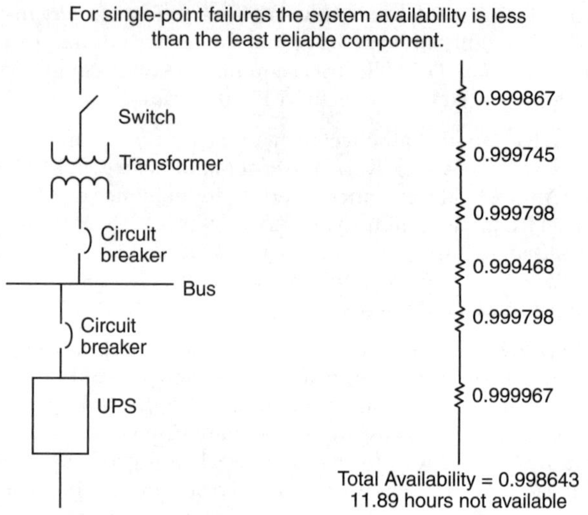

FIGURE A.30.4(c) Single Line Power System One Line and Availability Calculation.

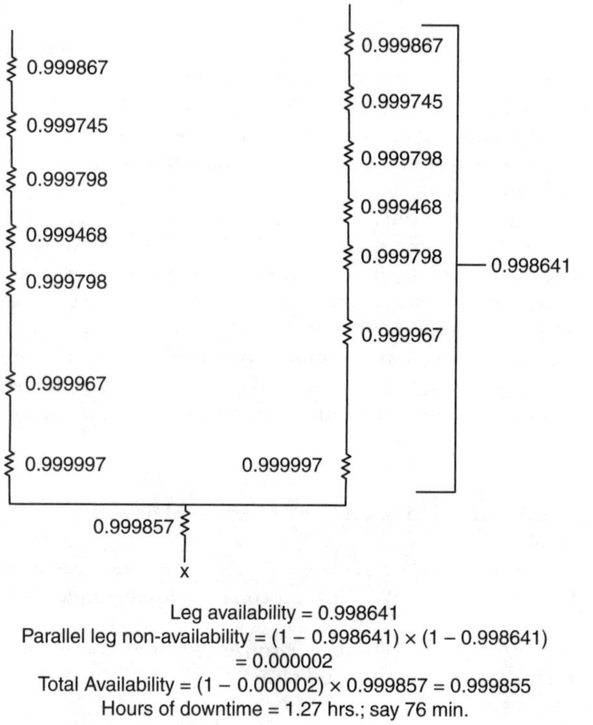

FIGURE A.30.4(e) Double Ended Substation Availability Calculation.

Annex B How to Instruct

This annex is not a part of the recommendations of this NFPA document but is included for informational purposes only.

B.1 Introduction. Training is basically a process for changing behavior. Behavioral changes are the product of new knowledge, reshaped attitudes, replaced skills, and newly acquired skills that express themselves or become observable as improved work techniques of the learner.

The trainer's function is to structure the instruction process in a manner that makes learning take place more effectively and in the shortest period of time.

B.2 The Shortcomings of Learning by Trial and Error. Trial-and-error learning is learning at random. It is slow and costly in terms of time and mistakes. It also is costly because it involves so much "unlearning" of incorrect practices and "relearning" after mistakes have been made.

Trial and error is the instructional process that continues to dominate industry. Its inefficiency is illustrated by the example of a newly hired maintenance electrician assigned to instrument circuit repair work until "he gets the feel of the plant and 'learns' his way around." The new electrician's assignment is to disconnect an instrument from the power source so an instrument technician can change out a defective chart drive motor. Consistent with apparent good safety practice but without consulting anyone, the electrician opens the switch that feeds power to the entire instrument panel. Loss of control of the process results in major product spoilage.

This example illustrates what can happen when people are put on jobs, simple or complex, without first being given organized instruction, either personally on the job or in groups off the job.

An even clearer illustration of the inefficiency of trial-and-error learning is the example of an inexperienced maintenance electrician who is charged with responsibility for motor trip-out troubleshooting but who receives no formal instruction on this subject. His first attempts include many blind alleys, such as going to the job location without the proper tools; a random inspection of the motor starter, the motor, and the driven load; or a random replacement of heater elements. As the number of his attempts to correct motor trip-outs increases, he learns to avoid many of the blind alleys and eventually comes up with a logical (to him) sequence of steps that will shorten his job time.

However, if he had been properly trained, the maintenance electrician could have performed the job correctly in a minimum amount of time because he would have had full knowledge of the task and confidence in his own abilities to perform it. The further benefit of training would have been less downtime, less material waste, and less chance of injury to himself and to other employees.

The justification for planned on- and off-the-job training, therefore, is to get better results in the form of greater job knowledge, greater skills, and better job attitudes toward such factors as quality, cost, and productivity and in the shortest amount of time.

The job of the instructor is to direct learning activities of trainees to avoid the blind alleys and mislearning that are inevitable with trial and error. This requires organized presentation.

B.3 Philosophy of Training. The organization of a training program for a new learning situation involves the following major tasks:

(1) Selection of experiences that will help the trainee learn what needs to be done
(2) Guiding the trainee's efforts toward the proper learning objectives
(3) Applying the trainee's past experience
(4) Avoiding failures, frustrations, and loss of interest because the trainee does not perceive the relationships between what is being taught and future activity

For the purpose of discussing how the instructor can organize the presentation of subject material, assume that motivation has been provided and that the trainee recognizes the need for the training and has a desire to learn.

Whenever a skill is being taught, the instructor is not only presenting facts but also forming attitudes. For example, when a trainee is learning how to make a relay adjustment, new information is being acquired. In addition, the trainee is forming attitudes and a mindset concerning the information presented as well as performance, precision standards, quality, safety, and equipment design. It is these attitudes and mindsets that determine how the employee will approach or handle the job.

B.4 The Four-Step Method of Instruction. A proven method of instruction is the "four-step method." These four steps are as follows:

(1) Preparation
(2) Presentation
(3) Application
(4) Observation

B.4.1 Step 1: Preparation.

B.4.1.1 Preparation of Subject Matter. A carefully laid out plan of action is necessary for the presentation of new information and skills. Mistakes made in presenting new material early in the teaching process might permanently confuse the trainee. To avoid teaching mistakes, the instructor should use a clearly worked-out subject content outline and a step-by-step breakdown of the operations to be covered during instruction.

B.4.1.2 Subject Content Outline. A carefully worked-out subject content outline is important to both the beginning instructor and the expert. The new instructor might not deal fully with all the steps of the explanation, whereas the expert might overlook steps that seem to be obvious. Both the new instructor and the expert should plan their presentations from the viewpoint of the trainee.

B.4.1.3 Breakdown of the Subject Matter. Instruction proceeds from the known to the unknown. It begins with the simple and proceeds to the complex.

Use of a step-by-step breakdown ensures that the instruction moves progressively through a job, presenting it as it should be done from start to completion.

Instruction is accomplished by making certain that each new step is thoroughly explained and demonstrated in proper order and that after each step the trainee understands what has been covered.

The process of instruction is a natural process, with each step falling logically into place.

The problems encountered in instruction are generally due to the instructor's failure to take the time beforehand to carefully develop each explanation so that the entire topic makes sense.

When the presentation has been carefully broken down so that each unit being taught is clear and logical, the major obstacle to successful training has been overcome.

B.4.1.4 Preparation of Trainees. The following four steps should be followed:

(1) *Put the trainees at ease.* The trainees should be receptive. Tensions should be minimized. This can be achieved by creating an atmosphere of personnel security. Trainees should be introduced, a friendly manner demonstrated, and the business at hand promptly introduced. The situation should be relieved by anticipating the questions that normally are raised by trainees, by clearly describing the objectives, by making the trainees aware of the advantages, and by letting them know how the program will affect them personally.
(2) *Develop favorable attitudes.* Attitude is a by-product of everything that occurs. The instructor will influence the shaping of the trainees' attitudes. Because attitude is a by-product, the development of a favorable attitude or outlook toward the program cannot be obtained by the simple process of talking about attitude directly. Instead, the instructor's responsibility is to do a good job of presenting the course,

pointing out what is going to be covered, and explaining how the program serves both the trainees' and the company's interest.

(3) *Find out what the trainees already know.* Individual interest and receptivity of trainees to the subject material can be determined by briefly reviewing the backgrounds of members of the training group. This will avoid duplication and provide the instructor with information that will reveal the gap between what members of the group already know and the material to be presented.

(4) *Preview material to be covered.* Having determined background knowledge already known to the group, the instructor should brief the trainees on the ground to be covered. This briefing need not come in the same order as outlined here. The important consideration is that at some point before getting into the body of the lesson the instructor should tell the trainees what is going to be covered during the period.

Preliminary groundwork is frequently looked on as a waste of time. But in training, it should be remembered that part of getting the job done is dealing first with the intangible assignment of psychologically preparing the trainee. Step 1 failure is the most common among new instructors. No lesson should be considered ready for presentation until specific measures to prepare the trainees have been developed.

B.4.2 Step 2: Presentation. The main points in a successful presentation follow.

B.4.2.1 Show How to Do the Job. The instructor should demonstrate the operation carefully and accurately. If the operation is difficult, two or three demonstrations of the operation should be made. The instructor should not lose sight of the fact that *showing is very important in teaching.* The instructor should demonstrate, or show how, before the trainee tries to do the job.

B.4.2.2 Tell and Explain the Operation. After the class has seen the job demonstrated, the instructor should tell how the job is performed. It is important that the instructor let the class learn by doing *only after they have had the necessary instruction. Trainees should never be put in the position of having to learn only by trial and error or by simple observation.* In other words, trainees should be *shown* and *told* exactly *what is expected and how to do it.* The details that should be remembered should be pointed out to the trainees.

B.4.2.3 Present Related Theory. Electrical maintenance workers might actually carry out the sequence of actions required to do a job without knowing the basic principles that underlie the action. They might not understand why they do what they do; however, they will be better technicians if they do know why. This makes the difference between mechanical, machine-like, unmotivated performance and purposeful, participative workmanship.

B.4.2.4 Direct the Attention of the Learners. Showing and telling require that the instructor direct the attention of the trainees to the job. Describing an operation, showing a picture, or demonstrating an action is not enough. The important details should be pointed out and emphasized by directing the attention of the trainees to them. Attention can be directed in a number of ways.

One method of directing attention is to point out the item. Such emphasis will usually be coupled with telling, with a question, or with a demonstration. Attention might also be directed by the use of graphic devices, sketches, diagrams or board drawings, mobiles, and colors in printed material and on charts.

Board work can be emphasized by use of colored chalks. Changing the voice, slowing down the rate of talking, pausing, and the hundreds of devices of showmanship that dramatize a point are all effective means for directing learners' attention.

B.4.3 Step 3: Application (Try-Out Performance). Application provides a checkpoint on what has been learned. It is accomplished by having the class members carry out or show back how the job or operation is done. There are four major reasons for Step 3:

(1) To repeat instructions
(2) To show the trainees that the job can be done by following the instructions as given
(3) To point out and to learn at which points the trainee might be experiencing difficulty
(4) To indicate to the instructor whether or not the instructions given in Steps 1 and 2 have been effective

Performing the physical steps to actually do a job does not test all the learning that should have been acquired. The instructor should check the trainees by additional means such as questioning, having them identify parts, asking them to summarize the steps verbally, and having them state reasons for functions.

B.4.3.1 Have the Trainees Explain and Perform Each Step. To keep mistakes to a minimum, the instructor should have the trainees do the following:

(1) Tell *what* they are going to do
(2) Tell *how* they are going to do it
(3) *Do* the job

Telling "what" and "how" should come in advance of doing the job. The trainees should carry out the necessary physical movements *after*, not before, the instructor is satisfied that they know how to do the job.

The instructor should have the trainees show how to do the job by the same method the instructor used in performing the operations. Because Step 3 is the trainees' first opportunity to actually apply what has been taught, it is important to avoid incorrect practices from the start.

B.4.3.2 Have the Trainees Do Simpler Parts of the Operation First. At this point, encouragement and success are important conditioners. Early successes are beneficial to learning, to remembering, and to building interest in future learning.

Trainees should be into the job with as few errors as possible. As the most expert member of the group, the instructor might have to assist the trainees by handling the more difficult parts the first time through.

B.4.3.3 Question the Trainees on Key Points. One of the training hazards encountered in Step 3 is the instructor's tendency to overlook slight omissions and details of the job that require explanation. *The instructor should never assume that the trainees understand what has been taught but should verify it by asking questions.* If there are omissions of details in the trainees' demonstrations and explanations, the instructor should raise questions to cover the details and have complete discussion of the points involved.

B.4.3.4 Make Corrections in a Positive and Impersonal Manner. It should be remembered that the trainees are in the psychological position of trying to do what the instructor wants.

The instructor should not lose sight of this and should not attempt to rush the learning or become impatient. In particular, the instructor should carefully consider each corrective step taken and praise good work, even if it is minor. Then the instructor should tell how some operations might have been performed more effectively. During a trainee's demonstration, it is sometimes better to permit minor mistakes to pass until the trainee has completed the explanation. Questions raised after the demonstration cause less interference and can be used effectively to get across the correct knowledge, methods, and points of view. If a trainee's mistakes are too frequent, the instructor can usually find the cause by going back to the instruction provided in Steps 1 and 2. *In other words, rather than attempt to explain mistakes made in Step 3 presentations as being due to the trainees' failure to learn, the instructor's own handling of the trainees up to Step 3 should be re-examined.* When the frequency of errors in the presentation step is high or when the same errors are being made by several trainees in the group, the cause usually can be traced to ineffective instruction in Step 1 or Step 2.

In summary, the instructor should observe the following basic rules to obtain better results and to build more favorable work-related attitudes:

(1) Make corrections in a *positive* manner.
(2) Make corrections in an *impersonal* manner.
(3) Focus attention on the *causes* of mistakes.
(4) Help the trainees to detect their own mistakes and make their own critiques.
(5) Correct with leading *questions.*
(6) Get every trainee into the act and provide as much practice under direct observation as possible in the time allotted.

After members of the training group have shown they understand and can perform the operation, and after the instructor is satisfied that a solid foundation of basic learning has been acquired, the group is ready to move to the final phase of instruction.

B.4.4 Step 4: Observation (Follow-Up and Performance Testing). The final step in the cycle of instruction is observation of the trainees. The instruction process up to this point is summarized in Table B.4.4.

Table B.4.4 Instruction Process Summary

Step	Purpose
Step 1: Preparation	Organization
Step 2: Presentation	Motivation, showing, and telling
Step 3: Application	Trainee demonstration

The purpose of Step 4 is to show what the trainees have learned by putting them in a work situation as nearly typical of the normal maintenance environment operations as possible.

Step 4 provides an opportunity for the trainees to practice and gain experience in phases of the job that the instructor has covered. Job knowledge is reinforced and job skills are acquired only by doing. Without practice, skills cannot be developed.

The guidance factors given in B.4.4.1 through B.4.4.3 are critical in Step 4.

B.4.4.1 Provide Close Follow-Up on the Job. When training is provided simultaneously to a group, it is practically impossible for the instructor to do an adequate job of follow-up on each trainee. Despite this, *prompt follow-up is the most important aspect of Step 4.* Unless the trainees put the techniques they have been taught into practice, instruction has no purpose.

It takes application to learn techniques. It takes correct application to learn correct techniques. Trainees, if left on their own, often develop incorrect ways of doing their jobs. Follow-up is the only means to prevent this. Responsibility for providing follow-up should definitely be assigned. Although it is common practice for the instructor to provide Step 4 follow-up, there are definite advantages in sharing follow-up responsibilities with the supervisors of the employees in training.

The training of maintenance electricians finds greater acceptance when there has been active line-supervision involvement. One way that this can be achieved is by using engineering and maintenance supervision as a pilot group before the program is presented to the trainees. Another common practice is to use engineers or maintenance supervisors as instructors, which provides a variety of benefits, the most important being a bond between the classroom and on-the-job performance. Also, inadequacies in training show up quickly, and on-the-job follow-up is efficiently implemented.

B.4.4.2 Provide Immediate Follow-Up on the Job. Heavy emphasis has been placed on follow-up, and the timing of follow-up is crucial. Unfortunately, trainees sometimes view training as having ended when the presentation phase is completed.

Follow-up is an easy function to put off. Its benefits are intangible, while daily maintenance demands are not. It is something supervisors might not be accustomed to, and other demands on their time get priority. Meanwhile, "wrong learning" multiplies. *Learning is learning, right or wrong.* Each error repeated is just that much more firmly instilled in the memory, which makes timing important. On-the-job follow-up should be phased out as performance demonstrates that correct methods and procedures have been learned and are being applied.

B.4.4.3 Maintain Performance Standards. Performance expectations should be high. There is no room for exceptions. If a quality standard is right, it should be observed in appraising trainee performance. If the standard is not right, it should be changed, not ignored.

Fault-free performance should be the training standard. Uniform results depend on uniform methods. High standards of equipment performance depend equally on high standards of equipment installation, operation, and maintenance.

Performance observation is the final filter in the developmental process. If the mesh is coarse, the product will be irregular. Trainees should not be graduated until they demonstrate capability using prescribed methods to obtain prescribed quality standards.

There might be times when many members of a training group exhibit inadequate understanding of maintenance practices or quality requirements. Re-instruction of the entire group might be the most economical means for bringing about the improvement desired in such instances. Two items of correction technique have immense bearing on the success of retraining. First, emphasis should be placed on what to do instead of concentrating on what was done incorrectly. Each correct detail should be commended. The step that is right should be emphasized. The operation should be commenced at that point and the next right step supplied. Then each phase of the operation should be repeated as it should be done. This is positive reinforcement.

Second, questions should be asked instead of statements made. Correct information should be drawn out instead of being supplied again. The purpose should be to establish a learning situation in which the trainees are active participants. Trainees should be encouraged to analyze their own performance. The goal is maximum trainee involvement.

B.5 Summary of the Instruction Process.

B.5.1 Instruction is the process of teaching trainees the knowledge, skills, and attitudes they need to do their jobs.

B.5.2 Instruction involves a variety of methods and techniques. The acquisition of knowledge, skills, and attitudes is the objective. How effectively instruction is organized and carried out determines the amount, rate, and permanence of new learning.

The industrial instructor's challenge is to develop ways to involve trainees and to discard the passive, lecture-based, nonparticipative methods inherited from the old-line techniques of academic institutions. Involvement is recommended if trainees are to acquire new information and practical skills effectively in the least amount of time. Equally important are the interest and desire to apply the new learning in the work situation.

B.5.3 Organized instruction is effective only when it is based on training methods that motivate the trainees.

B.5.3.1 Instruction should be presented so that it has practical meaning. The instructor should practice the following:

(1) Present practical applications.
(2) Use familiar experiences and words.
(3) Get the trainees to participate in the instruction.
(4) Use problem-solving discussions.
(5) Relate class work to on-the-job situations.

B.5.3.2 Instruction should be purposeful, that is, it should have a goal. To give purpose to training, the instructor should do the following:

(1) Make certain the reasons for the training are clear.
(2) Emphasize the benefits to the trainees.
(3) Point out the practical applications of what is being taught.
(4) Let the trainees know how they are doing.

B.5.4 Instructions should be organized in a way that generates active trainee participation. Participation can be increased by methods such as the following:

(1) Using models, mockups, graphs, charts, exhibits, and inspection tours of actual operations
(2) Using discussion and questions, having trainees prepare class materials, and encouraging trainee solutions of problems brought up in class
(3) Making specific assignments to trainees, providing individual practice, and having trainees research information

B.5.5 The instruction process can be broken into four steps.

B.5.5.1 Step 1: Preparation of the Trainees. The instructor should do the following:

(1) Develop motivation, reasons, advantages, and objectives
(2) Get the students interested in the training project
(3) Become familiar with what trainees already know about the operation
(4) State the job to be done, covering the whole job briefly

B.5.5.2 Step 2: Presentation (Present the Operation). The instructor should do the following:

(1) Tell, explain, show, and illustrate one step at a time, going from simple to complex
(2) Stress each key point
(3) Instruct clearly, completely, and patiently

B.5.5.3 Step 3: Application (Try-Out Performance). The instructor should do the following:

(1) Have trainees perform operations step by step
(2) Make certain that errors are corrected
(3) Have each trainee perform the operation again while explaining each key point

B.5.5.4 Step 4: Observation (Follow-Up). The instructor should do the following:

(1) Put trainees on their own
(2) Designate to whom trainees should go for help
(3) Establish definite arrangements for frequent checks
(4) Encourage discussions and questions
(5) Taper off follow-up

Annex C Bibliography

This annex is not a part of the recommendations of this NFPA document but is included for informational purposes only.

C.1 Introduction.

C.1.1 This bibliography lists some of the more widely recognized sources of maintenance and testing information. Because they are so numerous, many excellent textbooks by individual authors are not listed; information on them is available from the various publishers.

C.1.2 For those who are interested in implementing an effective EPM program or improving an existing one, a suitable reference library should be readily available. The size of the plant and the extent of its maintenance and servicing operations will determine the desired publications for the reference library.

C.1.3 The need to use the manufacturer's service manuals and instructions furnished with specific equipment or apparatus has been previously mentioned and cannot be overemphasized. Additionally, there are many sources of helpful information on general and specific maintenance, troubleshooting, test methods, test instruments, and their use. Some of these are available without cost, but most entail a nominal charge. Publishers of technical and trade magazines are another important source of pertinent literature. Some can provide, without charge, reprints of specific articles, or, for a nominal fee, a compilation of reprints of articles on a particular subject.

C.1.3.1 American Association of Textile Chemists and Colorists (AATCC). P.O. Box 12215, Research Triangle Park, NC 27709.

ANSI/AATCC-27, *Wetting Agents: Evaluation of Rewetting Agents,* 2009.

C.1.3.2 American Petroleum Institute (API). 1220 L St. NW, Washington, DC 20005-4070.

Guide for Inspection of Refinery Equipment, Chapter XIV, Electrical Systems.

C.1.3.3 American Society for Testing and Materials (ASTM International). 100 Barr Harbor Drive, P.O. Box C700, West Conshohocken, PA 19428-2959.

ASTM D 120, *Standard Specifications for Rubber Insulating Gloves,* 2009.

ANSI/ASTM D 664, *Standard Test Method for Acid Number of Petroleum Products by Potentiometric Titration,* 2011a.

ASTM D 1048, *Standard Specification for Rubber Insulating Blankets*, 2012.

ASTM D 1049, *Standard Specification for Rubber Insulating Covers*, (Rev. 2010).

ASTM D 1050, *Standard Specification for Rubber Insulating Line Hose*, 2005 (Rev. 2011).

ASTM D 1051, *Standard Specification for Rubber Insulating Sleeves*, 2008.

C.1.3.4 Crouse-Hinds. P.O. Box 4999, Syracuse, NY 13221.

Crouse-Hinds Code Digest, *Suggestions for Installation and Maintenance of Electrical Equipment for Use in Hazardous Areas*, 2011.

C.1.3.5 Factory Mutual Engineering Corporation. 1151 Boston-Providence Turnpike, Norwood, MA 02061.

Handbook of Industrial Loss Prevention, Chapter 32, 1968.

C.1.3.6 Institute of Electrical and Electronics Engineers (IEEE). 445 Hoes Lane, P.O. Box 1331, Piscataway, NJ 08855-1331.

ANSI/IEEE 67, *Guide for Operation and Maintenance of Turbine Generators*, 2005.

ANSI/IEEE 80, *Guide for Safety in AC Substation Grounding*, 2000.

ANSI/IEEE 141, *Recommended Practice for Electric Power Distribution for Industrial Plants (IEEE Red Book)*, 1993.

ANSI/IEEE 142, *Recommended Practice for Grounding of Industrial and Commercial Power Systems (IEEE Green Book)*, 2007.

ANSI/IEEE 241, *Recommended Practice for Electric Power Systems in Commercial Buildings (IEEE Gray Book)*, 1990.

ANSI/IEEE 242, *Recommended Practice for Protection and Coordination of Industrial and Commercial Power Systems (IEEE Buff Book)*, 2001.

ANSI/IEEE 315 (ANSI Y32.2-75), *Graphic Symbols for Electrical and Electronics Diagrams*, 1975 (Reaff. 1993).

ANSI/IEEE 400, *Guide for Making High-Direct-Voltage Tests on Power Cable Systems in the Field*, 2012.

ANSI/IEEE 432, *Guide for Insulation Maintenance for Rotating Electrical Machinery (5 HP to less than 10,000 HP)*, 1992.

ANSI/IEEE 446, *Recommended Practice for Emergency and Standby Power Systems for Industrial and Commercial Applications (IEEE Orange Book)*, 1995.

ANSI/IEEE 450, *Recommended Practice for Maintenance, Testing, and Replacement of Vented Lead-Acid Batteries for Stationary Applications*, 2010.

ANSI/IEEE 519, *Recommended Practice and Requirements for Harmonic Control in Electrical Power Systems*, 1992.

IEEE 602, *Recommended Practice for Electrical Systems in Health Care Facilities (IEEE White Book)*, 2007.

ANSI/IEEE 739, *Recommended Practice for Energy Management in Industrial and Commercial Facilities (IEEE Bronze Book)*, 1995.

IEEE C2, *National Electrical Safety Code*, Part 1, Rules for the Installation and Maintenance of Electric Supply Stations and Equipment, 2012.

IEEE C37.41, *Standard Design Tests for High-Voltage (>1000 V) Fuses, Fuse and Disconnecting Cutouts, Distribution Enclosed Single-Pole Air Switches, Fuse Disconnecting Switches, and Fuse Links and Accessories Used with These Devices*, 2008.

ANSI/IEEE C37.95, *Guide for Protective Relaying of Utility-Consumer Interconnections*, 2002.

ANSI/IEEE C37.96, *Guide for AC Motor Protection*, 2000.

IEEE C57.94, *Recommended Practice for Installation, Application, Operation and Maintenance of Dry-Type General Purpose Distribution and Power Transformers*, 1982 (Reaff. 1987).

ANSI/IEEE C57.106, *Guide for Acceptance and Maintenance of Insulating Oil in Equipment*, 2006.

IEEE C57.111, *Guide for Acceptance and Maintenance of Silicone Insulating Fluid and Its Maintenance in Transformers*, 1989.

ANSI/IEEE C57.121, *Guide for Acceptance and Maintenance of Less Flammable Hydrocarbon Fluid in Transformers*, 1998.

C.1.3.7 International Electrotechnical Commission (IEC). 3 rue de Varembé, P.O. Box 131, 1211 Geneva 20, Switzerland. (In the United States, IEC Publications are available from American National Standards Institute, ANSI.)

IEC No. 60417-DB-HS, *Graphical Symbols for Use on Equipment*, 2008.

C.1.3.8 Intertec Publishing Corp. 9800 Metcalf Avenue, P.O. Box 12901, Overland Park, KS 66282-2901.

EC&M Magazine.

C.1.3.9 McGraw-Hill Publishing Co. 1221 Avenue of the Americas, New York, NY 10020.

D. Beeman, *Industrial Power Systems Handbook.*

E. W. Boozer, *Motor Applications and Maintenance Handbook.*

C. I. Hubert, *Preventative Maintenance of Electrical Equipment.*

C.1.3.10 National Electrical Contractors Association (NECA). 3 Bethesda Metro Center, Suite 1100, Bethesda, MD 20814-5372.

Total Energy Management — A Practical Handbook on Energy Conservation and Management, Index No. 2095.

C.1.3.11 National Electrical Manufacturers Association (NEMA). 1300 North 17th Street, Suite 1847, Rosslyn, VA 22209.

NEMA 280, *Application Guide for Ground-Fault Circuit Interrupters* (see Section 7, Field Test Devices, and Section 8, Field Troubleshooting), 1990.

NEMA AB 3, *Molded Case Circuit Breakers and Their Application*, 2012.

NEMA AB 4, *Guidelines for Inspection and Preventive Maintenance of Molded-Case Circuit Breakers Used in Commercial and Industrial Applications*, 2009.

NEMA ICS 1.3, *Preventive Maintenance of Industrial Control and Systems Equipment*, 1986 (Reaff. 2009).

NEMA ICS 2.3, *Instructions for the Handling, Installation, Operation, and Maintenance of Motor Control Centers Rated Not More Than 600 Volts*, 1995 (Reaff. 2008).

NEMA ICS 7, *Industrial Control and Systems Adjustable — Speed Drives*, 2006.

ANSI/NEMA KS 3, *Guidelines for Inspection and Preventive Maintenance of Switches Used in Commercial and Industrial Applications*, 2010.

ANSI/NEMA MG 2, *Safety Standard for Construction and Guide for Selection, Installation and Use of Electric Motors and Generators* (see Section 8.3, Maintenance), 2001 (Reaff. 1–2007).

NEMA PB 1.1, *General Instructions for Proper Installation, Operation, and Maintenance of Panelboards Rated at 600 Volts or Less*, 2007.

NEMA PB 2.1, *General Instructions for Proper Handling, Installation, Operation, and Maintenance of Deadfront Distribution Switchboards Rated 600 Volts or Less,* 2007.

Evaluating Water-Damaged Electrical Equipment, 2011.

C.1.3.12 National Institute for Occupational Safety and Health (NIOSH). 4676 Columbia Parkway, Cincinnati, OH 45226.

Guidelines for Controlling Hazardous Energy During Maintenance and Servicing.

C.1.3.13 National Safety Council (NSC). 60611 1121 Spring Lake Drive, Itasca, IL 60143.

NSC 129.46, *Electrical Inspections Illustrated.*

NSC 635, *Lead-Acid Storage Batteries.*

C.1.3.14 U.S. Department of the Army. U.S. Army Corps of Engineers, 441 G Street NW, Washington, D.C. 20314-1000.

TM-5-682, *Electrical Safety, Facilities Engineering U.S. Army,* November 1999.

TM-5-683, *Electrical Interior, Facilities Engineering U.S. Army,* November 1995.

TM-5-684, *Electrical Exterior, Facilities Engineering U.S. Army,* November 1996.

TM-5-685, *Operation, Maintenance and Repair of Auxiliary Generators U.S. Army,* August 1996.

TM 5-686, *Power Transformer Maintenance and Acceptance Testing,* November 1998.

TM 5-688, *Foreign Voltages and Frequencies Guide,* November 1999.

TM 5-691, *Utility Systems Design Requirements for Command, Control, Communications, Computer, Intelligence, Surveillance, and Reconnaissance (C41SR) Facilities,* December 2000.

TM 5-692-1, *Maintenance of Mechanical and Electrical Equipment at Command, Control Communications, Computer, Intelligence, Surveillance, and Reconnaissance (C41SR) Facilities-Recommended Maintenance Practices,* April 2001.

TM 5-692-2, *Maintenance of Mechanical and Electrical Equipment at Command, Control Communications, Computer, Intelligence, Surveillance, and Reconnaissance (C41SR) Facilities-System Design Features,* April 2001.

C.1.3.15 Additional Addresses for Bibliography.

Chemical Rubber Co., 18901 Cranwood Parkway, Cleveland, OH 44128.

E. P. Dutton & Co., 201 Park Avenue S., New York, NY 10003.

Gale Research Co., 1400 Book Tower, Detroit, MI 48226.

Hayden Book Co., Inc., 50 Essex Street, Rochelle Park, NJ 07662.

National Fire Protection Association (NFPA), 1 Batterymarch Park, Quincy, MA 02169-7174.

Plant Engineering, 1350 E. Touhy Avenue, Des Plaines, IL 60018.

Prentice-Hall, Inc., Englewood Cliffs, NJ 07632.

Howard W. Sams Co., Inc., 4300 W. 62nd Street, Indianapolis, IN 46268.

TAB Books, Blue Ridge Summit, PA 17214.

Underwriters Laboratories Inc., 333 Pfingsten Road, Northbrook, IL 60062-2096.

U.S. Government Printing Office, Superintendent of Documents, Mail Stop: SSOP, Washington, DC 20402-9328.

Van Nostrand Reinhold Publishing Company, 135 W. 50th Street, New York, NY 10020

Westinghouse Electric Corp., Printing Division, 1 Stewart Station Drive, Trafford, PA 15085.

John Wiley & Sons, Inc., 605 Third Avenue, New York, NY 10016.

C.2 Power Quality References.

C.2.1 Harmonics.

Power Quality for the Utility Engineer, Electric Power Research Institute, 1988, p. VIII-12.

Van Wagner, et al., *Industrial Power Quality Case Study: General Motors Corporation Buick-Oldsmobile-Cadillac Group Powertrain Group — Livonia Engine,* Detroit Edison, March 1990, p. 43.

C.2.2 Sags and Swells.

Berutti, Al, and R. M. Waggoner, "Practical Guide to Quality Power for Sensitive Electronic Equipment," *EC&M,* Intertec Publishing, 1993. Based on materials originally written by John A. DeDad and editors of *EC&M.*

Chavousite, David M., "Selective Power Conditioning Keeps Plant Operating," *Power Quality Assurance,* July/August 1994, pp. 72–75.

Conrad, L., K. Little, and C. Grigg, "Predicting and Preventing Problems Associated with Remote Fault-Clearing Voltage Dips," *IEEE Transactions on Industry Applications,* January 1991, Vol. 27, pp. 167–172.

DeDad, John A., "Power Quality and Electronic Equipment Protection. What to Use, When to Use It," *EC&M,* January 1991, pp. 37–46.

Dorr, Douglas S., "Point of Utilization Power Quality Study Results," National Power Laboratory Power Quality Study, October 1994.

Douglas, John, "Power Quality," *EPRI Journal,* December 1993, pp. 8–15.

Dranetz Field Handbook for Power Quality Analysis, Dranetz Technologies, Inc., 1989.

IEEE 1159, *Recommended Practice on Monitoring Electric Power Quality,* 1995.

Kreiss, David, "Determining the Severity and Cause of Voltage Sags Using Artificial Intelligence," PQ Conference, October 1994.

Lamoree, Jeff, "How Utility Faults Impact Sensitive Customer Loads," *Electrical World,* April 1992, pp. 60–63.

Martzloff, Francois D., and Thomas M. Gruz, "Power Quality Site Surveys; Facts, Fiction, and Fallacies," *IEEE Transactions on Industry Applications,* November/December 1988, Vol. 24, No. 6.

McGrahaghan, Mark, David R. Mueller, and Marek J. Samothj,"Voltage Sags in Industrial Systems," *IEEE Transactions on Industry Applications,* March/April 1993, Vol. 29, No. 2, pp. 397–403.

Smith, Charles J. et al., "The Impact of Voltage Sags on Industrial Plant Loads."

"Voltage Sags in Industrial Systems," *IEEE Transactions on Industry Applications,* March/April 1993.

C.2.3 Noise.

Dranetz Field Handbook for Power Quality Analysis, 1991.

Federal Information Processing Standards Publication 94, U.S. Department of Commerce, 1983.

IEEE 100, *Standard Dictionary of Electrical and Electronic Terms*, 1996.

ANSI/IEEE 1100, *Recommended Practice for Powering and Grounding Electronic Equipment (IEEE Emerald Book)*, 2005.

C.2.4 Interruptions.

IEEE 1159, *Recommended Practice for Monitoring Electric Power Quality*, 2009.

IEEE 1250, *IEEE Guide for Identifying and Improving Voltage Quality in Power Systems*, 2011.

ANSI/NEMA C84.1, *Electric Power Systems and Equipment, Voltage Ratings (60 Hertz)*, 2006.

C.2.5 Interharmonics.

Gunther, Erich, Draft document on interharmonics.

IEC 60076-2, *Power Transformers – Part 2: Temperature Rise*, 1993-04-07.

Leiter, David, *Distributed Energy Resources*, U.S. Department of Energy for Fuel Cell Summit IV, Washington, DC, May 10, 2000.

MIL-STD-339, *Wiring and Wiring Devices for Combat and Tactical Vehicles, Selection and Installation of*, December 7, 1987, available from Defense Automated Printing Service (Customer Service), 700 Robbins Avenue, Building 4D, Philadelphia, PA 19111-5094.

RAC Publications, *Reliability Tool Kit*, page 12.

Annex D Informational References

D.1 Referenced Publications. The documents or portions thereof listed in this annex are referenced within the informational sections of this recommended practice and are not part of the recommendations of this document unless also listed in Chapter 2 for other reasons.

D.1.1 NFPA Publications. National Fire Protection Association, 1 Batterymarch Park, Quincy, MA 02169-7471.

NFPA 70®, National Electrical Code®, 2011 edition.

D.1.2 Other Publications.

D.1.2.1 ASTM Publications. ASTM International, 100 Barr Harbor Drive, P.O. Box C700, West Conshohocken, PA 19428-2959.

ASTM D 92, *Standard Test Method for Flash and Fire Points by Cleveland Open Cup Tester*, 2012.

ASTM D 1933, *Standard Specification for Nitrogen Gas as an Electrical Insulation Material*, 2003 (Rev. 2008).

D.1.2.2 IEEE Publications. Institute of Electrical and Electronics Engineers, 445 Hoes Lane, P.O. Box 1331, Piscataway, NJ 08855-1331.

ANSI/IEEE 43, *Recommended Practice for Testing Insulation Resistance of Rotating Machinery*, 2000.

IEEE 100, *Authoritative Dictionary of IEEE Standards Terms*, 2000.

ANSI/IEEE 315 (ANSI Y32.2-75), *Graphic Symbols for Electrical and Electronics Diagrams*, 1975 (Reaff. 1993).

IEEE 450, *Recommended Practice for Maintenance, Testing, and Replacement of Vented Lead-Acid Batteries for Stationary Applications*, 2010.

IEEE 1188, *Recommended Practice for Maintenance, Testing, and Replacement of Valve-Regulated Lead-Acid (VRLA) Batteries for Stationary Applications*, 2005.

IEEE C57.12.00, *General Requirements for Liquid-Immersed Distribution, Power, and Regulating Transformers*, 2010.

D.1.2.3 NETA Publications. InterNational Electrical Testing Association, P.O. Box 687, Morrison, CO 80465.

ANSI/NETA MTS, *Standard for Maintenance Testing Specifications for Electrical Power Distribution Equipment and Systems*, 2011.

D.1.2.4 Other Publications.

ANSI C107.1, *Guidelines for Handling and Disposal of Capacitor and Transformer-grade Askarels Containing Polychlorinated Biphenyls*, 1974.

D.2 Informational References.

ANSI/IEEE 1100, *Recommended Practice for Powering and Grounding Electronic Equipment (IEEE Emerald Book)*, 2005.

Power Quality for the Utility Engineer, Electric Power Research Institute, p. VIII-12, 1988.

Van Wagner et al., *Industrial Power Quality Case Study: General Motors Corporation Buick-Oldsmobile-Cadillac Group Powertrain Group — Livonia Engine*, Detroit Edison, March 1990, p. 43.

D.3 References for Extracts in Informational Sections. (Reserved)

Annex E Suggestions for Inclusion in a Walk-Through Inspection Checklist

This annex is not a part of the recommendations of this NFPA document but is included for informational purposes only.

E.1 General. These suggested items are directed toward minimizing day-to-day electrical hazards. The list is not complete, nor do the items necessarily appear in order of importance. It is presented as a guide for the preparation of a checklist that should be developed for each plant. Because of the similarity to the plant fire prevention inspection, both inspections can be carried out by the same personnel.

E.1.1 Flexible Cords (Including Those on Appliances). Heater-type cords are recommended for portable heating appliances, such as toasters, grills, and coffee makers. An inspection should be made for badly worn or frayed spots, splices (not permitted), improper type, or current-carrying capacity that is too small.

E.1.2 Plugs and Connectors. A check should be made for stray strands and loose terminals. They should be grounding type where required for specific appliances. The green conductor should be connected to the grounding terminal.

E.1.3 Extension Cords. Are extension cords used in place of permanent wiring, and are they of excessive length and of proper type? They should not pass through walls, partitions, or doors.

E.1.4 Multiple Current Taps. Are multiple current taps used because of too few receptacles? In particular, are they used in areas such as canteens, lunchrooms, and offices?

E.1.5 Appliances. Grills, toasters, and similar equipment should be permanently spaced from combustible material.

E.1.6 Heating Appliances. Where used with combustible material, such appliances generally require a signal light to indicate when they are "on."

E.1.7 Hot-Water Heaters. A check should be made for proper electrical protection. The combination temperature- and pressure-relief valve should be manually operated to be sure it is free and the drain line is clear. The setting should be visually checked.

E.1.8 Office Equipment. The condition of flexible cords, plugs, and connectors should be checked, and excessive use of extension cords and multiple current taps should be noted.

E.1.9 Receptacle Outlets. Grounding-type receptacles are generally required. Each receptacle should be checked for continuity of grounding connection, using a suitable test instrument. Are special receptacle configurations used for those supplying unusual voltages, frequencies, and so on? Are they well marked or identified? In particular, missing faceplates, receptacles showing signs of severe arcing, loose mounting, and so on, should be noted.

E.1.10 Portable Equipment (Tools, Extension Lamps, and Extension Cords). In the shop or tool room, a check should be made after each use for isolation between live parts and frame. The condition of cords and plugs should be noted. Is continuity maintained between the frame and the grounding pin of the plug? The green conductor should connect only to the plug grounding pin. On lamps, the condition of guards, shields, and so on, should be checked. See *NFPA 70, National Electrical Code* for portable hand lamps; metal-shell and paper-lined lampholders for hand lamps are not permitted.

E.1.11 Lighting Fixtures. All lighting fixtures should be labeled and grounded. See *NFPA 70, National Electrical Code,* for connection of electric-discharge lighting fixtures. These are permitted to be connected by suitable, three-conductor flexible cord where visible for its entire length and terminated at outer end in a grounding-type attachment plug or busway plug. No fixtures should be located close to highly combustible material. The location of fixtures with burned out bulbs or tubes; fixtures that are heavily coated with dust, dirt, or other material; and reflectors that are in need of cleaning should be noted.

E.1.12 Equipment Grounding. Where machinery or wiring enclosures are grounded through the conduit system, broken or loose connections at boxes and fittings, flexible connections, and exposed ground straps should be identified. Multiple bonding of conduit and other metallic enclosures to interior water piping systems, including sprinkler systems, is sometimes used as a precaution where building vibration is severe, even though a separate equipment-grounding conductor is run with the circuit conductors inside the conduit.

E.1.13 Yard Transformer Stations. The condition of transformers, fence, gates, and locks should be noted. Yard and equipment should be free of storage of combustible material, weeds, grass, vines, birds' nests, and so on. Localized overheating, indicated by conductor discoloration, should be watched for. Indication of excessive transformer temperature, pressure, or oil leakage should be noted.

E.1.14 Services. The condition of weatherheads and weatherhoods should be visually checked to determine that they remain in good condition. Birds' nests, rats' nests, and so on, should be eliminated. At the same time, the apparent condition of lightning arresters, surge capacitors, grounding conductors, and grounds should be determined. Are switches safely and readily accessible?

E.1.15 Switch Rooms and Motor Control Centers. Switch rooms and motor control centers should be clean, used for no other purpose, and free of storage of any kind, especially combustible material. Ventilation equipment should be in working condition and unobstructed. Any unusual noises or odors should be noticed and reported promptly. Metering equipment should be checked for high or low voltage and current and any indication of accidental grounding (ungrounded systems). Are switches and motor controllers properly identified as to function; are fire extinguishers in place, of suitable type, and charged?

E.1.16 Grouped Electrical Control Equipment (Such as Might Be Mounted on Walls). Is grouped electrical control equipment protected from physical damage and readily accessible? Are any equipment enclosures damaged, or do any have missing or open covers? Are any live parts exposed? Any condition that prevents quick or ready access should be reported.

E.1.17 Enclosures of Electrical Parts (Motor Control Equipment, Junction Boxes, Switches, etc.). Are covers secured in place? The location of broken or loose conduit, wiring gutters, and so on, should be reported. Missing dust caps should be replaced.

E.1.18 Hazardous (Classified) Location Equipment. All cover bolts should be in place and tight. Permanent markings should not be obstructed by paint. Joints between cover and case should be examined for signs of having been pried open in the removal of the cover. This might have damaged the mating surfaces of the joints. Excessive accumulations of dust and dirt should be noted for removal from all enclosures, including motors, which also should be examined for obstructed ventilation. The use of nonexplosionproof electric equipment, including lighting that might have been installed in the hazardous (classified) location area, should be noted and reported.

E.1.19 Emergency Equipment.

E.1.19.1 All exit lights should be functioning properly.

E.1.19.2 Emergency lights should all be in working condition. Periodic tests are recommended to ensure that emergency lights function when normal lighting is lost.

E.1.19.3 Emergency power supplies, such as batteries and engine-driven generators, normally receive scheduled tests. Records of periodic tests should be checked. Are fuel and cooling supplies for engine drives adequate? Are fire extinguishers in place, of proper type, and charged?

E.1.19.4 Alarm systems, such as for fire, intrusion, smoke detection, sprinkler water flow, and fire pumps, also receive periodic tests. Records of these tests should be checked to ensure that all signals are properly transmitted and that equipment is in good working condition.

Annex F Symbols

This annex is not a part of the recommendations of this NFPA document but is included for informational purposes only.

F.1 Figure F.1 contains some typical electrical symbols that are used on electrical power and control schematic drawings.

F.2 Figure F.2 contains some typical electrical symbols that are used on electrical control schematic drawings.

F.3 Figure F.3 contains some typical miscellaneous electrical symbols and tables that are used on electrical control schematics.

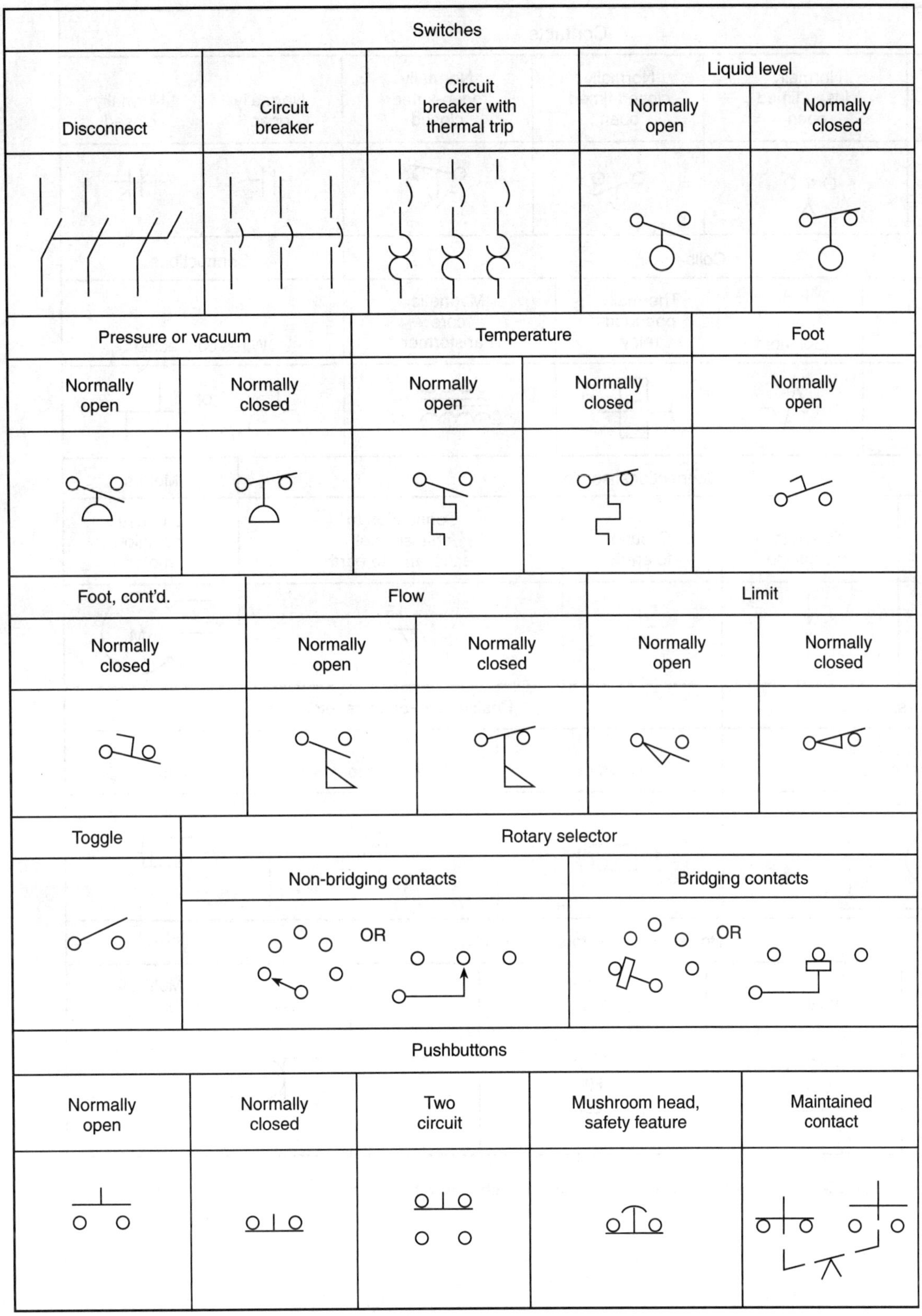

FIGURE F.1 Some Typical Electrical Symbols for Power and Control Schematics. (Courtesy of ANSI/IEEE 315, *Graphic Symbols for Electrical and Electronic Diagrams.*)

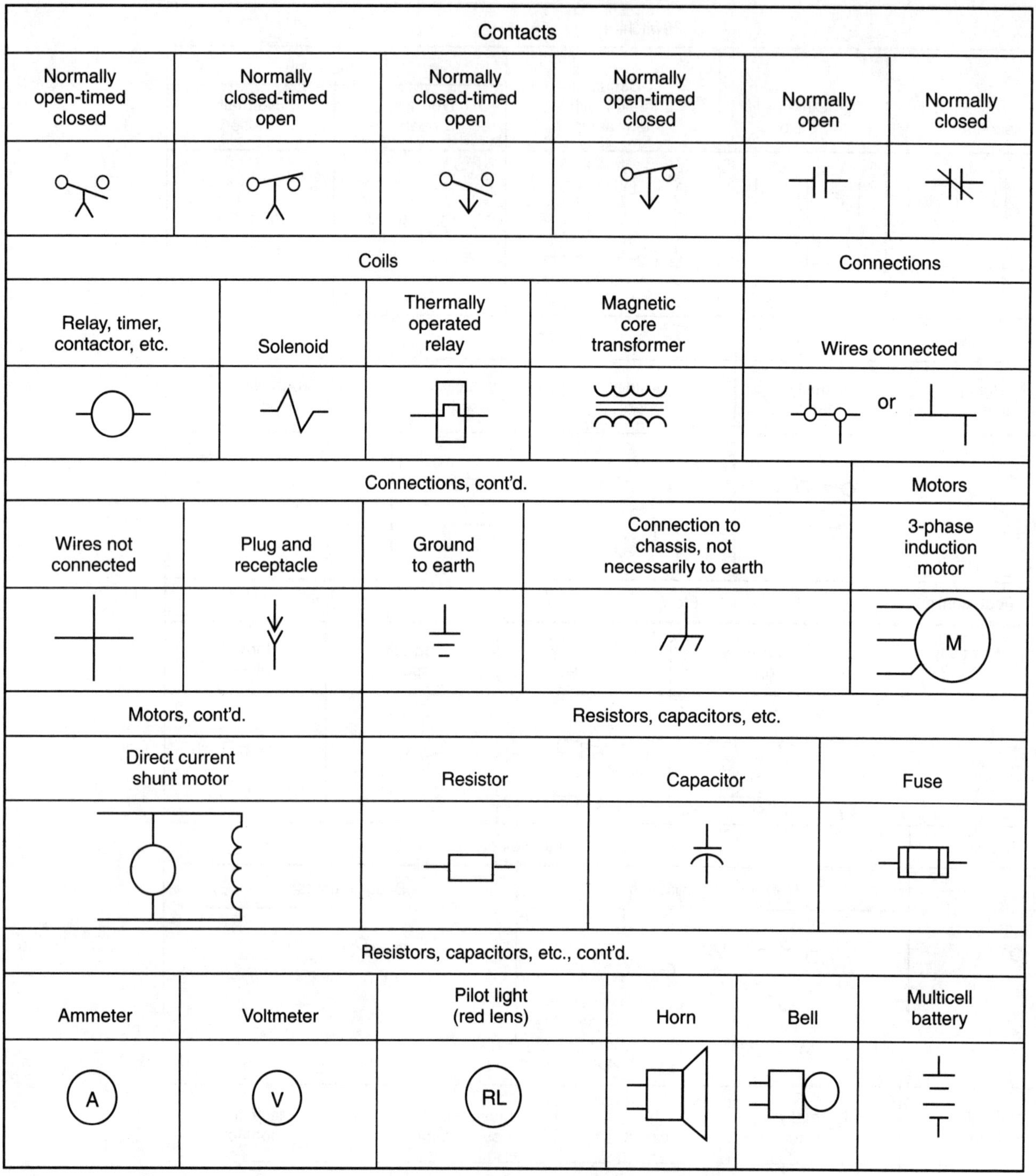

FIGURE F.2 Some Typical Electrical Symbols for Electrical Control Schematic Drawings.

Semiconductors			Miscellaneous
Rectifier diode	Rectifier bridge	Thermistor	Terminal
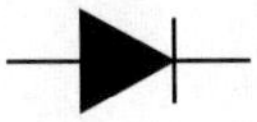	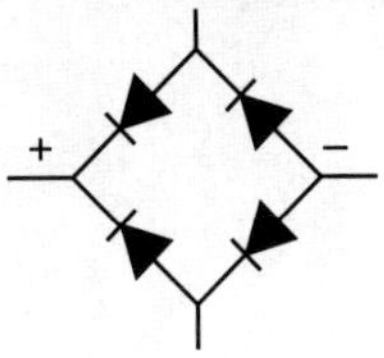		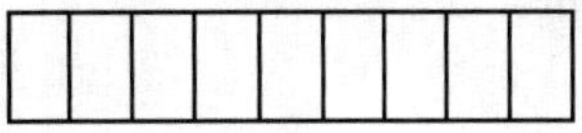

Miscellaneous, cont'd.

Table of contact operation for control switch

Detached contacts shown elsewhere on diagram

Contact	Position		
	1	2	3
A	X		
B		X	
C			X

X — Indicates contact closed

Miscellaneous, cont'd.

Table of contact operation for drum switch (sliding contact type)

Detached contacts shown elsewhere on diagram

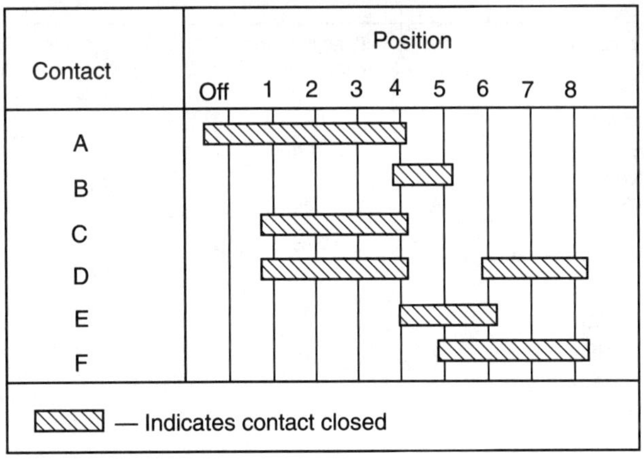

FIGURE F.3 Some Typical Miscellaneous Electrical Symbols.

Annex G Diagrams

This annex is not a part of the recommendations of this NFPA document but is included for informational purposes only.

G.1 Note that Annex G is presented to show use of symbols and should not be construed to indicate recommendations. Figure G.1 shows the use of some typical symbols in a single-line power distribution program.

G.2 Figure G.2 shows a wiring diagram for a reversing starter with control transformer.

G.3 Figure G.3 shows a power and control schematic for reversing starter with low-voltage remote pushbuttons. Forward, reverse, and stop connections are shown.

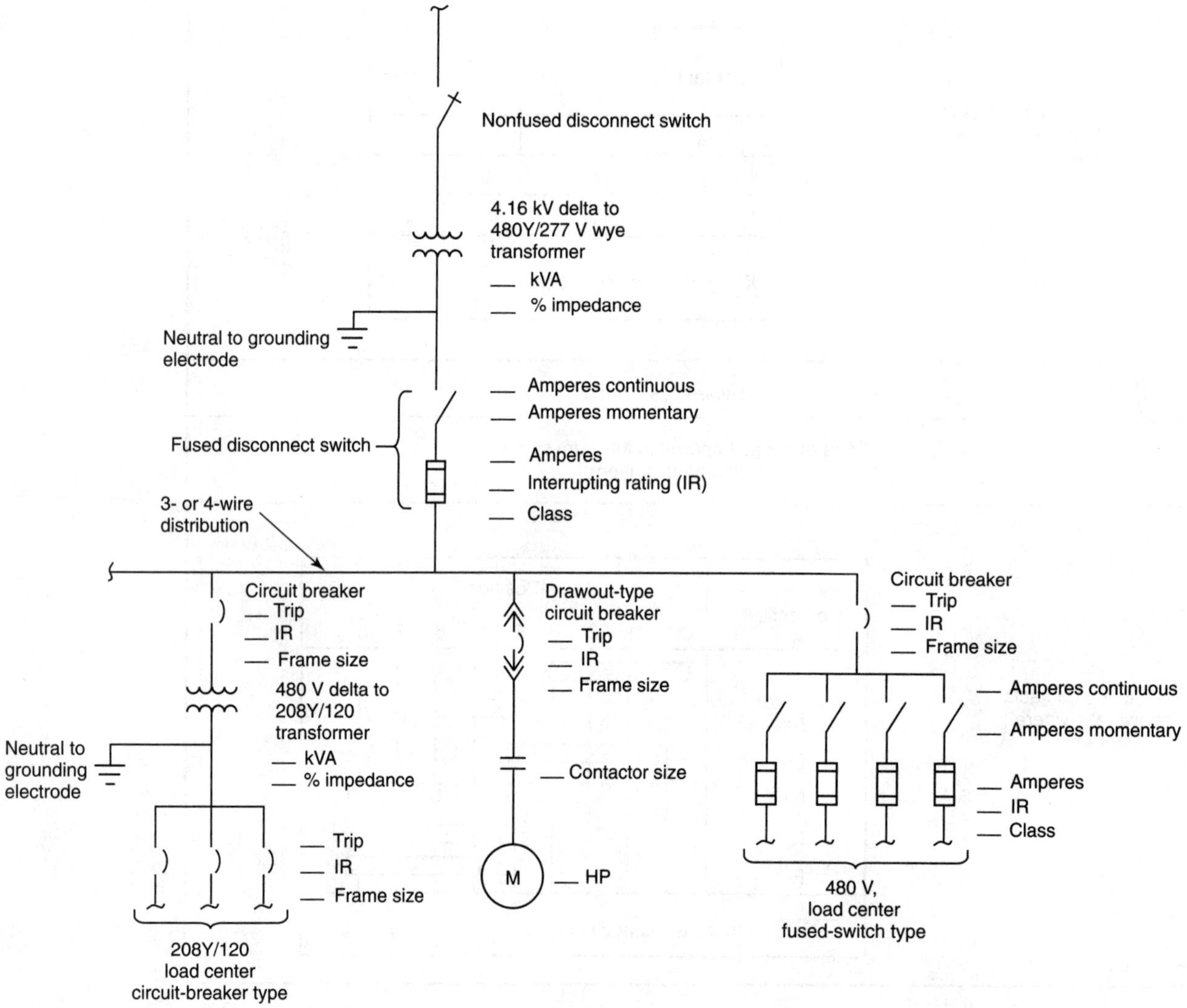

FIGURE G.1 Typical Use of Symbols in a Single-Line Power Distribution Program.

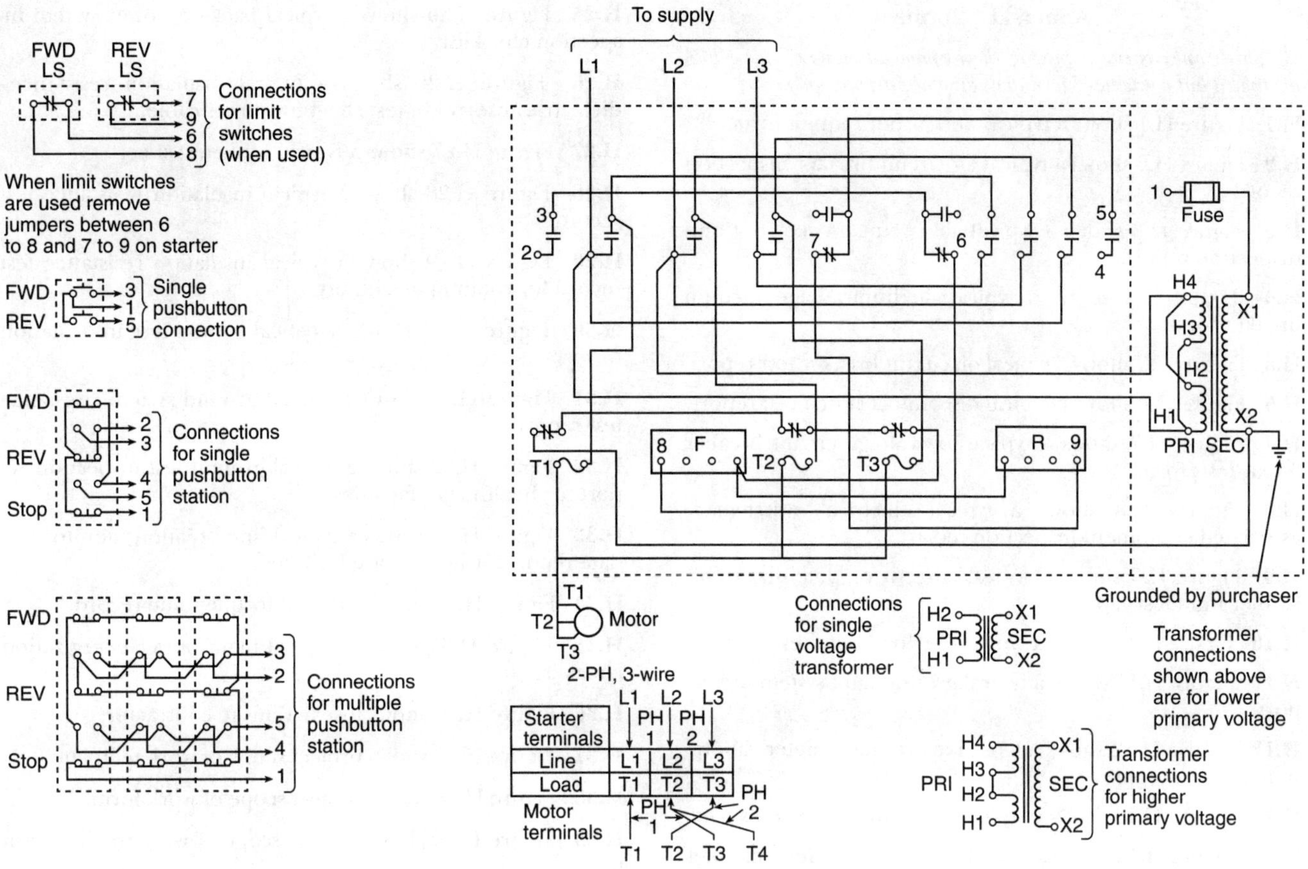

FIGURE G.2 Wiring Diagram for a Reversing Starter with Control Transformer.

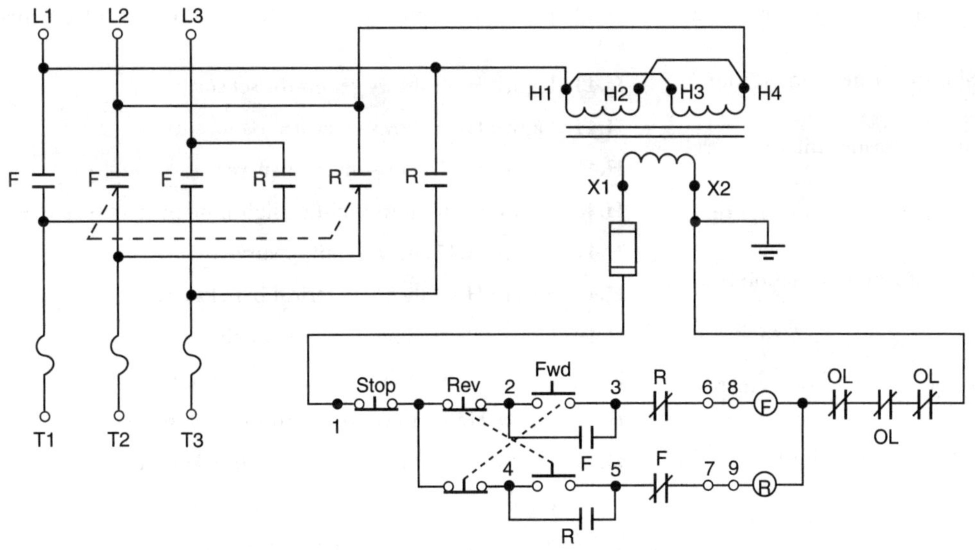

FIGURE G.3 Power and Control Schematic for Reversing Starter with Low-Voltage Remote Pushbuttons. Forward, reverse, and stop connections are shown.

Annex H Forms

This annex is not a part of the recommendations of this NFPA document but is included for informational purposes only.

H.1 Figure H.1 shows a typical work order request form.

H.2 Figure H.2 shows a typical air circuit breaker inspection record.

H.3 Figure H.3 shows a typical air circuit breaker test and inspection report.

H.4 Figure H.4 shows a typical medium-voltage vacuum breaker form.

H.5 Figure H.5 shows a typical oil circuit breaker test report.

H.6 Figure H.6 shows a typical disconnect switch test report.

H.7 Figure H.7 shows a typical low-voltage circuit breaker 5-year tests form.

H.8 Figure H.8 shows a typical electrical switchgear–associated equipment inspection record.

H.9 Figure H.9 shows a typical current or potential transformer ratio test report.

H.10 Figure H.10 shows a typical overload relay test report.

H.11 Figure H.11 shows a typical ground-fault system test report.

H.12 Figure H.12 shows a typical instrument/meter calibration and test report.

H.13 Figure H.13 shows a typical watt-hour meter test sheet.

H.14 Figure H.14 shows a typical panelboard/circuit breaker test report.

H.15 Figure H.15 shows a typical transformer test and inspection report.

H.16 Figure H.16 shows a typical transformer (dry type) inspection record.

H.17 Figure H.17 shows a typical transformer (liquid filled) inspection record.

H.18 Figure H.18 shows a typical transformer oil sample report.

H.19 Figure H.19 shows a typical transformer oil trending report.

H.20 Figure H.20 shows a typical transformer insulation resistance record.

H.21 Figure H.21 shows a typical battery record.

H.22 Figure H.22 shows a typical engine generator set inspection checklist.

H.23 Figure H.23 shows a typical automatic transfer switch form.

H.24 Figure H.24 shows a typical uninterruptible power supply system inspection checklist.

H.25 Figure H.25 shows a typical back-up power system inspection checklist.

H.26 Figure H.26 shows a typical insulation resistance–dielectric absorption test sheet for power cable.

H.27 Figure H.27 shows a typical cable test sheet.

H.28 Figure H.28 shows a typical insulation resistance test record.

H.29 Figure H.29 shows a typical insulation resistance test record for rotating machinery.

H.30 Figure H.30 shows a typical motor test information form.

H.31 Figure H.31 shows a typical ground system resistance test report.

H.32 Figure H.32 shows a typical ground test inspection report for health care facilities.

H.33 Figure H.33 shows a typical line isolation monitor test data report for health care facilities.

H.34 Figure H.34 shows a typical torque value record.

H.35 Figure H.35 shows a typical main power energization checklist.

H.36 Figure H.36 shows instructions to contractor.

H.37 Figure H.37 shows project scope of work template.

H.38 Figure H.38 shows project scope of work form.

H.39 Figure H.39 shows project scope of work modification form.

H.40 Figure H.40 shows cover and contents.

H.41 Figure H.41 shows point of contact.

H.42 Figure H.42 shows power distribution unit (PDU) survey.

H.43 Figure H.43 shows generator set survey.

H.44 Figure H.44 shows electrical panel survey.

H.45 Figure H.45 shows inverter survey.

H.46 Figure H.46 shows building lightning protection survey.

H.47 Figure H.47 shows rectifier survey.

H.48 Figure H.48 shows electrical panel survey.

H.49 Figure H.49 shows transfer switches survey.

H.50 Figure H.50 shows power transformers survey.

H.51 Figure H.51 shows uninterruptible power system survey.

H.52 Figure H.52 shows low voltage breaker data record.

H.53 Figure H.53 shows recloser data record.

H.54 Figure H.54 shows generator data record.

WORK ORDER REQUEST

Work Order No. — Craft

Plant Department

Directions to Requester: Complete Section I ONLY. Submit four copies to the Plant Department. Maintain last copy for your files. Prepare a separate request for each job. This request will be returned to you and becomes a work order only when approved and assigned a work order number by the Plant Department. Allow sufficient time for completion. Please TYPE your request.

I. To be completed by requester: Date ____ / ____ / ____

Summary of work request ____

Location of work: Room(s) ____ Building ____

Details of work request ____

Typical work order request form consists of five parts — includes copies for plant department (or plant engineer), data processing, receiving stores, requester, and requester's department. Work to be done is spelled out in detail.

Special time requirement: Date needed ____ / ____ / ____ Indicate reason ____

Department ____ Tel. ext. ____ ❑ Plan attached ❑ Info. attached

Authorized signature ____ Title ____ Approval if required ____

II. For plant department use only: Date Received ____ / ____ / ____

A. Your request has been ❑ Approved ❑ Disapproved ❑ Forwarded to ____ for action. Use the assigned work order number when referring to this request.

B. Instructions: ____

Job Estimates	Craft	Total Hours	Total Labor	Material	Grand Total
	Hours		$	$	$

Assigned to ____ Craft ____ ❑ Day ❑ Night

Foreman —

C. Completed per plant instructions? ❑ Yes ❑ No

Can recurrence be prevented? ❑ Yes ❑ No

If yes, indicate ____

Actual hours used	Tot. reg.	Tot. O/T	Tot. equiv. hrs.

Requester —

Completed per your request? ❑ Yes ❑ No

Plant and requester note variations ____

Date ____ Foreman's signature ____ Requester's signature ____

III. For data processing use only:

Dept.	Bldg.	Class	Category	Cause	Pay	O/T $

Total Labor $	+	Total Material $	=	Total $

Work description (alphabetic) ____

Work Order No. — Craft

Plant Department

 NFPA 70B

FIGURE H.1 Typical Work Order Request Form.

AIR CIRCUIT BREAKER INSPECTION RECORD

Plant ______________________ Date ______________________

Location ______________________ Serial No. ______________________

Mfr. ______________________ Type or Model ______________________

Drawout ❑ Non-drawout ❑ Switchboard ❑ Metal clad ❑

Rating: Volts ________ Amperes ________ Interrupting Amperes ________

Operation: Manual ❑ Electrical ❑ Remote Control ❑

Volts close ________ ac ❑ dc ❑ Volts trip ________ ac ❑ dc ❑

Protective Devices: Induction Relays ❑ Direct Trips ❑ Direct Trips ❑

CL Fuses ❑ TD Setting ________ Inst. Setting ________

Annual Inspection

Date										
Inspector's Initials										
	Aux.	Main	Aux.	Main	Aux.	Main	Aux.	Main	Aux.	Main
Contact Condition										
Good — Surface Smooth										
Fair — Minor Burns										
Poor — Burned and Pitted										
Contact Check										
Pressure (Good, Weak, Bad)										
Drawout Contacts										
Pressure (Good, Weak, Bad)										
Alignment (Good, Bad)										
Lubricate (Must Do — Use a No-Oxide Lubricant by Mfr.)										
Arcing Assemblies										
Clean and Check the Arc-Splitting Plates Surface Conditions										
Bushings										
Clean and Check Surface Condition										

Date					
Inspector's Initials					
Operating Mechanisms					
Checks					
Positive Close and Trip					
Bushing and Pin Wear					
Set Screws and Keepers					
Protective Devices					
Lubricate Wear Points					
Clean Pots and Replace Oil with Equipment Mfrs. Recommended Oil					
Insulation Condition					
Loose Connections					
Discolored Areas					
Corona Tracking					
Clean Surfaces					
Insulation Tests					
Phase to Phase (Megohm)					
Phase to Ground (Megohm)					
Test Operation					
Close and Trip					
Counter Reading (No. of Ops.)					
Electrical Load					
Peak Indicated Amperes					

Remarks (record action taken when indicated by inspection or tests):

____

____

Other repairs recommended:

____

____

 NFPA 70B

FIGURE H.2 Typical Air Circuit Breaker Inspection Record.

AIR CIRCUIT BREAKER TEST AND INSPECTION REPORT

Customer ____________ Date ____________ Work Order No. ____________
Address ____________ Air Temp. ____________ Rel. Humidity ____________
Breaker Owner/User ____________ Date Last Inspection ____________
Address ____________ Last Inspection Report No. ____________
Equipment Location ____________
Owner Identification ____________

Breaker Data:

Manufacturer ____________ Voltage ____________ Type ____________ Amperes ____________ Int. Rating ____________
Serial No. ____________ Type Oper. Mech. ____________ Age ____________ Other N.P. Data ____________

Test Data:

	Tank 1	Tank 2	Tank 3
Ins. Res. ____ kV, Megohms			
Contact Resistance, Microhms			
Closing Speed/Opening Speed			
Reference, P.F. Test Sheet No.			

Adjustments:

	Mfr's. Rec.	As Found	As Left
Arcing Contact Wipe			
Main Contact Gap			
Main Contact Wipe			
Latch Wipe			
Latch Clearance			
Contact Travel			
Prop Clearance			
Stop Clearance			

Inspection and Maintenance:

	Insp.	Dirty	Cleaned/ Lubed	See Remarks
Overall Cleanliness				
Insulating Members				
Mech. Connections				
Structural Members				
Cubicle				
Pri. Contact Fingers				
Shutter Mech.				
Relays				
Auxiliary Devices				
Racking Device				
Arc Chutes				
Blow Out Coil				
Puffers				
Liner				
Arc Runners				
Main Contacts				
Cubicle Wiring				
Breaker Wiring				
Heaters				
Panel Lights				
Bearings				
Contact Sequence				
Ground Connection				
Counter Reading				

Remarks: ____________

Inspections and Test by: ____________ Equipment Used: ____________ Sheet No.: ____________

 NFPA 70B

FIGURE H.3 Typical Air Circuit Breaker Test and Inspection Report.

MEDIUM VOLTAGE VACUUM BREAKER

SHEET NO. ________ OF ________

CUSTOMER ____________ DATE ____________ PROJECT NO. ____________
ADDRESS ____________ AIR TEMP. ____________ REL. HUMIDITY ____________
OWNER/USER ____________ DATE LAST INSPECTION ____________
ADDRESS ____________ LAST INSPECTION REPORT NO. ____________
EQUIPMENT LOCATION ____________
OWNER IDENTIFICATION ____________

BREAKER DATA:

Manufacturer ____________ Voltage ________ Type ____________ Amps ________ Age ________
Serial No. ____________ Type Oper Mech ____________ Int. Rating ____________ Other ____________

TEST DATA:

Ins Res @ ______ kV

Results In
Gigaohms ______
Megohms ______

A to G	B to G	C to G
A to B	B to C	C to A
A-L to L	B-L to L	C-L to L

Contact Resistance
Microhms - As Found
Microhms - As Left

A-L to L	B-L to L	C-L to L

HiPot Test @

ADJUSTMENTS:

	MFR'S. REC.	AS FOUND	AS LEFT
Erosion Indicator			
Main Contact Gap			

INSPECTION AND MAINTENANCE:

	INSP	NA	CLEAN LUBE	SEE REMARKS
Overall Cleanliness				
Insulating Members				
Mech. Connections				
Structural Members				
Cubicle				
Pri. Contact Fingers				
Shutter Mech.				
Relays				
Auxiliary Devices				
Racking Device				
Main Contacts				
Cubicle Wiring				
Breaker Wiring				
Heaters				
Panel Lights				
Bearings				

Contact Sequence		
Ground Connection		
Counter Reading		

Remarks ____________

Equipment Used ____________
Submitted By ____________

Courtesy of Northeast Electrical Testing NFPA 70B

FIGURE H.4 Typical Medium-Voltage Vacuum Breaker Form.

OIL CIRCUIT BREAKER TEST

SHEET NO. ________ OF ________

CUSTOMER ____________ DATE ____________ PROJECT NO. ____________
ADDRESS ____________ AIR TEMP. ____________ REL. HUMIDITY ____________
OWNER/USER ____________ DATE LAST INSPECTION ____________
ADDRESS ____________ LAST INSPECTION REPORT ____________
EQUIPMENT LOCATION ____________
OWNER IDENTIFICATION ____________

BREAKER DATA:

Manufacturer ____________ Voltage ________ Type ____________ Amps ________ Age ________
Serial No. ____________ Type Oper Mech ____________ Int. Rating ____________ Other ____________
Bushing Data ____________

TEST DATA:

Ins Res @ ______ kV

Results In
Gigaohms ______
Megohms ______

A to G	B to G	C to G
A to B	B to C	C to A
A-L to L	B-L to L	C-L to L

Contact Resistance

Microhms - As Found			
Microhms - As Left			
Reference PF Test Sheet			
______			

ADJUSTMENTS:

	MFR'S. REC.	AS FOUND	AS LEFT
Stop Clearance			
Contact Travel			
Overtravel			
Contact Wipe			
______			
Trip Roller			
Latch Wipe			
Latch Clearance			
Prop Wipe			
Prop Clearance			
Cut-off Switch			
AA Switch			
______			
______			
______			

INSPECTION AND MAINTENANCE:

	INSP	NA	CLEAN LUBE	SEE REMARKS
Tank Liners				
Insulating Members				
Oil Gauges				
Opening Spring				
Bushings				
Main Contacts				
Secondary Contacts				
Interrupters				
Linkage				
Dashpots				
Shutter Mechanism				
Elevating Mechanism				
Compressor Air Strainer				
Unload Valve				
Check Valve				
Compressor Belt				
Air Leaks				
Compressor Oil				
Gaskets				
Nuts, Bolts, Pins				
Closing Sequence				
Heater				
Oil Level				
Ground Connection				
Counter Reading As-Found				
Counter Reading As-Left				

Remarks ____________

Equipment Used ____________
Submitted By ____________

Courtesy of Northeast Electrical Testing NFPA 70B

FIGURE H.5 Typical Oil Circuit Breaker Test Report.

DISCONNECT SWITCH TEST REPORT

SHEET NO. ________ OF ________

CUSTOMER ____________________ DATE ____________________ PROJECT NO. ____________
ADDRESS ____________________ AIR TEMP. ____________________ REL. HUMIDITY ____________
OWNER/USER ____________________ DATE LAST INSPECTION ____________________
ADDRESS ____________________ LAST INSPECTION REPORT NO. ____________________
EQUIPMENT LOCATION ____________________
OWNER IDENTIFICATION ____________________

Fused ____________ Non-Fused ____________

SWITCH DATA:

Manufacturer ____________ Voltage ____________ Type ____________ Amps ____________ Age ____________

Serial No. ____________ Type Oper Mech ____________ Int. Rating ____________ Other ____________

TEST DATA:

Ins Res @ ______ kV

	A to G	B to G	C to G
Results In	A to B	B to C	C to A
Gigaohms ______			
Megohms ______			
Contact Resistance			
Microhms-As Found			
Microhms-As Left			

FUSE DATA:

	A	B	C
Manufacturer			
Size			
Catalog No.			
I.C. Rating			
Contact Resistance			
Microhms-As Found			
Microhms-As Left			

INSPECTION AND MAINTENANCE:

	INSP	NA	CLEAN LUBE	SEE REMARKS
Overall Cleanliness				
Insulating Members				
Mech. Connections				
Structural Members				
Cubicle				
Pri. Contact Fingers				
Main Contacts				
Arc Chutes				
Arc Runners				
Manual Operation				
Contact Lube				
Mechanical Lube				
Paint				
Door Interlock				
Ground Connection				

Remarks ____________________

Equipment Used ____________________
Test Technician ____________________

Courtesy of Northeast Electrical Testing — NFPA 70B

FIGURE H.6 Typical Disconnect Switch Test Report.

LOW-VOLTAGE CIRCUIT BREAKER 5-YEAR TESTS FORM

Plant ______________________ Date ______________________

Substation ______________ Feeder ______________ Load Reading ______________

Breaker Data

Mfr. ______________ Type ______________ Serial No. ______________

Trip Coil Rating ________ Amperes Characteristic ______________ Mfr's. Time Curve ______________

Trip Devices: Long Time Delay ❑ Short Time Delay ❑ Instantaneous Trip ❑

Time Delay Type: Oil Sucker Dashpot ❑ Air Bellows ❑ Air Orifice ❑ Oil Orifice ❑ Other ❑

Settings:

LT Delay — Amperes ____________ Adjustable Range ____________ Time Adjustable? Yes ❑ No ❑

ST Delay — Amperes ____________ Adjustable Range ____________ Time Adjustable? Yes ❑ No ❑

Instantaneous Trip — Amperes ______________________ Adjustable? Yes ❑ No ❑

Test Data

Date of Test / Inspector's Initials		Left Pole	Center Pole	Right Pole	Time Range from Curve
As Found Test (Trip Time in Seconds) % Pickup ______ Amperes ______					
Time Delay	(As Found — Amperes)				
Minimum Pickup (Nullify Time Delay)	(Adjusted — Amperes)				
Time Delay Tests (Trip Time in Seconds) % Pickup Amperes Long Time ______ ______					
______ ______					
Short Time ______ ______					
______ ______					
Resettable Delay	(Satisfactory)				
(___ % for ___ sec)	(Tripped)				
Instantaneous Trip	(As Found — Amperes)				
	(Adjusted — Amperes)				

Remarks (record unusual conditions, corrections, needed repairs, etc.; use separate form to record annual breaker inspection details):

____

____

 NFPA 70B

FIGURE H.7 Typical Low-Voltage Circuit Breaker 5-Year Tests Form.

ELECTRICAL SWITCHGEAR–ASSOCIATED EQUIPMENT INSPECTION REPORT

Plant ______________________ Date ______________________

Location ______________________ Serial No. ______________________

Mfr. ______________________ Year Installed ______________________

Rating: Volts ______________________ Bus Capacity Amperes ______________________

Type: Switchboard ❑ Indoor Metal Clad ❑ Outdoor Metal Clad ❑

Annual Inspection (Disregard items that do not apply.)

Date					
Inspector's Initials					
Switchboards					
Clean					
Check Wiring					
Inspect Panel Insulation					
Exposed Bus and Connections					
Clean and Check Porcelain					
Check Insulators for Cracks or Chips					
Check and Tighten Connections					
Inspect Potheads for Leaks					
Check for Environmental Hazards					
Test Insulation (Megohms)					
Metal Clad Enclosures					
Clean					
Check for Openings That Permit Dirt, Moisture and Rodent Entrance — Repair					
Check Hardware for Rust or Corrosion					
Paint Condition					
Check Heaters and Ventilators					
Metal Clad Bus and Connections					
Clean Insulators and Supports					
Check and Tighten Connections					
Check for Corona Tracking					
Inspect Potheads for Leaks					
Test Insulation (Megohms)					

Date					
Inspector's Initials					
Disconnect Switches					
Check Contact Surfaces					
Check Insulation Condition					
Lubricate per Mfr's. Instructions					
Test Operate					
Fuses and Holders					
Check Contact Surfaces					
Lubricate per Mfr's. Instructions					
Meters and Instruments					
Check Operation					
Test Meters per Eng. Std.					
Test Relays per Mfr's. Instructions					
Interlocks and Safety					
Check for Proper Operations					
Check Lightning Arresters					
Check Ground Detectors					
Check Equipment Grounds					
Station Battery					
Periodic Routine					
Maintenance is performed					

Remarks (record action taken when indicated by inspection or tests):

____

____

Recommendations:

____

____

____

 NFPA 70B

FIGURE H.8 Typical Electrical Switchgear–Associated Equipment Inspection Record.

CURRENT OR POTENTIAL TRANSFORMER RATIO TEST REPORT

SHEET NO. ________ OF ________

CUSTOMER ______________________ DATE ____________ PROJECT NO. ____________

ADDRESS ______________________ AIR TEMP. ____________ °C REL. HUMIDITY (%) ____________

OWNER/USER ______________________ DATE LAST INSPECTION ______________________

ADDRESS ______________________ LAST INSPECTION REPORT NO. ______________________

EQUIPMENT LOCATION ______________________

CIRCUIT IDENTIFICATION ______________________

LOCATION OF C.T. OR P.T. ______________________

C.T. OR P.T. IDENTIFICATION	C.T. OR P.T. SECONDARY TAPS	NAMEPLATE RATIO	APPLIED VOLTAGE OR CURRENT	MEASURED VOLTAGE OR CURRENT	PERCENT (%) ACCURACY	POLARITY PRIMARY	POLARITY SECONDARY
POLE #1 (A)	X1-X2						
BURDEN TEST							AMPS / VOLTS
SATURATION TEST							VOLTS / MA
MEGGER TEST							MEGOHMS
POLE #2 (B)	X1-X2						
BURDEN TEST							AMPS / VOLTS
SATURATION TEST							VOLTS / MA
MEGGER TEST							MEGOHMS
POLE #3 (C)	X1-X2						
BURDEN TEST							AMPS / VOLTS
SATURATION TEST							VOLTS / MA
MEGGER TEST							MEGOHMS

REMARKS ______________________

SUBMITTED BY ______________________

Courtesy of Northeast Electrical Testing NFPA 70B

FIGURE H.9 Typical Current or Potential Transformer Ratio Test Report.

OVERLOAD RELAY TEST REPORT

SHEET NO. ________ OF ________
TEST REPORT NO. ____________

CUSTOMER ____________ DATE ____________ PROJECT NO. ____________
ADDRESS ____________ AIR TEMP. ________ °C REL. HUMIDITY ____________
OWNER/USER ____________ DATE LAST INSPECTION ____________
ADDRESS ____________ LAST INSPECTION REPORT ____________
EQUIPMENT LOCATION ____________
OWNER IDENTIFICATION ____________

MOTOR PROTECTED ____________
MOTOR FLA ____________ MOTOR VOLTAGE ____________

OVERLOAD INFORMATION

OVERLOAD MANUFACTURER ____________ CATALOG NUMBER ____________
OVERLOAD RELAY HEATER COIL ____________ HEATER POSITION ____________
MANUFACTURERS CURVE NO. ____________ AMBIENT TEMP. ____________
FULL LOAD CURRENT AMPERES ____________ MIN. ____________ MAX. ____________

TEST RESULTS

PHASE	HEATER CURRENT	TEST CURRENT		TEST TIME
		PERCENTAGE	AMPS	SECONDS
PHASE 1				
PHASE 2				
PHASE 3				

STARTER INFORMATION

STARTER MANUFACTURER ____________
STARTER SIZE ____________ STARTER CATALOG NO. ____________
STARTER ____________ OTHER INFORMATION ____________

CONDUCTOR ____________ CONDUCTOR INSULATION ____________

DATE ____________

INSULATION RESISTANCE RESULTS:

Aø - GND ____________ Aø - Bø ____________
Bø - GND ____________ Bø - Cø ____________
Cø - GND ____________ Cø - Aø ____________

DATE ____________

MEGGER MOTOR ø-GND: (1/2 MIN) ____________ (1 MIN) ____________

DATE ____________

MEGGER MOTOR ø-GND W/CONDUCTOR INCLUDED: ____________ (1 MIN) ____________

REMARKS: ____________

EQUIPMENT USED ____________ SERIAL NUMBER ____________
QUALITY CONTROL REP. ____________ TITLE ____________
SUBMITTED BY ____________ TEST CREW ____________

Courtesy of Northeast Electrical Testing NFPA 70B

FIGURE H.10 **Typical Overload Relay Test Report.**

GROUND FAULT SYSTEM TEST

SHEET NO. ________ OF ________

CUSTOMER ________ DATE ________ PROJECT NO. ________
ADDRESS ________ AIR TEMP. ________ REL. HUMIDITY ________
OWNER/USER ________ DATE LAST INSPECTION ________
ADDRESS ________ LAST INSPECTION REPORT NO. ________
EQUIPMENT LOCATION ________
CIRCUIT IDENTIFICATION ________

FIELD DATA

MAIN OVERCURRENT DEVICE:

☐ CIRCUIT ☐ FUSED SWITCH

MANUFACTURER ________
TYPE ________
MODEL/CAT. # ________
CURRENT RATING ________
SYSTEM VOLTAGE ________
VOLTAGE RATING ________

GROUND FAULT SYSTEM:

☐ NEUT-GND STRAP ☐ ZERO SEQUENCE

MANUFACTURER ________
MODEL ________
CAT. NO. ________
PICK-UP RANGE ________
TIME RANGE ________
SENSOR/ C.T. ________

INSPECTION

CORRECT	INCORRECT	INSPECTION POINT	SIZE - REMARKS
		NEUT.-GRD LOCATION	
		CONTROL POWER	
		MONITOR OR TEST PANEL OPERATION	
		OTHER ________	

ELECTRICAL TESTS

1. BREAKER/SWITCH REACTION TIME (RT) ________ ☐ SEC. ☐ CYC.
2. PICK UP CURRENT ________ AMPS
3. PICK UP CURRENT MINUS 10% (________) A. ☐ TRIP ☐ NO TRIP
4. SHUNT TRIP COIL PICK-UP VOLTAGE ________ VOLTS
5. SYSTEM NEUTRAL INSULATION RESISTANCE TO GND ________ MEGOHMS
6. TIME-CURRENT CALIBRATION TESTS:

PRIMARY CURRENT AMPERE-TURNS	% PICKUP	TOTAL TIME	RT	RELAY TIME	MFG. TOLERANCE

REMARKS: ________

SUBMITTED BY: ________

Courtesy of Northeast Electrical Testing NFPA 70B

FIGURE H.11 Typical Ground-Fault System Test Report.

INSTRUMENT/METER CALIBRATION AND TEST REPORT

SHEET NO. ________ OF ________

CUSTOMER ________ DATE ________ PROJECT NO. ________
ADDRESS ________ AIR TEMP. ________ °C
OWNER/USER ________ DATE LAST INSPECTION ________
ADDRESS ________ LAST INSPECTION REPORT NO. ________
EQUIPMENT LOCATION ________
CIRCUIT IDENTIFICATION ________

LOCATION/FUNCTION OF INSTRUMENT/METER ________
TYPE ________ MANUFACTURER ________ MODEL ________
FULL SCALE ________ ACTUAL INPUT ________
P.T. RATIO ________ C.T. RATIO ________ CAL. WATTS ________

FULL SCALE CARDINAL POINTS							
BASIC RANGE CALCULATED VALUE							
STANDARD "AS FOUND"							
STANDARD "AS LEFT"							
"AS LEFT" ACCURACY (%)							

REMARKS ________

LOCATION/FUNCTION OF INSTRUMENT/METER ________
TYPE ________ MANUFACTURER ________ MODEL ________
FULL SCALE ________ ACTUAL INPUT ________
P.T. RATIO ________ C.T. RATIO ________ CAL. WATTS ________

FULL SCALE CARDINAL POINTS							
BASIC RANGE CALCULATED VALUE							
STANDARD "AS FOUND"							
STANDARD "AS LEFT"							
"AS LEFT" ACCURACY (%)							

REMARKS ________

SUBMITTED BY: ________ EQPT. USED: ________

Courtesy of Northeast Electrical Testing NFPA 70B

FIGURE H.12 Typical Instrument/Meter Calibration and Test Report.

WATT-HOUR METER TEST SHEET

SHEET NO. ________ OF ________

CUSTOMER ________ DATE ________ PROJECT NO. ________

ADDRESS ________ AIR TEMP. ________ °C

OWNER/USER ________ DATE LAST INSPECTION ________

ADDRESS ________ LAST INSPECTION REPORT NO. ________

EQUIPMENT LOCATION ________

CIRCUIT IDENTIFICATION ________

TEST LOCATION ________ CIRCUIT METERED ________

METER MANUFACTURER ________ TYPE ________ SER. NO. ________

VOLTS ________ AMPS ________ ________ PHASE ________ WIRE INTERVAL ________

C.T. RATIO ________ P.T. RATIO ________ TEST K ________ PRI. TEST K ________

	AS FOUND	AS LEFT		AS FOUND	AS LEFT
KWH REGISTER READING			POTENTIAL IND LAMPS		
DEMAND REGISTER READING			CHECK AND VERIFY REGISTER RATIO		
DISC R.P.M.			SYNCHRONOUS MOTOR		
WORM WHEEL MESH			CHECK KW PTR AGAINST KWH PTR		
MAGNET CLEANLINESS			CREEP CHECK		
MAGNET TIGHTNESS			TIME INTERVAL		

	ACCURACY CHECK (% REG.)		COIL BALANCE CHECK					
			COIL NO. 1		COIL NO. 2		COIL NO. 3	
TEST FUNCTION	AS FOUND	AS LEFT	AS FOUND	AS LEFT	AS FOUND	AS LEFT	AS FOUND	AS LEFT
LIGHT LOAD								
FULL LOAD								
POWER FACTOR								

REMARKS: ________

CUSTOMER REPRESENTATIVE ________ TITLE ________

TEST EQUIPMENT USED ________ SERIAL # ________

SUBMITTED BY ________ TEST ________

Courtesy of Northeast Electrical Testing NFPA 70B

FIGURE H.13 Typical Watt-Hour Meter Test Sheet.

PANELBOARD/CIRCUIT BREAKER TEST REPORT

SHEET NO. ________ OF ________

CUSTOMER ____________________ DATE ____________________ PROJECT NO. ____________
ADDRESS ____________________ AIR TEMP. ____________________ REL. HUMIDITY ____________
OWNER/USER ____________________ DATE LAST INSPECTION ____________________
ADDRESS ____________________ LAST INSPECTION REPORT ____________________
EQUIPMENT LOCATION ____________________
OWNER IDENTIFICATION ____________________

PANEL BUS INSULATION RESISTANCE IN MEGOHMS

A-G ________ B-G ________ C-G ________ A-B ________ B-C ________ A-C ________

PANEL BOARD RATINGS: AMPS: ____________ VOLTAGE: ____________

TEST VOLTAGE: ____________ MODEL NO: ____________ CATALOG ____________

MFG. ____________ CURVE NO. ____________ CURVE RANGE: ____________

MFG. ____________ CURVE NO. ____________ CURVE RANGE: ____________

MFG. ____________ CURVE NO. ____________ CURVE RANGE: ____________

MFG. ____________ CURVE NO. ____________ CURVE RANGE: ____________

CIRCUIT #	CKT. BKR. SIZE	TEST AMPS	TRIP TIME	INST. TRIP	CONTACT RESIS.	CIRCUIT #	CKT. BKR. SIZE	TEST AMPS	TRIP TIME	INST. TRIP	CONTACT RESIS.

REMARKS: ____________________

CUSTOMER REPRESENTATIVE ____________________ TITLE ____________
TEST EQUIPMENT ____________________ SERIAL # ____________
SUBMITTED BY ____________________

Courtesy of Northeast Electrical Testing

NFPA 70B

FIGURE H.14 Typical Panelboard/Circuit Breaker Test Report.

TRANSFORMER TEST AND INSPECTION REPORT

SHEET NO. ________ OF ________

CUSTOMER ______________________ DATE ______________________ PROJECT NO. ______________

ADDRESS ______________________ AIR TEMP. ______________ °C REL. HUMIDITY (%) ______________

OWNER/USER ______________________ DATE LAST INSPECTION ______________________

ADDRESS ______________________ LAST INSPECTION REPORT ______________________

EQUIPMENT LOCATION ______________________

OWNER IDENTIFICATION ______________________

NAMEPLATE INFORMATION:

MANUFACTURER ______________ KVA ______________ PHASE ______________ CYCLE ______________

SERIAL NO. ______________ TYPE ______________ CLASS ______________

PRI. VOLTAGE ______________ △ ☐ OR Y ☐ RATED CURRENT ______________ AMPERES

SEC. VOLTAGE ______________ △ ☐ OR Y ☐ RATED CURRENT ______________ AMPERES

COOLANT ☐ OIL ☐ ASKAREL ☐ AIR ☐ NITROGEN ☐ OTHER ______________

COOLANT CAPACITY ______________ TEMP. RISE (°C) ______________ IMPEDANCE (%) ______________

NO LOAD TAP CHANGER VOLTAGES ______________

GAUGES AND COUNTERS

TEMP. ______________ TEMP. RANGE ______________ RESET GAUGE ______________

PRESSURE ______________ OIL LEVEL ______________ TAP SETTING ______________

VISUAL INSPECTION

BUSHING ______________ CONNECTIONS ______________ PAINT ______________ OTHER ______________

LOAD TAP CHANGER ______________ LEAKS ______________

FANS & CONTROLS ______________ GAS REGULATOR ______________ GROUNDS ______________

WINDING INSULATION RESISTANCE TEST (MEGOHMS)

PRIMARY TO GROUND, SEC. GUARDED ______________ KVDC

SECONDARY TO GROUND, PRI. GUARDED ______________ KVDC

PRIMARY TO SECONDARY, GROUND GUARDED ______________ KVDC

EQUIPMENT USED ______________

30 SEC.	1 MIN.	10 MIN.	D.A.	P.I.

TURNS RATIO TEST ☐ ☐

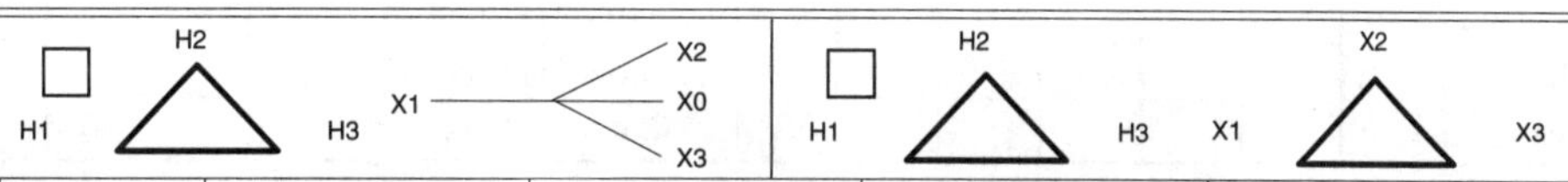

NAMEPLATE PRIMARY VOLTS	TAP POSITION	CONNECTION H___H___X___X___	CONNECTION H___H___X___X___	CONNECTION H___H___X___X___	CALCULATED RATIO
	A1				
	B2				
	C3				
	D4				
	E5				

REMARKS: ______________

EQUIPMENT USED ______________ SUBMITTED BY ______________

Courtesy of Northeast Electrical Testing NFPA 70B

FIGURE H.15 Typical Transformer Test and Inspection Report.

TRANSFORMER (DRY TYPE) INSPECTION RECORD

Plant ______________ Date ______________
Location ______________ Serial No. ______________
Year Purchased ______________ Year Installed ______________ Mfr. ______________
kVA ______________ Voltage ______________ Impedance ______________
Phase ______________ Taps ______________
Cooling System: Room Vent Fan ❑ Trans. Fan ❑ Gravity ❑

Annual Inspection

Date						Date					
Inspector's Initials						Inspector's Initials					
Electrical Load						Bushings					
						Cracks or Chips					
						Cleanliness					
Secondary Voltage						Equipment Ground					
No Load Volts						Check Connections					
Full Load Volts						Measured V					
Dust on Windings						Resistance					
Minor Collection						Temperature Alarms and Indicators					
Major Collection						Operation					
Cleaned						Accuracy					
Connections						Case Exterior					
Checked						Covers Intact					
Tightened						Paint Condition					
Cooling Systems						Lighting Arresters					
Fan Operation						Check Connections					
Filter Cleanliness						Check Bushings					
System Adequate											

Complete Internal Inspection

Report of Conditions Found:

Cooling System ______________

Coil Insulation ______________

Other ______________

Description of Work Performed:

____

____

Other Repairs Recommended: ______________

____

____

Shop or Contractor: ______________ Cost: ______________

 NFPA 70B

FIGURE H.16 Typical Transformer (Dry Type) Inspection Record.

TRANSFORMER (LIQUID FILLED) INSPECTION RECORD

Plant ______________________ Date ______________________

Location ______________________ Serial No. ______________________

Year Purchased ____________ Year Installed ____________ Mfr. ____________

kVA ____________ Voltage ____________ Taps ____________

Check type: Free Breathing ❑ Conservator ❑ Sealed ❑ Fan Cooled ❑

Phase ____________ Weight ____________ Impedance ____________

Insulating Fluid: Type ______________________ Gallons ______________________

Annual Inspection

Date						Date					
Inspector's Initials						Inspector's Initials					
Tank — Liquid Level						Exposed Bushings					
Normal						Cracks or Chips					
Below						Cleanliness					
Added Fluid											
Entrance Compartment Liquid Level						Equipment Ground Connection					
Normal						Good					
Below						Questionable					
Added Fluid						Tested					
Electrical Load						Temperature Indicator					
Peak Amperes						Highest Reading					
						Reset Pointer					
Secondary Voltage						Pressure–Vacuum Indicator					
Full Load						Pressure					
No Load						Vacuum					
Gaskets and Case Exterior						Ventilators, Dryers, Gauges, Filters, and Other Auxiliaries					
Liquid Leaks						Operation OK					
Paint Condition						Maint. Req'd.					

Remarks (record action when inspection data or tests are out of limits, etc.):

____

____

Reports of Conditions Found: ____

____

Description of Work Performed: ____

____

Other Repairs Recommended: ____

____

Shop or Contractor: ______________________ Cost: ____________

 NFPA 70B

FIGURE H.17 Transformer (Liquid Filled) Inspection Record.

TRANSFORMER OIL SAMPLE REPORT

TOTAL NO. OF SAMPLES ____
CUSTOMER ____ PROJECT NO. ____
LOCATION ____ DATE ____

LOCATION ____ ☐ OIL ____ SAMPLE NO. ____
IDENTIFICATION ____ ☐ ASKAREL ____
MFG. ____ ☐ NO GAUGE ____ PAINT ____ ☐ GOOD
SERIAL NO. ____ KVA ____ ☐ PRESSURE ____ ☐ POOR
CLASS ____ TYPE ____ ☐ VACUUM ____
INSUL. CLASS ____ PHASE ____ ☐ INDOOR ____ GASKETS ____ ☐ OK
VOLTAGE ____ ☐ OUTDOOR ____ ☐ LEAK
INSTR. BOOK ____ ☐ TEMP. GA. ____
AVG. DIELECTRIC ____ KV ____ BUSHINGS ____ ☐ OK
ACIDITY NO. ____ KOH ____ ☐ LEAK
ASTM COLOR NO. ____ LIQUID CAPACITY ____
PARTICLES ☐ YES ☐ NO TEMP ____ °C ____ OIL LEVEL ____ ☐ OK
WEATHER ____ ☐ LOW
RECOMMENDATIONS ____

LOCATION ____ ☐ OIL ____ SAMPLE NO. ____
IDENTIFICATION ____ ☐ ASKAREL ____
MFG. ____ ☐ NO GAUGE ____ PAINT ____ ☐ GOOD
SERIAL NO. ____ KVA ____ ☐ PRESSURE ____ ☐ POOR
CLASS ____ TYPE ____ ☐ VACUUM ____
INSUL. CLASS ____ PHASE ____ ☐ INDOOR ____ GASKETS ____ ☐ OK
VOLTAGE ____ ☐ OUTDOOR ____ ☐ LEAK
INSTR. BOOK ____ ☐ TEMP. GA. ____
AVG. DIELECTRIC ____ KV ____ BUSHINGS ____ ☐ OK
ACIDITY NO. ____ KOH ____ ☐ LEAK
ASTM COLOR NO. ____ LIQUID CAPACITY ____
PARTICLES ☐ YES ☐ NO TEMP ____ °C ____ OIL LEVEL ____ ☐ OK
WEATHER ____ ☐ LOW
RECOMMENDATIONS ____

LOCATION ____ ☐ OIL ____ SAMPLE NO. ____
IDENTIFICATION ____ ☐ ASKAREL ____
MFG. ____ ☐ NO GAUGE ____ PAINT ____ ☐ GOOD
SERIAL NO. ____ KVA ____ ☐ PRESSURE ____ ☐ POOR
CLASS ____ TYPE ____ ☐ VACUUM ____
INSUL. CLASS ____ PHASE ____ ☐ INDOOR ____ GASKETS ____ ☐ OK
VOLTAGE ____ ☐ OUTDOOR ____ ☐ LEAK
INSTR. BOOK ____ ☐ TEMP. GA. ____
AVG. DIELECTRIC ____ KV ____ BUSHINGS ____ ☐ OK
ACIDITY NO. ____ KOH ____ ☐ LEAK
ASTM COLOR NO. ____ LIQUID CAPACITY ____
PARTICLES ☐ YES ☐ NO TEMP ____ °C ____ OIL LEVEL ____ ☐ OK
WEATHER ____ ☐ LOW
RECOMMENDATIONS ____

EQUIPMENT USED ____ SUBMITTED BY ____

Courtesy of Northeast Electrical Testing NFPA 70B

FIGURE H.18 Typical Transformer Oil Sample Report.

TRANSFORMER OIL TRENDING REPORT

CUSTOMER ______________________ PROJECT NO. __________

LOCATION ______________________ DATE __________

LOCATION ______________ ☐ OIL __________ SAMPLE NO. __________

IDENTIFICATION ______________ ☐ ASKAREL __________

MFG. ______________ ☐ GAUGE __________ PAINT __________ ☐ GOOD

SERIAL NO. __________ KVA __________ ☐ PRESSURE __________ ☐ POOR

CLASS __________ TYPE __________ ☐ VACUUM __________

INSUL. CLASS __________ PHASE __________ ☐ INDOOR __________ GASKETS __________ ☐ OK

VOLTAGE ______________ ☐ OUTDOOR __________ ☐ LEAK

INSTR. BOOK ______________ ☐ TEMP. GA. __________

DIELECTRIC FLUID ANALYSIS

YR - JOB#	DIELE. (KV)	ACIDITY (mgKOH/g)	IFT (dynes/cm2)	COLOR	VISUAL	SPECIFIC GRAVITY	WATER (PPM)	POWER FACTOR (%)	PCB (PPM)

ACCEPTABLE DIELECTRIC TEST VALUES:

DIELECTRIC (ASTM D877) 30KV MIN. NEW OIL / 26KV MIN. USED OIL / 30KV MIN. NEW SILICONE/ 25kV MIN. USED SILICONE

ACID (ASTM D974) 0.03 mgKOH/g MAX. NEW OIL / 0.20mgKOH/g MAX. USED OIL / 0.1mgKOH/g MAX. NEW SILICONE/ 0.2mgKOH USED SILICONE

IFT (ASTM D971) 35 dynes/cm2 MIN. NEW OIL / 24 dynes/cm2 MIN. USED OIL / 31 dynes/cm2 MIN. SILICONE

COLOR (ASTM D1500) 1 MAX. NEW OIL / 4 MAX. USED OIL / CLEAR FOR SILICONE

WATER (ASTM D1533B) 25PPM MAX. NEW OIL / 35PPM MAX. USED OIL / 50PPM MAX. NEW SILICONE/ 100PPM MAX. USED SILICONE

DISSOLVED GAS ANALYSIS

YEAR	HYDROGEN (H2) (<100PPM)	OXYGEN (O2)	NITROGEN (N2)	METHANE (CH4) (<120PPM)	CARBON MONOXIDE (CO) (<350PPM)	CARBON DIOXIDE (CO2) (<2500PPM)	ETHYLENE (C2H4) (<50PPM)	ETHANE (C2H6) (<65PPM)	ACETYLENE (C2H2) (<1PPM)	TOTAL GAS CONTENT (%)	TOTAL COMBUST. GAS

REMARKS: ______________________

SUBMITTED BY ______________________

Courtesy of Northeast Electrical Testing NFPA 70B

FIGURE H.19 Typical Transformer Oil Trending Report.

TRANSFORMER INSULATION RESISTANCE RECORD

Plant ____________________ Date ____________

Scope: Power transformers of 150 kVA and greater capacity with primary voltage of 2300 volts or higher. Direct reading — recorded and plotted.

Transformer Serial No. ____________ Phase ____________

Location ____________ Instrument Used ____________

Equipment Included in Test ____________

I I*	Date	Primary to Ground	Secondary to Ground	Primary to Secondary	Internal Temp.	Ambient Temp.

*Inspector's Initials

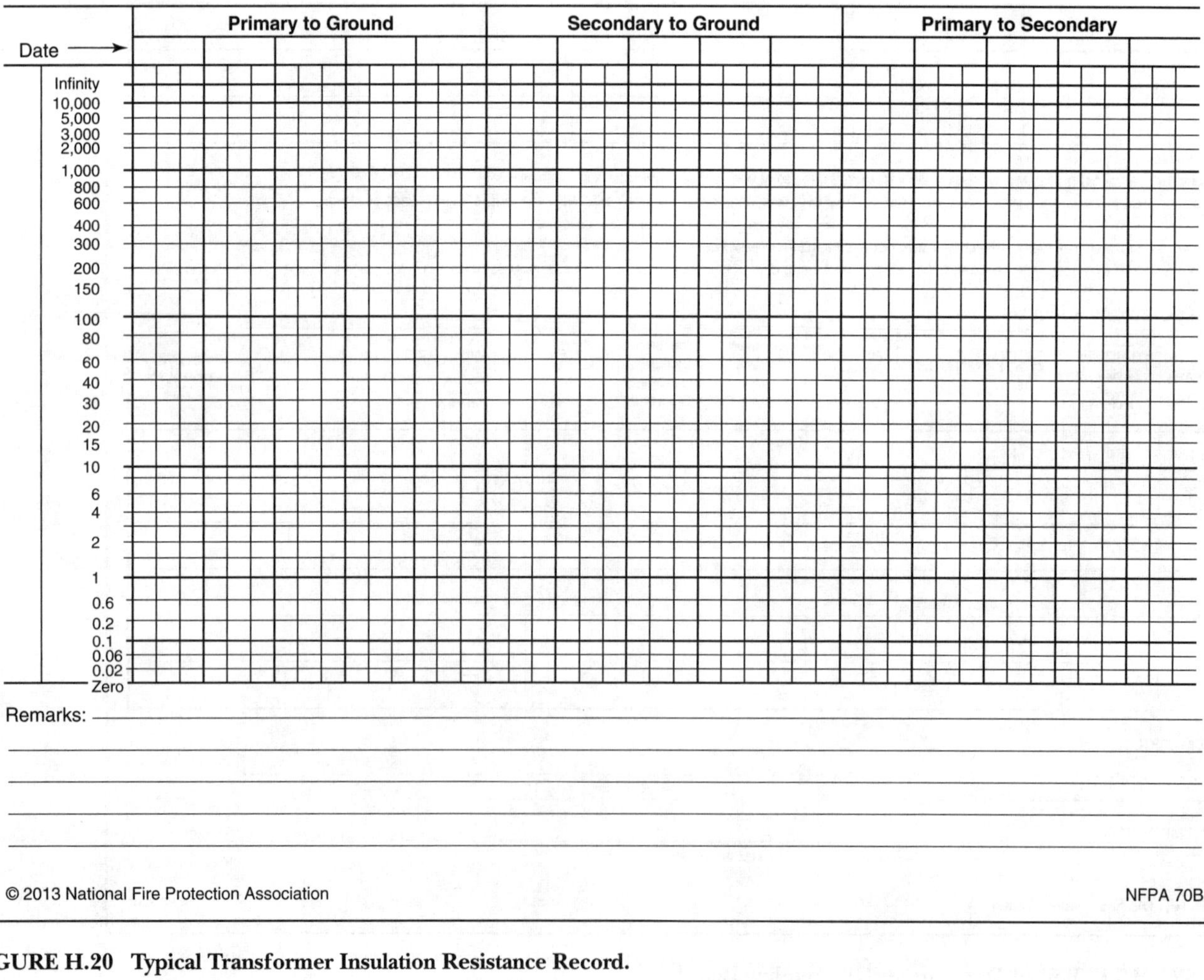

Remarks: ____________

 NFPA 70B

FIGURE H.20 Typical Transformer Insulation Resistance Record.

BATTERY RECORD

No. Cells ____________ Type ____________ Services ____________ Bldg. ____________

Note: Correct specific gravity readings for temperature.

Weekly Pilot Cell Readings: Cell No. ______

Date	Pilot Cell Specific Gravity	Pilot Cell Temp.	Bus Volts	Water Clean	Check Charger

Remarks: ____

Quarterly Cell Readings: Date ______

Cell	Specific Gravity	Volts	Cell	Specific Gravity	Volts
1			31		
2			32		
3			33		
4			34		
5			35		
6			36		
7			37		
8			38		
9			39		
10			40		
11			41		
12			42		
13			43		
14			44		
15			45		
16			46		
17			47		
18			48		
19			49		
20			50		
21			51		
22			52		
23			53		
24			54		
25			55		
26			56		
27			57		
28			58		
29			59		
30			60		

Bus Volts ____________ Current ____________

Battery Room Condition ____________

Cell Temps: 1 ________ 15 ________

30 ________ 45 ________ 60 ________

 NFPA 70B

FIGURE H.21 Typical Battery Record.

ENGINE GENERATOR INSPECTION

SHEET NO. ________ OF ________

CUSTOMER ______________________ DATE ______________ PROJECT NO. ______________
ADDRESS ______________________ AIR TEMP. ______________ REL. HUMIDITY ______________
OWNER/USER ______________________ DATE LAST INSPECTION ______________
ADDRESS ______________________ LAST INSPECTION REPORT ______________
EQUIPMENT LOCATION ______________________
CIRCUIT IDENTIFICATION ______________________

ENGINE TYPE: ☐ GASOLINE ☐ DIESEL ☐ GAS TURBINE

MAKE ______ MODEL ______ SERIAL NO. ______ KS # ______
KVA ______ KW ______ VOLTAGE ______ F.L.A. ______
RPM ______ HZ ______ HP ______ TECH. BULL. #______

1. ☐ Change oil and lube oil filters.
2. ☐ Remove unused oil from premises.
3. ☐ Change fuel oil elements.
4. ☐ Service crankcase breather.
5. ☐ Inspect air cleaner element, clean if required. If replacement is required, element(s) will be billed separately. Price of element(s) not included in contract price.
6. ☐ Check coolant level and maintain safe degree of protection. Engine mounted radiators only. (Remote radiators, cooling towers & heat exchangers serviced at user's request on a time and material basis.)
7. ☐ Check manifolds, brackets, mountings and flex connections.
8. ☐ Inspect fan belts, adjust if required.
9. ☐ Check pulley hub, bearings, lubricate if required.
10. ☐ Check operation of auxiliary water pump or fan motor.
11. ☐ Check operation of automatic louvers.
12. ☐ Repair minor fuel, coolant and lube oil leaks.
13. ☐ Check operation of jacket water heater(s).
14. ☐ Inspect generator, perform any routine maintenance as required.
    ☐ Megger
15. ☐ Inspect governor/actuator linkage.
16. ☐ Check battery electrolyte level and maintain to include:
    ☐ Temperature ☐ Specific Gravity ☐ Voltage
17. ☐ Check operation of charger and/or alternator.
18. ☐ Inspect fuel supply system for leaks or low level, inform owner of any discrepancies.
19. ☐ Drain condensation from day tank and check for any contamination. ONLY if day tank is equipped with a drain valve.
20. ☐ Check operation of transfer pump.
21. ☐ Check for correct generator output voltage & frequency, adjust if required.
22. ☐ Simulate & check operation of each safety shutdown and alarm device, relay type control panels only.
23. ☐ Check operation of generator control instrumentation; volts, amps, etc.
24. ☐ Test fault lamps & replace bulbs as required, panels with lamp test only.
25. ☐ Tank crankcase oil sample, owner to be notified of any discrepancies.
26. ☐ Submit report to owner
27. ☐ Auto start test.

REMARKS ____

SUBMITTED BY ______________________ EQUIPMENT USED ______________________

Courtesy of Northeast Electrical Testing NFPA 70B

FIGURE H.22 Typical Engine Generator Set Inspection Checklist.

AUTOMATIC TRANSFER SWITCH

SHEET NO. ________ OF ________

CUSTOMER ____________________ DATE ____________________ PROJECT NO. ____________________

ADDRESS ____________________ AIR TEMP. ____________________ REL. HUMIDITY ____________________

OWNER/USER ____________________ DATE LAST INSPECTION ____________________

ADDRESS ____________________ LAST INSPECTION REPORT NO. ____________________

EQUIPMENT LOCATION ____________________

OWNER IDENTIFICATION ____________________

Mfg. ____________________ Type: ____________________ Bul #: ____________________

Cat. # ____________________ Serial # ____________________ Voltage: ____________________

Amps: ____________________ Phase: ____________________ Op. Coil: ____________________

Inst. Bk: ____________________ Parts Bk. ____________________ Wire Diag: ____________________

Time Range "Transfer to Emergency" From ____________________ To ____________________

Time Range "Retransfer to Normal" From ____________________ To ____________________

TEST OPERATIONS

Transfer Time to Emergency As Found ____________________ As Left ____________________

Retransfer Time to Normal As Found ____________________ As Left ____________________

	NORMAL			EMERGENCY		
Contact Resistance in Microhms:	A ______	B ______	C ______	A ______	B ______	C ______
Voltage Drop in Millivolts:	A ______	B ______	C ______	A ______	B ______	C ______
Voltage Readings:	A-N ______	B-N ______	C-N ______	A-N ______	B-N ______	C-N ______
	A-B ______	B-C ______	C-A ______	A-B ______	B-C ______	C-A ______
Amperage Readings:	A ______	B ______	C ______	A ______	B ______	C ______

Undervoltage Relay:

	1V	2V	3V
Pickup	______	______	______
Dropout	______	______	______

________ Relay: Pickup ________ Dropout ________

________ Relay: Voltage Pickup ________ Dropout ________

Frequency Pickup ________ Dropout ________

Arc Chutes: ________

Contacts: ________

Megger: ________

Cleaned: ________

Lubrication: ________

Circuit Properly Tagged: ________

Bolted Connections: ________

Mechanical Operation: ________

Unusual Conditions: ________

Remarks: ____________________

Test Crew: ____________________

Courtesy of Northeast Electrical Testing NFPA 70B

FIGURE H.23 Typical Automatic Transfer Switch.

UNINTERRUPTIBLE POWER SUPPLY (UPS) SYSTEM INSPECTION CHECKLIST

For use of this form see TM 5-694: the proponent agency is COE.

SECTION A – CUSTOMER DATA

1. PLANT/BUILDING	2. LOCATION	3. JOB NUMBER
4. EQUIPMENT	5. CIRCUIT DESIGNATION	6. DATE *(YYYYMMDD)*
7. TEST EQUIPMENT		8. TESTED BY

SECTION B – VISUAL AND ELECTRICAL/MECHANICAL INSPECTION

9. CHECK POINT	COND*	NOTES	CHECK POINT	COND*	NOTES
COMPONENT INSPECTION/TESTING			ENERGIZE AND TEST SYSTEM		
INSTALLATION INSPECTION/TESTING			UTILITY TRIP TEST		
WIRING VISUAL VERIFICATION			LOADED TRANSFER TEST (NORMAL, EMERGENCY & RETURN)		
GENERATOR CONTROL FUNCTIONS			TIGHTNESS OF BOLTED CONNECTIONS		
LOADING UPS TEST			BATTERY DISCHARGE TEST		
DISCONNECT RECTIFIERS & INVERTERS SEPARATELY. DOES SYSTEM OPERATE CORRECTLY?			TEST ALL UPS DIAGNOSTIC FAULT INDICATORS		

SECTION C – ELECTRICAL TESTS**

10.	A-N	B-N	C-N	A-B	B-C	C-A	A	B	C	N	G
UPS INPUT											
	A-N	B-N	C-N	A-B	B-C	C-A	A	B	C	N	G
UPS OUTPUT											
	A-N	B-N	C-N	A-B	B-C	C-A	A	B	C	N	G
UPS SWITCHBOARD HARMONIC (THD)											

11. NOTES

* CONDITION: A = ACCEPTABLE; R = NEEDS REPAIR, REPLACEMENT OR ADJUSTMENT; C = CORRECTED; NA = NOT APPLICABLE
** NOTE VALUE AND PHASING

NFPA 70B

FIGURE H.24 Typical Uninterruptible Power Supply System Inspection Checklist.

BACK-UP POWER SYSTEM INSPECTION CHECKLIST

For use of this form see TM 5-694: the proponent agency is COE.

SECTION A – CUSTOMER DATA

1. PLANT/BUILDING	2. LOCATION	3. JOB NUMBER
4. EQUIPMENT	5. CIRCUIT DESIGNATION	6. DATE *(YYYYMMDD)*
7. TEST EQUIPMENT AND CALIBRATION DATE		8. TESTED BY

SECTION B – EQUIPMENT DATA

9. MANUFACTURER	10. STYLES/S.O.	11. VOLTAGE RATING	12. CURRENT RATING
13. EQUIPMENT CLASSIFICATION	14. FREQUENCY	15. WET BULB TEMPERATURE	16. DRY BULB TEMPERATURE

SECTION C – VISUAL AND ELECTRICAL/MECHANICAL INSPECTION

17. CHECK POINT	COND*	NOTES	CHECK POINT	COND*	NOTES
COMPONENT INSPECTION/TESTING			WIRING VISUAL VERIFICATION		
ENERGIZE AND TEST SYSTEM			UTILITY TRIP/GENERATOR BUILDING LOAD TEST		
INSTALLATION INSPECTION/TESTING			TIGHTNESS OF BOLTED CONNECTIONS		
GENERATOR CONTROLS AND FUNCTIONS			CHECK FOR PROPER SIZE BREAKER		
WIRING CONTINUITY TESTING			REFERENCE DRAWINGS		
WORKING CLEARANCE			PROPER PHASING CONNECTIONS AND COLOR CODE		
SWITCHGEAR CONTROL FUNCTIONS					
PERFORM AUTOMATIC TRANSFER SYSTEM (ATS) FUNCTIONS UNDER THE ADJACENT CONTROLLER	A. OPERATE NORMAL POWER				
	B. ALL GENERATORS OPERATE				
	C. GENERATORS 1 AND 2 OPERATE				
	D. GENERATORS 2 AND 3 OPERATE				
	E. GENERATORS 1 AND 3 OPERATE				
	F. RETURN TO NORMAL POWER AFTER EACH OF THE ABOVE TESTS				
	G. PARALLEL WITH UTILITY UPON RETURN TO NORMAL POWER (ITEMS B THROUGH E)				

SECTION D – ELECTRICAL TESTS

18. MEASUREMENT DESCRIPTION	VOLTAGE AND CURRENT MEASUREMENTS										
	VOLTAGE**						CURRENT**				
	A-N	B-N	C-N	A-B	B-C	C-A	A	B	C	N	G
	A-N	B-N	C-N	A-B	B-C	C-A	A	B	C	N	G

19. NOTES

1. CHECK FOR PROPER GROUNDING CONNECTIONS PRIOR TO ENERGIZING.

* CONDITION: A = ACCEPTABLE; R = NEEDS REPAIR, REPLACEMENT OR ADJUSTMENT; C = CORRECTED; NA = NOT APPLICABLE
** NOTE VALUE AND PHASING

NFPA 70B

FIGURE H.25 Typical Back-Up Power System Inspection Checklist.

INSULATION RESISTANCE–DIELECTRIC ABSORPTION TEST SHEET FOR POWER CABLE

Test No. ____________

____________ Company Date ____________

____________ Location Time ____________

Circuit	Circuit Length	Aerial Duct Burned	
Number of Conductors	Conductor Size AWG MCM (kcmil)	Belted Shielded	
Insulating Material	Insulating Thickness	Voltage Rating	Age
Pothead or Terminal Type		Location Indoors Outdoors	
Number and Type of Joints			
Recent Operating History			
		Mfr.	
State if Potheads or Terminals Were Guarded During Test			
List Associated Equipment Included in Test			
Misc. Information			

Test Data — Megohms

Part Tested					Test Made	Hours Days	After Shutdown
Grounding Time					Dry-Bulb Temp.		°F
Test Voltage					Wet-Bulb Temp.		°F
Test Connections	To Line	To Line	To Line	To Line	Dew Point		°F
	To Earth	To Earth	To Earth	To Earth	Relative Humidity		%
	To Guard	To Guard	To Guard	To Guard	Absolute Humidity		Gr./#
¼ minute					Equipment Temp.		°F (°C)
½ minute					How Obtained		
¾ minute							
1 minute							
2 minutes					"Megger" Inst.		
3 minutes					Serial No.		
4 minutes					Range		
5 minutes					Voltage		
6 minutes							
7 minutes							
8 minutes							
9 minutes							
10 minutes							
10:1 min. Ratio							

Remarks ____________

Tested by:

 NFPA 70B

FIGURE H.26 Typical Insulation Resistance–Dielectric Absorption Test Sheet for Power Cable.

CABLE TEST SHEET

Company ______________________ Date ____________ Job No. ____________

Test Location ____________ Circuit ____________ Air Temp. ______ °C % Hum. ____________

Test Type: Acceptance ❑ Periodic ❑ Special ❑ Date Last ____________ Sheet No. ________ Weather ____________

Cable Mfr.	
Rated kV	(Gnd., Ungnd.)
Oper. kV	(Gnd., Ungnd.)
Length	
Age	
No. Cond.	
No. Size	
Insul. Material	
Insul. Thickness	
Insul. Type	
Covering	
Installed in	
Factory Test kV	
% Factory Test kV	
Max. Test kV	
Voltmeter kV	

Time Min.	Volts kV	Leakage Current-Microamps Φ1 – ⊙	Φ2 – △	Φ3 – ⊡

Insulation Resistance Test 1 Minute @ 2.5 kV — 1000 Megohms			
Φ TO GND.			
Φ 1 TO Φ	—		
Φ 2 TO Φ	—	—	

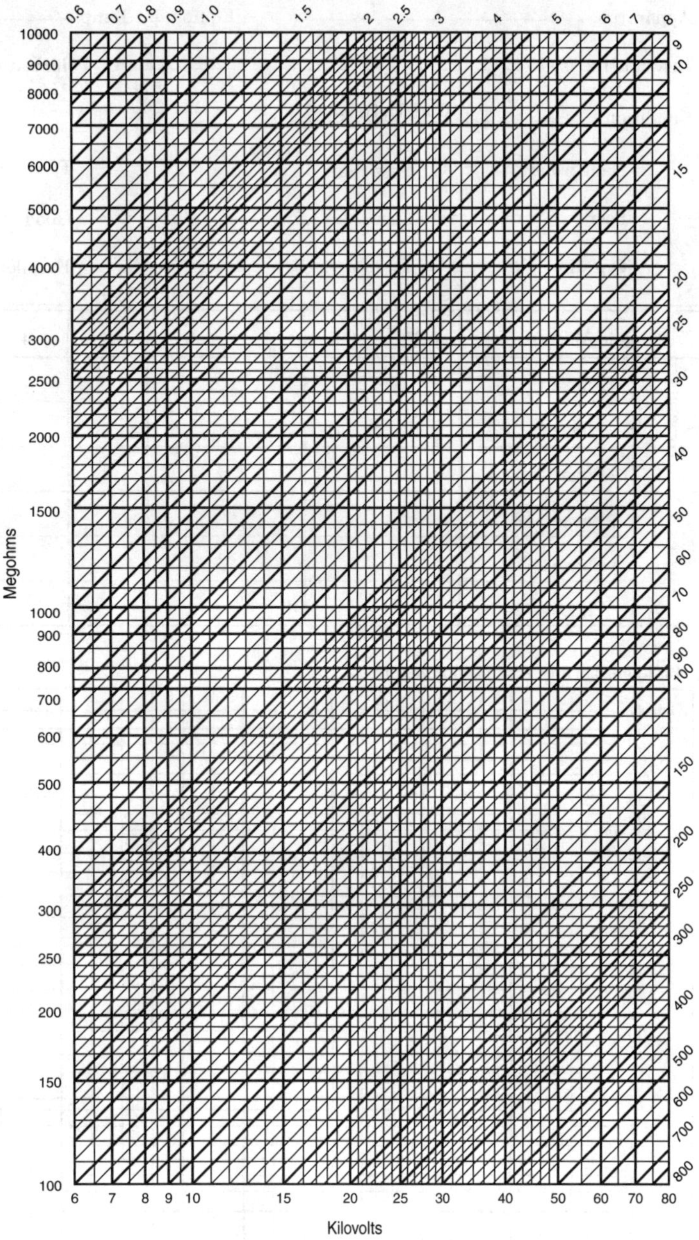

Authorization of Max.
Test kV and Verifying Voltmeter kV

____

Signature

Remarks: ______________________

____

____

____

Test Set No.: ____________ Tested by: ____________ Sheet No.: ____________

 NFPA 70B

FIGURE H.27 Typical Cable Test Sheet.

INSULATION RESISTANCE TEST RECORD

Date ______________________________

Scope: Dielectric Absorption Without Temperature Correction

Apparatus ______________________ Equipment Temp. __________ Ambient Temp. ______________

Instrument Used ______________________________ Polarization Index No. ______________________

Condition ________ 10:1 Min. Ratio

Dangerous --------- Less than 1

Poor -------------- Less than 1.5

Questionable ------- 1.5 to 2

Fair ------------- 2 to 3

Good ------------ 3 to 4

Excellent --------- Above 4

	Time in Minutes	0.25	0.5	1	2	3	4	5	6	7	8	9	10
To Ground	Phase 1												
	Phase 2												
	Phase 3												
Between Phases	Phase 1–2												
	Phase 2–3												
	Phase 3–4												

Plot the lowest group reading on graph.

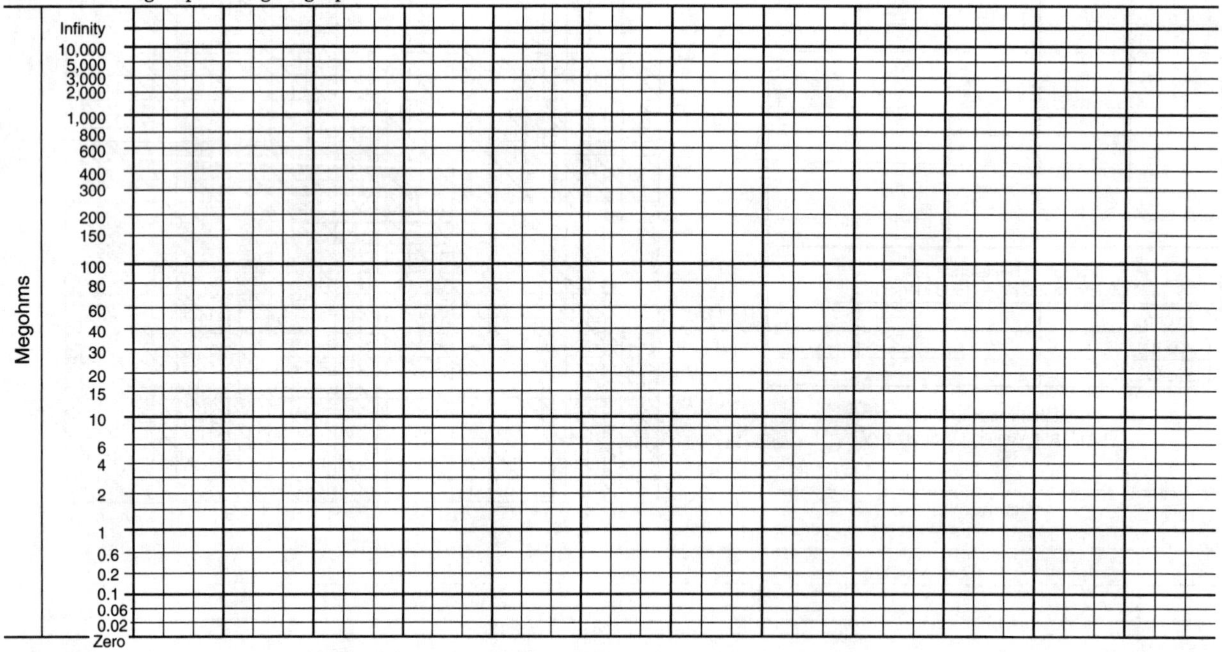

Tested by: ______________________________

 NFPA 70B

FIGURE H.28 Typical Insulation Resistance Test Record.

INSULATION RESISTANCE TEST RECORD FOR ROTATING MACHINERY

Reference: ANSI/IEEE 43, *Recommended Practice for Testing Resistance of Rotating Machinery*

Scope:

Dielectric Absorption — Temperature Corrected

ac machines 1000 kVA or more
dc machines 100 kW or more

Date ________________

Apparatus ________________ Voltage ________________ Rating ________________

Test Conditions:

List Associated Test Equipment Included in Test ________________

Winding Grounding Time ________________ Test Made ________________ Hours After Shutdown

Ambient Temperature ________________ °F Relative Humidity ________________ % Weather ________________

Equipment Temperature ________________ How Obtained ________________

Instrument ________________ Range ________________ Voltage ________________

Test Data:

Minutes	0.25	0.5	1.0	2.0	3.0	4.0	5.0	6.0	7.0	8.0	9.0	10.0
Reading												
Correction												

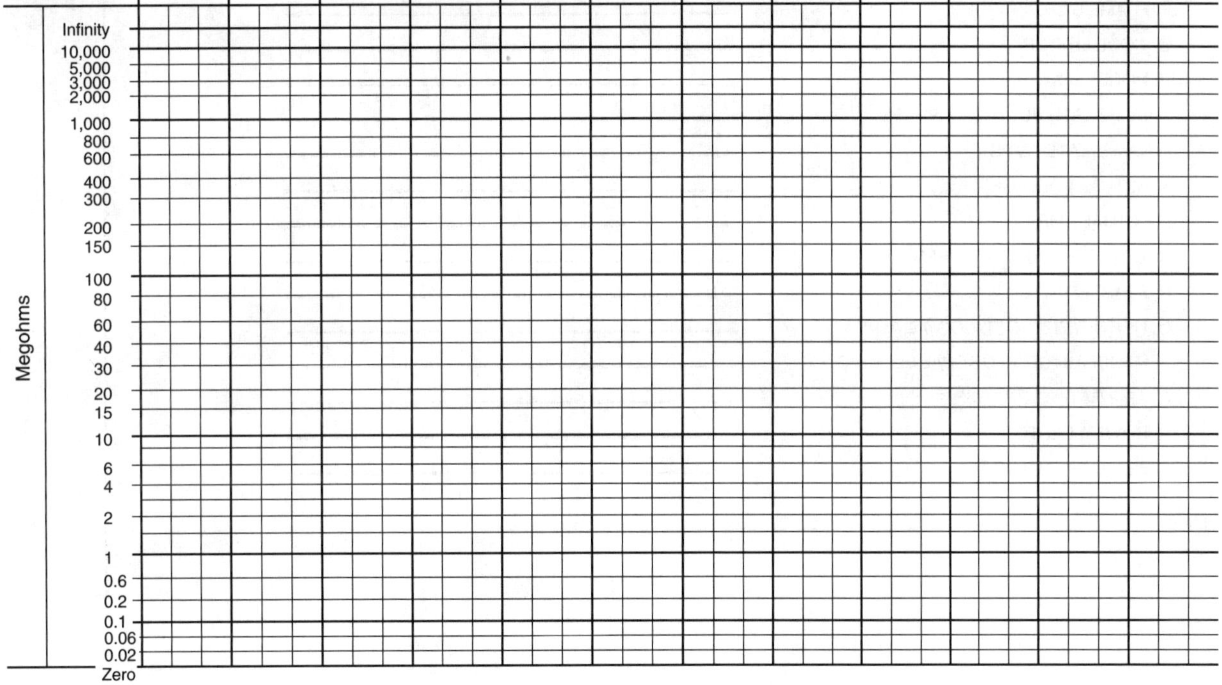

Polarization No. (10:1 min. ratio) ________________ Tested by ________________

Remarks: ________________

 NFPA 70B

FIGURE H.29 Typical Insulation Resistance Test Record for Rotating Machinery.

MOTOR TEST INFORMATION

SHEET NO. ________ OF ________
TEST REPORT NO. ____________

CUSTOMER ____________________ DATE ____________ PROJECT NO. ____________
ADDRESS ____________________ AIR TEMP. ____________ REL. HUMIDITY ____________
OWNER/USER ____________________ DATE LAST INSPECTION ____________
ADDRESS ____________________ LAST INSPECTION ____________
EQUIPMENT LOCATION ____________________
OWNER IDENTIFICATION ____________________

MOTOR TEST INFORMATION

INSULATION RESISTANCE TEST RESULTS AT ________ VDC IN MEGOHMS

30 SEC. ________
60 SEC. ________
10 MIN. ________
D.A. ________
P.I. ________

A. NAME & IDENTIFYING MARK OF MOTOR ____________
B. MANUFACTURER ____________
C. MODEL NUMBER ____________
D. SERIAL NUMBER ____________
E. RPM ____________
F. FRAME SIZE ____________
G. CODE LETTER ____________
H. HORSEPOWER ____________
I. NAMEPLATE VOLTAGE & PHASE ____________
J. NAMEPLATE AMPS ____________
K. ACTUAL VOLTAGE ____________
L. ACTUAL AMPS ____________
M. STARTER MANUFACTURER ____________
N. STARTER SIZE ____________
O. HEATER SIZE, CATALOG # & AMP ____________
P. MANUFACTURER OF DUAL ELEMENT ____________
Q. AMP RATING OF FUSE ____________
R. POWER FACTOR ____________
S. SERVICE FACTOR ____________

REMARKS: ____________________

TEST EQUIPMENT USED ____________________ SERIAL # ____________
SUBMITTED BY ____________________ TEST ____________

Courtesy of Northeast Electrical Testing NFPA 70B

FIGURE H.30 Typical Motor Test Information Form.

GROUNDING SYSTEM RESISTANCE TEST

SHEET NO. ________ OF ________

CUSTOMER ____________________ DATE ____________________ PROJECT NO. ____________________

ADDRESS ____________________ AIR TEMP. ____________________ REL. HUMIDITY ____________________

OWNER/USER ____________________ DATE LAST INSPECTION ____________________

ADDRESS ____________________ LAST INSPECTION REPORT NO. ____________________

LOCATION ____________________

SEASON	
SOIL TYPE	
SOIL CONDITION	
SINGLE ROD DEPTH	
MULTIPLE RODS (Y/N)	
LONGEST DIMENSION	
BURIED WIRE/STRIPS (Y/N)	
LONGEST DIMENSION	
DIST. TO AUX. ELECTRODE	
OTHER	

AUXILIARY POTENTIAL ELECTRODE

DISTANCE (FEET)	RESISTANCE (OHMS)

DISTANCE (FEET)

RESISTANCE (OHMS)

REMARKS ____________________

SUBMITTED BY ____________________ EQUIPMENT USED ____________________

Courtesy of Northeast Electrical Testing NFPA 70B

FIGURE H.31 Typical Ground System Resistance Test Report.

GROUND TEST INSPECTION REPORT — HEALTH CARE FACILITIES

SHEET NO. ________ OF ________

CUSTOMER ______________________ DATE ______________ PROJECT NO. ______________

ADDRESS ______________________ AIR TEMP. ______________ REL. HUMIDITY ______________

OWNER/USER ______________________ DATE LAST INSPECTION ______________

ADDRESS ______________________ LAST INSPECTION REPORT ______________

EQUIPMENT LOCATION ______________________

OWNER IDENTIFICATION ______________________

MAXIMUM TEST INTERVALS:
GENERAL CARE: 12 MONTHS
CRITICAL CARE: 6 MONTHS
WET LOCATIONS: 12 MONTHS

*NOTE: MAXIMUM READINGS PERMITTED:
20mV NEW CONSTRUCTION
40mV CRITICAL EXISTING CONSTRUCTION
500mV GENERAL CARE EXISTING CONSTRUCTION
0.1 ohm NEW CONSTRUCTION
0.2 ohm QUIET GROUNDS AND EXISTING CONSTRUCTION

ROOM NO.	DESCRIPTION (C) CRITICAL (G) GENERAL	VOLTAGE MEASUREMENT			IMPEDANCE MEASUREMENT	
		NUMBER OF RECEPTABLES	NUMBER OF OTHER	MAX READING (MILLIVOLTS)	NUMBER OF RECEPTACLES	MAX READING (OHMS)

REMARKS ______________________

CUSTOMER REPRESENTATIVE ______________________ TITLE ______________________

TEST EQUIPMENT ______________________ SERIAL # ______________________

SUBMITTED BY ______________________

Courtesy of Northeast Electrical Testing NFPA 70B

FIGURE H.32 Typical Ground Test Inspection Report— Health Care Facilities.

LINE ISOLATION MONITOR TEST DATA — HEALTH CARE FACILITIES

SHEET ________ OF ________

CUSTOMER ____________ DATE ____________ PROJECT NO. ____________

ADDRESS ____________ AIR TEMP. ________ °C REL. HUMIDITY ____________

OWNER/USE ____________ DATE LAST INSPECTION ____________

ADDRESS ____________ LAST INSPECTION REPORT NO. ____________

EQUIPMENT LOCATION ____________

CIRCUIT ____________

INSTRUMENT OR METER UNDER TEST

TYPE ____________ MANUFACTURER ____________ VOLTAGE ____________

SERIAL NO. ____________ MODEL NO. ____________

CATALOG NO. ____________ STYLE NO. ____________

TEST OPERATIONS

CAUTION: NO TEST EQUIPMENT NEEDED FOR THIS SECTION. REMOVE ALL PLUGS FROM MONITOR DURING THESE TESTS. PATIENT MUST NOT BE SUBJECTED TO HARMFUL TEST VOLTAGES.

		AS FOUND		AS LEFT	
1. AUDIBLE AND VISUAL INDICATORS	SELF TEST				
	SILENCE (MUTE)				
2. CHECK APPROPRIATE BOX IF INDICATOR IS OPERATIONAL	RED				
	GREEN				
	YELLOW				
3. LIM	MANUFACTURER'S SPECIFIED ALARM POINT		MA		MA
	METER READING		MA		MA

TEST OPERATIONS USING TEST EQUIPMENT:

TEST SET ____________

		AS FOUND	AS LEFT
4. LINE LEAKAGE TO GROUND	ONE mA		
	TWO mA		
	THREE mA		
	FOUR mA		
	FIVE mA		
5. ARE ALL BREAKERS OPERATIONAL AND CIRCUITS LABELED?			

REMARKS ____________

CUSTOMER REPRESENTATIVE ____________ TITLE ____________

TEST EQUIPMENT ____________ SERIAL # ____________

SUBMITTED BY ____________ TEST CREW ____________

Courtesy of Northeast Electrical Testing NFPA 70B

FIGURE H.33 Typical Line Isolation Monitor Test Data Report — Health Care Facilities.

TORQUE VALUE RECORD

SHEET NO. ________ OF ________

CUSTOMER ____________________ DATE ____________________ PROJECT NO. ____________________

ADDRESS ____________________ AIR TEMP. ____________________ REL. HUMIDITY ____________________

OWNER/USER ____________________ DATE LAST INSPECTION ____________________

ADDRESS ____________________ LAST INSPECTION REPORT NO. ____________________

EQUIPMENT LOCATION ____________________

OWNER IDENTIFICATION ____________________

GENERAL INFORMATION:

Equipment ID: ____________________ Performed By: ____________________

Location: ____________________ Torque Marked: ☐ Yes ☐ No Color: ____________________

Date Performed: ____________________ Verified By: ____________________

Torque Wrench Information: ☐ IN-LBS ☐ FT-LBS Verification Marked: ☐ Yes ☐ No Color: ____________________

Manufacturer: __________ Model: __________ Approved By: ____________________

TORQUE AND VERIFICATION

No.	No. of Items	Item Description/Location	Vendor Specification		NETA Specification		Torque		Note
			NO.	FT-LB / IN-LB	NO.	FT-LB / IN-LB	NO.	FT-LB / IN-LB	
				FT-LB		FT-LB		FT-LB	
				IN-LB		IN-LB		IN-LB	
				FT-LB		FT-LB		FT-LB	
				IN-LB		IN-LB		IN-LB	
				FT-LB		FT-LB		FT-LB	
				IN-LB		IN-LB		IN-LB	
				FT-LB		FT-LB		FT-LB	
				IN-LB		IN-LB		IN-LB	
				FT-LB		FT-LB		FT-LB	
				IN-LB		IN-LB		IN-LB	
				FT-LB		FT-LB		FT-LB	
				IN-LB		IN-LB		IN-LB	
				FT-LB		FT-LB		FT-LB	
				IN-LB		IN-LB		IN-LB	
				FT-LB		FT-LB		FT-LB	
				IN-LB		IN-LB		IN-LB	
				FT-LB		FT-LB		FT-LB	
				IN-LB		IN-LB		IN-LB	
				FT-LB		FT-LB		FT-LB	
				IN-LB		IN-LB		IN-LB	
				FT-LB		FT-LB		FT-LB	
				IN-LB		IN-LB		IN-LB	
				FT-LB		FT-LB		FT-LB	
				IN-LB		IN-LB		IN-LB	
				FT-LB		FT-LB		FT-LB	
				IN-LB		IN-LB		IN-LB	
				FT-LB		FT-LB		FT-LB	
				IN-LB		IN-LB		IN-LB	
				FT-LB		FT-LB		FT-LB	
				IN-LB		IN-LB		IN-LB	

REMARKS ____________________

CUSTOMER REPRESENTATIVE ____________________ TITLE ____________________

TEST EQUIPMENT ____________________ SERIAL # ____________________

SUBMITTED BY ____________________

Courtesy of Northeast Electrical Testing NFPA 70B

FIGURE H.34 Typical Torque Value Record.

MAIN POWER ENERGIZATION CHECKLIST

For use of this form see TM 5-694: the proponent agency is COE.

SECTION A – CUSTOMER DATA

1. PLANT/BUILDING	2. LOCATION	3. JOB NUMBER	
4. CIRCUIT DESIGNATION	5. CIRCUIT FED FROM	6. CIRCUIT FED TO	7. DATE *(YYYYMMDD)*
8. TEST EQUIPMENT TYPE/BRAND AND CALIBRATION DATE		9. TESTED BY	

SECTION B – VISUAL AND ELECTRICAL/MECHANICAL INSPECTION

10. CHECK POINT	COND*	NOTES	CHECK POINT	COND*	NOTES
COMPONENT INSPECTION/TESTING COMPLETED			VERIFY SWITCHGEAR CONTROL FUNCTIONS		
WIRING VISUAL VERIFICATION			CHECK FOR WORKING CLEARANCE		
VERIFY WIRING DIAGRAMS			ENERGIZE AND TEST SYSTEM		
VERIFY CIRCUIT SWITCHER CONTROL FUNCTIONS			TRANSFORMER TRANSFER CONTROL FUNCTIONS		
ENERGIZE AND TEST SYSTEM FOR ALL CONDITIONS			CHECK FOR UNUSUAL SOUNDS AFTER ENERGIZING		
CHECK BUSHING OR TERMINALS			CHECK ANCHORING OF TRANSFORMER SWITCHGEAR AND SWITCHES ENCLOSURE		
CHECK FOR REMOVAL OF PAINT OR HEAVY DENTS			CHECK FOR NORMAL/ABNORMAL SWITCHING OPERATION		

SECTION C – ELECTRICAL TESTS

11. MEASUREMENT DESCRIPTION	VOLTAGE AND CURRENT MEASUREMENTS										
	VOLTAGE**						CURRENT**				
	A-N	B-N	C-N	A-B	B-C	C-A	A	B	C	N	G
	A-N	B-N	C-N	A-B	B-C	C-A	A	B	C	N	G
	A-N	B-N	C-N	A-B	B-C	C-A	A	B	C	N	G
	A-N	B-N	C-N	A-B	B-C	C-A	A	B	C	N	G
	A-N	B-N	C-N	A-B	B-C	C-A	A	B	C	N	G

12. NOTES

* CONDITION: A = ACCEPTABLE; R = NEEDS REPAIR, REPLACEMENT OR ADJUSTMENT; C = CORRECTED; NA = NOT APPLICABLE
** NOTE VALUE AND PHASING

FIGURE H.35 Typical Main Power Energization Checklist.

Facility Identification **Job No.**

INSTRUCTIONS TO CONTRACTOR

Date: ________________

Contractor Name: ________________

Address: ________________

City, State, Zip Code: ________________

Subject: Project Title: ________________

Project No.: ________________

Enclosed is one complete set of the following bid documents covering work for the subject project.

1. Project Scope of Work, Dated: ________________
2. Proposal Forms for:
   General Maintenance of Electrical Power Equipment, and/or
   Infrared Surveying of Electrical Power Equipment, and/or
   Circuit Breaker Overhaul and Trip Unit Retrofit
3. Plant Electrical Power Equipment Documentation:
   - **Plant one line diagrams** **Dwg. Nos.** ____________, **Dated** ____________
   - **Plant layout drawings** **Dwg. Nos.** ____________, **Dated** ____________
   - **Plant equipment list** **Doc. No.** ____________, **Dated** ____________
   - **Short circuit analyses and time-current coordination studies**
   - **Equipment manufacturers' requirements will be available at the plant for your use.**

We would appreciate receiving a quote from you for this work on the enclosed Project Scope of Work in strict accordance to the quote documents. Please respond in writing if you do not intend to submit a quote for this project.

One original and copy(s) of your quote will be due not later than ________ at ________ local time.

Fax and send one original of your proposal to: ***(Enter Project Engineer's Name, Address, and Fax No. here)***

We welcome suggestions regarding changes in specifications and/or modifications in design or production methods that will aid in reducing costs without impairing quality or that will improve the quality, safety, and/or performance of the product on which you are quoting. However, your base bid price must be submitted on the basis of the bid documents. All voluntary alternates are to be presented as a separate price from the base bid.

Pre-Quote Walk Through (if applicable)

You are invited to attend a pre-quote walk through meeting scheduled for:

Time: ________________

Day: ________________

Date: ________________

Location: ________________

If you desire further information addressing the technical specifications or site visitation, please contact ***(project engineer)*** at ***(phone no.).***

Sincerely,

(Enter Project Engineer's Name and Location here)

 NFPA 70B

FIGURE H.36 Instructions to Contractor.

Facility Identification **Job No.**

PROJECT SCOPE OF WORK

Introduction

This Scope of Work document will define the maintenance activities to be included in this project. The contractor shall provide all labor, materials, and equipment necessary to perform requested electrical power equipment maintenance. All maintenance activities will be performed in accordance with the applicable Technical Specification and other pertinent portions of facility's Electrical Power Equipment Maintenance Manual. *The costs of premium time, if required, shall be included in this quote.*

If electrical power equipment is found to be within the Manufacturer's specifications for continued service and the required maintenance is unnecessary, clean, adjust, and reassemble the electrical power equipment and perform all Manufacturer recommended procedures for continued service.

If the contractor discovers that the electrical power equipment cannot be brought into compliance with the Manufacturer's specifications for continued service, advise the Owner of the "as found" condition and await further direction. *Save all component parts for Owner inspection.*

General Task Description

Schedule

Time to complete all tasks: ________________

Time equipment is available: ________________

Time/day to begin work: ________________

Time/day to complete work: ________________

 NFPA 70B (p. 1 of 2)

FIGURE H.37 Project Scope of Work Template.

Facility Identification **Job No.**

PROJECT SCOPE OF WORK *(continued)*

Specific Task Description

Contractor will perform the specific tasks described in this section.

General Maintenance

The following maintenance activities will be performed in accordance with the Technical Specifications for General Maintenance or Electrical Power Equipment.

Circuit Breaker Overhaul and Trip Unit Retrofit

The following maintenance activities will be performed in accordance with the Technical Specifications for Circuit Breaker Overhaul and Trip Unit Retrofit.

Infrared Surveying

The following maintenance activities will be performed in accordance with the Technical Specifications for Infrared Surveying of Electrical Power Equipment.

Work Not Included (Listed Owner furnished services)

Exceptions to Specifications

 NFPA 70B (p. 2 of 2)

FIGURE H.37 ***Continued***

PROJECT SCOPE OF WORK QUOTE FORM

Location: ______________ Project No. ______________ Purch. Req. No. ______________

Contractor: ______________ MBO No. ______________

The Contractor shall perform the work in strict accordance with the facility's Electrical Power Equipment Maintenance Manual EMM-1 General Maintenance Technical Specification, the Circuit Breaker Overhaul and Trip Unit Retrofit Technical Specification, the plant equipment list, single line diagrams and drawings, and the Project Scope of Work. All alternates shall be fully described with exceptions listed by Item Number, and priced separately. When applicable, General Maintenance Technical Specification (GMTS) codes shall be used on this Proposal Form as they are used in the Project Scope of Work.

Item No.	Maintenance Activity Description and Location of Equipment	GMTS Code	Quantity	Item/Unit Price	Item Total Price
				Page Total	

 NFPA 70B

FIGURE H.38 Project Scope of Work Form.

PROJECT SCOPE OF WORK MODIFICATION QUOTE FORM

Location: ____________ Project No. ____________ Purch. Req. No. ____________

Contractor: ____________ MBO No. ____________ Modification No. ____________

All modifications shall be fully described with reference to the original Item Number, and priced separately. The Contractor shall perform the work in strict accordance with the facility's Electrical Power Equipment Maintenance Manual EMM-1 General Maintenance Technical Specification, the Circuit Breaker Overhaul and Trip Unit Retrofit Technical Specification, the plant equipment list, single line diagrams and drawings, and the Project Scope of Work. When applicable, General Maintenance Technical Specification (GMTS) codes shall be used on this Proposal Form as they are used in the Project Scope of Work. Contractor shall not begin work until approved by the plant.

Item No.	Maintenance Activity Description and Location of Equipment	GMTS Code	Qty.	Labor	Mat'l	Item Total Price
					Page Total	

Submitted by: ____________ Date: ____________

Approved by: ____________ Date: ____________

 NFPA 70B

FIGURE H.39 Project Scope of Work Modification Form.

POWER QUALITY SURVEY DATA COLLECTION MANUAL

Installation: ____________________________

Location: ____________________________

Collection Date: ________________

Courtesy of U.S. Army Corps of Engineers NFPA 70B

FIGURE H.40 Cover and Contents.

POINTS OF CONTACT

Installation: ______________________________

Address: ______________________________

____

Names	Titles	Office Symbols	Phones

Contractor's Name: ______________________________

Address: ______________________________

____

Names	Titles	Office Names	Phones

Utilities: ______________________________

Address: ______________________________

____

Names	Titles	Office Names	Phones

Courtesy of U.S. Army Corps of Engineers NFPA 70B

FIGURE H.41 Points of Contact.

POWER DISTRIBUTION UNIT (PDU) SURVEY

Date: ____________

Installation: ____________ Location: ____________

Power Distribution Unit (PDU) Identification: ____________

Manufacturer: ____________ Model/Serial #: ____________

Size: ______ kVA 3-Phase: _____ Single-Phase: _____ Frequency: ______ Hz

Input Voltage Rating: ______ V Input Current Rating: ______ A Tap Changing Range: ______

Output Voltage Rating: ______ V Output Current Rating: ______ A

Measured Input Volts	$V_{(IN)A\text{-}B}$ = ______ V	$V_{(IN)B\text{-}C}$ = ______ V	$V_{(IN)C\text{-}A}$ = ______ V	
Input Voltage Harmonic Distortion	$V_{(IN)THD(A\text{-}B)}$ = ____ % 3rd = ______ __th = ______ __th = ______	$V_{(IN)THD(B\text{-}C)}$ = ____ % 3rd = ______ __th = ______ __th = ______	$V_{(IN)THD(A\text{-}C)}$ = ____ % 3rd = ______ __th = ______ __th = ______	
Measured Input Amps	$I_{(IN)A}$ = ______ A	$I_{(IN)B}$ = ______ A	$I_{(IN)C}$ = ______ A	
Input Current Harmonic Distortion	$I_{(IN)THD(A)}$ = ____ % 3rd = ______ __th = ______ __th = ______	$I_{(IN)THD(B)}$ = ____ % 3rd = ______ __th = ______ __th = ______	$I_{(IN)THD(C)}$ = ____ % 3rd = ______ __th = ______ __th = ______	
Measured Output Volts	$V_{(O)A\text{-}N}$ = ______ V $V_{(O)A\text{-}G}$ = ______ V	$V_{(O)B\text{-}N}$ = ______ V $V_{(O)B\text{-}G}$ = ______ V	$V_{(O)C\text{-}N}$ = ______ V $V_{(O)C\text{-}G}$ = ______ V	$V_{(O)N\text{-}G}$ = ______ V
Output Voltage Harmonic Distortion	$V_{(O)THD(A\text{-}N)}$ = ____ % 3rd = ______ __th = ______ __th = ______	$V_{(O)THD(B\text{-}N)}$ = ____ % 3rd = ______ __th = ______ __th = ______	$V_{(O)THD(C\text{-}N)}$ = ____ % 3rd = ______ __th = ______ __th = ______	$V_{(O)THD(N\text{-}G)}$ = ____ % 3rd = ______ __th = ______ __th = ______
Measured Output Amps	$I_{(O)A}$ = ______ A	$I_{(O)B}$ = ______ A	$I_{(O)C}$ = ______ A	$I_{(O)N}$ = ______ A
Output Current Harmonic Distortion	$I_{(O)THD(A)}$ = ____ % 3rd = ______ __th = ______ __th = ______	$I_{(O)THD(B)}$ = ____ % 3rd = ______ __th = ______ __th = ______	$I_{(O)THD(C)}$ = ____ % 3rd = ______ __th = ______ __th = ______	$I_{(O)THD(N)}$ = ____ % 3rd = ______ __th = ______ __th = ______

K Factor	Measured K Factor: ______ Nameplate K Factor: ______
Ground System	Neutral bus of PDU is connected to ground? ❑ Yes ❑ No Ground bus of PDU is connected to upstream switchgear/switchboard/panel ground bus? _____ to building metal frame? _____ to raised floor frame? _____ Ground current measurement: ______ A Ground resistance measurement: ______ Ω
Temperature	Transf winding temperature range: ______ °F Bus temperature range: ______ °F CBs having temperature higher that 90°F: ______
Power Factor	PF: ______ Displacement Power Factor (DPF): ______

Courtesy of U.S. Army Corps of Engineers NFPA 70B (p. 1 of 2)

FIGURE H.42 Power Distribution Unit (PDU) Survey.

POWER DISTRIBUTION UNIT (PDU) SURVEY *(continued)*

Deficiencies found	
Problems in the past	
Customer's concerns	
Notes	

Courtesy of U.S. Army Corps of Engineers NFPA 70B (p. 2 of 2)

FIGURE H.42 ***Continued***

GENERATOR SET SURVEY

Date: ______________

Installation: ______________________________ Location: ______________________________

Number of generator sets at this location: ________

Generator Set #1

Physical Conditions: ❑ Good condition ❑ Damage ❑ Not in use ❑ Need repair
❑ Old ❑ Corrosion ❑ Need maintenance ❑ Other: ______________

- Designed for: ❑ Prime operation ❑ Standby operation ❑ Emergency operation

Engine Data:

- Manufacturer: ______________________
- Model/Type: ______________________ Rated Voltage: ______________________
- Rated hp (or kW): ______________________ Rated Current: ______________________
- Power Factor: ______________________ Frequency: ______________________

Generator Data:

- Manufacturer: ______________________
- Model/Type: ______________________
- Generated Voltages: __________ V Generated Frequencies: __________ Hz
- Rated kVA: ______________ Rated kW: ______________
- Rated Currents: __________ A Efficiency Factor: ______________
- Winding Connection (D/W/GW): __________ Power Factor: ______________

Batteries

- ❑ Good condition ❑ Leakage ❑ Need maintenance ❑ Dead ❑ Other: ______________
- Measured Voltages: __________ V Measured Temperatures: __________ °F

Generator Set #2

Physical Conditions: ❑ Good condition ❑ Damage ❑ Not in use ❑ Need repair
❑ Old ❑ Corrosion ❑ Need maintenance ❑ Other: ______________

- Designed for: ❑ Prime operation ❑ Standby operation ❑ Emergency operation

Engine Data:

- Manufacturer: ______________________
- Model/Type: ______________________ Rated Voltage: ______________________
- Rated hp (or kW): ______________________ Rated Current: ______________________
- Power Factor: ______________________ Frequency: ______________________

Generator Data:

- Manufacturer: ______________________
- Model/Type: ______________________
- Generated Voltages: __________ V Generated Frequencies: __________ Hz
- Rated kVA: ______________ Rated kW: ______________
- Rated Currents: __________ A Efficiency Factor: ______________
- Winding Connection (D/W/GW): __________ Power Factor: ______________

Batteries

- ❑ Good condition ❑ Leakage ❑ Need maintenance ❑ Dead ❑ Other: ______________
- Measured Voltages: __________ V Measured Temperatures: __________ °F

Courtesy of U.S. Army Corps of Engineers NFPA 70B (p. 1 of 2)

FIGURE H.43 Generator Set Survey.

GENERATOR SET SURVEY *(continued)*

Generator Set #3

Physical Conditions: ❑ Good condition ❑ Damage ❑ Not in use ❑ Need repair
❑ Old ❑ Corrosion ❑ Need maintenance ❑ Other: ____________________

- Designed for: ❑ Prime operation ❑ Standby operation ❑ Emergency operation

Engine Data:

- Manufacturer: ____________________
- Model/Type: ____________________ Rated Voltage: ____________________
- Rated hp (or kW): ____________________ Rated Current: ____________________
- Power Factor: ____________________ Frequency: ____________________

Generator Data:

- Manufacturer: ____________________
- Model/Type: ____________________
- Generated Voltages: __________ V Generated Frequencies: __________ Hz
- Rated kVA: __________ Rated kW: __________
- Rated Currents: __________ A Efficiency Factor: __________
- Winding Connection (D/W/GW): __________ Power Factor: __________

Batteries

- ❑ Good condition ❑ Leakage ❑ Need maintenance ❑ Dead ❑ Other: __________
- Measured Voltages: __________ V Measured Temperatures: __________ °F

Generator Operation:

- Can these generators run in parallel with the utility power sources? ❑ Yes ❑ No
- The generators are being used as: ❑ Backup source ❑ Peak shaving ❑ Prime source
- Are the generators properly protected against overload? ❑ Yes ❑ No abnormal conditions? ❑ Yes ❑ No or reverse power flow (if generators can run in parallel with utility source)? ❑ Yes ❑ No
- Can the generators automatically start? ❑ Yes ❑ No and automatically shut off? ❑ Yes ❑ No
- How many times did generator fail to start or break down (with unknown reason) during the last few years? _____

Maintenance:

- Does the generator operation log book exist and is it up to date? ❑ Yes ❑ No
- How often does the generator run for maintenance? ______ times per week/month, ❑ with loads or ❑ without loads.
- How long did the generator run during each maintenance period? ______ minutes
- How often is the generator fuel system checked? ______ times per week/month

Generator Grounding System:

- ❑ Solidly grounded ❑ High resistance ❑ Low resistance ❑ Reactance
- Measured ground impedance in ohms: _______
- Is the generator's neutral bus connected to ground? ❑ Yes ❑ No
- Is the generator frame connected to ground? ❑ Yes ❑ No

Notes:

____

____

____

____

____

Courtesy of U.S. Army Corps of Engineers NFPA 70B (p. 2 of 2)

FIGURE H.43 ***Continued***

ELECTRICAL PANEL SURVEY

Date: ____________

Installation: ______________________ Location: ______________________

Panel Identification: ______________________

Manufacturer Name: ______________________ Panel Type/Model: ______________________

Voltage Rating: ________ V Current Rating: ________ A Phases: ________ # of Wires: ________

Main Breaker: Type/Model: ____________ Rating: ________ A Adjustable Setting Range: ________

Measured Voltages	$V_{A\text{-}N}$ = ______ V $V_{A\text{-}G}$ = ______ V	$V_{B\text{-}N}$ = ______ V $V_{B\text{-}G}$ = ______ V	$V_{C\text{-}N}$ = ______ V $V_{C\text{-}G}$ = ______ V	$V_{N\text{-}G}$ = ______ V
Voltage Sine Waves	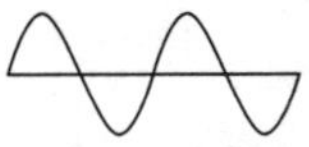 $V_{A\text{-}N}$	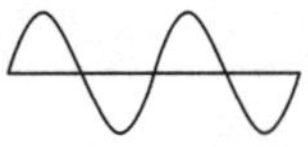 $V_{B\text{-}N}$	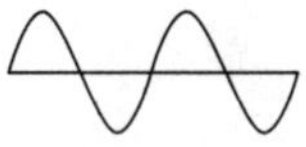 $V_{C\text{-}N}$	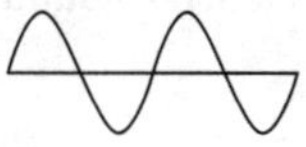 $V_{N\text{-}G}$
Harmonic Voltage Distortion	$V_{THD(A\text{-}N)}$ = ____ % 3rd = ______ % __th = ______ % __th = ______ %	$V_{THD(B\text{-}N)}$ = ____ % 3rd = ______ % __th = ______ % __th = ______ %	$V_{THD(C\text{-}N)}$ = ____ % 3rd = ______ % __th = ______ % __th = ______ %	$V_{THD(N\text{-}G)}$ = ____ % 3rd = ______ % __th = ______ % __th = ______ %
Measured Currents	I_A = ________ A	I_B = ________ A	I_C = ________ A	I_N = ________ A
Current Sine Waves	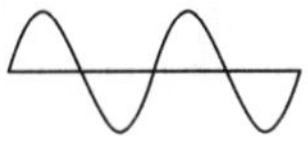 I_A	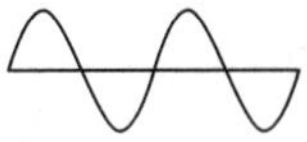 I_B	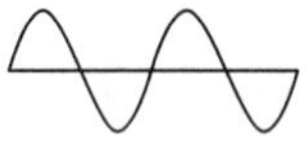 I_C	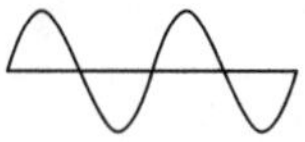 I_N
Harmonic Current Distortion	$I_{THD(A)}$ = ______ % 3rd = ______ % __th = ______ % __th = ______ %	$I_{THD(B)}$ = ______ % 3rd = ______ % __th = ______ % __th = ______ %	$I_{THD(C)}$ = ______ % 3rd = ______ % __th = ______ % __th = ______ %	$I_{THD(N)}$ = ______ % 3rd = ______ % __th = ______ % __th = ______ %
Power Factor	PF: ______________ Displacement Power Factor (DPF): ______________			
Grounding System	Ground bus isolated from frame?____ or bonded to frame?____ Ground by metal conduits?____ or by ground conductors?____ Ground bus bonded to neutral bus? ❑ Yes ❑ No Each branch circuit has separated neutral ❑ Yes ❑ No and ground conductor? ❑ Yes ❑ No Ground current measurement: ______ A Ground resistance measurement: ______ Ω Sketch the existing grounding system (on the back sheet) when it is necessary.			
Temperature	Bus temperature range: ______ °F Conductor temperature range: ______ °F CBs having temperature higher that 90°F: ______________________			
Entrance Conductor	Phases: ________ MCM Number of conductors per phase: ________			
Sizes	Neutral: ________ MCM Number of conductors per phase: ________ Ground: ________ MCM Number of conductors per phase: ________			
Lightning Protection	Manufacturer: ______________ Type: ________ Voltage rating: ________			

 NFPA 70B (p. 1 of 2)

FIGURE H.44 Electrical Panel Survey.

ELECTRICAL PANEL SURVEY *(continued)*

Other Circuit Breakers in the Panel	Circuit breaker rating: _________ A 3 Ph or Single:_____ How many CB:_____ Circuit breaker rating: _________ A 3 Ph or Single:_____ How many CB:_____ Circuit breaker rating: _________ A 3 Ph or Single:_____ How many CB:_____ Circuit breaker rating: _________ A 3 Ph or Single:_____ How many CB:_____ Circuit breaker rating: _________ A 3 Ph or Single:_____ How many CB:_____ Circuit breaker rating: _________ A 3 Ph or Single:_____ How many CB:_____ Circuit breaker rating: _________ A 3 Ph or Single:_____ How many CB:_____ Circuit breaker rating: _________ A 3 Ph or Single:_____ How many CB:_____
Deficiencies Found	
Problems in the Past	
Customer's Concerns	
Notes	

 NFPA 70B (p. 2 of 2)

FIGURE H.44 ***Continued***

INVERTER SURVEY

Date: ______________

Installation: ______________ Location: ______________

Inverter Identification: ______________

Number of inverters at this location: ________ Do they all have the same size and characteristics? ❑ Yes ❑ No

Manufacturer: ______________ Model: ________ Type: ________ Phases: _____ Wires: _____

Input Voltages	Rated: ________ Measured: ________
Input Currents	Rated: ________ Measured: ________
Output Voltages	Rated: ________ Measured: ________
Output Currents	Rated: ________ Measured: ________
kVA	Rated: ________ Measured input: ______ Measured output: ______ PF: ______
kW	Rated: ________ Measured input: ______ Measured output: ______ DPF: ______
Conductor Sizes	Phases: ( )______ Neutral: ( )______ Ground: ( )______
Measured Temperatures	Bus temperature range: ________°F Conductor temperature range: ________°F
Grounding System	Terminal of inverter bonded to ground? ❑ Yes ❑ No Ground bus of inverter bonded to the frame? ❑ Yes ❑ No Ground current measurement: ______ A Ground resistance measurement: ______ Ω Sketch the existing grounding system (on the back sheet) when it is needed.
Batteries	Manufacturer: ______________ Model: __________ Type: __________ Cell voltage: ______________ Number of cells: ______________ Total battery voltages: ________V Total battery currents: ________A Conductor sizes: Phases: ( ) ______ Neutral: ( ) ______ Ground: ( ) ______ Physical conditions: ❑ Damage ❑ Corrosion ❑ Leakage Fluid fill level: ________ Proper mounting: ________ Proper clearance: ________ Battery rack condition: ____________ Battery rack grounded? ❑ Yes ❑ No Battery bank terminal grounded? ❑ Yes ❑ No Battery fluid specific gravity last checked: ______________

Measured Cell Battery Voltages and Fluid Temperatures

Cell #____ _____V _____°F	Cell #____ _____V _____°F	Cell #____ _____V _____°F	Cell #____ _____V _____°F	Cell #____ _____V _____°F	Cell #____ _____V _____°F	Cell #____ _____V _____°F
Cell #____ _____V _____°F	Cell #____ _____V _____°F	Cell #____ _____V _____°F	Cell #____ _____V _____°F	Cell #____ _____V _____°F	Cell #____ _____V _____°F	Cell #____ _____V _____°F
Cell #____ _____V _____°F	Cell #____ _____V _____°F	Cell #____ _____V _____°F	Cell #____ _____V _____°F	Cell #____ _____V _____°F	Cell #____ _____V _____°F	Cell #____ _____V _____°F
Cell #____ _____V _____°F	Cell #____ _____V _____°F	Cell #____ _____V _____°F	Cell #____ _____V _____°F	Cell #____ _____V _____°F	Cell #____ _____V _____°F	Cell #____ _____V _____°F
Cell #____ _____V _____°F	Cell #____ _____V _____°F	Cell #____ _____V _____°F	Cell #____ _____V _____°F	Cell #____ _____V _____°F	Cell #____ _____V _____°F	Cell #____ _____V _____°F

 NFPA 70B (p. 1 of 2)

FIGURE H.45 Inverter Survey.

INVERTER SURVEY *(continued)*

Deficiencies Found	
Problems in the Past	
Notes	

 NFPA 70B (p. 2 of 2)

FIGURE H.45 ***Continued***

BUILDING LIGHTNING PROTECTION SURVEY

Date: ___________

Installation: ___________ Building #: ___________

Physical Conditions	❑ Good Condition ❑ Rust/Corrosion ❑ Damaged
Roof	Materials: ❑ Metal ❑ Non-metal Types: ❑ Flat ❑ Gable ❑ Hip ❑ Gambrel ❑ Intermediate Ridges ❑ Domed ❑ Shed
Air Terminals	Size (diameter): ________ inches Height: ________ feet ________ inches Material: ❑ Copper ❑ Copper Alloys ❑ Aluminum Approximate distance between two consecutive air terminals: ________ feet Air terminals securely mounted on appropriate bases? ❑ Yes ❑ No Air terminal bases are of the same material as the air terminals? ❑ Yes ❑ No Air terminal bases properly fastened/anchored to the roof? ❑ Yes ❑ No An air terminal at each corner of the roof? ❑ Yes ❑ No Two paths for currents to flow (to ground) at each air terminal? ❑ Yes ❑ No
Lightning Conductors	Sizes: ________ AWG Material: ❑ Copper ❑ Copper Alloys ❑ Aluminum Approximate distance between two consecutive lightning conductors: ________ feet Interconnected lightning conductors properly bonded together? ❑ Yes ❑ No Any sharp bend curves (less than 8 inch radius and 90° angle)? ❑ Yes ❑ No Lightning conductor securely fastened every 4 feet? ❑ Yes ❑ No
Down Conductors	Size: ________ AWG Material: ❑ Copper ❑ Copper Alloys ❑ Aluminum Are the conductors electrically continuous running down to the ground? ❑ Yes ❑ No Approximate distance between two consecutive down conductors: ________ feet At least 2 down conductors installed at opposite corners of the building? ❑ Yes ❑ No Total number of down conductors installed: ________ Average resistance measurement at down conductors: ________ Ω
Objects on the Roof	Metal object has a thickness less than 3⁄16 inch? ❑ Yes ❑ No Metal object is directly bonded to lightning conductors or through an air terminal to lightning conductors? ❑ Yes ❑ No The bonding surface has a contact area of not less than 3 sq-inches? ❑ Yes ❑ No Non-metal objects on the roof? ❑ Yes ❑ No Are they protected with air terminals? ❑ Yes ❑ No Does each air terminal provide a two-way path to the ground? ❑ Yes ❑ No For non-metal object, is the distance from the farthest corner of the object to the air terminal less than 2 feet? ❑ Yes ❑ No
Grounding System	Separate grounding loop for lightning protection system? ❑ Yes ❑ No Is the grounding loop for lightning protection system bonded to the electrical grounding system? ❑ Yes ❑ No Average ground resistance measurement (at the location where it is connected to electrical grounding system): ________ Ω

 NFPA 70B (p. 1 of 2)

FIGURE H.46 Building Lightning Protection Survey.

BUILDING LIGHTNING PROTECTION SURVEY *(continued)*

Deficiencies Found	
Problems in the Past	
Sketch the roof floor plan and mark down the location of air terminals, cross-roof lightning conductors, down conductors, and distances between them.	

 NFPA 70B (p. 2 of 2)

FIGURE H.46 ***Continued***

RECTIFIER SURVEY

Date: ____________

Installation: ______________________ Location: ______________________

Rectifier Identification: ______________________

Number of units at this location: ________ Do they all have the same size and characteristics? ❑ Yes ❑ No

Manufacturer: ______________ Model: ________ Type: ________ Phases: _____ Wires: _____

Input Voltages	Rated: ________ Measured: ________
Input Currents	Rated: ________ Measured: ________
Output Voltages	Rated: ________ Measured: ________
Output Currents	Rated: ________ Measured: ________
kVA	Rated: ________ Measured input: ______ Measured output: ______ PF: ______
kW	Rated: ________ Measured input: ______ Measured output: ______ DPF: ______
Conductor Sizes	Phases: ________ Neutral: ________ Ground: ________ Number of conductors per phase: ________
Measured Temperatures	Bus temperature range: ________ °F Conductor temperature range: ________ °F
Grounding System	Terminal of rectifier bonded to ground? ❑ Yes ❑ No Ground bus of rectifier bonded to the frame? ❑ Yes ❑ No Ground current measurement: ______ A Ground resistance measurement: ______ Ω Sketch the existing grounding system (on the back sheet) when it is needed.
Batteries	Manufacturer: ________ Model: ________ Type: ________ Cell voltage: ________ Number of cells: ________ Total battery voltages: ________ V Total battery currents: ________ A Conductor sizes: Phases: ( ) ______ Neutral: ( ) ______ Ground: ( ) ______ Physical conditions: ❑ Damage ❑ Corrosion ❑ Leakage Fluid fill level: ________ Proper mounting: ________ Proper clearance: ________ Battery rack condition: ________ Battery rack grounded? ❑ Yes ❑ No Battery bank terminal grounded? ❑ Yes ❑ No Battery fluid specific gravity last checked: ________

Measured Battery Cell Voltages and Fluid Temperatures

Cell #____ _____V _____°F	Cell #____ _____V _____°F	Cell #____ _____V _____°F	Cell #____ _____V _____°F	Cell #____ _____V _____°F	Cell #____ _____V _____°F	Cell #____ _____V _____°F
Cell #____ _____V _____°F	Cell #____ _____V _____°F	Cell #____ _____V _____°F	Cell #____ _____V _____°F	Cell #____ _____V _____°F	Cell #____ _____V _____°F	Cell #____ _____V _____°F
Cell #____ _____V _____°F	Cell #____ _____V _____°F	Cell #____ _____V _____°F	Cell #____ _____V _____°F	Cell #____ _____V _____°F	Cell #____ _____V _____°F	Cell #____ _____V _____°F
Cell #____ _____V _____°F	Cell #____ _____V _____°F	Cell #____ _____V _____°F	Cell #____ _____V _____°F	Cell #____ _____V _____°F	Cell #____ _____V _____°F	Cell #____ _____V _____°F
Cell #____ _____V _____°F	Cell #____ _____V _____°F	Cell #____ _____V _____°F	Cell #____ _____V _____°F	Cell #____ _____V _____°F	Cell #____ _____V _____°F	Cell #____ _____V _____°F

 NFPA 70B (p. 1 of 2)

FIGURE H.47 Rectifier Survey.

RECTIFIER SURVEY *(continued)*

Deficiencies Found	
Problems in the Past	
Customer's Concerns	
Notes	

 NFPA 70B (p. 2 of 2)

FIGURE H.47 ***Continued***

ELECTRICAL PANEL SURVEY

Date: ____________

Installation: ______________________ Location: ______________________

Panel Identification: ______________________

Manufacturer Name: ______________________ Panel Type/Model: ______________________

Voltage Rating: ________ V Current Rating: ________ A Phases: ________ # of Wires: ________

Main Breaker: Type/Model: ____________ Rating: ________ A Adjustable Setting Range: ________

Measured Voltages	V_{A-N} = ______ V V_{A-G} = ______ V	V_{B-N} = ______ V V_{B-G} = ______ V	V_{C-N} = ______ V V_{C-G} = ______ V	V_{N-G} = ______ V
Voltage Sine Waves	V_{A-N}	V_{B-N}	V_{C-N}	V_{N-G}
Harmonic Voltage Distortion	$V_{THD(A-N)}$ = ____ % 3rd = ______ % __th = ______ % __th = ______ %	$V_{THD(B-N)}$ = ____ % 3rd = ______ % __th = ______ % __th = ______ %	$V_{THD(C-N)}$ = ____ % 3rd = ______ % __th = ______ % __th = ______ %	$V_{THD(N-G)}$ = ____ % 3rd = ______ % __th = ______ % __th = ______ %
Measured Currents	I_A = ________ A	I_B = ________ A	I_C = ________ A	I_N = ________ A
Current Sine Waves	I_A	I_B	I_C	I_N
Harmonic Current Distortion	$I_{THD(A)}$ = ______ % 3rd = ______ % __th = ______ % __th = ______ %	$I_{THD(B)}$ = ______ % 3rd = ______ % __th = ______ % __th = ______ %	$I_{THD(C)}$ = ______ % 3rd = ______ % __th = ______ % __th = ______ %	$I_{THD(N)}$ = ______ % 3rd = ______ % __th = ______ % __th = ______ %

Power Factor	PF: ____________ Displacement Power Factor (DPF): ____________
Grounding System	Ground bus isolated from frame?____ or bonded to frame?____ Ground by metal conduits?____ or by ground conductors?____ Ground bus bonded to neutral bus? ❑ Yes ❑ No Each branch circuit has separated neutral ❑ Yes ❑ No and ground conductor? ❑ Yes ❑ No Ground current measurement: ______ A Ground resistance measurement: ______ Ω Sketch the existing grounding system (on the back sheet) when it is necessary.
Temperature	Bus temperature range: ______ °F Conductor temperature range: ______ °F CBs having temperature higher that 90°F: ____________
Entrance Conductor	Phases: ________ MCM Number of conductors per phase: ________
Sizes	Neutral: ________ MCM Number of conductors per phase: ________ Ground: ________ MCM Number of conductors per phase: ________
Lightning Protection	Manufacturer: ____________ Type: ________ Voltage rating: ________

 NFPA 70B (p. 1 of 2)

FIGURE H.48 Electrical Panel Survey.

ELECTRICAL PANEL SURVEY *(continued)*

Other Circuit Breakers in the Panel	CB rating: _________ A How many CB: _________ Conductor sizes: _________ CB rating: _________ A How many CB: _________ Conductor sizes: _________ CB rating: _________ A How many CB: _________ Conductor sizes: _________ CB rating: _________ A How many CB: _________ Conductor sizes: _________ CB rating: _________ A How many CB: _________ Conductor sizes: _________ CB rating: _________ A How many CB: _________ Conductor sizes: _________ CB rating: _________ A How many CB: _________ Conductor sizes: _________ CB rating: _________ A How many CB: _________ Conductor sizes: _________
Deficiencies Found	
Problems in the Past	
Customer's Concerns	
Notes	

 NFPA 70B (p. 2 of 2)

FIGURE H.48 ***Continued***

TRANSFER SWITCHES SURVEY

Date: ______________

Installation: ______________ Location: ______________

Transfer Switch Identification: ______________

Manufacturer Name: ______________ Model/Type: ______________ Serial #: ______________

Voltage Rating: __________V Current Rating: __________A Fuse sizes: ______________

Automatic or Manual: ______________ Phases: ________ # of Poles:_____ # of Wires:_____

Measured Voltages	$V_{A\text{-}G}$ = ________V $V_{A\text{-}N}$ = ________V	$V_{B\text{-}G}$ = ________V $V_{B\text{-}N}$ = ________V	$V_{C\text{-}G}$ = ________V $V_{C\text{-}N}$ = ________V	$V_{N\text{-}G}$ = ________V
Voltage Sine Waves	$V_{A\text{-}G}$	$V_{B\text{-}G}$	$V_{C\text{-}G}$	$V_{N\text{-}G}$
Harmonic Voltage Distortion	$V_{THD(A\text{-}G)}$ = ____% 3rd = ______% 5th = ______% __th = ______%	$V_{THD(B\text{-}G)}$ = ____% 3rd = ______% 5th = ______% __th = ______%	$V_{THD(C\text{-}G)}$ = ____% 3rd = ______% 5th = ______% __th = ______%	$V_{THD(N\text{-}G)}$ = ____% 3rd = ______% 5th = ______% __th = ______%
Measured Currents	I_A = ________A	I_B = ________A	I_C = ________A	I_N = ________A
Current Sine Waves	I_A	I_B	I_C	I_N
Harmonic Current Distortion	$I_{THD(A)}$ = ______% 3rd = ______% 5th = ______% __th = ______%	$I_{THD(B)}$ = ______% 3rd = ______% 5th = ______% __th = ______%	$I_{THD(C)}$ = ______% 3rd = ______% 5th = ______% __th = ______%	$I_{THD(N)}$ = ______% 3rd = ______% 5th = ______% __th = ______%

Grounding System	Neutral bus exists? ❑ Yes ❑ No Ground bus exists? ❑ Yes ❑ No Neutral bus bonded to ground bus at the transfer switch? ❑ Yes ❑ No Neutral conductors just run through transfer switch? ❑ Yes ❑ No Ground bus bonded to the frame? ❑ Yes ❑ No Ground bus connected to upstream source ground? ❑ Yes ❑ No Ground bus connected to downstream load ground? ❑ Yes ❑ No Ground resistance measurement: ________ Ω Ground current measurement: ________A Sketch the existing grounding system (on the back sheet) when it is necessary.
Temperatures	Bus temperatures: ________°F Conductor temperatures: ________°F
Power Factor	PF: ______________ Displacement Power Factor (DPF): ______________
Conductor Sizes	Normal source: Phases: ( )_____ Neutral: ( )_____ Ground: ( )_____ Emerg/Standby source: Phases: ( )_____ Neutral: ( )_____ Ground: ( )_____ Load side: Phases: ( )_____ Neutral: ( )_____ Ground: ( )_____

 NFPA 70B (p. 1 of 2)

FIGURE H.49 Transfer Switches Survey.

TRANSFER SWITCHES SURVEY *(continued)*

Operation Scheme of Transfer Switch	Make before break? ❑ Yes ❑ No Break before make? ❑ Yes ❑ No Time delay available? ❑ Yes ❑ No
Deficiencies Found	
Problems in the Past	
Customer's Concerns	
Notes	

 NFPA 70B (p. 2 of 2)

FIGURE H.49 ***Continued***

POWER TRANSFORMERS SURVEY

Date:________

Installation:________ Location:________

Transformer Identification:________

Type: ❑ Isolation ❑ Shielded Isolation ❑ Dry ❑ Oil ❑ Pad or Pole Mounted

Number of units:______ kVA Rating of each unit:______ Phases:______ Impedance:______ %

Load Tap Changing: ❑ Automatic ❑ Manual ❑ None

Cooling System (O/A/FA/etc.): Nameplate Power Factor: ______ Nameplate "k" Factor: ______

Measured Power Factor:______ Measured Displacement Power Factor:______ Measured "k" Factor:______

	High Voltage Sides				*Low Voltage Sides*			
Voltage Rating	______ V				______ V			
Current Rating	______ A				______ A			
Wiring Connection	(D/Y/GY):______				*(D/Y/GY):*______			
Measured Voltages	$V_{A\text{-}B}$	$V_{B\text{-}C}$	$V_{A\text{-}C}$		$V_{a\text{-}n}$	$V_{b\text{-}n}$	$V_{c\text{-}n}$	$V_{n\text{-}g}$
Harmonic Voltages	THD							
	3rd							
	5th							
	_th							
	_th							
Measured Currents	I_A	I_B	I_C		I_a	I_b	I_c	I_n
Harmonic Currents	THD							
	3rd							
	5th							
	_th							
	_th							
Conductor Sizes	Phases:____ # of Conductors/phase:____ Neutral:____ # of Conductors/phase:____ Ground:____ # of Conductors: ____				*Phases:____ # of Conductors/phase:____* *Neutral:____ # of Conductors/phase:____* *Ground:____ # of Conductors:____*			

 NFPA 70B (p. 1 of 2)

FIGURE H.50 Power Transformers Survey.

POWER TRANSFORMERS SURVEY *(continued)*

Temperature	Winding temperature range: _________°F Bus temperature range: _________°F Enclosure temperature: _________°F
Physical Conditions	❑ Good condition ❑ Damage ❑ Corrosion ❑ Fluid leakage Proper mounting: _______ Proper clearance: _______ Need maintenance: _______ Sight of burning/overheat (color change): _______
Grounding System	Ground bus connected to transformer frame? ❑ Yes ❑ No Ground bus connected to upstream source ground? ❑ Yes ❑ No Ground bus connected to downstream load ground? ❑ Yes ❑ No Ground bus connected to building metal frame? ❑ Yes ❑ No Ground current measurement: _________A Ground resistance measurement: _________ Ω
Deficiencies Found	
Problems in the Past	
Customer's Concerns	
Notes	

 NFPA 70B (p. 2 of 2)

FIGURE H.50 ***Continued***

UNINTERRUPTIBLE POWER SYSTEM SURVEY

Date: ____________

Installation: ____________ Location: ____________

UPS System Identification: ____________

Number of modules: ________ Do these modules have the same sizes and characteristics? ❑ Yes ❑ No

Module #1

Manufacturer: ____________

Model/Type: ____________

Frequencies: Input: ________ Hz Output: ________ Hz

Power Factor: Input: ________ Output: ________

Wiring Connection: Input: ❑ 3-Phases/3 Wires ❑ 3-Phases/4 Wires ❑ Single-Phase

Output: ❑ 3-Phases/3 Wires ❑ 3-Phases/4 Wires ❑ Single-Phase

Grounding System: Ground Current: ________ A Ground Resistance: ________ Ω

Input Voltages: Rated: ______ V Measured $V_{A\text{-}B}$ = ______ V $V_{B\text{-}C}$ = ______ V $V_{A\text{-}C}$ = ______ V

Input Currents: Rated: ______ A Measured I_A = ______ A I_B = ______ A I_C = ______ A

DC Link Voltages: Rated: ______ V Measured $V_{A\text{-}B}$ = ______ V $V_{B\text{-}C}$ = ______ V $V_{A\text{-}C}$ = ______ V

DC Link Currents: Rated: ______ A Measured I_A = ______ A I_B = ______ A I_C = ______ A

Output Voltages: Rated: ______ V Measured $V_{A\text{-}B}$ = ______ V $V_{B\text{-}C}$ = ______ V $V_{A\text{-}C}$ = ______ V

Output Currents: Rated: ______ A Measured I_A = ______ A I_B = ______ A I_C = ______ A

kVA: Rated: ______ Measured Input: ______ ______ ______ Measured Output: ______ ______ ______

kW: Rated: ______ Measured Input: ______ ______ ______ Measured Output: ______ ______ ______

Module #2 (If they are of different sizes/characteristics)

Manufacturer: ____________

Model/Type: ____________

Frequencies: Input: ________ Hz Output: ________ Hz

Power Factor: Input: ________ Output: ________

Wiring Connection: Input: ❑ 3-Phases/3 Wires ❑ 3-Phases/4 Wires ❑ Single-Phase

Output: ❑ 3-Phases/3 Wires ❑ 3-Phases/4 Wires ❑ Single-Phase

Grounding System: Ground Current: ________ A Ground Resistance: ________ Ω

Input Voltages: Rated: ______ V Measured $V_{A\text{-}B}$ = ______ V $V_{B\text{-}C}$ = ______ V $V_{A\text{-}C}$ = ______ V

Input Currents: Rated: ______ A Measured I_A = ______ A I_B = ______ A I_C = ______ A

DC Link Voltages: Rated: ______ V Measured $V_{A\text{-}B}$ = ______ V $V_{B\text{-}C}$ = ______ V $V_{A\text{-}C}$ = ______ V

DC Link Currents: Rated: ______ A Measured I_A = ______ A I_B = ______ A I_C = ______ A

Output Voltages: Rated: ______ V Measured $V_{A\text{-}B}$ = ______ V $V_{B\text{-}C}$ = ______ V $V_{A\text{-}C}$ = ______ V

Output Currents: Rated: ______ A Measured I_A = ______ A I_B = ______ A I_C = ______ A

kVA: Rated: ______ Measured Input: ______ ______ ______ Measured Output: ______ ______ ______

kW: Rated: ______ Measured Input: ______ ______ ______ Measured Output: ______ ______ ______

 NFPA 70B (p. 1 of 2)

FIGURE H.51 Uninterruptible Power System Survey.

UNINTERRUPTIBLE POWER SYSTEM SURVEY *(continued)*

Batteries

Manufacturer:__________ Model:__________ Type:__________
Number of battery banks:__________ Number of cells per bank:__________
Measured total voltage of each bank:______ ______ ______ ______ ______ V
Measured total current of each bank:______ ______ ______ ______ ______ A
Battery rack condition:__________ Battery rack properly grounded? ❑ Yes ❑ No
Battery bank terminal grounded? ❑ Yes ❑ No
Battery bank switch (3 or 4 poles):__________ Switch properly grounded? ❑ Yes ❑ No
Conductor sizes: Phases: ( ) ______ Neutral: ( ) ______ Ground: ( ) ______
Ground current measurement:______ A Ground resistance measurement:______ Ω
Batteries properly mounted? ❑ Yes ❑ No Proper ventilation? ❑ Yes ❑ No
Fluid level checked:__________ Battery fluid specific gravity last checked:__________
Battery physical conditions: ❑ Damage ❑ Corrosion ❑ Leakage ❑ Need maintenance

Measured Battery Cell Voltages and Fluid Temperatures

Cell #____ ____V ____°F	Cell #____ ____V ____°F	Cell #____ ____V ____°F	Cell #____ ____V ____°F	Cell #____ ____V ____°F	Cell #____ ____V ____°F	Cell #____ ____V ____°F
Cell #____ ____V ____°F	Cell #____ ____V ____°F	Cell #____ ____V ____°F	Cell #____ ____V ____°F	Cell #____ ____V ____°F	Cell #____ ____V ____°F	Cell #____ ____V ____°F
Cell #____ ____V ____°F	Cell #____ ____V ____°F	Cell #____ ____V ____°F	Cell #____ ____V ____°F	Cell #____ ____V ____°F	Cell #____ ____V ____°F	Cell #____ ____V ____°F
Cell #____ ____V ____°F	Cell #____ ____V ____°F	Cell #____ ____V ____°F	Cell #____ ____V ____°F	Cell #____ ____V ____°F	Cell #____ ____V ____°F	Cell #____ ____V ____°F
Cell #____ ____V ____°F	Cell #____ ____V ____°F	Cell #____ ____V ____°F	Cell #____ ____V ____°F	Cell #____ ____V ____°F	Cell #____ ____V ____°F	Cell #____ ____V ____°F
Cell #____ ____V ____°F	Cell #____ ____V ____°F	Cell #____ ____V ____°F	Cell #____ ____V ____°F	Cell #____ ____V ____°F	Cell #____ ____V ____°F	Cell #____ ____V ____°F

Deficiencies Found

____
____
____
____

Notes

____
____
____
____
____

 NFPA 70B (p. 2 of 2)

FIGURE H.51 ***Continued***

LOW VOLTAGE BREAKER DATA RECORD

Site: ______________________ Date: ______________ Page: ______________

No.	Location	Breaker					Relay									
		Mfgr.	Type	Rated Voltage	Frame Rating	Interr. Time	Mfgr.	Type	Long Time Delay		Short Time Delay		Instant	Ground	
									Range	Bands Avail.	Range	Bands Avail.	Range	Range	Bands Avail.

 NFPA 70B

FIGURE H.52 Low Voltage Breaker Data Record.

RECLOSER DATA RECORD

Site: ______________ Date: ______________ Page: ______________

No.	Location	CT Ratio	Nominal Voltage	Mfgr.	Type	BIL	Continuous Current Rating	Interruption Rating	Minimum Trip	Hydrau or Electro	Operation Sequence Available	Reclosing Times Available	Tripping Curves Available

 NFPA 70B

FIGURE H.53 Recloser Data Record.

GENERATOR DATA RECORD

Site: ______ Date: ______ Page: ______

No.	Location	Type	kVA Rating	Generated Volts	Rated Current	Speed in RPM	Wiring Connections	Subtrans. Impedance	Ground Impedance	Power Factor

 NFPA 70B

FIGURE H.54 Generator Data Record.

Annex I NEMA Configurations

This annex is not a part of the recommendations of this NFPA document but is included for informational purposes only.

I.1 Figure I.1 shows the typical NEMA configurations for general-purpose nonlocking plugs and receptacles.

I.2 Figure I.2 shows the typical NEMA configurations for locking plugs and receptacles.

DESCRIPTION		NEMA NUMBER	15 AMPERE		20 AMPERE		30 AMPERE		50 AMPERE		60 AMPERE	
			RECEPTACLE	PLUG	RECEPTACLE	PLUG	RECEPTACLE	PLUG	RECEPTACLE	PLUG	RECEPTACLE	PLUG
2-POLE 2-WIRE	125V	1	1-15R	1-15P POLARIZED / NON POLAR.		2-20P MATES WITH 5-20R		1-30P MATES WITH 5-30R				
	250V	2		2-15P MATES WITH 6-15R	2-20R	2-20P	2-30R	2-30P				
	277V AC	3										
	600V	4										
2-POLE 3-WIRE GROUNDING	125V	5	5-15R	5-15P	5-20R	5-20P	5-30R	5-30P	5-50R	5-50P		
	125V	5ALT			5ALT-20R							
	250V	6	6-15R	6-15P	6-20R	6-20P	6-30R	6-30P	6-50R	6-50P		
	250V	6ALT			6ALT-20R							
	277V AC	7	7-15R	7-15P	7-20R	7-20P	7-30R	7-30P	7-50R	7-50P		
	347V AC	24	24-15R	24-15P	24-20R	24-20P	24-30R	24-30P	24-50R	24-50P		
	480V AC	8										
	600V AC	9										
3-POLE 3-WIRE	125 / 250V	10			10-20R	10-20P	10-30R	10-30P	10-50R	10-50P		
	3 Ø 250V	11	11-15R	11-15P	11-20R	11-20P	11-30R	11-30P	11-50R	11-50P		
	3 Ø 480V	12										
	3 Ø 600V	13										
3-POLE 4-WIRE GROUNDING	125 / 250V	14	14-15R	14-15P	14-20R	14-20P	14-30R	14-30P	14-50R	14-50P	14-60R	14-60P
	3 Ø 250V	15	15-15R	15-15P	15-20R	15-20P	15-30R	15-30P	15-50R	15-50P	15-60R	15-60P
	3 Ø 480V	16										
	3 Ø 600V	17										
4-POLE 4-WIRE	3 Ø Y 120 / 280V	18	18-15R	18-15P	18-20R	18-20P	18-30R	18-30P	18-50R	18-50P	18-60R	18-60P
	3 Ø Y 277 / 480V	19										
	3 Ø Y 347 / 600V	20										
4-POLE 5-WIRE GROUNDING	3 Ø Y 120 / 208V	21										
	3 Ø Y 277 / 480V	22										
	3 Ø Y 347 / 600V	23										

Note: Blank spaces reserved for future configurations.

FIGURE I.1 NEMA Configurations for General-Purpose Nonlocking Plugs and Receptacles.

DESCRIPTION		NEMA NUMBER	15 AMPERE RECEPTACLE	15 AMPERE PLUG	20 AMPERE RECEPTACLE	20 AMPERE PLUG	30 AMPERE RECEPTACLE	30 AMPERE PLUG	50 AMPERE RECEPTACLE	50 AMPERE PLUG	60 AMPERE RECEPTACLE	60 AMPERE PLUG
2–POLE 2–WIRE	125V	1	L 1–15R	L 1–15P								
2–POLE 2–WIRE	250V	2			L 2–20R	L 2–20P						
2–POLE 2–WIRE	277V AC	3										
2–POLE 2–WIRE	600V	4										
2–POLE 3–WIRE GROUNDING	125V	5	L 5–15R	L 5–15P	L 5–20R	L 5–20P	L 5–30R	L 5–30P	L 5–50R	L 5–50P	L 5–60R	L 5–60P
2–POLE 3–WIRE GROUNDING	250V	6	L 6–15R	L 6–15P	L 6–20R	L 6–20P	L 6–30R	L 6–30P	L 6–50R	L 6–50P	L 6–60R	L 6–60P
2–POLE 3–WIRE GROUNDING	277V AC	7	L 7–15R	L 7–15P	L 7–20R	L 7–20P	L 7–30R	L 7–30P	L 7–50R	L 7–50P	L 7–60R	L 7–60P
2–POLE 3–WIRE GROUNDING	347V AC	24			L 24–20R	L 24–20P						
2–POLE 3–WIRE GROUNDING	480V AC	8			L 8–20R	L 8–20P	L 8–30R	L 8–30P	L 8–50R	L 8–50P	L 8–60R	L 8–60P
2–POLE 3–WIRE GROUNDING	600V AC	9			L 9–20R	L 9–20P	L 9–30R	L 9–30P	L 9–50R	L 9–50P	L 9–60R	L 9–60P
3–POLE 3–WIRE	125 / 250V	10			L 10–20R	L 10–20P	L 10–30R	L 10–30P				
3–POLE 3–WIRE	3 Ø 250V	11	11–15R	11–15P	L 11–20R	L 11–20P	L 11–30R	L 11–30P				
3–POLE 3–WIRE	3 Ø 480V	12			L 12–20R	L 12–20P	L 12–30R	L 12–30P				
3–POLE 3–WIRE	3 Ø 600V	13					L 13–30R	L 13–30P				
3–POLE 4–WIRE GROUNDING	125 / 250V	14			L 14–20R	L 14–20P	L 14–30R	L 14–30P	L 14–50R	L 14–50P	L 14–60R	L 14–60P
3–POLE 4–WIRE GROUNDING	3 Ø 250V	15			L 15–20R	L 15–20P	L 15–30R	L 15–30P	L 15–50R	L 15–50P	L 15–60R	L 15–60P
3–POLE 4–WIRE GROUNDING	3 Ø 480V	16			L 16–20R	L 16–20P	L 16–30R	L 16–30P	L 16–50R	L 16–50P	L 16–60R	L 16–60P
3–POLE 4–WIRE GROUNDING	3 Ø 600V	17					L 17–30R	L 17–30P	L 17–50R	L 17–50P	L 17–60R	L 17–60P
4–POLE 4–WIRE	3 Ø Y 120 / 208V	18			L 18–20R	L 18–20P	L 18–30R	L 18–30P				
4–POLE 4–WIRE	3 Ø Y 277 / 480V	19			L 19–20R	L 19–20P	L 19–30R	L 19–30P				
4–POLE 4–WIRE	3 Ø Y 347 / 600V	20			L 20–20R	L 20–20P	L 20–30R	L 20–30P				
4–POLE 5–WIRE GROUNDING	3 Ø Y 120 / 208V	21			L 21–20R	L 21–20P	L 21–30R	L 21–30P	L 21–50R	L 21–50P	L 21–60R	L 21–60P
4–POLE 5–WIRE GROUNDING	3 Ø Y 277 / 480V	22			L 22–20R	L 22–20P	L 22–30R	L 22–30P	L 22–50R	L 22–50P	L 22–60R	L 22–60P
4–POLE 5–WIRE GROUNDING	3 Ø Y 347 / 600V	23			L 23–20R	L 23–20P	L 23–20R	L 23–20P	L 23–50R	L 23–50P	L 23–60R	L 23–60P

Note: Blank spaces reserved for future configurations.

FIGURE I.2 NEMA Configurations for Locking Plugs and Receptacles.

Annex J Primary Contact Matrix

This annex is not a part of the recommendations of this NFPA document but is included for informational purposes only.

J.1 Primary Contact Matrix.

PRIMARY CONTACT MATRIX

		Primary Contacts	Other
Facility Manager			
Maintenance Manager			
Information Technology (IT) Manager			
Hazardous Material (Haz Mat) Team			
Site Security			
Medical Response Team			
Fire Response Team			
Other Responders			
System Affected			
Power delivery systems	Generator(s)		
	Power equipment failure (cables, switchgear, circuit breakers, SCADA)		
	UPS systems		
HVAC systems	Make-up water not available		
	Primary components fail (pump, filter, etc.)		
	Make water support equipment (pipes, valves)		
	HVAC controls		
	HVAC zones control (pneumatic)		
	CPU monitored/ controlled		
	HVAC bldg. zone controls		
Building security and egress systems	Video monitoring, personnel entry/exit surveillance systems, exit/egress lighting		
	Fire sensors, pathogens detection-response systems		
	Fire suppression activation/failure		
	Egress system failure (doors, elevation)		

 NFPA 70B (1 of 2)

FIGURE J.1 Primary Contact Matrix.

PRIMARY CONTACT MATRIX *(continued)*

		Primary Contacts	Other
Potential emergency/ failures from natural/ human actions	Utility supply		
	Floods (nature), floods (broken piping)		
	Snow		
	Fire (wild and facility)		
Meteorological hazards	Windstorms (hurricanes, tornadoes, cyclones)		
	Temperature extremes		
	Lightning strike/surge events		
Biological events	"Normal" diseases, exotic diseases		
	Animal/insect infestations		
Accidental (human or equipment caused)	Hazards (chemical, radiological, biological) spill or release into environment (air, water)		
	Internal site explosions		
	Fuel/resource outages		
	Internal communications equipment failure, human communication failure		
	HVAC equipment/controls failure		
	CPU server failure		
	Terminal/video readout failure		
	Security equipment failure		
	Fire suppression equipment failure		
	Human/animal sickness		
Accidental external utility related disruptions	Utility power disruptions		
	Domestic water utility disruptions/pollution/ contaminations		
	Fuel gases interruptions		
	External communications failure		

 NFPA 70B (2 of 2)

FIGURE J.1 ***Continued***

Annex K Long-Term Maintenance Guidelines

This annex is not a part of the recommendations of this NFPA document but is included for informational purposes only.

K.1 Introduction. This annex deals specifically with the maintenance of equipment that, by nature of its application, necessitates long intervals between shutdowns. It should be stressed that environmental or operating conditions of a specific installation should be considered and might dictate a different frequency of maintenance than suggested in this annex.

Maintenance guidelines are presented in Sections K.2 through K.4.

K.2 Medium-Voltage Equipment. Table K.2(a) through Table K.2(k) address equipment that should be considered items with long-term maintenance intervals, including the following equipment and techniques:

(1) Cables, terminations, and connections
(2) Liquid-filled transformers
(3) Dry-type transformers
(4) Metal-clad switchgear
(5) Circuit breakers
(6) Metal-enclosed switches
(7) Buses and bus ducts
(8) Protective relays
(9) Automatic transfer control equipment
(10) Circuit breaker overcurrent trip devices
(11) Fuses
(12) Lightning arresters

K.3 Medium- and Low-Voltage Equipment. The medium- and low-voltage equipment that should be considered items for long-term maintenance intervals are outside overhead electric lines.

Table K.3 shows medium- and low-voltage equipment, outside overhead electric lines; maintenance of equipment subject to long intervals between shutdowns — electrical distribution.

K.4 Low-Voltage Equipment. Table K.4(a) through Table K.4(k) address equipment that should be considered items for long-term maintenance intervals. This includes the following equipment and techniques:

(1) Low-voltage cables and connections
(2) Dry-type transformers
(3) Switchgear
(4) Drawout-type circuit breakers
(5) Buses and bus ducts
(6) Panelboards
(7) Protective relays
(8) Automatic transfer control equipment
(9) Circuit breaker overcurrent trip devices
(10) Fuses
(11) Lighting arresters

Table K.2(a) Medium-Voltage Equipment, Cables, Terminations, and Connections: Maintenance of Equipment Subject to Long Intervals Between Shutdowns — Electrical Distribution

Type	Inspections, Maintenance, and Tests	Typical Frequency and Remarks
Solid Dielectric (Chapter 19)	**Inspections (while energized) (19.2.1):**	Annually.
	Conduit entrances (19.4).	Observe for deformation due to pressure and for bends with radius less than minimum allowed.
	Poles and supports.	Same as above.
	0+Binder tape terminations (aerial cables) (19.3).	Same as above.
	Ends of trays (19.4).	Same as above.
	Splices (19.2.3).	Same as above.
	Terminations (stress cones and potheads) (19.2.3, 19.2.5).	Same as above plus dirt, tracking, water streaks, chipped porcelain, shield ground connections (where visible), and adequate clearances from grounded metal parts. Observe for continuity.
	Fireproofing (where required) (19.2.3).	Make certain loads are within cable ampacity rating.
	Loading.	
Varnished Cambric Lead Covered and Paper Insulated Lead Covered	**Inspections (while energized) (19.2.1):**	
	Same as above.	Same as above.
	Lead sheath (19.2.3).	Observe for cracks or cold wipe joints, often indicated by leakage of cable oil or compound.
All Types	**Major Maintenance and Test (deenergized) (7.1, 19.2.1):**	3–6 years.
	Complete inspection same as above.	Same as above.
	Clean and inspect porcelain portions of potheads (19.2.5, 15.1.2.1).	For cracks and chips.
	Clean and inspect stress cones and leakage sections (19.2.3, 15.2.14).	For soundness of stress cones. X-ray or disassemble, if soft spots are detected. For surface tracking.
	Check plastic jackets for longitudinal shrinkage from splices and terminations.	Jacket shrinkage might have damaged shielding tapes or stress cones.

Table K.2(a) ***Continued***

Type	Inspections, Maintenance, and Tests	Typical Frequency and Remarks
	Check integrity of shield grounding (19.2.3).	Observe ground connections for stress cones. Suggest checking electrical continuity of shielding tape.
	Check general condition of cable (19.2.3).	Does insulating material appear to have been damaged by overheating?
	Observe connectors for overheating (19.2.5, 15.1.3, 15.2.15).	Discoloration or oxidation indicates possible problem. Check bolts for tightness, if accessible. If connectors are insulated with tape, deterioration or charring of tape is indicative of overheated connector, caused by loose bolts, etc. Infrared survey while conductors are energized and loaded to at least 40 percent of ampacity might be beneficial to detect overheated connections. Use good-quality infrared scanning equipment.
	Test cable insulation with high potential dc (19.5, 11.9.1).	Disconnect cables from equipment and provide corona protection on ends. Ground other conductors not being tested. Record leakage current in microamperes at each test voltage level. Record temperature and relative humidity.
	Determine condition of cable insulation (11.9.2.6).	Interpret test results, considering length of cable, number of taps, shape of megohm or leakage current curve, temperature and relative humidity.
	Reconnect cables to equipment.	Tighten connectors adequately.
	Aluminum conductors.	Make certain that connectors of the proper type are correctly installed. Use Belleville washers when bolting aluminum cable lugs to equipment. Advisable to determine conductivity of connection using microhmmeter or determine voltage drop under test load conditions.

Table K.2(b) Medium-Voltage Equipment, Liquid-Filled Transformers: Maintenance of Equipment Subject to Long Intervals Between Shutdowns — Electrical Distribution

Type	Inspections, Maintenance, and Tests	Typical Frequency and Remarks
Oil and Askarel Sealed Tank, Conservator and Gas Sealed Systems (Chapter 21)	**Inspections (while energized):**	
	Top liquid temperature (21.2.4).	Weekly to monthly. Record findings. Current temperature and highest indicated. Reset drag needle; 80°C (176°F) nominal max. permitted.
	Head space pressure (sealed-tank type) (21.2.5.2).	Should vary under changes in loading and ambient temperature. If gauge remains at zero, gauge is broken or leak exists in tank head space, which permits transformer to breathe and allows entrance of moisture.
	Nitrogen pressure (pressurized-tank type).	Check nitrogen bottle pressure and pressure in transformer head space.
	Liquid level in tanks (21.2.5.1).	Should be between min. and max. marks on gauge.
	Liquid levels in oil-filled bushings (if so equipped).	Should be between min. and max. marks on gauge.
	Evidence of oil leaks (21.2.7.4).	From tanks, fittings, cooling tubes, and bushings.
	Automatic load tap changer mechanism.	General condition; note and record number of operations.
	Tests (while energized):	
	Oil — draw sample and test in laboratory (21.2.8).	Annually for normal service transformers, biannually for rectifier and arc furnace transformers. Dielectric strength, acidity, and color. If dielectric is low, determine water content.
	Askarel — draw sample and test in laboratory (21.2.8). (Observe EPA regulations for handling and disposal.)	Same frequency as for oil. Dielectric strength, acidity, color, and general condition. If dielectric is low determine water content.
	Comprehensive liquid tests.	3–6 years. In addition to above, tests include interfacial tension, water content, refractive index power factor at 25°C (77°F) and 100°C (212°F) (20.9.3.2) corrosive sulfur (askarel), and inclusion of cellulose material.
	Dissolved gas content in liquid of transformers in critical service or in questionable condition as might be indicated by above liquid tests (11.11.10).	6 years or as conditions indicate. Draw sample in special container furnished by test laboratory. Spectrophotometer test detects gases in oil caused by certain abnormal conditions in transformer. A series of tests on samples drawn over a period of time might be necessary to determine if abnormal condition exists and to determine problem. Devices are available for installation on transformers to collect gases to be tested for combustibility to determine if internal transformer problem exists.
	Major Maintenance and Testing (deenergized) (7.1, 21.2.7.2):	3–6 years or more often if above tests indicate.
	Make above tests well in advance of scheduled shutdown.	Determine possible problems that require attention.

Table K.2(b) ***Continued***

Type	Inspections, Maintenance, and Tests	Typical Frequency and Remarks
	Inspect pressure-relief diaphragm for cracks or holes or mechanical pressure-relief device for proper operation (21.2.7.3).	Replace if defective. Possible cause of pressure in sealed-type transformers remaining at zero.
	Pressure test with dry nitrogen the head space areas of sealed-type transformers if pressure gauge remains at zero and pressure relief device is satisfactory.	Apply liquid along seams, etc., to locate leaks. Make necessary repairs.
	Clean bushings and inspect surfaces (21.2.7.3).	Consider application of silicone grease in badly contaminated areas; should be removed and reapplied at maximum 2-year intervals, preferably 1 year.
	Inspect load tap changer mechanism and contact.	Follow manufacturer's instructions on maintenance and number of operations between contact replacements.
	Paint tank as required.	Wire-brush and prime rust spots. Finish paint.
	Check ground system connections (21.2.7.5). Perform turns ratio test (11.11.2).	In each tap position; as an acceptance test and after major repairs.
	Perform power factor tests (disconnect from equipment) (11.9.3.2).	Windings, bushings, and insulating liquid.
	Consider making winding/tap changer resistance tests.	Use microhmmeter in each tap position to detect abnormally high contact resistance.
	Make undercover inspection through manholes (provide positive protection to prevent entrance of moisture) (11.2.7). This inspection might not be necessary at 6-year intervals unless tests indicate problems.	6-year frequency should definitely be considered for rectifier and arc furnace transformers. Inspect for moisture or rust under cover, water on horizontal surfaces under oil, tap changer contacts (insofar as possible), trash, oil sludge deposits, loose bracing, and loose connections.
	Consider high-potential dc tests (11.9.2.6) (11.5 through 11.8).	dc in excess of 34 kV can polarize liquid and thereby increase leakage currents.
	If above inspections and/or tests indicate possible internal problems, it might be necessary to transport transformer to shop to untank the core and coil assembly for cleaning, inspecting, testing and making repairs as found necessary.	
	Filtering insulating liquid (deenergize transformer and ground windings).	Frequency as required. Remove moisture by heating and pumping liquid through cellulose filters, a centrifuge or a vacuum dehydrator. Thoroughly clean hose and filtering equipment before switching from oil to askarel or vice versa (21.2.1.2). Observe ANSI C107.1 for handling and disposal of askarel.
	Re-refining insulating oil (deenergize transformer and ground windings).	Frequency as required. Filter through fuller's earth to remove polar compounds and acids. Add dibutylparacresol to replace oxidation inhibitors.
	Refilling transformer with insulating liquid (21.2.7.7, 21.2.7.8).	Refill under partial vacuum if transformer tank is so designed. Follow manufacturer's instructions. Always test insulating liquid for dielectric strength (min. 26 kV for oil) prior to pumping into transformer and pump through filter (min. 30 kV askarel).
	Special Testing (deenergized):	To test phase-to-phase and turn-to-turn insulation (200 Hz to 300 Hz for 7200-volt cycles).
	Induced potential test (11.11.2.1).	Proof test.
	ac high potential test (11.9.3.1)	Proof test.

Table K.2(c) Medium-Voltage Equipment, Dry-Type Transformers: Maintenance of Equipment Subject to Long Intervals Between Shutdowns — Electrical Distribution

Type	Inspections, Maintenance, and Tests	Typical Frequency and Remarks
Ventilated (indoors) (21.1, 21.3)	**Inspections (while energized):**	Weekly to monthly. Record findings.
	Operating temperature (21.3.4).	Current temperature and highest indicated. Reset drag needle. 150°C (302°F) is max. operating temperature for transformers rated 80°C (176°F) rise. 220°C (428°F) is max. operating temperature for transformers rated 150°C (302°F) rise.
	Cleanliness of screens located over or behind ventilation louvers in enclosure (21.3.5.1).	Clogged screens restrict ventilation and thereby increase operating temperature of core and coil assembly. If dust and lint are on outside of screens, vacuum screens without deenergizing transformer. If dust and lint are on inside, transformer should be deenergized and enclosure sides removed to clean screens.
	Ventilating fan operation (if so equipped).	Check operation of fans with control switch in "Manual" position. Do not operate fans continuously with switch in "Manual"; leave in "Automatic" so temperature detectors will operate fans at temperatures above specified levels. Also check alarm contacts for proper operation at excessive temperature levels.
	Room ventilation (21.3.5.1)	Adequate ventilation system to admit and exhaust air. Air streams should not be directed toward upper vent louvers in transformer enclosure because doing so would restrict ventilation inside transformer and cause overheating.
	Evidence of condensation and water leaks in room (21.3.5.1).	Inspect top of transformer. Make necessary corrections.
	Major Maintenance (deenergized) (7.1, 21.3.5):	3–6 years, more often if required.
	Remove enclosure covers and clean vent louvers and screens (21.3.5.5).	
	Clean insulators, core, and windings (21.3.5.5, 21.3.5.6).	Use bottle of dry nitrogen with pressure regulator, hose and small nozzle to blow off dust. Restrict pressure to 207 kPa (30 psi max). Clean with soft bristle brush as required.
	Inspect following components:	
	Interphase barriers (21.3.5.5).	Should not touch windings.
	Wedges and clamping rings (21.3.5.5).	For proper clamping of windings; tighten as required.
	Primary and secondary buses and conductors (15.1.3, 21.3.5.5).	For tightness of connections.
	Porcelain insulators (15.1.2).	For chips, cracks, and water streaks.
	Insulating materials (15.2.11 through 15.2.15, 21.3.5.5).	For surface tracking.
	Windings (15.2.15, 21.3.5.5, 21.3.5.6).	For damage to insulation, including overheating.
	Tap connections (21.3.5.5).	For tightness and correctness to provide proper voltage.
	Core assembly.	For loose or dislocated laminations, for localized or general overheating, and for integrity of ground strap, which is *only* place where core assembly is permitted to be grounded.
	Ventilating channels between core and windings and between windings (21.3.5.6).	For clogging with lint, dust, or tape used to hold spacers, etc., in place during assembly; clean as required to allow proper air flow.

Table K.2(c) *Continued*

Type	Inspections, Maintenance, and Tests	Typical Frequency and Remarks
	Space heaters for proper operation.	Used to keep windings dry when transformer is deenergized.
	Temperature detectors.	For proper location and proper support of leads.
	Temperature indicators.	For accuracy and operation of fan and alarm contacts at proper temperatures.
	Cooling fans.	For free turning and proper operation.
	Testing (deenergized) (11.1, 11.4 through 11.8):	3–6 years, more often if required.
	Turns-ratio test (11.11.2).	In each tap position as an acceptance test and after major repairs.
	Polarization index (PI) test (11.11.2, 11.11.9, 11.9).	Use 1000-volt insulation resistance tester. Low PI results often indicate moisture in winding; investigate cause and satisfactorily dry transformer before making high potential dc test and returning transformer to service.
	High-potential dc test (11.11.10, 11.9.2).	Record leakage currents in microamperes, temperature, and relative humidity.
	Special Testing (deenergized):	
	Induced potential test (11.11.2).	To test phase-to-phase and turn-to-turn insulation (200 Hz to 300 Hz for 7200-volt cycles). Proof test.
	ac high-potential test (11.9.3).	Proof test.

Table K.2(d) Medium-Voltage Equipment, Metal-Clad Switchgear: Maintenance of Equipment Subject to Long Intervals Between Shutdowns — Electrical Distribution

Type	Inspections, Maintenance, and Tests	Typical Frequency and Remarks
Indoor (Chapter 15)	**Inspections (while energized):**	3–6 months.
	Open external doors and inspect components:	
	Fronts of circuit breakers.	Record number of operations.
	Protective and control relays (15.9.7).	Wiring and connections, not internals.
	Auxiliary devices, wiring, and terminal blocks (15.4.6).	Proper indicating lights should light.
	Space heaters (15.2.8).	Operate continuously to overcome possible malfunction of thermostats. Consider installation of ammeters in heater supply circuits to monitor full load current of heaters on each circuit to ensure that all are operating.
	Ventilation (15.2.9).	Ventilation louvers should be open.
	Insulators and insulating materials (15.2.11 through 15.2.15).	
	Cable terminations (19.1 through 19.4).	Observe stress cones and leakage sections annually for cleanliness and tracking.
	Batteries (15.9.4).	
	Also inspect for following conditions:	
	Loading.	Record loads.
	Cleanliness (15.2.10).	Moderate amount of dry nonconductive dust not harmful.
	Dryness (15.2.6, 15.2.7).	Evidence of condensation or water leaks.
	Rodents and reptiles (15.2.5).	
	Overheating of parts (15.2.15).	Discoloration or oxidation indicates possible problem.
	Tracking on insulating surfaces (15.2.14).	Take necessary corrective action.

(continues)

Table K.2(d) *Continued*

Type	Inspections, Maintenance, and Tests	Typical Frequency and Remarks
	Major Maintenance or Overhaul:	3–6 years, depending on ambient conditions.
	Deenergize (7.1). **Verify that no parts of the power or control circuitry are energized by "back feed" from alternate power or control sources.** Completely clean, inspect, tighten, and adjust all components (15.4.1):	Follow manufacturer's maintenance instructions.
	Structure and enclosure (15.2.4, 15.2.5).	Wire-brush and prime rust spots. Finish paint.
	Ventilating louvers and air filters (15.2.9).	Clean or replace filters as required.
	Buses, splices, and bolts (15.1.3, 15.2.15).	Check bolts for manufacturer's recommended torque. If inaccessible, check insulating tape, boot, or compound box over bus splices for heat deterioration due to loose bolts, etc.
	Insulators and insulating materials (15.1.2, 15.2.11 through 15.2.15).	Clean and inspect for surface tracking.
	Circuit breakers (15.4 through 15.6).	Refer to oil and air circuit breaker sections.
	Breaker disconnect studs and finger clusters (15.4.3.8).	Lubricate, unless manufacturer's instructions specify that they should not be lubricated.
	Drawout breaker racking mechanisms (15.1.7).	Alignment and ease of operation.
	Cable terminations and connections (19.1 through 19.4).	Clean and inspect for surface tracking. Check connections for tightness.
	Meters (15.9.7).	Test for accuracy.
	Controls, interlocks, and closing power rectifiers (15.9.8).	Make functional tests. Check voltages.
	CTs, PTs, and control power transformers (15.9.5).	
	Fuseclips and fuses (18.2).	Check clips for adequate spring pressure.
	Grounding (15.1.5, 15.9.9).	Proper fusing.
	Components and conditions in above block.	Make necessary repairs.
	Testing (7.1, Chapter 11):	3–6 years, depending on ambient conditions.
	Test buses, breakers, PTs, CTs, and cables with high-potential dc.	Record leakage currents in microamperes (19.5, 11.9.2.6).
	Calibrate and test protective relays (11.12).	Refer to protective relays section.
	Functionally trip breakers with relays (11.12.2).	Preferably, inject test current into CT and relay circuits.
	Test conductivity of aluminum cable connections (15.1.3).	Use microhmmeter or determine voltage drop under test load conditions.
	Test wiring for controls, meters and protective relays for insulation resistance (11.9.2.3).	1000-volt dc for control wiring. 500-volt dc for meters and relays.
Outdoor	**Inspections (while energized):**	1–3 months.
	Same as for indoor gear except:	
	Special emphasis on evidence of condensation and water leaks (15.2.4, 15.2.6, 15.2.7).	Rust spots on underside of metal roof indicative of condensate.
	Special emphasis on space heater operation (15.2.8).	
	Ventilating louvers and air filters (15.2.9).	Clean or replace air filters as required.
	Major Maintenance or Overhaul:	3 years, more often if conditions require.
	Deenergize (7.1). **Verify that no parts of the power or control circuitry are energized by "back feed" from alternate power or control sources.** Same as for indoor gear.	Follow manufacturer's maintenance instructions.
	Testing (7.1, Chapter 11):	
	Same as for indoor gear.	3 years, more often if conditions require.

Table K.2(e) Medium-Voltage Equipment, Circuit Breakers: Maintenance of Equipment Subject to Long Intervals Between Shutdowns — Electrical Distribution

Type	Inspections, Maintenance, and Tests	Typical Frequency and Remarks
Air-break, Drawout Type (15.4)	**Inspection and Maintenance (withdrawn from switchgear and deenergized) (7.1):**	Max. of 3 years or at manufacturer's maximum number of operations since previous maintenance, whichever occurs first. Immediately after breaker opens to interrupt a serious fault. Follow manufacturer's maintenance instructions. If breaker is stored-energy closing type, follow manufacturer's safety precautions, determine that closing springs are discharged, or mechanism is blocked to prevent personal injury. Keep hands away from contacts and mechanism while test operating breaker (15.4.1.1).
	Remove arc chutes. Inspect, adjust and clean where necessary:	
	Main contacts (15.4.3).	For pitting, spring pressure, overheating, alignment, overtravel, or wipe; adjust or replace accordingly.
	Arcing contacts (15.4.3.2).	For alignment, overtravel, or wipe and for arc erosion; adjust or replace accordingly.
	Moving parts and linkages (15.4.5.1 through 15.4.5.3).	For freedom of movement.
	Closing mechanism (15.4.5).	For quick and positive closing action.
	Tripping mechanism (15.4.5).	For freedom of movement and reliability to open breaker contacts.
	Interlocks and safety devices (15.4.6.2, 15.9.8).	Functionally test to prove proper operation.
	Primary disconnect finger clusters (15.4.3.8).	For proper adjustment and spring pressure; lubricate, unless manufacturer's instructions specify that they should not be lubricated.
	Secondary disconnect contacts (15.4.3.8).	For alignment and spring pressure. Lubricate.
	Closing and trip coils (15.4.6.1).	General condition and evidence of overheating.
	Spring charging motor and mechanism (stored energy type) (15.4.6.1).	Proper operation. Oil leaks from gear motor.
	Shunt trip device (15.4.6.1).	For freedom of movement. Functionally test.
	Undervoltage trip device.	For freedom of movement. Functionally test.
	Auxiliary contacts.	For proper operation with closing and opening of breaker.
	Closing (x and y) relays (electrically operated breakers).	Contact erosion. Dress or replace as required.
	Current transformers (15.2.11, 15.9.5.2).	General condition. Check nameplate ratio.
	Connection bolts (11.4.1 through 11.4.3).	Check for tightness.
	Structure or frame.	For proper alignment and loose or broken parts.
	Fuses and mountings (18.1, 18.2).	General condition and tightness.
	Frame-grounding device.	Connect before and disconnect after primary fingers.
	Position indicators (15.4.6.2, 15.9.6.2).	For proper operation.
	Auxiliary wiring.	General condition and tightness of terminal screws.
	Arc chutes (15.4.4).	For broken parts, missing arc splitters, and amount of metal spatter and burning on interior surfaces. Snuffer screens should be clean. Repair or replace as necessary.

(continues)

Table K.2(e) *Continued*

Type	Inspections, Maintenance, and Tests	Typical Frequency and Remarks
	Operation counter.	For proper operation. Record number of operations.
	Insulators and insulating materials (15.2.11, 15.2.13, 15.2.14, 15.4.2).	For cracks, breaks, corona, tracking, and overheating.
	Breaker auxiliary devices (15.4.6).	Make necessary repairs.
	Testing (withdrawn from switchgear and deenergized) (7.1, 11.5 through 11.8):	Max. of 3 years, etc., same as preceding block.
	Insulation (11.9.1, 11.9.2.3, 11.9.2.4, 11.9.2.6).	High-potential test each main contact with breaker open and all other main contacts and frame grounded. Record results. Use 1000-volt megohmmeter on auxiliary devices, controls, and associated wiring.
	Contact conductivity or resistance (11.16.1.2.2)	Use microhmmeter or determine voltage drop under test load conditions.
	System Testing (breaker installed):	
	Electrically operated breaker.	After above maintenance and testing have been satisfactorily completed, install electrically operated breaker in proper switchgear cell and rack it into "Test" position, or when test stand (station) is provided, connect breaker control contacts to same with cord and plug provided with breaker. Operate closing control devices to ensure that breaker closes and latches without trip-free operations. Operate trip control devices to ensure that breaker trips open in a reliable manner (15.4.6.3). Functionally test all electrical interlock and safety devices. After satisfactorily passing all operational tests, the breaker can be racked into the "Connected" position and placed in normal service.
Oil-immersed, Drawout Type (15.6)	**Inspection and Maintenance (withdrawn from switchgear and deenergized) (7.1):**	Max. of 3 years or at manufacturer's maximum number of operations since last previous maintenance, whichever occurs first; also immediately after breaker opens to interrupt a serious fault. Follow manufacturer's maintenance instructions.
	Lower oil tank. Inspect, adjust, and clean where necessary:	
	Main contacts (15.6.3).	For pitting, spring pressure, overheating, alignment, overtravel, or wipe. Adjust or replace accordingly.
	Arc-quenching assemblies (15.6.4).	For alignment, overtravel, or wipe and for arc erosion. Adjust or replace accordingly.
	Moving parts and linkages (15.4.5.1 through 15.4.5.3).	For freedom of movement.
	Closing mechanism (15.4.5).	For quick and positive closing action.
	Tripping mechanism (15.4.5).	For freedom of movement and reliability to open breaker contacts.
	Interlocks and safety devices (15.4.6.2, 15.9.8).	Functionally test to prove proper operation.
	Primary disconnect finger clusters (15.4.3.8).	For proper adjustment and spring pressure. Lubricate, unless manufacturer's instructions specify that they should not be lubricated.

Table K.2(e) ***Continued***

Type	Inspections, Maintenance, and Tests	Typical Frequency and Remarks
	Secondary (control) disconnect contacts (15.4.3.8).	For alignment and spring pressure. Lubricate.
	Closing and trip coils (15.4.6.1).	General condition and evidence of overheating.
	Shunt trip device (15.4.6.1).	For freedom of movement. Functionally test.
	Undervoltage trip device.	For freedom of movement. Functionally test.
	Bushings (15.6.2.1).	Cracked and chipped porcelain. Condition of surfaces.
	Auxiliary contacts.	For proper operation with closing and opening of breaker.
	Closing (x and y) relays (electrically operated breakers).	Contact erosion. Dress or replace as required.
	Current transformers (15.2.11, 15.9.5.2).	General condition. Check nameplate ratio.
	Connection bolts (11.4.1 through 11.4.3).	Check for tightness.
	Inspection and Maintenance (withdrawn from switchgear and deenergized):	
	Structure or frame.	For proper alignment and loose or broken parts.
	Fuses and mountings (18.1, 18.2).	General condition and tightness.
	Frame-grounding device.	Connect before and disconnect after primary fingers.
	Position indicators (15.4.6.2, 15.9.6.2).	For proper operation.
	Auxiliary devices (15.6.6).	
	Auxiliary wiring.	General condition and tightness of terminal screws.
	Arc quenchers (15.6.4).	For broken and missing parts and amount of metal spatter and burning on interior surfaces. Repair or replace as necessary.
	Operation counter.	For proper operation. Record number of operations.
	Insulators and insulating materials (15.6.2.1).	For cracks, breaks, and chips.
	Insulating oil (15.6.2.2 through 15.6.2.4, 11.11.8, 11.19).	For level, general condition, dielectric strength, and acidity. Make necessary repairs.
	Testing (withdrawn from switchgear and deenergized) (7.1, 11.5 through 11.8):	Max. of 3 years, etc., same as preceding block.
	Insulation (11.9.1, 11.9.2.3, 11.9.2.4, 11.9.2.6).	High-potential test each main contact with breaker open and all other main contacts and frame grounded. Use 1000-volt insulation resistance tester on auxiliary devices and controls and associated wiring. Test oil for dielectric strength. Clean tank and breaker mechanism. Filter oil or replace as required (15.6.2.4).
	Contact conductivity or resistance (11.16.2.2.2).	Use microhmmeter or determine voltage drop under test load conditions.
	dc high-potential and/or power-factor test (11.9.1, 11.9.2, 11.9.3.2).	Record results.
	Overcurrent trip devices (electromechanical type) on breakers so equipped (15.4.6.4).	Pass specified currents from current test set through coils of trip devices to open breaker contacts within time limits according to manufacturer's or specially designed time–current coordination curves. Adjust trip devices as required to accomplish desired results.

(continues)

Table K.2(e) *Continued*

Type	Inspections, Maintenance, and Tests	Typical Frequency and Remarks
		Test set should be equipped with cycle counter for accuracy of instantaneous trip tests. Record results.
	System Testing (breaker installed): Electrically operated breaker.	After above maintenance and testing have been satisfactorily completed, connect electrically operated breaker to switchgear or test stand control wiring by means of the test cord and plug. Operate closing control devices to ensure that breaker closes and latches without trip-free operations. Operate trip control devices to ensure that breaker trips open in a reliable manner (15.4.6.3). Functionally test all electrical interlock and safety devices. After satisfactorily passing all operational tests, the breaker can be placed in its switchgear cell, racked into the "Connected" position, and placed in normal service.
Air-break and Oil-immersed, Fixed Type (15.4.1.2)	**Maintenance and Testing:** Open all disconnect switches to isolate main contacts from electrical supply and load wiring (7.1). **Verify that no parts of the power or control circuitry are energized by "back feed" from alternate power or control sources.** Open closing and tripping power switches to deenergize control devices and wiring (15.1.4.1, 15.1.4.3).	Same frequency as similar drawout-type breaker in preceding blocks. Use adequate safety procedures.
	Perform maintenance and test work in accordance with applicable portions of preceding sections.	Follow manufacturer's instructions.
	Close switches to restore closing and tripping power. Functionally test controls and protective relays for proper operation of breaker.	
Pneumatically and Hydraulically Operated Type (Usually Fixed, Outdoor Type)	**Inspection (while energized):** Check for proper air or hydraulic pressure in storage tank for closing mechanism. Operate motor-driven compressor. Check interior of control cabinet for evidence of water leaks and condensation (15.2.6, 15.2.7). Check space heater for proper operation (15.2.6, 15.2.8).	Monthly. Follow manufacturer's instructions.
	Check machined parts of mechanism for rust spots.	Should be covered with thin coat of lubricant.
	Check operation counter.	Record number of operations.
	Check control battery (15.9.4).	
	Check oil gauges on high-voltage bushings and breaker tanks (15.6.6).	For proper oil level.
	Porcelain bushings (15.6.2.1).	For cracks, chips, and breaks.
	Insulating oil (15.6.2.2 through 15.6.2.4, 11.11.8, 11.19).	For level, general condition, dielectric strength, and acidity.

Table K.2(e) *Continued*

Type	Inspections, Maintenance, and Tests	Typical Frequency and Remarks
	Check oil level in compressor crank case. Inspect control wiring for evidence of damage. Inspect breaker tanks for evidence of oil leaks. Inspect breaker tanks for rust spots.	Make necessary repairs.
	Maintenance and Testing (while deenergized) (7.1): Same as applicable portions of preceding sections plus: Complete check of pneumatic or hydraulic operating mechanism. Power factor test. On some breakers, timing of contact closing and opening might be required (11.16.1.2.7). Measure contact resistance. Measure contact penetration. Measure resistance of internal resistors. Check lever systems, stops, and adjustments. Check dashpot or shock absorber operation. Inspect contact interrupting plates. Inspect gaskets, joints, conduit, and tank fittings. Check pressure switch operation. Check for loose bolts, tightness of joints, etc.	Follow manufacturer's instructions and proper safety procedures. Record results. Use circuit breaker time–travel analyzer or electronic timer. Make necessary repairs.
Vacuum and Gas-Filled Type (11.5)	**Inspections, Maintenance, and Testing:** Follow manufacturer's instructions.	Under certain conditions, high-potential testing can cause x-ray emission from vacuum bottles. Use manufacturer's recommended safety precautions.

Table K.2(f) Medium-Voltage Equipment, Metal-Enclosed Switches: Maintenance of Equipment Subject to Long Intervals Between Shutdowns — Electrical Distribution

Type	Inspections, Maintenance, and Tests	Typical Frequency and Remarks
Indoor Air (15.1.4, 15.3, 15.7)	**Inspections:** Observe components visible through inspection windows (if provided): Switch contacts (15.2.15). Auxiliary devices, wiring, and terminal blocks. Fuseclips and fuses (18.2). Insulators and insulating materials (15.1.2, 15.2.11 through 15.2.15). Space heaters (15.2.8).	6 months. Operate continuously to overcome possible malfunction of thermostats. Consider installation of ammeters in heater supply circuits to monitor full load current of heaters on each circuit to ensure that all are operating.
	Cable terminations (19.1 through 19.4). Adequate grounding (15.1.5, 15.9.9).	Observe stress cones and leakage sections annually for cleanliness and tracking.
	Also observe conditions: Loading.	Record loads if gear is equipped with ammeters.

(continues)

Table K.2(f) ***Continued***

Type	Inspections, Maintenance, and Tests	Typical Frequency and Remarks
	Cleanliness (15.2.10).	Moderate amount of dry nonconductive dust not harmful.
	Dryness (15.2.7).	Evidence of condensation or water leaks.
	Rodents and reptiles.	
	Overheating of parts (15.2.15).	Discoloration or oxidation indicates possible problem. Infrared survey might be beneficial. Use good-quality infrared scanning equipment. Switches should be loaded to at least 40 percent of rating while being scanned.
	Tracking on insulating surfaces (15.2.14).	Make necessary repairs.
	Major Maintenance or Overhaul:	3–6 years, depending on ambient conditions.
	Deenergize (7.1). **Verify that no parts of the power or control circuitry are energized by "back feed" from alternate power or control sources.** Completely clean, inspect, tighten and adjust all components (15.4.1):	Follow manufacturer's maintenance instructions.
	Structure and enclosure (15.2.4, 15.2.5).	Wire-brush and prime rust spots. Finish paint.
	Ventilating louvers and air filters (15.2.9).	Clean or replace air filters as required.
	Buses, splices, and bolts (15.1.3, 15.2.15).	Check bolts for manufacturer's recommended torque.
	Insulators and insulating materials (15.1.2, 15.2.10 through 15.2.15).	Clean and inspect for surface tracking.
	Space heaters (15.2.7).	Operate continuously to overcome possible malfunction of thermostats. Consider installation of ammeters in heater supply circuits to monitor full load current of heaters on each circuit to ensure that all are operating. The ammeters permit frequent check of heater operation while switches are energized.
	Main switch blades and contacts (15.2.15, 15.4.1, 15.4.3).	Use safety precautions if switch is stored energy type. Lubricate.
	Arcing switch blades and contacts (15.4.3.2).	Dress or replace if arc eroded. Do not lubricate.
	Arc chutes or interrupter devices (15.4.4).	Check for condition, alignment, and proper operation.
	Switch-operating mechanism and linkage (15.4.5).	Adjust for adequate contact closure and overtravel. Lubricate.
	Switch/fuse door and other interlocks (15.9.8).	Make functional check for proper operation sequence.
	Switch disconnect studs and finger clusters (if switch is drawout type) (15.4.3.7).	Lubricate, unless manufacturer's instructions specify that they should not be lubricated.
	Cable terminations and connections (19.1 through 19.4).	Clean and inspect for surface tracking. Check connections for tightness.
	Meters (15.9.7).	Check for accuracy.
	Fuseclips and fuses (18.2).	Check clips for adequate spring pressure. Proper fuse rating.
	Grounding (15.1.5, 15.9.9).	
	Potential and control power transformers (15.9.5).	
	Current transformers (15.9.5).	
	Auxiliary devices (15.4.6).	
	Components and conditions in preceding blocks.	Make necessary repairs.

Table K.2(f) *Continued*

Type	Inspections, Maintenance, and Tests	Typical Frequency and Remarks
	Testing (7.1, Chapter 20):	3–6 years, depending on ambient conditions.
	Test buses, breakers, PTs, CTs, and cables with high-potential dc.	Record leakage currents in microamperes (19.5, 11.9.2.6).
	Calibrate and test protective relays (11.12).	Refer to protective relays section.
	Functionally open electrically operated type switches with protective relays (11.12.2).	
	Test conductivity of switch contacts and aluminum cable connections.	Use microhmmeter or determine voltage drop under test load conditions.
	Test wiring for controls, meters, and protective relays for insulation resistance (11.9.2.3).	1000-volt megohmmeter for control wiring. 500-volt megohmmeter for meters and relays.
Outdoor Air	**Inspections (while energized):**	1–3 months.
	Same as for indoor gear except:	
	Special emphasis on evidence of condensation and water leaks (15.2.4, 15.2.6, 15.2.7).	Rust spots on underside of metal roof indicative of condensate.
	Special emphasis on space heater operation (15.2.8).	
	Ventilating louvers and air filters for cleanliness (15.2.9).	Clean or replace air filters as required.
	Major Maintenance or Overhaul:	3 years, more often if conditions require.
	Deenergize (7.1). **Verify that no parts of the power or control circuitry are energized by "back feed" from alternate power or control sources.** Same as for indoor gear (15.4.1).	
	Testing (7.1, Chapter 11):	
	Same as for indoor gear.	3 years, more often if conditions require.
Oil and Gas	**Exterior Inspection:**	Annually.
	Check oil level and gas pressure in switch.	
	Take oil or gas sample.	Test as recommended by manufacturer.
	Check for evidence of leakage.	Repair if necessary.
	Inspect exterior of switch for corrosion.	Paint as required.
	Major Maintenance or Overhaul:	After 500 operations.
	Deenergize (7.1, 15.4.1). **Verify that no parts of the power or control circuitry are energized by "back feed" from alternate power or control sources.**	
	Drain or vent insulating medium.	
	Check gaskets for cracks and flexibility.	Replace where necessary.
	Check cable entrances for mechanical damage or tracking.	
	Inspect all mechanical and electrical connections for tightness.	
	Clean switch interior.	
	Refill.	Use new or reconditioned oil or gas.
	Testing:	
	Actuate each operating mechanism.	Check for proper operation.
	Test with dc high-potential tester.	

Table K.2(g) Medium-Voltage Equipment, Buses and Bus Ducts: Maintenance of Equipment Subject to Long Intervals Between Shutdowns — Electrical Distribution

Type	Inspections, Maintenance, and Tests	Typical Frequency and Remarks
Indoor	**Inspections (while energized):**	3–6 months.
	Open buses:	
	Condition of bus conductors (15.1.3).	
	Evidence of overheated joints (15.1.3, 15.2.15).	Discoloration or oxidation indicates possible problem. Infrared survey might be beneficial. Use good-quality infrared scanning equipment. Buses should be loaded to at least 40 percent of capacity while being scanned.
	Condition of insulators and insulated sleeving (15.1.2).	Cleanliness and breaks.
	Clearance from grounded metal surfaces and above floor.	
	Guards and caution signs.	Where required.
	Loading.	Make certain load is within ampacity rating.
	Bus duct (covers in place):	
	Condition of enclosures (15.2.4).	
	Evidence of water drips on enclosure.	Investigate and correct immediately.
	Adequate grounding (15.1.5, 15.9.9).	
	Loading.	Make certain load is within ampacity rating.
	Maintenance and Testing (deenergized):	1–6 years, depending on conditions.
	Deenergize (7.1). **Verify that no parts of the power or control circuitry are energized by "back feed" from alternate power or control sources.**	
	Open buses:	
	Check for evidence of overheated joints (15.1.3, 15.2.15).	Discoloration or oxidation of bare joints indicates possible problem. Charred tape or cover over insulated joint indicates problem.
	Check connection bolts for tightness where not covered (15.1.3).	Torque according to manufacturer's installation instructions.
	Clean and inspect insulators (15.1.2).	For cracks, chips, breaks, and surface tracking.
	Clean and inspect insulated sleeving over buses, if provided.	For cracks, breaks, properly taped joints, and surface tracking.
	dc high-potential test (11.9.2).	Record results.
	Bus duct (covers removed):	
	Clean and check condition of sleeving over buses (15.2.1.3, 15.2.11 through 15.2.14).	For cracks, breaks, properly insulated joints, and surface tracking. Make necessary repairs.
	Clean and inspect insulators (15.1.2).	For cracks, chips, breaks and surface tracking or burning.
	Check for evidence of internal moisture (15.2.7).	From water leaks or condensation.
	Check for proper ventilation (15.2.9).	All ventilating louvers should be open.
	Check for proper space heater operation (15.2.8).	Operate continuously to overcome possible malfunction of thermostats. Consider installation of ammeters in heater supply circuits to monitor full load current of heaters on each circuit to ensure that all are operating. The ammeters permit frequent check of heater operation while buses are energized.
	Check space heater wiring.	For proper clearances from buses.
	Check condition of enclosure (15.2.4).	Close all unused holes.
	Check grounding connections (15.1.5).	For tightness.
	Check integrity of barriers.	
	dc high-potential test (11.9.2.6).	Record results.

Table K.2(g) *Continued*

Type	Inspections, Maintenance, and Tests	Typical Frequency and Remarks
Outdoor	**Inspections (while energized):**	3–6 months.
	Open buses:	
	Same as for indoor buses	
	Bus duct (covers in place):	
	Condition of enclosure (15.2.4):	Enclosure should be weatherproof type.
	Adequate grounding (15.1.5, 15.9.9).	
	Loading.	Make certain load is within ampacity rating.
	Maintenance and Testing (deenergized):	3–6 years.
	Deenergize (7.1). **Verify that no parts of the power or control circuitry are energized by "back feed" from alternate power or control sources.**	
	Open buses: Same as for indoor buses.	
	Bus duct (covers removed): Same as for indoor bus duct plus following:	
	Check condition of cover gaskets (15.2.6, 15.2.7).	For deterioration, breaks, and omissions.
	Check operation of space heaters (15.2.8).	Operate continuously to overcome the possible malfunction of thermostats. Consider installation of ammeters in heater supply circuits to monitor full load current of heaters on each circuit to ensure that all are operating. The ammeters permit frequent check of heater operation while buses are energized.
	Check enclosure ventilating louvers (15.2.9).	Clean or replace air filters, as necessary. Check for ability to exclude insects, rodents, reptiles, and metal rods.

Table K.2(h) Medium-Voltage Equipment, Protective Relays: Maintenance of Equipment Subject to Long Intervals Between Shutdowns — Electrical Distribution

Type	Inspections, Maintenance, and Tests	Typical Frequency and Remarks
Induction Disk Relays (drawout type) (11.12)	**Inspection, Cleaning, Maintenance, Calibration, and Testing (while associated circuit breaker is closed and supplying load):**	2–3 years, more often where dust, moisture, corrosion, vibration, or wide temperature variation is present. Follow manufacturer's instructions for type of relay and test set.
	Brush or blow dust off top edge of relay cover and remove cover.	Use caution to not accidentally cause relay contacts to close, which would trip associated circuit breaker and shut down load.
	Remove relay disconnect device or open relay trip circuit switch and then open supply circuit switches in relay case.	Remove only one relay from service at a time. Leave other relays in service to provide protection for circuit.
	Release locking mechanisms, withdraw relay from case, and place on workbench in clean area adjacent to test equipment.	Handle with care to avoid damage to delicate mechanism.
	Clean mechanism with soft, long-bristle brush or very light air pressure from hose.	
	Tighten all screws and nuts. Inspect for broken or defective connections.	Do not overtighten. Repair defective connections.
	Inspect closely for dust and iron filings clinging to magnet and in air gap, which might restrict rotation of disk.	Thoroughly clean to remove dust and foreign matter.
	Inspect for correct alignment of disk and proper clearances from mechanism, magnet, etc.	Make necessary adjustments to provide proper clearances so disk does not drag on mechanism or magnet.

(continues)

Table K.2(h) ***Continued***

Type	Inspections, Maintenance, and Tests	Typical Frequency and Remarks
	Burnish contact surfaces and inspect contacts for burning and pitting.	Use relay contact burnishing tool. Replace badly burned or pitted contacts.
	Inspect disk restraint spring.	For proper shape, tension, and possible damage from overheating due to excessive current, flow through same.
	Record"as-found" time lever setting and temporarily set time lever adjustment on position 10. Turn disk with thumb until relay contacts close.	Check to detect if disk binds at any point. Examine spring and contacts in closed position. Make necessary adjustments, repairs, or replacements.
	Release disk and allow it to reset until contact bracket is resting against "full-open" stop device.	Watch disk movement to detect if disk slows down or stops before contact bracket rests against the "full-open" stop device. If so, make necessary adjustments, repairs, or replacement.
	Reset time lever to "as-found" position or to the desired new position, if different one is specified on the applicable coordination or instruction sheet.	Refer to applicable time/current coordination sheet or setting instructions, which are necessary to obtain correct coordination with associated upstream or downstream protective devices.
	Check current tap plug for correct position and tightness. Reposition, if necessary, to agree with setting specified on applicable coordination sheet.	Same as preceding item.
	If precise relay operation accuracy is required, it might be necessary to test some types of relays in their own cases in switchgear or in an unwired duplicate case located on workbench.	For "in-case" testing, connect relay test set leads to proper type relay test jack and insert it into relay disconnect contacts, being careful to make connections in accordance with applicable instructions and diagrams. If relay is in its own case, make certain that it is disconnected from the switchgear wiring to the relay case by means of open switches in the case or by wiring to only the relay side of the test jack. Make certain that switchgear side of test has jumpers installed to short out the CT circuits.
	Some relays can be satisfactorily tested outside of case.	Place withdrawn relay on a clean workbench and connect test leads to proper terminals using alligator clips.
	Select test points from applicable coordination or instruction sheet and calculate amount of current or voltage to be applied to relay operating coil. Adjust test set to apply proper values of test current or voltage.	Consider switchgear PT or CT ratios in calculating proper amount of test current or voltage. Pass test current through current coils. Apply test voltage across potential coils.
	Test relay pickup point by applying test set voltage or current (determined from coordination curve) at which disk begins to turn very slowly.	After disk begins to turn slowly, lower test current or voltage slightly and check if disk stops turning and rests. Make necessary adjustments or repairs. Record results.
	Connect desired relay contact to timer circuit of test set. Adjust test set for current or voltage specified to test time contacts of relay. Push "Initiate" button on test set and check actual time required for relay time contacts to close.	When contacts close, test set automatically removes applied test current or voltage, and its accurate timer will stop.
	Compare actual time with time specified on coordination curve.	If actual time is close enough to the specified time to satisfy the required coordination accuracy, record results and proceed to next step. If not, readjust time lever accordingly and repeat test. Continue until the desired timing accuracy is attained.

Table K.2(h) *Continued*

Type	Inspections, Maintenance, and Tests	Typical Frequency and Remarks
	If relay is equipped with instantaneous current attachment, adjust test set for current or voltage specified for testing same on coordination sheet. Apply and check accurate timer for time required for instantaneous contacts to close. Adjust instantaneous setting to close contacts at current value specified on coordination sheet.	After desired results are obtained, reduce test current slightly and check that relay contacts do not close at value below that specified. If they do, further adjustment and retesting at the specified current will be required until the close and no-close results are within the allowed tolerances.
	Test seal-in contacts for closing at specified values.	
	Check target flags for proper operation each time relay contacts close. Leave all targets in "dropped" position.	Reset target flags and check that moderate vibration does not cause a false operation.
	Record"as-found" and "as-left" settings, test current values and operating results, and maintenance and corrective action taken.	
	Clean and inspect interior of relay case located in switchgear.	Use soft-bristle brush or stream of low-pressure dry, clean air, being careful to not open current-shorting contacts or short potential contacts. Do not short trip circuit terminals, etc.
	Clean glass or plastic window in relay cover and check target reset mechanism for free movement.	Use cleaning materials that will not damage plastic.
	Insert relay into its case in switchgear and secure locking devices. Insert connection device or close switches inside relay case.	Observe that disk does not begin to rotate. If it does, remove connection device before contacts close and trip breaker, investigate reason, and make necessary correction.
	Replace relay cover and secure fastenings. Operate target reset mechanism to determine that targets reset properly. Seal relay cover to discourage unauthorized entry.	On relays with individual switches, close current or potential switches and observe that disk does not rotate before closing trip circuit switch.
Induction Disk Relays (nondrawout type) (11.12)	**Inspection, Cleaning, Maintenance, Calibration, and Testing:**	Same frequency and remarks as for drawout-type relays. Make certain that the connections to the test jack are correct before inserting test jack into test receptacle.
	Same general procedures as for drawout type, except relays cannot be easily removed from their cases and switchgear. A test receptacle is usually provided in the switchgear adjacent to each relay to facilitate testing. A suitable isolating test jack is connected to the relay test set and inserted into the test receptacle. This disconnects the relay contacts from the breaker trip circuit in the switchgear and connects the test set current or potential leads to the proper operating coils in the relay.	
All Types (11.12)	**General Maintenance and Functional Testing (switchgear deenergized and associated breakers out of service) (7.3):**	3–6 years to coincide with major switchgear maintenance.
	Same as above, except checking condition of wiring and terminals.	Check wiring for condition of conductors and insulation. Check terminals for tightness.
	Functionally test by closing associated circuit breaker and injecting proper value of test current into associated CT circuit or applying proper value of test voltage to associated potential wiring after disconnecting same from its supply PTs.	Check to determine that contacts of proper relays close and that associated breaker trips open. If not, determine cause and make necessary corrections.

Table K.2(i) Medium-Voltage Equipment, Automatic Transfer Control Equipment: Maintenance of Equipment Subject to Long Intervals Between Shutdowns — Electrical Distribution

Type	Inspections, Maintenance, and Tests	Typical Frequency and Remarks
Indoor and Outdoor	**Inspections (while energized):**	3–6 months.
	Protective, sensing, timing, and control relays.	For condition of contacts.
	Control wiring and terminals.	General condition.
	Control power batteries (15.9.4).	
	Enclosure (15.2.4, 15.2.6, 15.2.7, 15.2.10).	Cleanliness and evidence of condensation and leaks.
	Space heaters (outdoor enclosures) (15.2.8).	Operate continuously to overcome possible malfunction of thermostats. Consider installation of ammeter in heater supply circuit for frequent monitoring of full load current of heaters to ensure that all are operating. The ammeters permit frequent check of heater operation while buses are energized.
	Maintenance and Testing (while deenergized) (7.1):	3 years, more often if conditions require.
	Clean enclosure, relays, control devices, etc. (15.9.7).	
	Clean, inspect, and burnish contacts.	
	Test and calibrate protective relays (11.12).	Refer to protective relays section.
	Tighten terminals.	
	Test circuits and devices insulation.	Use 500-volt dc insulation resistance tester.
	Maintain enclosure.	Wire-brush and prime rust spots. Finish paint.
	Functionally test by placing selector switch in manual position and operating control switches to open and close associated circuit breakers.	Remove breakers from service.
	Functionally test by placing selector switch in automatic position and simulating conditions that should cause controls to operate associated breakers to effect transfer of power.	Remove breakers from service.

Table K.2(j) Medium-Voltage Equipment, Fuses: Maintenance of Equipment Subject to Long Intervals Between Shutdowns — Electrical Distribution

Type	Inspections, Maintenance, and Tests	Typical Frequency and Remarks
All Types (Chapter 18)	**Visual Inspections (while energized):**	3–6 months. Binoculars might be necessary to inspect fuses on overhead lines.
	Evidence of contact overheating.	Discoloration or oxidation indicates possible problem. Infrared survey might be beneficial. Use good-quality infrared scanning equipment. Fuses should be loaded to at least 40 percent of rating while being scanned.
	Cracked, chipped, or broken insulation of fuse barrels and mounting insulators (15.1.2, 18.2.3.1).	
	Cleanliness of insulation surfaces.	
	Overload.	Steady load should not exceed E rating of fuse.
	Proper oil level in barrel of oil-filled type.	

Table K.2(j) ***Continued***

Type	Inspections, Maintenance, and Tests	Typical Frequency and Remarks
	Maintenance (while deenergized) (7.1, 18.2.2): Remove fuses from mountings and inspect for:	3–6 years, depending on ambient conditions.
	Cleanliness (15.1.2, 18.2.3.1).	Clean insulating and contact surfaces.
	Cracked, chipped, or broken insulation (18.2.3.1).	Replace defective insulation.
	Evidence of overheating and arc erosion on fuse ferrules and spring clips on mountings (18.2.3.2).	Replace defective parts.
	Tension of spring clips and pressure against contact surfaces (18.2.3.3).	Replace weakened or annealed spring clips.
	Tightness of connections (18.2.3.3).	
	Ampere rating agreement with specified rating (18.1.4).	Disassemble refill-type fuses and check nameplate information on refill unit. Check contact surfaces for evidence of overheating. Reassemble and tighten securely.
	Interrupting rating adequacy for fault capability of system on which fuse is installed (18.1.4).	
	Testing (while deenergized): Mounting insulators can be dc high-potential tested (11.9.2.6).	3–6 years. High-potential testing of fuse mountings is not a standard maintenance practice.

Table K.2(k) Medium-Voltage Equipment, Lightning Arresters: Maintenance of Equipment Subject to Long Intervals Between Shutdowns — Electrical Distribution

Type	Inspections, Maintenance, and Tests	Typical Frequency and Remarks
All Types	**Visual Inspection (while energized):**	3–6 months. Binoculars might be necessary to inspect arresters on overhead lines.
	Cleanliness of porcelain surfaces (15.9.2.1). Cracked, chipped, or broken porcelain (15.1.2). Disconnected line or ground connections.	
	Maintenance (while deenergized) (7.1): Clean porcelain surfaces (15.9.2.1).	3–6 years, depending on ambient conditions. Consider application of silicone grease in badly contaminated areas; should be removed and reapplied at max. 2-year, preferably 1-year intervals. Wire-brush and prime rust spots. Finish paint.
	Check tightness of line and ground connections. Inspect nameplate data for voltage rating suitability for system voltage and grounding. Clean internal porcelain surfaces of nonsealed arresters if test results indicate contamination present.	
	Testing (while deenergized) (7.1):	3–6 years.
	Power factor test (15.9.2.2).	Record results.
	Test insulation resistance (15.9.2.2).	Record results. Compare resistances of all arresters of same rating and type, which should be approximately the same.

Table K.3 Medium- and Low-Voltage Equipment, Outside Overhead Electric Lines: Maintenance of Equipment Subject to Long Intervals Between Shutdowns — Electrical Distribution

Equipment	Inspections, Maintenance, and Tests	Typical Frequency and Remarks
Wood Poles	**Inspect from Ground Level for:** Leaning. Washout. Splitting. Bird damage. Lightning damage.	4–6 months. Binoculars usually required.
Wood Crossarms	Twisting. Splitting. Decay. Loose or missing braces. Loose pins. Surface tracking or burning (15.2.14).	4–6 months. Binoculars usually required.
Insulators and Bushings	Cracks (require careful inspection) (15.1.2.1). Chips or bad breaks (15.1.2.1). Unscrewed from pin. Leaning at bad angle. Cleanliness (15.1.2.1).	4–6 months. Binoculars usually required. If atmosphere is contaminated, annual or biannual cleaning and coating of porcelain insulator surfaces with silicone grease might be necessary. This can be done with lines energized by using special equipment, materials, and trained personnel.
Lightning Arresters	Cracked, chipped, and broken insulators (15.1.2.1, 15.9.2.1). Ground connection (15.1.5, 15.9.9). Cleanliness.	4–6 months. Binoculars usually required. If atmosphere is contaminated, annual or biannual cleaning and coating of porcelain insulator surfaces with silicone grease might be necessary. This can be done with lines energized by using special equipment, materials, and trained personnel.
Guys and Anchors	Broken strands. Corrosion. Looseness and slippage. Loose clamps. Excessive tension. Anchor eye above ground. Adequate clearance from conductors. Insulators properly located.	4–6 months. Binoculars usually required.
Conductors	Off insulator and resting on crossarms. Broken strands. Blisters or burned spots. Excessive or uneven sagging. Loose connections (15.1.3, 15.2.15, 11.17). Horizontal and vertical clearances. Trees that touch or can fall across conductors.	4–6 months. Binoculars usually required.
Hardware	Looseness. Corrosion.	4–6 months. Binoculars usually required.

Table K.3 ***Continued***

Equipment	Inspections, Maintenance, and Tests	Typical Frequency and Remarks
Switches and Fuses	General condition. Broken arcing horns. Bent or misaligned arms.	4–6 months. Binoculars are usually required.
Connections	Evidence of overheating (15.1.3, 15.2.15, 11.17).	4–6 months. Binoculars usually required. Infrared survey can be beneficial.
Ground Wires	Breaks (15.1.5). Attachment to pole.	Report on all conditions that require correction. Make necessary repairs.
All Poles	**Climbing or Bucket Truck Inspection for Detailed Inspection of Foregoing Items and Conditions**	3–5 years. Tighten hardware and make necessary repairs and replacements. Wire-brush, prime, and finish-paint rusted areas of metal poles.
Wood Poles and Crossarms	**Ground-Line Inspection and Preservative Treatment:**	8–10 years in southern areas, 10–15 years in northern areas.
	Sound pole with hammer to 1.83 m (6 ft) above ground.	
	Excavate to 0.46 m–0.51 m (18 in.–20 in.) belowground, wire-brush, inspect for surface decay.	Cut out moderate decay pockets.
	Test bore to determine internal decay; if found, determine extent.	If not too extensive, inject preservative fluid, and plug holes.
	Apply preservative to external surface from 0.51 m (20 in.) belowgrade to 0.15 m (6 in.) above.	
	Wrap treated area with protective film and backfill excavation.	If decay is excessive, reinforce or replace pole.
	Aboveground Inspection and Preservative Treatment:	
	Sound pole with hammer. Bore hollow areas and inject preservative fluid; plug holes.	
	Bore pole 0.15 m (6 in.) above bolts and inject preservative fluid; plug holes.	
	Inspect crossarms for decay pockets.	
	Apply preservative treatment.	Replace crossarm if decay is extensive.
	Inspect roof for decay. Apply preservative and cover.	If decay is present but does not extend below top crossarm, cut off pole to sound wood, treat with preservative, and install cover.
	Inspect all wood for termites.	Treat if not excessive.
	Tighten pole hardware.	
	Inspect for bird (woodpecker) damage.	Fill holes with compound if pole is not excessively weakened by damage. Weakened areas might be reinforced.
Current-Carrying Parts	**Thermal Scanning or Infrared Inspection** (11.17):	5–8 years, depending on ambient conditions.
	Scan all conductors, connectors, switches, fuses, etc., with special thermal detecting equipment to locate hot spots caused by loose connectors and bad contacts.	Conductors should be loaded to at least 40 percent of the rating while being scanned. Use good-quality infrared scanning equipment. Small gun-type thermal detectors not usually effective at overhead line distances and require too much time. Make repairs or replacements as indicated by results.

Table K.4(a) Low-Voltage Equipment, Low-Voltage Cables and Connections: Maintenance of Equipment Subject to Long Intervals Between Shutdowns — Electrical Distribution

Type	Inspections, Maintenance, and Tests	Typical Frequency and Remarks
Solid Dielectric, Elastomeric, PVC, etc. (Chapter 19)	**Inspections (while energized)** (19.2.1):	Annually.
	Conduit entrances (19.4).	Observe for deformation due to pressure and for bends with radius less than minimum allowed.
	Poles and supports.	Same as above.
	Binder tape terminations (aerial cables) (19.3).	Same as above.
	Ends of trays (19.4).	Same as above.
	Splices (19.2.2).	Same as above.
	Terminal lugs and connectors (19.2.2, 19.2.4).	Observe for evidence of overheating. Infrared survey might be beneficial. Use good-quality infrared scanning equipment. Conductors should be loaded to at least 40 percent capacity while being scanned. Discoloration or oxidation indicates possible problem.
	Fireproofing (where required) (19.2.2).	Observe for continuity.
	Loading.	Make certain loads are within cable ampacity rating.
Varnished Cambric Lead Covered	**Inspections (while energized)** (19.2.1):	
	Same as preceding block.	Same as preceding block.
	Lead sheath (19.2.2).	Observe for cracks or cold wipe joints often indicated by leakage of cable oil or compound.
All Types	**Major Maintenance and Testing (deenergized):**	3–6 years.
	Deenergize (7.1, 19.2.1). **Verify that no parts of the power or control circuitry are energized by "back feed" from alternate power or control sources.**	
	Complete inspection same as preceding blocks.	Same as preceding blocks.
	Clean porcelain of potheads (19.2.4) (15.1.2.1).	Inspect for cracks and chips.
	Check general condition of cable.	Does insulating material appear to have been damaged by overheating?
	Observe lugs and connectors for overheating (15.1.3, 15.2.15, 19.2.4).	Discoloration or oxidation indicates possible problem. Check bolts for tightness.
	Test cable insulation with high-potential dc (19.5, 11.9.1, 11.9.2, 11.9.2.6).	Disconnect cables from equipment. Record leakage current in microamperes at each test voltage level. Record temperature and relative humidity.
	As an alternative, test cable insulation resistance (11.9.2.3).	Use 2500-volt or 5000-volt megohmmeter.
	Determine condition of cable insulation (11.9.2.6).	Interpret test results, considering length of cable, number of taps, temperature and relative humidity.
	Reconnect cables to equipment.	Tighten connectors adequately.
	Aluminum conductors.	Make certain that connectors are of the proper type and correctly installed. Use Belleville washers when bolting aluminum cable lugs to equipment. Advisable to determine conductivity of connection using microhmmeter or to determine voltage drop under test load conditions.

Table K.4(b) Low-Voltage Equipment, Dry-Type Transformers: Maintenance of Equipment Subject to Long Intervals Between Shutdowns — Electrical Distribution

Type	Inspections, Maintenance, and Tests	Typical Frequency and Remarks
Ventilated and Nonventilated	**Inspections (while energized):**	Monthly.
	Operating temperature (21.3.4).	Odor of overheated insulation in vicinity of ventilated transformer might be best indication of problem. 150°C (302°F) is max. operating temperature for transformers rated 80°C (176°F) rise. 220°C (428°F) is max. operating temperature for transformers rated 150°C (302°F) rise. Nonventilated enclosure of latter might normally be too hot to touch.
	Cleanliness of ventilation louvers in enclosure and excessive accumulation of dust on top of nonventilated enclosure (21.3.5.1 through 21.3.5.6).	Clogged louvers restrict ventilation and thereby increase operating temperature of core and coil assembly. Vacuum louvers without deenergizing transformer if dust and lint are on outside. If dust and lint are inside, transformer should always be deenergized and enclosure sides removed to clean louvers, etc. Clean excessive accumulation of dust off top of nonventilated enclosure.
	Area ventilation and temperature (21.3.5.1).	If ambient temperature exceeds maximum allowed, transformer should be derated accordingly.
	Loading.	Make certain loading is within rating of transformer.
	Major Maintenance (deenergized):	3–6 years, more often if required.
	Deenergized (7.1) (21.3.5.4). **Verify that no parts of the power or control circuitry are energized by "back feed" from alternate power or control sources.**	
	Remove enclosure covers and clean vent louvers (21.3.5.1, 21.3.5.2).	
	Clean insulators, core, and windings (21.3.5.4, 21.3.5.6).	Use bottle of dry air or nitrogen with pressure regulator, hose, and small nozzle to blow off dust. Restrict pressure to 207 kPa (30 psi) max. Clean with soft-bristle brush as required.
	Inspect following components:	
	Wedges and clamping rings (21.3.5.5).	For proper clamping of windings. Tighten as required.
	Primary and secondary buses and conductors (15.1.3, 21.3.5.5).	For tightness of connections and evidence of excessive heat.
	Porcelain insulators (15.1.2).	For chips, cracks, and water streaks.
	Insulating materials (15.2.11, 15.2.15, 21.3.5.5).	For breaks and damage due to excessive heat.
	Windings (15.2.15, 21.3.5.5, 21.3.5.6).	For damage to insulation, including overheating.
	Tap connections (21.3.5.5).	For tightness and correctness to provide proper voltage.
	Core assembly.	For loose or dislocated laminations, localized or general overheating, and integrity of ground strap, which is only place where core assembly is permitted to be grounded.
	Ventilating channels between core and windings and between windings (21.3.5.6).	For clogging with lint, dust, or tape used to hold spacers, etc., in place during assembly. Clean as required to allow proper air flow.

(continues)

Table K.4(b) *Continued*

Type	Inspections, Maintenance, and Tests	Typical Frequency and Remarks
	Testing (deenergized):	3–6 years. More often if required.
	Deenergize (7.1, 21.3.5.4). **Verify that no parts of the power or control circuitry are energized by "back feed" from alternate power or control sources.**	
	Polarization index (PI) test (11.11.2, 11.11.3, 11.11.9).	Use 1000-volt megohmmeter. Low PI results often indicate moisture in winding. If so, investigate cause and satisfactorily dry transformer before making high-potential dc test and returning transformer to service.
	High-potential dc test (11.11.10, 11.9.2.6).	Record leakage currents in microamperes, temperature, and relative humidity.
	As an alternative, test transformer with 1000-volt, 2500-volt, or 5000-volt megohmmeter.	Use 1000, 2500, or 5000 volts dc, depending on size and voltage rating of transformer.

Table K.4(c) Low-Voltage Equipment, Switchgear: Maintenance of Equipment Subject to Long Intervals Between Shutdowns — Electrical Distribution

Type	Inspections, Maintenance, and Tests	Typical Frequency and Remarks
Indoor (Chapter 15)	**Inspections (while energized):**	3–6 months.
	Open doors and inspect components:	Detect overheating.
	Fronts of circuit breakers.	
	Protective and control relays (if used) (15.9.7).	Control wiring, not internals.
	Auxiliary devices, wiring, and terminal blocks (15.4.6).	Proper indicating lamps should light.
	Insulators and insulating materials (15.1.2, 15.2.11 through 15.2.15).	
	Cable connections (15.2.15).	Observe for evidence of overheating. Infrared survey might be beneficial. Use good-quality infrared scanning equipment. Conductors should be loaded to at least 40 percent of capacity while being scanned.
	Batteries (if used) (15.9.4).	
	Also inspect for following conditions:	
	Loading.	Record loads. Make certain loads are within ampacity ratings of breakers and their overcurrent trip coils.
	Cleanliness (15.2.10).	Moderate amount of dry nonconductive dust not harmful.
	Dryness (15.2.6, 15.2.7).	Evidence of condensation or water leaks.
	Rodents and reptiles (15.2.5).	
	Overheating of parts (15.2.15).	Discoloration or oxidation indicates possible problem. Infrared survey might be beneficial. Use good-quality infrared scanning equipment. Components should be loaded to at least 40 percent of capacity while being scanned. Make necessary repairs.

Table K.4(c) *Continued*

Type	Inspections, Maintenance, and Tests	Typical Frequency and Remarks
	Major Maintenance or Overhaul:	3–6 years, depending on ambient conditions.
	Deenergize (7.1). **Verify that no parts of the power or control circuitry are energized by "back feed" from alternate power or control sources.** Completely clean, inspect, tighten, and adjust all components (15.4.1).	Follow manufacturer's instructions.
	Structure and enclosure (15.2.4, 15.2.5).	Wire-brush and prime rust spots. Finish paint.
	Ventilating louvers (15.2.9).	Clean.
	Buses, splices, and bolts (15.1.3, 15.2.15).	Check bolts for manufacturer's recommended torque.
	Insulators and insulating materials (15.1.2, 15.2.11 through 15.2.15).	Clean and inspect for cracks.
	Circuit breakers (15.4 through 15.6).	Refer to circuit breaker section.
	Breaker disconnect studs (15.4.3.7).	Inspect for pitting and evidence of overheating. Lubricate, unless manufacturer's instructions specify that they should not be lubricated.
	Breaker disconnect finger clusters (15.4.3.7).	Inspect for proper adjustment and spring pressure and overheating. Inspect retainer rings for stress cracks in corners. Lubricate, unless manufacturer's instructions specify that they should not be lubricated.
	Cable connections (15.1.3, 15.2.15, 19.1 through 19.4).	Inspect for evidence of overheating. Check for tightness. Use Belleville washers on aluminum lugs.
	Drawout breaker racking mechanisms (15.1.7).	Alignment and ease of operation.
	Meters (15.9.7).	Test for accuracy.
	Controls, interlocks, and closing power rectifiers (15.9.8).	Make functional tests. Check voltages.
	CTs, PTs, and control power transformers (15.9.5).	
	Fuse clips and fuses (18.2).	Check clips for adequate spring pressure. Proper fuse rating.
	Grounding (15.1.5, 15.9.9).	
	Components and conditions in above block.	Make necessary repairs.
	Testing (deenergized) (Chapter 11):	
	Deenergize (7.1). **Verify that no parts of the power or control circuitry are energized by "back feed" from alternate power or control sources.**	3–6 years, depending on ambient conditions.
	Test buses, breakers, PTs, CTs, wiring, and cables for insulation resistance (11.9.2.3).	2500-volt dc on buses, breakers, and 600-volt equipment. 1000-volt dc on control wiring. 500-volt dc on meters and protective relays.
	Calibrate and test protective relays (11.12).	Refer to protective relays section.
	Functionally trip breakers with relays (11.12.2).	Preferably, inject test current into CT and relay circuits.
	Calibrate and test overcurrent trip devices (15.4.6.4, 15.4.6.5, 11.10).	Use high-current test equipment and adjust trips to operate in accordance with manufacturer's and specially prepared time–current coordination curves. Adjust for proper conformance.
	Test conductivity of aluminum cable connections (15.1.3, 11.10.5.1.5).	Use microhmmeter or determine voltage drop under test load conditions. Use Belleville washers when bolting aluminum cable lugs to equipment.

(continues)

Table K.4(c) ***Continued***

Type	Inspections, Maintenance, and Tests	Typical Frequency and Remarks
Outdoor (Chapter 15)	**Inspections (while energized):** Same as for indoor gear plus:	1–3 months.
	Space heaters (15.2.8).	Operate during cool weather.
	Special emphasis on condensation and water leaks (15.2.4, 15.2.6, 15.2.7).	Rust spots on underside of roof indicative of condensation.
	Air filters behind ventilating louvers (15.2.9).	Clean or replace as required.
	Major Maintenance or Overhaul: Deenergize (7.1). **Verify that no parts of the power or control circuitry are energized by "back feed" from alternate power or control sources.** Same as for indoor gear (15.4.1).	3 years, more often if conditions require. Follow manufacturer's maintenance instructions.
	Testing (deenergized) (Chapter 11): Deenergize (7.1). **Verify that no parts of the power or control circuitry are energized by "back feed" from alternate power or control sources.** Same as for indoor gear.	3 years, more often if conditions require.

Table K.4(d) Low-Voltage Equipment, Drawout-Type Circuit Breakers: Maintenance of Equipment Subject to Long Intervals Between Shutdowns — Electrical Distribution

Type	Inspections, Maintenance, and Tests	Typical Frequency and Remarks
Air-Break (15.4)	**Inspection and Maintenance (withdrawn from switchgear and deenergized):**	Max. of 3 years or at manufacturer's maximum number of operations since previous maintenance, whichever occurs first. Immediately after breaker opens to interrupt a serious fault. Follow manufacturer's maintenance instructions. If breaker is stored-energy closing type, follow manufacturer's safety precautions, determine that closing springs are discharged or mechanism is blocked to prevent personal injury. Keep hands away from contacts and mechanism while test-operating breaker (15.4.1.1).
	Remove arc chutes. Inspect, adjust, and clean where necessary:	
	Main contacts (15.4.3).	For pitting, spring pressure, overheating, alignment, overtravel, or wipe. Adjust or replace accordingly.
	Arcing contacts (15.4.3.2).	For alignment, overtravel, or wipe and for arc erosion. Adjust or replace accordingly.
	Moving parts and linkages (15.4.5.1 through 15.4.5.3).	For freedom of movement.
	Closing mechanism (15.4.5).	For quick and positive closing action.
	Tripping mechanism (15.4.5).	For freedom of movement and reliability to open breaker contacts.
	Interlocks and safety devices (15.4.6.2, 15.9.8).	Functionally test to prove proper operation.
	Primary disconnect finger clusters (15.4.3.8).	For proper adjustment and spring pressure. Lubricate, unless manufacturer's instructions specify that they should not be lubricated.
	Secondary disconnect contacts (15.4.3.8).	For alignment and spring pressure. Lubricate.
	Closing and trip coils (15.4.6.1).	General condition and evidence of overheating.

Table K.4(d) ***Continued***

Type	Inspections, Maintenance, and Tests	Typical Frequency and Remarks
	Spring charging motor and mechanism (stored-energy type) (15.4.6.1).	Proper operation. Oil leaks from gear motor.
	Shunt trip device (15.4.6.1).	For freedom of movement. Functionally test.
	Undervoltage trip device.	For freedom of movement. Functionally test.
	Anti-single-phase or blown fuse lockout devices (fused breakers only).	General condition. Functionally test with proper voltage to trip and lock out breaker.
	Auxiliary contacts.	For proper operation with closing and opening of breaker.
	Closing (x and y) relays (electrically operated breakers).	Contact erosion. Dress or replace as required.
	Current transformers (15.2.11, 15.9.7.2).	General condition. Check nameplate ratio.
	Connection bolts (16.4.1).	Check for tightness.
	Structure or frame (16.4.1, 16.4.2, 16.4.3).	For proper alignment and loose or broken parts.
	Fuses and mountings (18.1).	General condition and tightness.
	Frame grounding device.	Connect before and disconnect after primary fingers.
	Position indicators (15.4.6.2, 15.9.6.2).	For proper operation.
	Auxiliary wiring (15.4.6).	General condition and tightness of terminal screws.
	Arc chutes (15.4.4).	For broken parts, missing arc splitters, and amount of metal spatter and burning on interior surfaces. Snuffer screens should be clean. Repair or replace as necessary.
	Operation counter (if so equipped).	For proper operation. Record number of operations.
	Insulators and insulating materials (15.2.11, 15.2.15).	For cracks, breaks, and overheating.
	Breaker auxiliary devices (15.4.6).	Make necessary repairs.
	Testing (withdrawn from switchgear and deenergized) (7.1, 11.5 through 11.8):	Max. of 3 years, etc., same as preceding block.
	Test insulation resistance (11.9.1 through 11.9.2.4).	2500-volt megohmmeter on each main contact with breaker open and all other main contacts and frame grounded. 1000-volt megohmmeter on auxiliary devices and controls and associated wiring, except solid-state trip devices.
	Contact conductivity or resistance (11.10.5.1.5).	Use microhmmeter or determine voltage drop under test load conditions.
	Overcurrent (OC) trip devices (electromechanical, series type) (15.4.6.4).	Pass specified currents from high-current test set through coils of series type OC trip devices. Trip devices should open breaker contacts within time limits according to manufacturer's or specially designed time–current coordination curves. Adjust trip devices as required to accomplish desired results. Test set should be equipped with cycle counter for accuracy of short-time and instantaneous trip tests. Record results.
	Overcurrent trip devices (electromechanical, 5 amp CT type).	Test 5 amp, type OC trip devices in similar manner using reduced current proportional to ratio of CTs in switchgear that normally supply current to the OC trip coils. Record results.

(continues)

Table K.4(d) *Continued*

Type	Inspections, Maintenance, and Tests	Typical Frequency and Remarks
	Overcurrent trip devices (solid-state type) (15.4.6.5).	Use manufacturer's instructions and test set specifically designed for solid-state trip device being tested, or use primary injection from high-current test set. Adjust trip device settings to obtain desired tripping times and currents to conform with applicable coordination curves. Record results. Do not use megohmmeter insulation resistance tester on solid-state trip devices or associated wiring.
	System Testing (breaker installed):	
	Electrically operated breaker	After preceding maintenance and testing have been satisfactorily completed, install electrically operated breaker in proper switchgear cell and rack it into "Test" position. Operate closing control devices to ensure that breaker closes and latches without trip-free operations. Operate trip control devices to ensure that breaker trips open in a reliable manner (8.4.6.3). Functionally test all electrical interlock and safety devices. After satisfactorily passing all operational tests, the breaker can be racked into the "Connected" position and placed in normal service.
Oil-Immersed	**Inspection and Maintenance (withdrawn from switchgear and deenergized):**	Max. of 3 years or at manufacturer's maximum number of operations since last previous maintenance, whichever occurs first. Immediately after breaker opens to interrupt a fault. Follow manufacturer's maintenance instructions.
	Lower oil tank. Inspect, adjust, and clean where necessary:	
	Main contacts.	For pitting, spring pressure, overheating, alignment, overtravel, or wipe. Adjust or replace accordingly.
	Arcing contacts.	For alignment, overtravel, or wipe and for arc erosion. Adjust or replace accordingly.
	Moving parts.	For freedom of movement.
	Closing mechanism.	For quick and positive closing action.
	Tripping mechanism.	For freedom of movement and reliability to open breaker contacts.
	Mechanical interlocks.	Functionally test to prove proper operation.
	Primary disconnect finger clusters.	For proper adjustment and spring pressure. Lubricate, unless manufacturer's instructions specify that they should not be lubricated.
	Secondary (control) disconnect contacts.	For alignment and spring pressure. Lubricate.
	Closing and trip coils.	General condition and evidence of overheating.
	Shunt trip device.	For freedom of movement. Functionally test.
	Undervoltage trip device.	For freedom of movement. Functionally test.
	Bushings.	Cracked and chipped porcelain. Condition of surfaces.

Table K.4(d) ***Continued***

Type	Inspections, Maintenance, and Tests	Typical Frequency and Remarks
	Auxiliary contacts.	For proper operation with closing and opening of breaker.
	Closing (x and y) relays (electrically operated breakers).	Contact erosion. Dress or replace as required.
	Current transformers.	General condition. Check nameplate ratio.
	Connection bolts.	Check for tightness.
	Structure or frame.	For proper alignment and loose or broken parts.
	Fuses and mountings.	General condition and tightness.
	Frame grounding device.	Connect before and disconnect after primary fingers.
	Position indicators.	For proper operation.
	Auxiliary wiring.	General condition and tightness of terminal screws.
	Arc quenchers.	For broken and missing parts and amount of metal spatter and burning on interior surfaces. Repair or replace as necessary.
	Operation counter (if so equipped).	For proper operation. Record number of operations.
	Insulators and insulating materials.	For cracks, breaks, and chips. Make necessary repairs.
	Testing (withdrawn from switchgear and deenergized):	Max. of 3 years, etc., same as preceding block.
	Insulation.	2500-volt megohmmeter on each main contact with breaker open and all other main contacts and frame grounded. 1000-volt megohmmeter on auxiliary devices and controls and associated wiring.
	Contact conductivity.	Use microhmmeter or determine voltage drop under test load conditions.
	Overcurrent trip devices (electromechanical type).	Pass specified currents from current test set through coils of trip devices to open breaker contacts within time limits according to manufacturer's or specially designed time–current coordination curves. Adjust trip devices as required to accomplish desired results. Test set should be equipped with cycle counter for accuracy of instantaneous trip tests. Record results.
	System Testing (breaker installed): Electrically operated breaker.	After above maintenance has been satisfactorily completed, connect electrically operated breaker to switchgear or test stand control wiring by means of the test cord and plug. Operate closing control devices to ensure that breaker closes and latches without trip-free operations. Operate trip control devices to ensure that breaker trips open in a reliable manner. Functionally test all electrical interlock and safety devices. After satisfactorily passing all operational tests, the breaker can be placed in its switchgear cell, racked into the "Connected" position, and placed in normal service.

Table K.4(e) Low-Voltage Equipment, Buses and Bus Ducts: Maintenance of Equipment Subject to Long Intervals Between Shutdowns — Electrical Distribution

Type	Inspections, Maintenance, and Tests	Typical Frequency and Remarks
Indoor	**Inspections (while energized):**	3–6 months.
	Open buses:	
	Condition of bus conductors (15.1.3).	
	Evidence of overheated joints (15.1.3).	Discoloration or oxidation indicates possible problem. Infrared survey can be beneficial. Use good-quality infrared scanning equipment. Buses should be loaded to at least 40 percent of capacity while being scanned.
	Condition of insulators (15.1.2.1).	Cleanliness and breaks.
	Clearance from grounded metal surfaces and above floor.	
	Loading.	Make certain load is within ampacity rating.
	Bus duct (covers in place):	
	Condition of enclosure (15.2.4, 20.4.3).	
	Evidence of water drips on enclosure.	
	Security of switches attached to plug-in type bus duct.	
	Adequate grounding (15.1.5, 15.9.9).	
	Loading.	Make certain load is within ampacity rating.
	Maintenance and Tests (deenergized):	3–6 years.
	Deenergize (7.1). **Verify that no parts of the power or control circuitry are energized by "back feed" from alternate power or control sources.**	
	Open buses:	
	Check connection bolts for tightness (15.1.3).	Torque according to manufacturer's installation instructions.
	Clean insulators and inspect (15.1.2).	Check for cracks, chips, and breaks.
	Test insulation resistance (11.9.2.3).	2500-volt megohmmeter, if suitable. If not, 1000-volt megohmmeter is satisfactory.
	Bus duct (covers removed):	
	Check condition of bus conductors (15.1.3).	
	Check for evidence of overheated joints (15.1.3, 15.2.15).	Discoloration or oxidation indicates possible problem.
	Check connection bolts for tightness (15.1.3).	Torque according to manufacturer's installation instructions.
	Check switches attached to plug-in type bus duct (15.1.5, 15.9.9).	For condition of contacts, operating mechanism, fuse clips, fuses, and load cables. Make necessary repairs.
	Check ground connections (20.4.3).	For tightness.
	Check condition of enclosure.	Close all unused holes.
	Check for proper ventilation (15.2.9).	All ventilating louvers should be open.
	Check for evidence of internal moisture (15.2.7).	From water leaks or condensation.
	Clean and inspect insulators (15.1.2).	Check for cracks, chips, and breaks.
	Test insulation resistance (11.9.2.3).	Manufacturer usually permits 1000-volt dc test for 1 minute.
Outdoor	**Inspections (while energized):**	3–6 months.
	Open buses:	
	Same as for indoor open buses.	
	Bus duct (covers in place):	
	Condition of enclosure (15.2.4).	Enclosure should be weatherproof type.
	Adequate grounding (15.1.5, 15.9.9).	
	Loading.	Make certain load is within ampacity rating.
	Maintenance and Testing (deenergized):	3–6 years.

Table K.4(e) *Continued*

Type	Inspections, Maintenance, and Tests	Typical Frequency and Remarks
	Deenergize (7.1). **Verify that no parts of the power or control circuitry are energized by "back feed" from alternate power or control sources.**	
	Open buses:	
	Same as for indoor open buses.	
	Bus duct (covers removed):	
	Check condition of bus conductors (15.1.3).	
	Check for evidence of overheated joints (15.1.3, 15.2.15).	Discoloration or oxidation indicates possible problem.
	Check connection bolts for tightness (15.1.3).	Torque according to manufacturer's installation instructions.
	Check ground connections (15.1.5, 15.9.9).	For tightness.
	Check condition of enclosure (15.2.4, 20.4.3).	Close all unused holes. Wire-brush and prime rust spots. Finish paint.
	Check for evidence of internal moisture (15.2.7).	From water leaks or condensation. Correct.
	Clean and inspect insulators (15.1.2).	For cracks, chips, and breaks.
	Test insulation resistance (11.9.2.3).	Manufacturer usually permits 1000-volt dc test for 1 minute.
	Check operation of space heaters (15.2.8).	Operate during cool weather.
	Check enclosure ventilating louvers (15.2.9)	All ventilating louvers should be open. Should exclude insects, rodents, reptiles, and metal rods.

Table K.4(f) Low-Voltage Equipment, Panelboards: Maintenance of Equipment Subject to Long Intervals Between Shutdowns — Electrical Distribution

Type	Inspections, Maintenance, and Tests	Typical Frequency and Remarks
Fused Switch Type	**Inspections (while energized):**	3–6 months.
	Switches for overheating (16.5.4).	Feel front of each switch to detect overheating. Arrangements should be made to shut down any overheated switch to determine cause and repair or replace.
	Portion of enclosure over supply cable terminals for overheating (16.4.1, 16.4.2).	Feel enclosure. If overheated, remove cover and inspect supply cables and terminals for evidence of overheating. Discoloration or oxidation indicates possible problem.
	Loading.	Constant load on switch should not exceed 80 percent of switch nameplate rating, unless switch is rated for 100 percent continuous operation.
	Enclosure for general condition (16.3).	Arrange to have unused conduit knockout holes plugged with knockout closures.
	Ground connections (15.1.5).	For integrity.
	Directory.	For accuracy of loads served.
	Evidence of water dripping on or striking NEMA 1 enclosure.	Stop water leaks.
	Cleaning, Inspection, and Maintenance (deenergized) (7.1, 16.5.2):	3–6 years.
	Clean interior of enclosure and switches (16.3).	
	Inspect fuses for overheating (11.7, 18.1).	Refer to low-voltage fuse section.
	Inspect fuseclips for overheating and weakness (18.1.2).	Replace weak or burned clips.

(continues)

Table K.4(f) *Continued*

Type	Inspections, Maintenance, and Tests	Typical Frequency and Remarks
	Inspect fuses for proper ampere rating for cable size and for interrupting adequacy for fault current capability of supply system (18.1.4).	Fuse ampere rating should not exceed NEC ampacity of cables. Fuse interrupting rating should exceed fault current available from system.
	Check connection bolts for tightness (15.2.15).	Do not overtighten and strip threads.
	Check set screws in all cable connectors for tightness (15.1.3).	AL/CU set screw-type connectors tend to loosen. Set screws in many old-style AL/CU connectors are unplated aluminum on which threads tend to gall and cause set screws to bind before they tighten sufficiently against cables. Replace this connector, if necessary.
	Open and close switches. Inspect contact surfaces and operating mechanism (16.5.5).	Repair or replace burned contacts. Make certain switch contacts close fully.
	Inspect all insulating materials (15.2.11, 15.2.15).	For cracks, breaks, cleanliness, and thermal damage.
	Inspect arc chutes.	For broken parts and missing arc splitters.
	Check door/switch mechanical interlocks (16.11.2).	That switch door cannot be opened when switch handle is in "On" position, unless interlock defeater mechanism is operated. That switch handle cannot be thrown to "On" position while switch door is open, unless interlock defeat mechanism is operated.
	Check padlock devices (16.11.2.2).	That switch handle cannot be thrown to "On" position with padlock in lockout device.
	Check door latches.	That doors do not open when operating handle is in "On" position.
	Check directory for accuracy.	
	Enclosure (16.3).	Wire-brush and prime rust spots. Finish paint.
	Test supply cables, switches, and load cables for insulation resistance (11.9.2.3).	2500-volt megohmmeter preferred, 1000-volt megohmmeter acceptable.
Molded-Case Circuit-Breaker Type (Chapter 17, 11.10)	Inspections and maintenance similar to fused switch type except: Circuit breakers usually cannot be opened for inspection and maintenance. Breakers usually operate at a higher temp. Breakers usually contain no fuses. Breakers often not equipped with external door or operating handle other than handle integral with breaker. Breaker overcurrent trip operation can be tested with high-current test set (11.10).	3–6 years.

Table K.4(g) Low-Voltage Equipment, Protective Relays: Maintenance of Equipment Subject to Long Intervals Between Shutdowns — Electrical Distribution

Type	Inspections, Maintenance, and Tests	Typical Frequency and Remarks
Induction Disk (11.12)	Same as for medium-voltage protective relays.	Refer to medium-voltage protective relays sections.

Table K.4(h) Low-Voltage Equipment, Automatic Transfer Control Equipment: Maintenance of Equipment Subject to Long Intervals Between Shutdowns — Electrical Distribution

Type	Inspections, Maintenance, and Tests	Typical Frequency and Remarks
Indoor and Outdoor	Same as for medium-voltage auto transfer control equipment.	Refer to medium-voltage auto transfer equipment section.

Table K.4(i) Low-Voltage Equipment, Circuit Breaker Overcurrent Trip Devices: Maintenance of Equipment Subject to Long Intervals Between Shutdowns — Electrical Distribution

Type	Inspections, Maintenance, and Tests	Typical Frequency and Remarks
Series and 5 Amp Type (11.10.6.3)	Same as OC trip item in low-voltage drawout circuit breaker section.	Refer to low-voltage drawout circuit breaker section.

Table K.4(j) Low-Voltage Equipment, Fuses: Maintenance of Equipment Subject to Long Intervals Between Shutdowns — Electrical Distribution

Type	Inspections, Maintenance, and Tests	Typical Frequency and Remarks
Cartridge-and-Plug Type (18.1)	Same as for medium-voltage fuses except: Inspect for discoloration or charring of fiber barrel ends adjacent to ferrules (15.2.15).	Refer to medium-voltage fuse section. This can be done while fuses are energized. Indicates probability of loose contact between fuse ferrules or blades and spring clips.
	Check for assembly rightness of renewable fuse ferrules on barrels (15.1.3, 18.1.2).	Looseness can possibly cause overheating of ferrules and fiber barrel ends.
	Check for constant moderate overload on circuit supplied by fuses.	Fuses should not be continuously loaded beyond 80 percent of ampere rating.

Table K.4(k) Low-Voltage Equipment, Lightning Arresters: Maintenance of Equipment Subject to Long Intervals Between Shutdowns — Electrical Distribution

Type	Inspections, Maintenance, and Tests	Typical Frequency and Remarks
Indoor and Outdoor Type (15.9.2)	Same as for medium-voltage lightning arresters except:	Refer to medium-voltage lightning arrester section. Use 500-volt megohmmeter.
	Test insulation resistance. Arresters in metal containers.	Inspect conductors for damage where they enter container.

Annex L Maintenance Intervals

This annex is not a part of the recommendations of this NFPA document but is included for informational purposes only.

L.1 Introduction. This annex provides, in Table L.1, an initial guideline for maintenance intervals for equipment. It should be stressed that environmental or operating conditions of a specific installation should be considered and might dictate a different frequency of maintenance than suggested in this annex *(see 8.2.4)*. Chapter 12 and Annex K deal specifically with the maintenance of equipment that, by nature of its application, necessitates long intervals between shutdowns. It should be noted that maintenance, inspection, and test methods for equipment that can operate for long periods are essentially the same as for equipment that might be shut down frequently. However, the recommended work should be performed with more care and diligence to obtain the desired reliability for service to loads that can operate continuously for months or years.

Table L.1 Interval Guidelines

Item/Equipment	Task/Function	Interval	Reference
Substations (Outdoor)	Infrared scanning	Annually	11.17
Insulators	Visual inspection	4–6 months	15.1.2.1
	Corona detection	4–6 months	15.1.2.2
	Electrical tests	As indicated by other PM	11.9
Conductors	Visual inspection of connections	4–6 months	15.1.3
	Check connections for tightness	As indicated by other PM	15.1.3
Air-disconnecting switches	Visual inspection	4–6 months	15.1.4.2
	Operation check	Annually	15.1.4.3
	Contact inspection	Annually	15.1.4.3

(continues)

Table L.1 ***Continued***

Item/Equipment	Task/Function	Interval	Reference
Grounding equipment	Visual inspection	Annually	15.1.5
	Check connections for tightness	1–2 years	15.1.5
	Electrical test	3 years	11.13.1, 11.13.2, 11.13.3
Enclosures	Security/operational check	1–3 months	15.1.6
Switchgear Assemblies	Infrared scanning	Annually	11.17
Enclosures	Security/operational check		15.2.5
	Outdoor	1–3 months	
	Indoor	6 months	
	Visual inspection		15.2.6 through 15.2.7.2
	Outdoor	1–3 months	
	Indoor	6 months	
Ventilation	Visual inspection	1–3 months	15.2.9
Space heaters	Operational check	Annually	15.2.8
Insulation	Visual inspection/clean	Annually	15.2.11 through 15.2.15.3
	Electrical tests	2 years	11.9
Air Circuit Breakers, Medium Voltage			
Insulation	Visual inspection/clean	Annually	15.4.2
	Electrical tests	3 years	11.9
Contacts	Visual inspection/clean	Annually	15.4.3.5
	Adjust	Annually	15.4.3.6
	Electrical test	3 years	11.16.1.2.2, 11.16.1.2.7, 11.9.3.2
Arc interrupters	Visual inspection/clean	Annually	15.4.4.3, 15.4.4.4
	Electrical test	3 years	15.4.4.4
	Air-puffer operational check	Annually	15.4.4.5
Operating mechanism	Visual inspection	Annually	15.4.5.2
	Operational check/adjustment	Annually	15.4.5.2
Trip device circuit	Operational check	Annually	15.4.6.3
Air Circuit Breakers, Low Voltage	Visual inspection/clean/adjust	Annually	15.4, 11.9, 11.10.6
	Electrical tests	3 years	
Vacuum Circuit Breaker	Visual inspection/clean/adjust	Annually	15.4
	Contact checks/vacuum integrity	3 years	15.5.1, 15.5.2
	Electrical tests	3 years	11.9, 11.10.6
Oil Circuit Breaker	General inspections and tests	3 years	11.9, 11.16.4.2.5
Bushings	Visual inspection/clean	3 years	15.6.2.1
Oil	Dielectric breakdown test	Annually	15.6.2.3, 11.9
	level	Annually	15.6.6
Contacts	Resistance check	3 years	15.6.3
	Visual inspection	3 years	15.6.3
Interrupter Switches	*See Air Circuit Breakers, Medium Voltage*		15.7
Surge Arresters	Visual inspection		15.9.2.1
	Outdoor	3–6 months	
	Indoor	Annually	
	Electrical test	3–6 years	15.9.2.2
Capacitors	Visual inspection	3–6 months	15.8.3.4
	Fuse check	3–6 months	15.8.3.3

Table L.1 *Continued*

Item/Equipment	Task/Function	Interval	Reference
Stationary Batteries and Chargers	Visual inspection/clean	Monthly	15.9.4.4 through 15.9.4.4.13
	Torque intercell connectors	Annually	15.9.4.4.11
	Pilot cell measurements	Monthly	11.14.2.2
	All lead–acid cell-specific gravity	Quarterly	11.14.2.1
	Capacity test	1–5 years	11.14.2.3
	Sample connection resistances	Quarterly	11.14.2.4
	Infrared scanning	Annually	11.14.2.5, 11.17
Protective Relays	Cleaning, calibration, and function tests		15.9.7.3, 11.12
	Electromechanical	1–2 years	
	Solid state	3 years	
Supervisory Control and Data Acquisition (SCADA)			
Electrical/electronic Systems	Lamp test/verify indicators	Monthly	12.14.11.1, 12.14.11.2
	Inspect enclosures for dirt, heat, water	Monthly	8.7.1, Table L.1
	Physically exercise valves and actuators	6 months	(Reserved for future)
	Actuate switches	6 months	(Reserved for future)
	Run PLC diagnostics	6 months	22.4.4
	Calibrate sensors and transmitters	Annually	11.7
	Calibrate actuators	Annually	11.7
	Calibrate meters	Annually	
	Test batteries	6 months	15.9.4.6
	Test automatic control sequences	Annually	(Reserved for future)
	Verify alarms	Annually	15.9.6.1, 12.4.11, 15.9.6
Pneumatic system/components	Check regulators and filters	Monthly	(Reserved for future)
	Inspect tubing and piping	Monthly	(Reserved for future)
	Actuate pressure switches	6 months	(Reserved for future)
	Physically exercise valves and actuators	6 months	(Reserved for future)
	Calibrate switches and sensors	Annually	(Reserved for future)
	Calibrate pressure gauges	Annually	(Reserved for future)
	Calibrate thermometers	Annually	(Reserved for future)
Power and Distribution Transformers			
Liquid filled	Current and voltage readings	Weekly–monthly	21.2.2.2, 21.2.3
	Temperature readings	Weekly–monthly	21.2.4
	Liquid level check	Weekly–monthly	21.2.5.1
	Pressure/vacuum gauge readings	Weekly–monthly	21.2.5.2
	Liquid analysis	Annually	21.2.8
	Comprehensive liquid tests	Annually	11.11.9, 11.19
	Insulation test	3–5 years	21.2.9, 11.9
	Turns-ratio test	3–5 years	21.2.9, 11.11.2
	Fault gas analysis	Annually	21.2.9, 11.11.9
	Dissolved-gas-in-oil analysis	Annually	11.11.10
Dry type	Cleaning, inspection, and testing	2 years	11.9, 11.11.2
Power Cables	Visual inspection	Yearly	19.2
	Electrical testing	1–3 years	19.5, 11.9.2.4
Motor Control Equipment	Infrared scanning	Annually	11.17
Enclosures	Visual inspection/clean	Annually	16.2.1 through 16.3.5
Bus bar, wiring, and terminal connections	Check connections for tightness and proper torque	2 years	16.4.2
	Visual inspection of insulators	Annually	16.4.4
	Visual inspection of wiring	Annually	16.4.5
	Electrical tests	2 years	11.9.2.3

(continues)

Table L.1 *Continued*

Item/Equipment	Task/Function	Interval	Reference
Disconnects	Visual inspection/clean	Annually	16.5.3
	Operation check	Annually	16.5.5
Contactors	Visual inspection/clean	Annually	16.8.2.1 through 16.8.3
Motor overload relays, nonthermal type	Check connections for tightness	2 years	16.9.4
	Cleaning, calibration, and function tests	3 years	16.9.1.3, 16.9.2
Electrical interlocks	Inspection	Annually	16.11.1.2
Mechanical interlocks	Inspection	Annually	16.11.2
Electronic Equipment	Inspection	Annually	22.4.1
	Cleaning	3 years	22.4.2
	Adjustments/calibration	3–5 years	22.4.3
Molded-Case Circuit Breakers	Visual inspection/clean	3 years	17.7 through 17.11
	Mechanical test	2 years	17.11
	Electrical test	3–5 years	11.10.5
Fuses, 1000 Volts or Less			
Fuse terminals and fuseclips	Visual inspection	3 years	18.1.2
	Clip contact pressure	3 years	18.1.3
	Cleaning of contact surfaces	3 years	18.1.3
Fuses	Visual inspection for discoloration and damage	3 years	18.1.3
Fuses, Over 1000 Volts			
Insulators	Visual inspection/cleaning	3 years	18.2.3.1
Fuse terminals and fuseclips	Inspection of contact surfaces	3 years	18.2.3.2
Fuses	Visual inspection for corrosion	3 years	18.2.3.2
	Terminal connections and hardware	3 years	18.2.3.3
	Fuse tubes	3 years	18.2.3.3
Fuses, Vented Expulsion Type	Visual inspection of seals	3 years	18.2.3.4
Rotating Equipment	Vibration analysis	Continuously to 6-month intervals	26.7
Stator and rotor windings	Visual and mechanical inspection, cleaning	Annually	8.7, 25.3, 25.6
	Electrical testing	Annually	11.20
Brushes, collector rings, and commutators	Visual and mechanical inspection	Annually	25.4
Bearings, sleeved	Oil level check	Weekly–monthly	25.5
	Drain, flush, and lubricate	Annually	
Waste-packed	Re-oil, check air gap	1000 hours	25.5.2.3
Ball and roller	Inspection and lubrication	Per manufacturer	25.5.3
Kingsbury thrust bearings	Drain, flush, and lubricate	Per manufacturer	25.5.4
Wiring Devices			
Attachment plugs, cord connector bodies	Inspection	Monthly and when used	24.2.1 through 24.6
Receptacles	Inspection	Monthly and when used	24.3.1
	Operation check	Monthly and when used	24.3.2
General-use snap switches	Operation check	When used	24.5.2 through 24.5.4
Pin-and-sleeve devices, heavy-duty industrial-type plugs, cord connectors, and receptacles	Inspections, cleaning, and checks	Monthly and when used	24.8
Portable Electric Tools	Inspections/cleaning	Monthly and when used	29.1.3, 29.3.1, 29.3.2
	Lubrication	Per manufacturer	29.3.3
	Electrical tests	Quarterly	29.7

Table L.1 ***Continued***

Item/Equipment	Task/Function	Interval	Reference
Low-Voltage Busway	Infrared scanning	Annually	20.4.2.1
	Visual inspection	Annually	20.4.3.2
	Electrical test	2 years	20.4.8
Uninterruptible Power Supply Systems	Infrared scanning	Annually	11.17, 20.3.8
	Visual inspection	Quarterly	28.3.8
	Routine maintenance	6 months	28.3.8
	System tests	2 years	28.5
	Battery tests	*See Stationary Batteries and Chargers.*	15.9.4
UPS support standby generator	Test run, exercise	Monthly	28.3.5

Annex M Equipment Storage and Maintenance During Construction

This annex is not a part of the recommendations of this NFPA document but is included for informational purposes only.

M.1 Introduction. Preferably, all types of electrical equipment should be stored in a clean, heated building affording good physical protection and providing controlled access to prevent unauthorized tampering with the equipment. However, equipment can be stored in other inside and outside environments with proper provisions to satisfy the following general recommendations and the recommendations specified in the particular equipment sections.

M.1.1 Before storage, when equipment is received, it should be inspected for shipping damage, and reports should be made as recommended to recover repair or replacement costs from the carrier in the event damage was sustained. In some cases, visual inspection might indicate a need to test for concealed damage before the equipment is removed from the carrier vehicle.

M.1.2 Covers are recommended unless storage conditions specified in M.1.1 exist. Canvas tarpaulins or the equivalent are preferred over other coverings because they provide better humidity control and enclosure scuff protection.

M.1.3 The manufacturer's shipping skids should be left on the equipment to provide structural support until the equipment is set in its final resting place.

M.1.4 Insulation tests should be conducted and test values recorded when the equipment is received. Periodic tests are recommended in the following sections for particular types of equipment. In all cases, insulation should be retested prior to start-up, with sufficient time provided for any necessary dry-out or repair prior to energizing.

M.1.5 Regular inspections should be made to check the general effectiveness of equipment storage provisions, and improvements should be made as indicated.

M.2 Equipment. Where storage conditions specified in Section M.1 are not available, indoor or outdoor storage should comply with the following paragraphs.

M.2.1 Switchgear, Switchboards, Motor Control, and Other Control Equipment.

M.2.1.1 Metal-enclosed equipment should be stored in the upright position. Good ventilation of the shelter and protection from dirt, moisture, and physical damage should be provided.

M.2.1.2 Space heaters furnished with the equipment should be connected to a continuous source of power of the proper rating.

> **CAUTION:** Where space heaters are supplied from auxiliary power transformers, care should be taken that low-voltage heater circuits are properly isolated before power-source connection to prevent inadvertent energizing of the auxiliary transformer and associated high-voltage primary wiring.

M.2.1.3 Temporary heaters or lamp banks should be used where space heaters are not furnished to maintain temperature at a level approximately 12°C (10°F) above ambient.

M.2.1.4 In humid locations, such as in the tropics, it might be necessary to remove the equipment from shipping cases to permit adequate ventilation and to avoid mildew.

M.2.1.5 Oil-immersed circuit breakers, starters, and similar items that are shipped dry should be stored indoors or should be filled with insulating liquid as soon as they are received on site. Units filled with liquid can be stored outdoors if raised above grade to prevent any damage from surface water and if a shed roof and tarpaulin siding (or equivalent) are provided.

M.2.1.6 Insulation resistance values of such parts as operating coils should be spot-checked every 6 weeks. If any readings are low, the affected parts should be dried out before they are placed in operation.

M.3 Busway and Associated Fittings.

M.3.1 Busway sections and fittings preferably should be stored in a heated building that has adequate air circulation and is protected from dirt, water, and physical damage. Where this is not possible, sections and fittings should be stored in a clean, dry shelter that has provisions for maintaining temperature uniformity necessary to prevent condensation.

M.3.2 If busway sections and associated fittings are stored outdoors, they should be securely covered for protection from weather and dirt. Temporary electrical heating should be installed beneath the cover to prevent condensation. At least 0.0283 watt/m^3 (3 watts/ft^3) is adequate for the average environment.

M.3.3 Weatherproof busway should be treated exactly the same as indoor busway until after it is installed. It is not weatherproof until completely and properly installed.

M.4 Motors and Generators.

M.4.1 Indoor storage should be provided for all motors and generators except motors designed for outdoor use, which can be stored outdoors without protective covering. These generally are explosionproof motors, totally enclosed motors, and totally enclosed motors with integral coolers. However, in special cases where they are designed for indoor use only, they should be stored indoors. Other motors can be stored outdoors if protective covering that permits good ventilation is furnished.

M.4.2 For motors and mechanical equipment with motors such as motor-operated valve actuators stored outdoors without protective cover, the following should be observed:

(1) All enclosure openings not intended to be open during operation of the equipment, such as conduit and cable entrances in terminal boxes, should be closed with watertight plugs. Temporary shipping plugs should be replaced with permanent storage plugs.
(2) All motors and mechanical equipment with motors should be stored in their normal operating position (e.g., vertical motors in an upright position with their shaft extension downward).

M.4.3 If space heaters are furnished in the units, they should be connected to a continuous supply of power of the proper rating.

M.4.4 Insulation resistance values of each winding should be measured and should be recorded for future reference. The first set of values should be reasonably consistent with the factory insulation resistance measurement values. These measurements should be taken as soon as possible after a unit arrives at the site.

M.4.5 Brushes should be removed from brush holders and should be stored in a dry, warm place where condensation will not occur.

M.4.6 After installation at the final service location, motors with oil-lubricated bearings and ac motors and generators with collector rings should be protected as follows:

(1) *Oil-lubricated bearings.* All internal surfaces of bearing housing should be coated with rust preventative. Vent and drain connections should be plugged, capped, or blinded, as applicable, using steel fittings. The internal surfaces should be recoated (by fill, fill and drain, slushing, spraying, or rotation, as appropriate) at 1-month intervals.
(2) *Collector rings.* Applied protective coatings should be examined and renewed if not intact.

M.4.7 Every 3 months, insulation resistance values of each winding of units rated 2300 volts and higher should be measured and recorded. Temperature and weather conditions should be recorded at time of reading. If resistance is low and cables have been connected, the cables should be disconnected and the measurements repeated. If resistance of winding insulation only is low, leads in a unit's terminal box should be dried out by removing the cover and exposing them to dry, clear weather or by placing an electric lamp or heater in the terminal box. If these measures do not result in acceptable insulation resistance values, the windings should be dried out by an approved method until acceptable values are obtained.

M.4.8 Six weeks before start-up, insulation resistance values of each winding of all units should be measured and recorded. Temperature and weather conditions should be recorded at time of reading. If resistance is low, M.4.7 should be followed.

M.4.9 If grease-lubricated units are on site more than 1 year from the date of shipment from the factory without having been operated, the bearing grease should be inspected. If there has been any visible deterioration of the lubricating properties of the grease, the grease should be cleaned out and the bearings replaced per the manufacturer's recommendations.

M.4.10 Immediately before start-up, the following should be performed:

(1) Insulation values of all units with cables connected should be measured and recorded. If readings are low, the units should be dried out before starting.
(2) Protective coatings should be cleaned from collector ring surfaces.
(3) Surfaces of commutators should be cleaned per manufacturer's instructions.

M.5 Transformers.

M.5.1 Indoor storage should be provided for all transformers except the following:

(1) Transformers intended for outdoor installation can be stored outdoors without protective covering.
(2) Large indoor units can be stored outdoors if raised above grade to prevent any damage from surface water and if a shed roof and tarpaulin siding (or equivalent) are provided.

M.5.2 Ventilated dry-type units should have the same storage conditions as indoor switchgear.

M.5.3 Drums of insulating liquid stored outdoors should be laid on their sides with the large bung downward. Drums should be placed so that the large bung is at about a 45-degree angle from the bottom center position, to minimize contamination by moisture or other liquids.

M.5.4 Transformers Filled with Insulating Liquid. If a transformer is shipped with its main tank filled with insulating liquid (except for expansion space), the level of the liquid and the ambient temperature should be measured and recorded when the unit arrives on site and every month thereafter. If the level falls, leaks should be repaired and insulating liquid added to keep the level within tolerances.

M.5.5 Gas Under Pressure. If a transformer is shipped with its main tank filled with insulating liquid and blanketed with gas under pressure or filled with gas under pressure, the gas pressure and the ambient temperature should be measured and recorded when the unit arrives on site and every month thereafter. If the pressure falls, leaks should be repaired and gas added to keep the pressure within tolerances.

M.5.6 Primary disconnect switches should be handled per the recommendations for switchgear, motor control, and control equipment.

M.6 Cables.

M.6.1 Reels of paper-insulated, lead-sheathed cable should be rotated 90 degrees every 2 weeks. Sealed ends should be checked for leaks and patched if necessary.

M.6.2 Low-pressure, gas-filled cable should be handled as follows:

(1) Gas pressure should be measured and recorded when the cable is received at the site and every month thereafter. The pressure should be between 34.48 kPa (5 psig) and 89.64 kPa (13 psig). If falling pressure indicates a leak in the cable, a cylinder of dry nitrogen should be connected to the cable to maintain pressure until the leak is located and sealed.

(2) Nitrogen used to maintain pressure during storage, if needed, should be in accordance with ASTM D 1933, *Standard Specification for Nitrogen Gas as an Electrical Insulating Material.* Where available, nitrogen type III is preferred. Manufacturer's recommendations should be followed during installation and operation of any nitrogen cylinders.

M.6.3 Physical protection from vehicles or striking objects is recommended for solid dielectric types (XLP or EPR) on reels. Cable ends should be sealed at the factory and maintained until cables are terminated properly.

M.7 Storage Batteries.

M.7.1 All batteries should be stored indoors, in a dry place.

M.7.2 Batteries that have been shipped dry and charged should have the seals inspected when they are received at the site. Any seals that are damaged should be renewed per the manufacturer's instructions.

M.7.3 Lead–acid batteries that have been shipped wet should be handled as follows:

(1) Electrolyte level should be inspected when batteries are received at the site. Electrolyte should be added to the proper level, if any has been lost.
(2) Three months after date of shipment from the factory, and every three months thereafter, batteries should be given a freshening charge to restore the voltage to 2.15 volts per cell and the specific gravity to 1.21 at 25°C (77°F). The charging rate should not exceed the manufacturer's recommended value; batteries should not be overcharged.
(3) Other type batteries that have been shipped wet should be handled per the manufacturer's instructions.

Annex N Reliability Centered Maintenance

This annex is not a part of the recommendations of this NFPA document but is included for informational purposes only.

N.1 Definitions. These definitions are referenced in several reliability publications and the formulas can be verified in MIL-STD-339, *Wiring and Wiring Devices for Combat and Tactical Vehicles, Selection and Installation of,* or in IEEE 100, *Authoritative Dictionary of IEEE Standards Terms.*

N.1.1 Availability. The probability that a system or product will be available to perform its intended mission or function when called upon to do so at any point in time. It can be measured in one of several ways.

N.1.1.1 Function of Uptime. Availability can be considered as the percent of total time that a system is available. It is measured using Equation 1 (note that the period of time over which this measure of availability is made must be defined). Downtime includes administrative time and delays, as well as time for maintenance and repair.

$$\text{Availability} = \frac{\text{Uptime}}{\text{Downtime} + \text{Uptime (Total time)}} \quad [1]$$

N.1.1.2 Operational Availability.

N.1.1.2.1 Another equation for availability directly uses parameters related to the reliability and maintainability characteristics of the item as well as the support system. Equation 2 reflects this measure.

$$\text{Operational Availability} = \frac{\text{Mean Time Between Maintenance (MTBM)}}{\text{Mean Downtime} + \text{MTBM}} \quad [2]$$

N.1.1.2.2 In Equation 2, MTBM includes all maintenance required for any reason, including repairs of actual design failures, repairs of induced failures, cases where a failure cannot be confirmed, and preventive maintenance.

N.1.1.3 Inherent Availability. When only maintenance required to correct design failures is counted and the effects of the support system are ignored, the result is inherent availability, which is given by Equation 3.

$$\text{Inherent Availability} = \frac{\text{Mean Time Between Failures (MTBF)}}{\text{Mean Time to Repair} + \text{MTBF}} \quad [3]$$

N.1.2 RCM Maintenance. Those activities and actions that directly retain the proper operation of an item or restore that operation when it is interrupted by failure or some other anomaly. (Within the context of RCM, proper operation of an item means that the item can perform its intended function.) These activities and actions include removal and replacement of failed items, repair of failed items, lubrication, servicing (includes replenishment of consumables such as fuel), and calibrations. Other activities and resources are needed to support maintenance. These include spares, procedures, labor, training, transportation, facilities, and test equipment. These activities and resources are usually referred to as logistics. Although some organizations might define maintenance to include logistics, it is used in this section in the more limited sense and does not include logistics.

N.1.2.1 Corrective Maintenance. Actions required to restore a failed item to proper operation. Restoration is accomplished by removing the failed item and replacing it with a new item, or by fixing the item by removing and replacing it with a new item, or by fixing the item by removing and replacing internal components or by some other repair action.

N.1.2.2 Preventive Maintenance. Scheduled activities based on an interval to ensure safety, reduce the likelihood of operational failures, and obtain as much useful life as possible from an item.

N.1.2.3 Condition-Based Maintenance. Actions performed on the basis of observed wear or on predicting when the risk of failure is excessive.

N.1.2.3.1 Some items exhibit wear as they are used. If the probability of failure can be related to a measurable amount of wear, it might be possible to prescribe how much wear can be tolerated before the probability of failure reaches some unacceptable level. If so, then this point becomes the criterion for removal or overhaul. Measurement can be done using a variety of techniques depending on the characteristic being measured. The temperature of electrical equipment, for example, can be measured using infrared thermography.

N.1.2.3.2 In predictive maintenance, a given operating characteristic of the item, current, or temperature, for example, is trended and compared with the known "normal" operating levels. An acceptable range is established with either upper and lower limits or some maximum or minimum level. As long as the trend data remain inside the acceptable values, any variation is considered to be normal deviation due to variances in materials, operating environment, and so forth. When the trend line intersects the "unacceptable" limit line, preventive maintenance is required to avoid a failure in the future. The limits are based on knowledge of the normal operating characteristics and the level of risk of failure that is acceptable.

N.1.3 Reliability. The probability that an item will perform its intended function(s) without failure for a specified time under stated conditions.

N.1.4 Reliability-Centered Maintenance (RCM). A logical, structured framework for determining the optimum mix of applicable and effective maintenance activities needed to sustain the operational reliability of systems and equipment while ensuring their safe and economical operation and support.

N.2 Benefits of RCM.

N.2.1 Reduced Costs. Savings have been achieved by industries for equipment when going from a traditional to an RCM-based PM program. It is important to note that these costs savings were achieved with no reduction in safety.

N.2.2 Increased Availability. For many systems, availability is of primary importance. The level of availability achieved in actual use of a product is a function of how often it fails and how quickly it can be restored to operation. The latter, in turn, is a function of how well the product was designed to be maintainable, the amount of PM required, and the logistics resources and infrastructure that have been put in place to support the product. RCM directly contributes to availability by reducing PM to that which is essential and economic.

N.3 Relationship of RCM to Other Disciplines.

N.3.1 Reliability. Much of the analysis needed for reliability provides inputs necessary for performing an RCM analysis. The fundamental requirement of the RCM approach is to understand the failure characteristics of an item. As used herein, failure characteristics include the consequences of failure, and whether or not the failure manifests itself and, if it does, how. Reliability is measured in different ways, depending on one's perspective: inherent reliability, operational reliability, mission (or functional) reliability, and basic (or logistics) reliability. RCM is related to operational reliability.

N.3.1.1 Inherent Versus Operational Reliability. From a designer's perspective, reliability is measured by "counting" only those failures that are design-related. When measured in this way, reliability is referred to as "inherent reliability." From a user's or operator's perspective, any event that causes the system to stop performing its intended function is a failure event. These events include all design-related failures that affect the systems' function. Also included are maintenance-induced failures, no-defect-found events, and other anomalies that might have been outside the designer's contractual responsibility or technical control. This type of reliability is called "operational reliability."

N.3.1.2 Mission-Critical or Functional Reliability Versus Basic or Logistics Reliability. Any failure that causes the product to fail to perform its function or critical mission is counted in "mission-critical reliability." Redundancy improves mission-critical reliability. Consider a case where one part of a product has two elements in parallel where only one is needed (redundant). If a failure of one element of the redundant part of the product fails, the other continues to function, allowing the product to do its job. Only if both elements fail will a mission-critical failure occur.

N.3.1.3 Basic Reliability. In "basic" reliability, all failures are counted, whether or not a mission-critical or functional failure has occurred. This measure of reliability reflects the total demand that will eventually be placed on maintenance and logistics.

N.3.1.3.1 Safety. RCM specifically addresses safety and is intended to ensure that safety is never compromised.

N.3.1.3.2 Environmental Concerns. In the past several years, environmental concerns and issues involving regulatory bodies have been accorded importance in the RCM approach for some items that are equal (or nearly so) to safety. Failures of an item that can cause damage to the environment or that result in some federal or state law being violated can pose serious consequences for the operator of the item. So the RCM logic can be modified to specifically address environmental or other concerns.

N.3.1.3.3 Maintainability. RCM is a method for prescribing PM that is effective and economical. Whether or not a given PM task is effective depends on the reliability characteristics of the item in question. Whether or not a task is economical depends on many factors, including how easily the PM tasks can be performed. Ease of maintenance, corrective or preventive, is a function of how well the system has been designed to be maintainable. This aspect of design is called maintainability. Providing ease of access, placing items requiring PM where they can be easily removed, providing means of inspection, designing to reduce the possibility of maintenance-induced failures, and other design criteria determine the maintainability of a system.

N.4 Supporting Data. Data are critical to the success of an RCM analysis. Since conducting an RCM analysis requires an extensive amount of information, and much of this information is not available early in the design phase, RCM analysis for a new product cannot be completed until just prior to production. The data fall into four categories: failure characteristics, failure effects, costs, and maintenance capabilities and procedures. Table N.4 illustrates reliability and maintainability information crucial to an RCM analysis.

Table N.4 Reliability and Maintainability Information for RCM Analysis

Calculated Data	Formula for Calculation
Ao, Operational Availability	Ao = MTBM/(MTBM+MDT)
Ai, Inherent Availability	Ai = MTBF/(MBTF+MTTR)
R(t), Reliability (for time interval t)	$R(t) = e^{-\lambda t}$
MTBF, Mean Time Between Failures (h)	MTBF = Tp/Tf
BTTR, Mean Time To Repair (h)	MTTR = Rdt/Tf
MTTM, Mean Time To Maintain (h)	MTTM = Mdt/Tma
MDT, Mean Downtime (h)	MDT = (Rdt + Rlt + Mdt)/Tde
Probability of satisfactory start, prob_s_s	Prob_s_s = total_start/total_attempt
Probability of failure to start, prob_f_s	Prob_f_s = total_fail_start/total_attempt
Hrdt/Year, Hours Downtime per Year	Hrdt/Year = (1 – Ao) × 8760

N.5 Reliability, Inherent Availability, and Operational Availability Data. Table N.5 is provided to help you understand and properly apply the data categories in your analysis. The summary information calculated from the individual equipment records is also included. Calculation formulas for each category are given in Table N.4. These definitions are referenced in several reliability publications, and the formulas can be verified in MIL-STD-339 or in the IEEE standard definition publication.

Table N.5 Reliability, Inherent Availability, and Operational Availability Data

Roll Up Report by Category, Class, and Item				
CATEGORY[a]	CLASS[b]	Reliability[c]	Inherent Availability[d]	Operational Availability[e]
Accumulator		0.993467721	0.999993849	0.999884828
	Accumulator, Pressurized.	0.993913727	0.999992102	0.999841861
	Accumulator, Unpressurized.	0.992345933	0.999998246	0.999992983
Air Compressor		0.964395571	0.999966392	0.999377084
	Air Compressor, Electric.	0.926805720	0.999919556	0.999207149
	Air Compressor, Fuel.	0.989726301	0.999996935	0.999487902
Air Dryer		0.997716217	0.999998695	0.999926162
	Air Dryer, All Types.	0.997716217	0.999998695	0.999926162
Air Handling Unit		0.989056337	0.999997032	0.999875595
	Air Handling Unit, Non-humid wo/Drive.	0.989056337	0.999997032	0.999875595
Arrester		0.998679474	0.999999397	0.999999397
	Arrester, Lightning.	0.998679474	0.999999397	0.999999397
Battery		0.993006248	0.999990299	0.999969547
	Battery, Gel Cell-Sealed, Strings.	0.980061731	0.999995402	0.999967422
	Battery, Lead Acid, System.	0.992563514	0.999972627	0.999968207
	Battery, Nickel-Cadmium.	0.999399558	0.999999292	0.999971403
Blower		0.999825378	1.000000000	0.999960812
	Blower, wo/Drive.	0.999825378*	1.000000000	0.999960812
Boiler		0.878642210	0.999360697	0.995132436
	Boiler, Hot Water, Gravity and Circulated.	0.959008598	0.999985268	0.999501894
Steam		0.842870823	0.999064090	0.993057393
	Boiler, Steam, High Pressure.	0.928026957	0.999619462	0.991492148
	Boiler, Steam, Low Pressure.	0.719936234	0.998154400	0.995621239
Bus Duct		0.999696290	1.000000000	1.000000000
	Bus Duct, All types, (100 ft).	0.999696290*	1.000000000	1.000000000
Cabinet Heaters		0.999897930	0.999999994	0.999978224
	Cabinet Heaters, Forced Air Flow, Steam or Hot Water.	0.999897930	0.999999994	0.999978224
Cable		0.998149212	0.999998818	0.999987869
Above Ground		0.999509398	0.999999527	0.999998357
	Cable, Above Ground, In Conduit, ≤600V, Per 1000 ft.	0.999932074	0.999999938	0.999990264
	Cable, Above Ground, In Conduit, >600V ≤5kV, Per 1000 ft.	0.999463225	0.999999476	0.999998707
	Cable, Above Ground, No Conduit, ≤600V, Per 1000 ft.	0.999879838	0.999999966	0.999999904
	Cable, Above Ground, No Conduit, >600V ≤5kV, Per 1000 ft.	0.999244433	0.999999655	0.999999655
	Cable, Above Ground, Trays, ≤600V, Per 1000 ft.	0.968468243*	1.000000000	1.000000000
	Cable, Above Ground, Trays, >600V ≤5kV, Per 1000 ft.	0.997171966*	1.000000000	1.000000000
Aerial		0.988381339	0.999997295	0.999997259
	Cable, Aerial, ≤15kV, Per Mile.	0.953928762	0.999990218	0.999990218
	Cable, Aerial, >15kV, Per Mile.	0.995896395	0.999998806	0.999998762

(continues)

Table N.5 ***Continued***

Roll Up Report by Category, Class, and Item				
CATEGORY[a]	**CLASS[b]**	**Reliability[c]**	**Inherent Availability[d]**	**Operational Availability[e]**
Below Ground		0.994225869	0.999995527	0.999928197
	Cable, Below Ground, Duct, ≤600V, Per 1000 ft.	0.999875009	0.999999766	0.999999697
	Cable, Below Ground, Duct, >600V ≤5kV, Per 1000 ft.	0.987125021*	1.000000000	1.000000000
	Cable, Below Ground, In Conduit, ≤600V, Per 1000 ft.	0.997994901	0.999997428	0.999991686
	Cable, Below Ground, In Conduit >600V ≤5kV, per 1000 ft.	0.997646877	0.999995779	0.999987126
	Cable, Below Ground, Insulated, >5kV, Per 1000 ft.	0.980031515	0.999988193	0.999674546
	Cable, Below Ground, Insulated, ≤600V, Per 1000 ft.	0.973653295	0.999976836	0.999976836
Insulated		0.992748496	0.999998338	0.999998338
	Cable, Insulated, DC, Per 100 ft.	0.992748496	0.999998338	0.999998338
Cable Connection		0.999629261	0.999999968	0.999999968
Capacitor Bank		0.839937440	0.999954142	0.999942075
	Capacitor Bank, Power Factor Corrector, (in kVAR).	0.839937440	0.999954142	0.999942075
Charger		0.992621004	0.999999577	0.999986472
	Charger, Battery.	0.992621004	0.999999577	0.999986472
Chiller		0.888515818	0.999829779	0.997620632
	Chiller, Absorption.	0.841986658	0.999769437	0.995132437
	Chiller, Centrifugal, 600–1000 Tons.	0.955142622	0.999923928	0.997604888
	Chiller, Reciprocating, Closed, w/Drive, 50–200 Tons.	0.879941865	0.999809524	0.998734968
	Chiller, Reciprocating, Open, wo/Drive, 50–200 Tons.	0.826705884	0.999775088	0.999312485
	Chiller, Rotary, 600–1000 Tons.	0.986993503	0.999964132	0.996197991
	Chiller, Screw, >300 Tons.	0.956286690	0.999510164	0.996566046
Circuit Breaker, 600V		0.999996752	0.999999582	0.999983888
3 Phase, Fixed		0.999996551	0.999999899	0.999992732
	Circuit Breaker, 600V, 3 Phase, Fixed, Including molded case, ≤600 amp, Normally Closed, Trp. Ckt. Incl.	0.999984307*	1.000000000	0.999997443
	Circuit Breaker, 600V, 3 Phase, Fixed, Including molded case, ≤600 amp, Normally Open, Trp. Ckt. Incl.	0.999887215	0.999999760	0.999990187
	Circuit Breaker, 600V, 3 Phase, Fixed, Including molded case, >600 amp, Normally Closed, Trp. Ckt. Incl.	0.999994218*	1.000000000	0.999992509
	Circuit Breaker, 600V, 3 Phase, Fixed, Including molded case, >600V ≤5kV.	0.996576534	0.999985320	0.999880051
Drawout (Metal Clad)		0.998892235	0.999999605	0.999837990
	Circuit Breaker, 600V, Drawout (Metal Clad), <600 amp, Normally Closed, Trp. Ckt. Incl.	0.999792091	0.999999858	0.999798004
	Circuit Breaker, 600V, Drawout (Metal Clad), <600 amp, Normally Open, Trp. Ckt. Incl.	0.997456731	0.999998256	0.999860901
	Circuit Breaker, 600V, Drawout (Metal Clad), >600 amp, Normally Closed, Trp. Ckt. Incl.	0.998150509	0.999999894	0.999954301
	Circuit Breaker, 600V, Drawout (Metal Clad), >600 amp, Normally Open, Trp. Ckt. Incl.	0.994487152	0.999998738	0.999927638

Table N.5 ***Continued***

Roll Up Report by Category, Class, and Item				
CATEGORY[a]	**CLASS**[b]	**Reliability**[c]	**Inherent Availability**[d]	**Operational Availability**[e]
Vacuum		0.980129686	0.999975385	0.999852780
	Circuit Breaker, 5kV, Vacuum, <600 amp, Normally Closed, Trp. Ckt. Incl.	0.997191564	0.999997432	0.999960511
	Circuit Breaker, 5kV, Vacuum, <600 amp, Normally Open, Trp. Ckt. Incl.	0.998887668*	1.000000000	0.999983060
	Circuit Breaker, 5kV, Vacuum, >600 amp, Normally Closed, Trp. Ckt. Incl.	0.976752059	0.999960259	0.999619774
	Circuit Breaker, 5kV, Vacuum, >600 amp, Normally Open, Trp. Ckt. Incl.	0.961020019	0.999957368	0.999854272
Compressor		0.986548811	0.999986587	0.999865676
	Compressor, Refrigerant, >1 Ton.	0.995193627	0.999998075	0.999907183
	Compressor, Screw Type.	0.946328222	0.999931777	0.999667651
Condensers		0.900083857	0.999913810	0.999583534
	Condensers, Double Tube.	0.973573588	0.999992357	0.999758971
	Condensers, Propeller Type Fans/Coils, DX.	0.733621551	0.999734138	0.999393134
	Condensers, Shell and Tube.	0.998878743*	1.000000000	0.999614286
Control Panel		0.994698171	0.999998908	0.999800824
	Control Panel, Generator, wo/Switchgear.	0.988952766	0.999997330	0.999980962
	Control Panel, HVAC/Chillers/AHUs, wo/Switchgear.	0.999848787*	1.000000000	0.999982209
	Control Panel, Switchgear Controls.	0.980568763	0.999997149	0.998160003
Convectors		0.999913016	1.000000000	0.999998481
	Convectors, Fin Tube Baseboard, Electric.	0.999582861*	1.000000000	0.999999626
	Convectors, Fin Tube Baseboard, Steam or Hot Water.	0.999890105*	1.000000000	0.999998180
Cooling Tower		0.968333522	0.999702865	0.997170520
	Cooling Tower, Atmospheric Type, wo/Fans, Motors, Pumps, Valves, etc.	0.928543791	0.999247479	0.994184363
	Cooling Tower, Evaporative Type, wo/Fans, Motors, Pumps, Valves, etc.	0.994195540	0.999988924	0.999046330
Damper Assembly		0.999971953	0.999999975	0.999990131
	Damper Assembly, Motor.	0.999966919*	1.000000000	0.999989337
	Damper Assembly, Pneumatic.	0.999277503	0.999999835	0.999994555
Diesel Engine		0.589772164	0.998540049	0.993985981
Generator Packaged		0.775917369	0.999329810	0.997272882
	Diesel Engine Generator, Packaged, 250kW-1.5MW, Continuous.	0.558396351	0.998287624	0.996927250
	Diesel Engine Generator, Packaged, 250kW-1.5MW, Standby.	0.883822868	0.999742312	0.997409685
Unpackaged		0.317735957	0.996759289	0.986574653
	Diesel Engine Generator, Unpackaged, 750kW-7MW, Continuous.	0.162719469	0.994801067	0.980739869
	Diesel Engine Generator, Unpackaged, 750kW-7MW, Standby.	0.531004159	0.998262059	0.991052357
Drive		0.978172315	0.999958316	0.999925947
	Drive, Adjustable Speed.	0.978172315	0.999958316	0.999925947
Evaporator		0.995968933	0.999993228	0.999908962
Coil		0.995812835	0.999992633	0.999899263
	Evaporator, Coil, Direct Expansion.	0.995812835	0.999992633	0.999899263
Shell Tube		0.997036799	0.999997290	0.999975270
	Evaporator, Shell Tube, Direct Expansion.	0.997036799	0.999997290	0.999975270

(continues)

Table N.5 ***Continued***

Roll Up Report by Category, Class, and Item				
CATEGORY[a]	**CLASS[b]**	**Reliability[c]**	**Inherent Availability[d]**	**Operational Availability[e]**
Fan		0.987559807	0.999971610	0.999351118
	Fan, Centrifugal.	0.981021428	0.999946483	0.999770440
	Fan, Propeller/Disc.	0.989640193	0.999957798	0.999093547
	Fan, Tubeaxial.	0.989938879	0.999990870	0.999055744
	Fan, Vaneaxial.	0.996408668*	1.000000000	1.000000000
Filter		0.999898973	1.000000000	0.999903911
	Filter, Electrical Tempest.	0.998510134*	1.000000000	1.000000000
Mechanical		0.999891630	1.000000000	0.999896927
	Filter, Mechanical, Air Regulator Set.	0.999840000*	1.000000000	0.999981949
	Filter, Mechanical, Fuel Oil.	0.999271146*	1.000000000	0.999910729
	Filter, Mechanical, Lube Oil.	0.999377566*	1.000000000	0.999554311
Fuse		0.997969725	1.000000000	1.000000000
	Fuse, >5kV ≤15kV.	0.999341365*	1.000000000	1.000000000
	Fuse, 0-5kV.	0.998627456*	1.000000000	1.000000000
Gas Turbine Generator		0.647849145	0.998890863	0.990692798
Packaged		0.587787144	0.998689955	0.989043771
	Gas Turbine Generator, Packaged, 750kW-7MW, Continuous.	0.177710554	0.994598022	0.983584136
	Gas Turbine Generator, Packaged, 750kW-7MW, Standby.	0.829472916	0.999868149	0.990615770
Unpackaged		0.994155201	0.999775158	0.997950995
	Gas Turbine Generator, Unpackaged, 750kW-7MW, Continuous.	0.994155201	0.999775158	0.997950995
Gauge		0.999042094	1.000000000	0.999999785
	Gauge, Fluid Level.	0.999042094*	1.000000000	0.999999785
Heat Exchanger		0.989034610	0.999997303	0.998935596
	Heat Exchanger, Boiler System, Steam.	0.971835048	0.999998369	0.997231137
	Heat Exchanger, Lube Oil.	0.996596565	0.999995330	0.999740960
	Heat Exchanger, Water To Water.	0.996130029*	1.000000000	0.999861134
Heater		0.947826981	0.999984168	0.994164558
	Heater, Electric, Lube/Fuel Oil Or Jacket.	0.947826981	0.999984168	0.994164558
Humistat		0.984575905	0.999998226	0.999998226
	Humistat, Assembly.	0.984575905	0.999998226	0.999998226
Inverters		0.995190512	0.999985691	0.999598793
	Inverters, All Types.	0.995190512	0.999985691	0.999598793
Meter		0.998913484	0.999993988	0.999993961
	Meter, Electric.	0.999635167	0.999999958	0.999999958
	Meter, Fuel.	0.946014073	0.999543853	0.999543853
	Meter, Water.	0.999621152	0.999999870	0.999999697
Motor Generator Set		0.975052652	0.999978501	0.993070544
	Motor Generator Set, 3 Phase, 400 Hz.	0.995075131	0.999995491	0.999628032
	Motor Generator Set, 3 Phase, 60 Hz.	0.957963867	0.999963722	0.987366458
Motor Starter		0.999147052	0.999995416	0.999944527
	Motor Starter, ≤600V.	0.998167781*	1.000000000	0.999984223
	Motor Starter, >600V.	0.996875738	0.999991427	0.999909983
Motor, Electric		0.999032041	0.999973300	0.999930849
	Motor, Electric, DC.	0.985531708	0.999031729	0.998182336
Induction		0.981918899	0.999992950	0.999724259
	Motor, Electric, Induction, ≤600V.	0.988992708	0.999998736	0.999957372
	Motor, Electric, Induction, >600V.	0.974689985	0.999986993	0.999484292
Single Phase		0.999980411	0.999999987	0.999988267
	Motor, Electric, Single Phase, ≤5 amp.	0.999979878*	1.000000000	0.999996192
	Motor, Electric, Single Phase, >5 amp.	0.998550210	0.999999503	0.999696847
Synchronous		0.998653401	0.999978284	0.999857033
	Motor, Electric, Synchronous, ≤600V.	0.996555656*	1.000000000	0.999777580
	Motor, Electric, Synchronous, >600V.	0.991366824	0.999964367	0.999907948

Table N.5 ***Continued***

Roll Up Report by Category, Class, and Item				
CATEGORY[a]	CLASS[b]	Reliability[c]	Inherent Availability[d]	Operational Availability[e]
Motor, Mechanical		0.195448823	0.999809717	0.998810724
Diesel		0.904562026	0.999953538	0.991433654
	Motor, Mechanical, Diesel.	0.904562026	0.999953538	0.991433654
Gas		0.161029030	0.999791533	0.999743425
	Motor, Mechanical, Gas.	0.161029030	0.999791533	0.999743425
Pipe		0.981888041	0.999994337	0.999991952
	Pipe, Flex, Non-Reinforced, >4 in.	0.985560776	0.999994466	0.999990038
	Pipe, Flex, Reinforced, >4 in.	0.977618384	0.999994186	0.999994186
Piping		0.999960899	0.999998770	0.999676366
Refrigerant		0.999954550	0.999999430	0.999990919
	Piping, Refrigerant, <1 in.	0.999925556*	1.000000000	0.999993884
	Piping, Refrigerant, <2 in.	0.997181886	0.999996564	0.999986684
	Piping, Refrigerant, >2 in.	0.999822269*	1.000000000	1.000000000
	Piping, Refrigerant, 1-3 in.	0.993176045	0.999993747	0.999895362
Water		0.999720116	0.999994706	0.997739077
	Piping, Water, ≤2 in.	0.998834378*	1.000000000	1.000000000
	Piping, Water, >12 in.	0.939385452*	1.000000000	1.000000000
	Piping, Water, >2 ≤4 in.	0.979679275	0.999966994	0.999966994
	Piping, Water, >4 ≤8 in.	0.998103531*	1.000000000	1.000000000
	Piping, Water, >8 ≤12 in.	0.999374866*	1.000000000	0.994961083
Pressure Control		0.993091820	0.999995568	0.999938101
	Pressure Control, Assembly.	0.993091820	0.999995568	0.999938101
Pressure Regulator		0.999163441	1.000000000	0.999993069
Hot Gas		0.999163441	1.000000000	0.999993069
	Pressure Regulator, Hot Gas.	0.999163441*	1.000000000	0.999993069
Pump		0.993705867	0.999994889	0.999826613
Centrifugal		0.994206434	0.999995523	0.999903450
	Pump, Centrifugal, Integral Drive.	0.992515450	0.999993654	0.999897429
	Pump, Centrifugal, wo/Drive.	0.995791244	0.999997272	0.999909083
	Pump, Positive Displacement.	0.991821538	0.999992500	0.999537023
Radiators		0.987545587	0.999977760	0.999934189
	Radiators, Small Tube.	0.987545587	0.999977760	0.999934189
Rectifiers		0.995540658	0.999991837	0.998972976
	Rectifiers, All Types.	0.995540658	0.999991837	0.998972976
Sending Unit		0.999566658	0.999999536	0.999999258
Air Velocity		0.998867884	0.999998707	0.999997599
	Sending Unit, Air Velocity.	0.998867884	0.999998707	0.999997599
	Sending Unit, Pressure.	0.997916028	0.999997883	0.999997089
	Sending Unit, Temperature.	0.999980697*	1.000000000	1.000000000
Software Con. ADAS Sys.		0.642221250	0.999854564	0.999658784
	Software Con. ADAS Sys., ≤1000 Acquisition Points.	0.777690112	0.999954199	0.999888246
	Software Con. ADAS Sys., >1000 Acquisition Points.	0.428800729	0.999644282	0.999174503
Strainer		0.999943310	1.000000000	0.999916767
	Strainer, Coolant.	0.998861684*	1.000000000	0.999333463
	Strainer, Duplex Fuel/Lube Oil.	0.995679886*	1.000000000	0.999861421
	Strainer, Fuel Oil.	0.998766615*	1.000000000	0.999924447
	Strainer, Lube Oil.	0.999529759*	1.000000000	0.999881981
Water		0.999926442	1.000000000	0.999960363
	Strainer, Water, ≤4 in.	0.999920044*	1.000000000	0.999999893
	Strainer, Water, >4 in.	0.999081068*	1.000000000	0.999505864

(continues)

Table N.5 ***Continued***

Roll Up Report by Category, Class, and Item				
CATEGORY[a]	**CLASS[b]**	**Reliability[c]**	**Inherent Availability[d]**	**Operational Availability[e]**
Switch		0.993744427	0.999996988	0.999960651
Automatic Transfer		0.950118163	0.999976051	0.999857315
	Switch, Automatic Transfer, >600 amp., ≤600V.	0.968631015	0.999994046	0.999809981
	Switch, Automatic Transfer, 0-600 amp., ≤600V.	0.917774618	0.999943753	0.999942269
Disconnect		0.999846881	0.999999966	0.999961037
	Switch, Disconnect, Enclosed, ≤600V.	0.999394569*	1.000000000	0.999938186
	Switch, Disconnect, Enclosed, >5kV.	0.998257804	0.999999801	0.999939288
	Switch, Disconnect, Enclosed, >600V ≤5kV.	0.997942528*	1.000000000	0.999867230
	Switch, Disconnect, Fused, DC, >600 amp., ≤600V.	0.999408178*	1.000000000	1.000000000
	Switch, Disconnect, Fused, DC, 0-600 amp., ≤600V.	0.999367257*	1.000000000	0.999987568
	Switch, Electric, On/Off Breaker Type, Non-knife., ≤600V.	0.999358198	0.999999927	0.999999780
Float		0.997716932	0.999999478	0.999985388
	Switch, Float, Electric.	0.997716932	0.999999478	0.999985388
Manual Transfer		0.999129111	1.000000000	0.999966262
	Switch, Manual Transfer, ≤600 amp., ≤600V.	0.997919138*	1.000000000	0.999952908
	Switch, Manual Transfer, >600 amp., ≤600V.	0.998503402*	1.000000000	0.999975863
	Switch, Oil Filled, ≥5kV.	0.998241979*	1.000000000	0.999996849
Static		0.997748999	0.999996656	0.999919287
	Switch, Static, >1000 amp., ≤600V.	0.996326697	0.999989918	0.999739539
	Switch, Static, >600 ≤1000 amp., ≤600V.	0.992336720	0.999998244	0.999994731
	Switch, Static, 0-600 amp. ≤600V.	0.998950665*	1.000000000	0.999999648
Switchgear		0.991916417	0.999974462	0.999585725
Bare Bus		0.989863408	0.999968286	0.999579123
	Switchgear, Bare Bus, ≤600V, All Cabnets, Ckt. Bkrs. Not Included.	0.990554799	0.999992098	0.999455269
	Switchgear, Bare Bus, >5kV, All Cabnets, Ckt. Bkrs. Not Included.	0.982216877	0.999995342	0.999839597
	Switchgear, Bare Bus, >600V ≤5kV, All Cabnets, Ckt. Bkrs. Not Included.	0.997007868	0.999872746	0.999607036
Insulated Bus		0.999613608	0.999989619	0.999601929
	Switchgear, Insulated Bus, ≤600V, All Cabnets, Ckt. Bkrs. Not Included.	0.998420947*	1.000000000	0.999468794
	Switchgear, Insulated Bus, >5kV, All Cabnets, Ckt. Bkrs. Not Included.	0.995913049	0.999982547	0.999626621
	Switchgear, Insulated Bus, >600V ≤5kV, All Cabnets, Ckt. Bkrs. Not Included.	0.996224761	0.999996546	0.999696028
Tank		0.995965564	0.999991636	0.999971186
Day		0.994810377	0.999997030	0.999974756
	Tank, Day, Genset Fuel.	0.994810377	0.999997030	0.999974756
Fuel		0.993549151	0.999955673	0.999872929
	Tank, Fuel.	0.993549151	0.999955673	0.999872929
Receiver		0.997280535	0.999997824	0.999996891
	Tank, Receiver, Air.	0.997280535	0.999997824	0.999996891
Water		0.996377265	0.999999793	0.999989539
	Tank, Water.	0.996377265	0.999999793	0.999989539
Thermostat		0.998319168	0.999999398	0.999997565
	Thermostat, Radiator.	0.998319168	0.999999398	0.999997565

Table N.5 ***Continued***

Roll Up Report by Category, Class, and Item

CATEGORY[a]	CLASS[b]	Reliability[c]	Inherent Availability[d]	Operational Availability[e]
Transducer		0.999978470	0.999999933	0.999998552
Flow		0.996713345	1.000000000	0.999986736
	Transducer, Flow.	0.996713345*	1.000000000	0.999986736
Pressure		0.997477750	0.999999423	0.999987243
	Transducer, Pressure.	0.997477750	0.999999423	0.999987243
Temperature		0.998242572	0.999999950	0.999999026
	Transducer, Temperature.	0.998242572	0.999999950	0.999999026
Transformer, Dry		0.999953743	0.999995817	0.999971899
Air Cooled		0.999882198	1.000000000	0.999944571
	Transformer, Dry, Air Cooled, ≤500kVA.	0.999775100*	1.000000000	0.999995570
	Transformer, Dry, Air Cooled, >1500kVA ≤3000kVA.	0.999393210*	1.000000000	0.999745124
	Transformer, Dry, Air Cooled, >500kVA ≤1500kVA.	0.999582527*	1.000000000	0.999987102
Isolation		0.997166548	0.999993113	0.999989567
	Transformer, Dry, Isolation, Delta Wye, <600V.	0.997166548	0.999993113	0.999989567
Transformer, Liquid		0.994797669	0.999950735	0.998990580
Forced Air		0.989259891	0.999836759	0.996601877
	Transformer, Liquid, Forced Air, ≤10,000kVA.	0.992879584	0.999797696	0.990915913
	Transformer, Liquid, Forced Air, ≤5,000kVA.	0.987452327	0.999994736	0.999987215
	Transformer, Liquid, Forced Air, >10,000kVA ≤50,000kVA.	0.994329760	0.999065253	0.985856760
Non-Forced Air		0.997113141	0.999998203	0.999985412
	Transformer, Liquid, Non-Forced Air, ≤3000kVA.	0.998891114	0.999999367	0.999996102
	Transformer, Liquid, Non-Forced Air, >10000kVA ≤50000kVA.	0.982624792	0.999987813	0.999893406
	Transformer, Liquid, Non-Forced Air, >3000kVA ≤10000kVA.	0.994771048	0.999999402	0.999985038
UPS		0.999078297	0.999998349	0.999951289
Rotary		0.995983397	1.000000000	0.999895500
	UPS, Rotary.	0.995983397*	1.000000000	0.999895500
Small Computer Room Floor		0.990661925	0.999997858	0.999967870
	UPS, Small Computer Room Floor.	0.990661925	0.999997858	0.999967870
Valve		0.999995192	0.999999568	0.999977752
3-way		0.999727982	1.000000000	0.999987577
	Valve, 3-way, Diverting/Sequencing.	0.999257278*	1.000000000	0.999999501
	Valve, 3-way, Mixing Control.	0.999570876*	1.000000000	0.999980689
Ball		0.999807822	0.999999957	0.999999204
	Valve, Ball, N.C.	0.999516658*	1.000000000	0.999998106
	Valve, Ball, N.O.	0.998749718	0.999999929	0.999999929
Butterfly		0.998692271	0.999999513	0.999995506
	Valve, Butterfly, N.C.	0.991788585	0.999996931	0.999990199
	Valve, Butterfly, N.O.	0.999965510*	1.000000000	0.999996507
Check		0.999742108	0.999999971	0.999980199
	Valve, Check.	0.999742108	0.999999971	0.999980199
Control		0.999937125	0.999999943	0.999996490
	Valve, Control, N.C.	0.999922211	0.999999929	0.999997478
	Valve, Control, N.O.	0.999832761*	1.000000000	0.999992325
Expansion		0.999742991	1.000000000	1.000000000
	Valve, Expansion.	0.999742991*	1.000000000	1.000000000

(continues)

Table N.5 *Continued*

Roll Up Report by Category, Class, and Item

CATEGORY[a]	CLASS[b]	Reliability[c]	Inherent Availability[d]	Operational Availability[e]
Gate		0.999827547	0.999999888	0.999999642
	Valve, Gate, N.C.	0.999421886	0.999999934	0.999998647
	Valve, Gate, N.O.	0.999872337	0.999999883	0.999999752
Globe		0.999980570	1.000000000	0.999921533
	Valve, Globe, N.C.	0.999975654*	1.000000000	0.999901776
	Valve, Globe, N.O.	0.999903788*	1.000000000	0.999999612
Plug		0.990331504	0.999997992	0.999997984
	Valve, Plug, N.C.	0.986191497	0.999997832	0.999997819
	Valve, Plug, N.O.	0.996093704	0.999998213	0.999998213
Reducing		0.998490771	1.000000000	0.999972616
	Valve, Reducing, Makeup Water.	0.998490771*	1.000000000	0.999972616
Relief		0.998671145	0.999999696	0.999994763
	Valve, Relief.	0.998671145	0.999999696	0.999994763
Suction		0.998214603	0.999998521	0.999994094
	Valve, Suction.	0.998214603	0.999998521	0.999994094
Valve Operator		0.992808232	0.999991177	0.999971677
	Valve Operator, Electric.	0.990159307	0.999979209	0.999934083
Hydraulic		0.915817948	0.999969884	0.999601804
	Valve Operator, Hydraulic.	0.915817948	0.999969884	0.999601804
Pneumatic		0.995224402	0.999998361	0.999997541
	Valve Operator, Pneumatic.	0.995224402	0.999998361	0.999997541
Voltage Regulator		0.964377637	0.999690405	0.999644857
	Voltage Regulator, Static.	0.964377637	0.999690405	0.999644857
Water Cooling Coil		0.999577258	0.999999879	0.999993176
Fan Coil Unit		0.999577258	0.999999879	0.999993176
	Water Cooling Coil, Fan Coil Unit.	0.999577258	0.999999879	0.999993176

[a]Represents the category of the item, for example, Boiler.

[b]Represents the Class of the item, for example, Boiler, Hot Water.

[c]This column represents the probability of the item failing in 1 year.

[d]Inherent Availability considers down time as a result of failure. No maintenance down time is considered. Ref: RAC publications, Reliability Tool Kit, page 12.

[e]Operational Availability considers down time as a result of maintenance and failure. Ref: RAC publications, Reliability Tool Kit, page 12.

N.6 FMECA Procedure as Part of an RCM Program.

N.6.1 Part of an effective RCM program is to determine the failure modes effects and conduct criticality analysis (FMECA) of all systems, determine the risk priority based on the product of the severity level of a component, failure occurrence level, and detection level.

N.6.2 Determine the failure modes associated with each system (e.g., chilled water supply can have no water flow or degraded flow). Assign a failure mechanism to each failure mode (e.g., degraded flow can be the result of leaky gasket, low supply voltage to motor) and determine the failure effects on system (e.g., no effect, decrease in chiller water temperature). Severity levels are assigned along with probability of failure and a risk priority is determined. This provides for greater emphasis and funding to be assigned to systems that have a greater risk of failure. Therefore systems with higher risk priority would receive more preventive and predictive maintenance than systems with lower risk priorities.

N.6.3 Risk priority is classified with a number, risk priority number (RPN). This is equal to the product of severity level of a component, occurrence, and detection level as noted below with the sum of RPNs for each component within a critical system:

$$\text{sum}\frac{{}^{i}S\ (RPN)n;\ \text{where } RPN=O\times S\times D\ (\text{Occurrence} \times \text{Severity} \times \text{Detection})}{n=1}$$

N.6.4 The purpose of preventive maintenance is not to prevent every component failure from occurring but to prevent the system operational failure. Critical components/subsystems that compromises system operation should receive a high degree of preventive and predictive maintenance. These are critical components or subsystems. A component/subsystem that represents a single point failure that does not compromise the system would receive less preventive and predictive maintenance or even just run to failure.

N.6.5 There are several FMECA methods that can be used to categorize components and subsystems. This depends on how much data is available for the particular systems. A basic block diagram of the RCM process is shown in Figure N.6.5.

N.6.5.1 Define the system: Identify each systems indenture levels. This identifies each system functional item and its associated

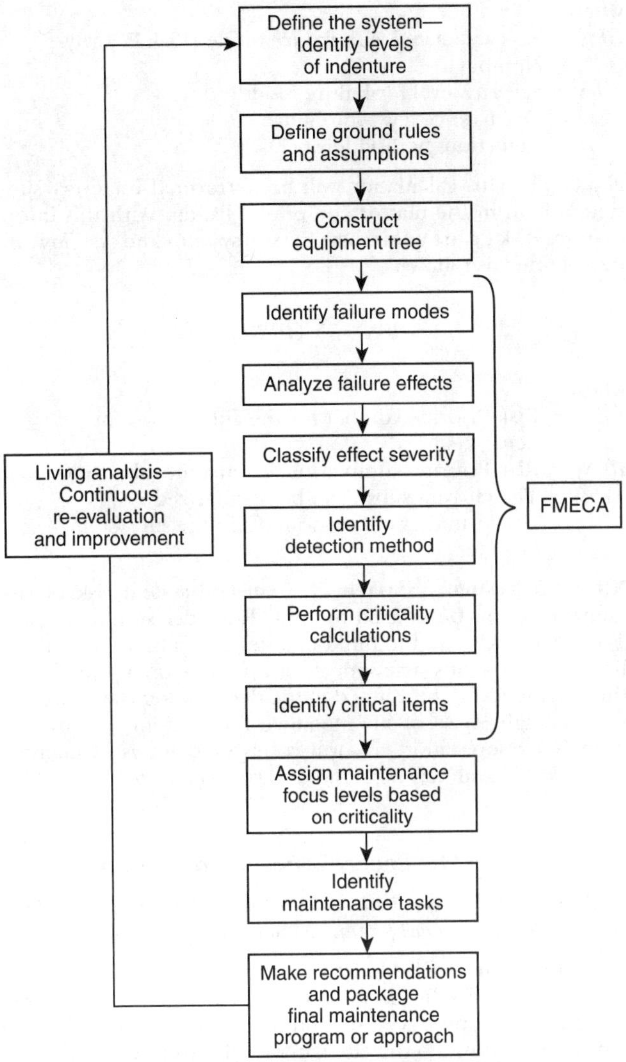

FIGURE N.6.5 Basic Block Diagram of the RCM Process.

failure modes for each functional output. These would be considered your different maintenance areas of concern.

N.6.5.2 Define ground rules and assumptions: The ground rules apply to mission system/equipment, analysis methods (what do we wish to prevent main power outage, operating time during mission stage, source of data).

N.6.5.3 Construct equipment tree. This is a block diagram of operation between indenture levels (function items) that provides different types of failure modes and effects.

N.6.5.4 Identify failure modes.

N.6.5.5 Analyze failure effects.

N.6.5.6 Classify effect severity

(1) Identify detection method.
(2) Perform criticality calculations
(3) Identify critical items.
(4) Assign maintenance focus based on criticality
(5) Identify maintenance tasks.
(6) Make recommendations and package final maintenance program or approach.

N.6.6 Example of FMECA.

N.6.6.1 Detection Method.

N.6.6.1.1 When system controls, automation configurations, and system safeguards are unknown, Detection Method Level can be assumed to be 1. This assumes and stresses that, for a mission critical facility, all item and system level function losses should and will be apparent.

N.6.6.1.2 Although this is an acceptable approach for initial analysis and demonstration purposes, it should be understood that the presence, or absence, of detection method in a systems has a direct effect on the risk associated with the operation of that system. Therefore, consideration of detection method will provide more accurate and resolute analysis results and recommendations. Furthermore, an understanding of current detection method provisions, along with results of an analysis which considered detection method and component level failure modes, can and should be utilized to make recommendations on future detection method provisions.

N.6.6.2 Occurrence.

N.6.6.2.1 Equipment specific PREP database availability numbers will provide indication of failure frequency. These metrics will help to provide less subjective item and system risk assessments. However, they must be adjusted to account for system redundancy, and ranked into discrete occurrence levels to be used in qualitative equipment criticality calculations.

N.6.6.2.2 By design and purpose, a redundant system is more reliable and less vulnerable than a single point, with respect to system function and mission requirements. Therefore, the occurrence level for a single point function must be weighted to reflect the operation, presumed reliability, and severity of loss of function of the redundant component system as accurately as possible.

N.6.6.2.3 The following formula is used to calculate the adjusted availability of a given subsystem due to a level of component or subsystem redundancy.

$$Ai^1 = \sum_{k=m}^{n} \frac{n!}{k!(n-k)!}(Ai)^k (1-Ai)^{(n-k)}$$

where:

Ai = Initial inherent component availability
Ai^1 = Adjusted redundant component availability level
m = Minimum number of components needed
n = Number of components available
k = Current component in redundant system being analyzed

N.6.6.2.4 With availability metrics representative of system configuration now available, component availability is ranked to provide discrete subsystem occurrence levels, as shown in Table N.6.6.2.4.

N.6.6.3 Severity.

N.6.6.3.1 It is also important to consider the concept of failure severity. Severity pertains to and ranks the consequences of system level failure mode effects. For example, a highly probable failure may occur for a subsystem of a piece of critical equipment without severe consequences.

Table N.6.6.2.4 Component Availability Rankings

Availability (nines)	Occurrence Rank	Occurrence Description
≥0.999999999	1	Almost Never
0.99999999	2	Remote
0.9999999	3	Very Slight
0.999999	4	Slight
0.99999	5	Low
0.9999	6	Medium
0.999	7	Moderately High
0.99	8	High
0.9	9	Very High
0	10	Almost Certain

N.6.6.3.2 Severity rankings used are as shown in Table N.6.6.3.2.

N.6.6.4 RPN Calculations and Ranking Methods for Flexible Analysis.

N.6.6.4.1 Severity, occurrence, and detection method levels are then utilized to produce a subsystem risk assessment as follows:

$$RPN = O \times S \times D$$

where:

RPN = Risk associated with failure mode (Risk Priority Number)
S = Severity level for failure mode
O = Occurrence level for failure mode
D = Detection method level (1)

N.6.6.4.2 This calculation will be performed for every subsystem item in the master equipment listing. With this information, Risk Priority Numbers for sub-systems and systems can be obtained as follows:

$$RPN_s = \sum_{n=1}^{j} (RPNc)n$$

where:

RPN_s = Risk Priority Number for the current system being analyzed
RPN_c = Risk Priority Number for the current subsystem
n = The current subsystem being analyzed
j = Total number of components in the subsystem or system

N.6.6.4.3 Results — System X. Item and system risk assessments can now be utilized to apply RCM decision logic (see Table N.6.6.4.3), and to build maintenance tasking program. Items and systems assessed to be of high operational risk should, especially, be applied to the decision logic and should receive high levels of maintenance focus. Items having extremely low operation risk will receive low levels of maintenance focus, and may be allowed to run to failure.

Table N.6.6.3.2 Severity Rankings

Ranking	Effect	Comment
1	None	No reason to expect failure to have any effect on Safety, Health, Environment, or Mission
2	Very Low	Minor disruption to mission
3	Low	Minor disruption to mission
4	Low to Moderate	Moderate disruption to mission
5	Moderate	Moderate disruption to mission
6	Moderate to High	Moderate disruption to mission
7	High	High disruption to mission
8	Very High	High disruption to mission
9	Hazard	Extremely high disruption to mission
10	Hazard	Extremely high disruption to mission

Annex O Energy Efficiency of Motors

This annex is not a part of the recommendations of this NFPA document but is included for informational purposes only.

O.1 Introduction. The federal Energy Policy Act (EPAct) of 1992 mandated efficiency levels for many general purpose electric motors produced or sold after October of 1997. Motors manufactured to the efficiency levels prescribed by EPAct are "energy efficient motors." In June of 2001 NEMA introduced a motor efficiency level that went above the efficiency levels of EPAct. The higher efficiency ratings are termed "NEMA Premium™." Table O.1(a) and Table O.1(b) provide efficiency levels for EPAct and NEMA Premium open drip proof and totally enclosed motors.

O.2 Electric Motor Efficiency. Efficiency is a function of electrical power input, power factor, losses and mechanical power output. The formulas for efficiency are:

Efficiency = (power output)/(power input) = (power input – losses)/(power input)

Table N.6.6.4.3 Example of Risk Priority Number Calculation

Facility Identifier	Equipment Type	Parent System	M	N	PREP ID	A	A'	O' Ranked	S	RPN
A-1	A	X	1	2	13	0.999988924	0.9999999999	1	1	9
A-2	A	X	1	2	13	0.999988924	0.9999999999	1	9	9
B-1	B	X	1	4	163	0.999993654	1.00000000000	1	9	9
B-2	B	X	1	4	163	0.999993654	1.00000000000	1	9	9
B-3	B	X	1	4	163	0.999993654	1.00000000000	1	9	9
B-4	B	X	1	4	163	0.999993654	1.00000000000	1	9	9

Table O.1(a) EPAct and NEMA Premium Efficiencies for Open Drip-Proof Motors

	Open Drip-Proof (ODP)					
	1200 RPM		1800 RPM		3600 RPM	
HP	EPACT	NEMA Premium	EPACT	NEMA Premium	EPACT	NEMA Premium
1	80.0	82.5	82.5	85.5	NA	77.0
1.5	84.0	86.5	84.0	86.5	82.5	84.0
2	85.5	87.5	84.0	86.5	84.0	85.5
3	86.5	88.5	86.5	89.5	84.0	85.5
5	87.5	89.5	87.5	89.5	85.5	86.5
7.5	88.5	90.2	88.5	91.0	87.5	88.5
10	90.2	91.7	89.5	91.7	88.5	89.5
15	90.2	91.7	91.0	93.0	89.5	90.2
20	91.0	92.4	91.0	93.0	90.2	91.0
25	91.7	93.0	91.7	93.6	91.0	91.7
30	92.4	93.6	92.4	94.1	91.0	91.7
40	93.0	94.1	93.0	94.1	91.7	92.4
50	93.0	94.1	93.0	94.5	92.4	93.0
60	93.6	94.5	93.6	95.0	93.0	93.6
75	93.6	94.5	94.1	95.0	93.0	93.6
100	94.1	95.0	94.1	95.4	93.0	93.6
125	94.1	95.0	94.5	95.4	93.6	94.1
150	94.5	95.4	95.0	95.8	93.6	94.1
200	94.5	95.4	95.0	95.8	94.5	95.0
250	95.4	95.4	95.4	95.8	94.5	95.0
300	95.4	95.4	95.4	95.8	95.0	95.4
350	95.4	95.4	95.4	95.8	95.0	95.4
400	—	95.8	95.4	95.8	95.4	95.8
450	—	96.2	95.8	96.2	95.8	95.8
500	—	96.2	95.8	96.2	95.8	95.8

Motor efficiency is a measure of the effectiveness with which electrical energy is converted to mechanical energy. Motor efficiency varies with the load. Most motors manufactured for use in North America are rated in horsepower, and are built to dimensions based on English units, according to NEMA frame standards. The rest of the world rates motor output in kilowatts (kW), and builds them to dimensions based on metric units, per International Electrotechnical Commission (IEC) standards.

O.3 Cost Savings with Increased Efficiency. The annual energy cost of a motor can be calculated using the following formula:

Annual Energy Cost = horsepower × 0.746 × hours of operation × electricity rate ÷ motor efficiency

Continuing the analysis, the following formula evaluates the financial impact due to the difference in motor efficiency levels.

$$D = 0.746 \times HP \times C \times N \times \frac{(100 - 100)}{E_1 \quad E_2}$$

where:

D = Difference in operating cost ($/year)
HP = Motor horsepower
C = Electricity rate ($/kWh)
N = Annual operating hours
E_1 = Nominal efficiency of lower efficiency motor %
E_2 = Nominal efficiency of higher efficiency motor %

An important point to remember about the relationship of efficiency and load is that efficiency varies only slightly over a wide range of output. That is because for induction motors of normal design the equality of load-dependent and load-independent losses will occur between 60 percent and 85 percent load (typically around 75 percent). Therefore, maximum efficiency usually occurs at that point, rather than at rated horsepower. Although efficiency typically drops rapidly as the load decreases below 50 percent, the actual energy wasted — the losses themselves — is much lower than at full load. Those losses represent energy cost without benefit. If they are small, their cost is small, despite the poor efficiency percentage. In other words, a large percentage of a small amount of power is still a small amount. In the extreme, the lowest possible efficiency for any motor is zero — at no load. But the total power used at that point is usually negligibly small.

O.4 Energy Efficient Motor Terminology. In the 1990s the terms "energy efficient" and "standard" as applied to motors were rather easy to understand. "Energy efficient" motors were those that met the EPAct requirements, and motors built to pre-EPAct levels were considered "standard."

With the introduction of NEMA Premium the terminology changed and has become more confusing. Most, but not all, now term high efficiency motors as those that meet the levels of NEMA Premium, and consider EPAct level efficiency motors as "standard."

To avoid the confusion associated with the terms associated with motor efficiency, it is best to make efficiency comparisons based on the nameplate efficiency.

Table O.1(b) EPAct and NEMA Premium Efficiencies for Totally Enclosed Fan-Cooled Motors

	Totally Enclosed Fan-Cooled (TEFC)					
	1200 RPM		1800 RPM		3600 RPM	
HP	EPACT	NEMA Premium	EPACT	NEMA Premium	EPACT	NEMA Premium
1	80.0	82.5	82.5	85.5	75.5	77.0
1.5	85.5	87.5	84.0	86.5	82.5	84.0
2	86.5	88.5	84.0	86.5	84.0	85.5
3	87.5	89.5	87.5	89.5	85.5	86.5
5	87.5	89.5	87.5	89.5	87.5	88.5
7.5	89.5	91.0	89.5	91.7	88.5	89.5
10	89.5	91.0	89.5	91.7	89.5	90.2
15	90.2	91.7	91.0	92.4	90.2	91.0
20	90.2	91.7	91.0	93.0	90.2	91.0
25	91.7	93.0	92.4	93.6	91.0	91.7
30	91.7	93.0	92.4	93.6	91.0	91.7
40	93.0	94.1	93.0	94.1	91.7	92.4
50	93.0	94.1	93.0	94.5	92.4	93.0
60	93.6	94.5	93.6	95.0	93.0	93.6
75	93.6	94.5	94.1	95.4	93.0	93.6
100	94.1	95.0	94.5	95.4	93.6	94.1
125	94.1	95.0	94.5	95.4	94.5	95.0
150	95.0	95.8	95.0	95.8	94.5	95.0
200	95.0	95.8	95.0	96.2	95.0	95.4
250	95.4	95.8	95.0	96.2	95.0	95.8
300	95.4	95.8	95.4	96.2	95.4	95.8
350	95.4	95.8	95.4	96.2	95.4	95.8
400	—	95.8	95.4	96.2	95.4	95.8
450	—	95.8	95.4	96.2	95.4	95.8
500	—	95.8	95.8	96.2	95.4	95.8

O.5 Example of Electric Motor Efficiency. To evaluate motor efficiency, consider the following example: 10 horsepower motor with an input of 8.3 kilowatts (kW). To convert motor HP to kW, multiply the HP times 0.746. The motor output power is therefore 10 × 0.746, or 7.46 kW. With 8.3 kW input, the losses are 8.3 – 7.46, or 0.84 kW. Using the first of the above formulas, the efficiency, with 8.3 kW input would be:

Efficiency = 7.46/8.3 = 0.898, or 89.8%

And using the second formula:

Efficiency = (8.3 – 0.84)/8.3 = 0.898, or 89.8%

If the motor efficiency were improved by a 20 percent reduction in losses, the efficiency in the example would change as follows:

Original losses = 0.84 kW

20% of 0.84 kW = 0.84 × 0.2 = 0.17 kW

New losses = 0.84–0.17 = 0.67 kW

Input power = output power plus losses = 7.46 + 0.67 = 8.13 kW

Applying the first formula above for efficiency:

Efficiency = 7.46/8.13 = 0.918, or 91.8%

The efficiency improvement is 91.8 – 89.8, or 2.0 percent. Or, a 20 percent reduction in losses equates to a 2 percent efficiency improvement in this example.

O.6 Electric Motor Losses. Five internal losses exist in a squirrel cage induction motor. Three of them are "load-dependent," varying quite closely as the square of the load or the load current. These include the I^2R loss in the stator winding ("copper loss"); the I^2R loss in the rotor cage ("slip loss"); and the stray load loss (involving numerous components in various parts of the machine). The two remaining losses are generally considered "load-independent" or "constant" losses — the core or iron loss (which will decrease slightly with increasing load) and the friction and windage loss. The relative proportions of these individual losses will vary with motor speed and size.

The difference between input and output makes up the loss component. Generally, the fundamental motor design rules mean that motor efficiency increases with horsepower (HP) rating. Bear in mind though, that the total losses also increase with nameplate horsepower. As can be seen in Table O.6, gains in efficiency do not offset the increased magnitude of losses as the motor horsepower increases.

Table O.6 Higher Motor Horsepower Results in Higher Losses, Even with Increased Efficiency

Motor HP	Efficiency %	Loss Watts
1	88	100
10	90	830
100	93	5600

O.7 Example of Cost Savings with Increased Efficiency. Using the following formula, the annual energy cost of a motor can be calculated.

Annual Energy Cost = horsepower × 0.746 × hours of operation × electricity rate ÷ motor efficiency

Example: A factory has a 100 HP motor that operates at full load every day of the year. The motor efficiency is 0.941. The electric rate is $0.09/kWh. We can calculate the annual energy costs (AEC) by using the formula above:

$$\text{AEC} = \frac{100\ \text{HP} \times 0.746 \times 8760\ \text{h/yr} \times \$0.09/\text{kWh}}{0.941} = \$62{,}502$$

Financial impact due to higher efficiencies:

Applying the following formula, the financial impact due to the difference in motor efficiency levels can be evaluated.

$$D = 0.746 \times HP \times C \times N \times \frac{(100 - 100)}{E_1 \quad E_2}$$

where:

D = Difference in operating cost ($/year)
HP = Motor horsepower
C = Electricity rate ($/kWh)
N = Annual operating hours
E_1 = Nominal efficiency of lower efficiency motor %
E_2 = Nominal efficiency of higher efficiency motor %

Example Data:

HP = 40 hp open enclosure, 4 pole motor
C = Electricity rate $.08/kWh
N = 4000 annual operating hours
E_1 = Nominal efficiency 93.0% NEMA Premium Efficient motor
E_2 = Nominal efficiency 94.1% NEMA Premium Efficient motor

$$D = 0.746 \times 40 \times 0.08 \times 4000 \times \frac{(100 - 100)}{93.0 \quad 94.1} = \$120.02$$

Annex P Identification of Transformers by Cooling Class

This annex is not a part of the recommendations of this NFPA document but is included for informational purposes only.

P.1 Introduction. Transformers manufactured after 2000, and ones manufactured according to IEC 60076-2: 1993 are identified on the nameplate by the cooling class. The following text is extracted from IEEE Standard C57.12.00-2000, *IEEE Standard General Requirements for Liquid-Immersed Distribution, Power, and Regulating Transformers.*

P.1.1 The identification letters are:

(1) First letter: Internal cooling medium in contact with the windings:

O: mineral oil or synthetic insulating liquid with fire point ≤300°C

K: insulating liquid with fire point > 300°C

L: insulating liquid with no measurable fire point

Fire point — The lowest temperature at which a specimen will sustain burning for 5 s. (ASTM D 92-2005, "Cleveland Open Cup" test method.)

(2) Second letter: Circulation mechanism for internal cooling medium:

N: *natural* convection flow through cooling equipment and in windings

F: *forced* circulation through cooling equipment (i.e., coolant pumps), natural convection flow in windings (also called nondirected flow)

D: forced circulation through cooling equipment, *directed* from the cooling equipment into at least the main windings

(3) Third letter: External cooling medium:

A: air

W: water

(4) Fourth letter: Circulation mechanism for external cooling medium:

N: natural convection

F: forced circulation [fans (air cooling), pumps (water cooling)]

P.1.2 In a transformer with forced, nondirected cooling, (second code letter F), the rates of coolant flow through all the windings vary with the loading, and are not directly controlled by the pumps. The pumped oil flows freely inside the tank and is not forced to flow through the windings.

P.1.3 In a transformer designated as having forced directed coolant circulation (second code letter D), the rate of coolant flow through the main windings is determined by the pumps and not by the loading. A minor fraction of the coolant flow through the cooling equipment may be directed outside the main windings to provide cooling for core and other parts. Regulating windings and/or other windings having relatively low power may also have nondirected coolant circulation.

P.2 Transformers with More than One Power Rating. A transformer may be specified with more than one power rating (also referred to as cooling stages). The transformer nameplate lists the rated power and cooling class designation for each rating. The ratings are listed in order of increasing power. The cooling class designations are normally listed in order with a diagonal slash separating each one.

P.2.1 For example, a transformer having set of fans which may be put in service as desired at high loading would be identified as ONAN/ONAF. This indicates the coolant circulation is by natural convection only.

P.2.2 For example, a transformer having coolant circulation by natural convection only at base loading would be identified as ONAN/OFAF. However, the transformer has cooling equipment with pumps and fans to increase the power-carrying capacity at high loading.

P.3 Cooling Class Designations. Cooling class designations used in IEEE Standard C57.12.00-1993, *General Requirements for Liquid-Immersed Distribution, Power, and Regulating Transformers,* and in previous revisions, and the corresponding new designations used in the 2000 revision of this standard are provided in Table P.3.

Table P.3 Cooling Class Designation

Present Designations	Previous Designations
ONAN	OA
ONAF	FA
ONAN/ONAF/ONAF	OA/FA/FA
ONAN/ONAF/OFAF	OA/FA/FOA
ONAN/ODAF	OA/FOA
ONAN/ODAF/ODAF	OA/FOA/FOA
OFAF	FOA
OFWF	FOW
ODAF	FOA
ODWF	FOW

Index

-F-

-G-

-H-

-I-

-L-

-M-

-N-

-P-

-Q-

-R-

-S-

-T-

-U-

-V-

-W-

Cou/W 2 3 4 5 6 16 15 14

Sequence of Events Leading to Issuance of This NFPA Committee Document

Step 1: Call for Proposals

- Proposed new Document or new edition of an existing Document is entered into one of two yearly revision cycles, and a Call for Proposals is published.

Step 2: Report on Proposals (ROP)

- Committee meets to act on Proposals, to develop its own Proposals, and to prepare its Report.
- Committee votes by written ballot on Proposals. If two-thirds approve, Report goes forward. Lacking two-thirds approval, Report returns to Committee.
- Report on Proposals (ROP) is published for public review and comment.

Step 3: Report on Comments (ROC)

- Committee meets to act on Public Comments to develop its own Comments, and to prepare its report.
- Committee votes by written ballot on Comments. If two-thirds approve, Report goes forward. Lacking two-thirds approval, Report returns to Committee.
- Report on Comments (ROC) is published for public review.

Step 4: Technical Report Session

- "*Notices of intent to make a motion*" are filed, are reviewed, and valid motions are certified for presentation at the Technical Report Session. ("Consent Documents" that have no certified motions bypass the Technical Report Session and proceed to the Standards Council for issuance.)
- NFPA membership meets each June at the Annual Meeting Technical Report Session and acts on Technical Committee Reports (ROP and ROC) for Documents with "certified amending motions."
- Committee(s) vote on any amendments to Report approved at NFPA Annual Membership Meeting.

Step 5: Standards Council Issuance

- Notification of intent to file an appeal to the Standards Council on Association action must be filed within 20 days of the NFPA Annual Membership Meeting.
- Standards Council decides, based on all evidence, whether or not to issue Document or to take other action, including hearing any appeals.

Committee Membership Classifications

The following classifications apply to Technical Committee members and represent their principal interest in the activity of the committee.

M *Manufacturer:* A representative of a maker or marketer of a product, assembly, or system, or portion thereof, that is affected by the standard.

U *User:* A representative of an entity that is subject to the provisions of the standard or that voluntarily uses the standard.

I/M *Installer/Maintainer:* A representative of an entity that is in the business of installing or maintaining a product, assembly, or system affected by the standard.

L *Labor:* A labor representative or employee concerned with safety in the workplace.

R/T *Applied Research/Testing Laboratory:* A representative of an independent testing laboratory or independent applied research organization that promulgates and/or enforces standards.

E *Enforcing Authority:* A representative of an agency or an organization that promulgates and/or enforces standards.

I *Insurance:* A representative of an insurance company, broker, agent, bureau, or inspection agency.

C *Consumer:* A person who is, or represents, the ultimate purchaser of a product, system, or service affected by the standard, but who is not included in the *User* classification.

SE *Special Expert:* A person not representing any of the previous classifications, but who has a special expertise in the scope of the standard or portion thereof.

NOTES:
1. "Standard" connotes code, standard, recommended practice, or guide.
2. A representative includes an employee.
3. While these classifications will be used by the Standards Council to achieve a balance for Technical Committees, the Standards Council may determine that new classifications of members or unique interests need representation in order to foster the best possible committee deliberations on any project. In this connection, the Standards Council may make appointments as it deems appropriate in the public interest, such as the classification of "Utilities" in the National Electrical Code Committee.
4. Representatives of subsidiaries of any group are generally considered to have the same classification as the parent organization.

Submitting Public Input / Public Comment through the Electronic Submission System (e-Submission):

As soon as the current edition is published, a Standard is open for Public Input.

Before accessing the e-Submission System, you must first sign-in at www.NFPA.org. *Note: You will be asked to sign-in or create a free online account with NFPA before using this system*:

a. Click in the gray Sign In box on the upper left side of the page. Once signed-in, you will see a red "Welcome" message in the top right corner.
b. Under the Codes and Standards heading, Click on the Document Information pages (List of Codes & Standards), and then select your document from the list or use one of the search features in the upper right gray box.

OR

a. Go directly to your specific document page by typing the convenient short link of www.nfpa.org/document#, (Example: NFPA 921 would be www.nfpa.org/921) Click in the gray Sign In box on the upper left side of the page. Once signed in, you will see a red "Welcome" message in the top right corner.

To begin your Public Input, select the link The next edition of this standard is now open for Public Input (formally "proposals") located on the Document Information tab, the Next Edition tab, or the right-hand Navigation bar. Alternatively, the Next Edition tab includes a link to Submit Public Input online

At this point, the NFPA Standards Development Site will open showing details for the document you have selected. This "Document Home" page site includes an explanatory introduction, information on the current document phase and closing date, a left-hand navigation panel that includes useful links, a document Table of Contents, and icons at the top you can click for Help when using the site. The Help icons and navigation panel will be visible except when you are actually in the process of creating a Public Input.

Once the First Draft Report becomes available there is a Public comment period during which anyone may submit a Public Comment on the First Draft. Any objections or further related changes to the content of the First Draft must be submitted at the Comment stage.

To submit a Public Comment you may access the e-Submission System utilizing the same steps as previous explained for the submission of Public Input.

For further information on submitting public input and public comments, go to: http://www.nfpa.org/publicinput

Other Resources available on the Doc Info Pages

Document information tab: Research current and previous edition information on a Standard

Next edition tab: Follow the committee's progress in the processing of a Standard in its next revision cycle.

Technical committee tab: View current committee member rosters or apply to a committee

Technical questions tab: For members and Public Sector Officials/AHJs to submit questions about codes and standards to NFPA staff. Our Technical Questions Service provides a convenient way to receive timely and consistent technical assistance when you need to know more about NFPA codes and standards relevant to your work. Responses are provided by NFPA staff on an informal basis.

Products/training tab: List of NFPA's publications and training available for purchase.

Community tab: Information and discussions about a Standard

Information on the NFPA Standards Development Process

I. Applicable Regulations. The primary rules governing the processing of NFPA standards (codes, standards, recommended practices, and guides) are the NFPA *Regulations Governing the Development of NFPA Standards (Regs)*. Other applicable rules include NFPA *Bylaws*, NFPA *Technical Meeting Convention Rules*, NFPA *Guide for the Conduct of Participants in the NFPA Standards Development Process*, and the NFPA *Regulations Governing Petitions to the Board of Directors from Decisions of the Standards Council.* Most of these rules and regulations are contained in the *NFPA Standards Directory*. For copies of the *Directory*, contact Codes and Standards Administration at NFPA Headquarters; all these documents are also available on the NFPA website at "www.nfpa.org."

The following is general information on the NFPA process. All participants, however, should refer to the actual rules and regulations for a full understanding of this process and for the criteria that govern participation.

II. Technical Committee Report. The Technical Committee Report is defined as "the Report of the responsible Committee(s), in accordance with the Regulations, in preparation of a new or revised NFPA Standard." The Technical Committee Report is in two parts and consists of the First Draft Report and the Second Draft Report. (See *Regs* at 1.4)

III. Step 1: First Draft Report. The First Draft Report is defined as "Part one of the Technical Committee Report, which documents the Input Stage." The First Draft Report consists of the First Draft, Public Input, Committee Input, Committee and Correlating Committee Statements, Correlating Input, Correlating Notes, and Ballot Statements. (See *Regs* at 4.2.5.2 and Section 4.3) Any objection to an action in the First Draft Report must be raised through the filing of an appropriate Comment for consideration in the Second Draft Report or the objection will be considered resolved. [See *Regs* at 4.3.1(b)]

IV. Step 2: Second Draft Report. The Second Draft Report is defined as "Part two of the Technical Committee Report, which documents the Comment Stage." The Second Draft Report consists of the Second Draft, Public Comments with corresponding Committee Actions and Committee Statements, Correlating Notes and their respective Committee Statements, Committee Comments, Correlating Revisions, and Ballot Statements. (See *Regs* at Section 4.2.5.2 and 4.4) The First Draft Report and the Second Draft Report together constitute the Technical Committee Report. Any outstanding objection following the Second Draft Report must be raised through an appropriate Amending Motion at the Association Technical Meeting or the objection will be considered resolved. [See *Regs* at 4.4.1(b)]

V. Step 3a: Action at Association Technical Meeting. Following the publication of the Second Draft Report, there is a period during which those wishing to make proper Amending Motions on the Technical Committee Reports must signal their intention by submitting a Notice of Intent to Make a Motion. (See *Regs* at 4.5.2) Standards that receive notice of proper Amending Motions (Certified Amending Motions) will be presented for action at the annual June Association Technical Meeting. At the meeting, the NFPA membership can consider and act on these Certified Amending Motions as well as Follow-up Amending Motions, that is, motions that become necessary as a result of a previous successful Amending Motion. (See 4.5.3.2 through 4.5.3.6 and Table1, Columns 1-3 of *Regs* for a summary of the available Amending Motions and who may make them.) Any outstanding objection following action at an Association Technical Meeting (and any further Technical Committee consideration following successful Amending Motions, see *Regs* at 4.5.3.7 through 4.6.5.3) must be raised through an appeal to the Standards Council or it will be considered to be resolved.

VI. Step 3b: Documents Forwarded Directly to the Council. Where no Notice of Intent to Make a Motion (NITMAM) is received and certified in accordance with the Technical Meeting Convention Rules, the standard is forwarded directly to the Standards Council for action on issuance. Objections are deemed to be resolved for these documents. (See *Regs* at 4.5.2.5)

VII. Step 4a: Council Appeals. Anyone can appeal to the Standards Council concerning procedural or substantive matters related to the development, content, or issuance of any document of the Association or on matters within the purview of the authority of the Council, as established by the *Bylaws* and as determined by the Board of Directors. Such appeals must be in written form and filed with the Secretary of the Standards Council (See *Regs* at 1.6). Time constraints for filing an appeal must be in accordance with 1.6.2 of the *Regs*. Objections are deemed to be resolved if not pursued at this level.

VIII. Step 4b: Document Issuance. The Standards Council is the issuer of all documents (see Article 8 of *Bylaws*). The Council acts on the issuance of a document presented for action at an Association Technical Meeting within 75 days from the date of the recommendation from the Association Technical Meeting, unless this period is extended by the Council (See *Regs at* 4.7.2). For documents forwarded directly to the Standards Council, the Council acts on the issuance of the document at its next scheduled meeting, or at such other meeting as the Council may determine (See *Regs* at 4.5.2.5 and 4.7.4).

IX. Petitions to the Board of Directors. The Standards Council has been delegated the responsibility for the administration of the codes and standards development process and the issuance of documents. However, where extraordinary circumstances requiring the intervention of the Board of Directors exist, the Board of Directors may take any action necessary to fulfill its obligations to preserve the integrity of the codes and standards development process and to protect the interests of the Association. The rules for petitioning the Board of Directors can be found in the *Regulations Governing Petitions to the Board of Directors from Decisions of the Standards Council* and in 1.7 of the *Regs*.

X. For More Information. The program for the Association Technical Meeting (as well as the NFPA website as information becomes available) should be consulted for the date on which each report scheduled for consideration at the meeting will be presented. For copies of the First Draft Report and Second Draft Report as well as more information on NFPA rules and for up-to-date information on schedules and deadlines for processing NFPA documents, check the NFPA website (www.nfpa.org/aboutthecodes) or contact NFPA Codes & Standards Administration at (617) 984-7246.